Suzuki
Samurai/Sidekick/ X-90/Vitara & Geo/Chevrolet Tracker Automotive Repair Manual

by Bob Henderson and John H Haynes
Member of the Guild of Motoring Writers

Models covered:

All Suzuki Samurai/Sidekick/X-90/Vitara and Geo/Chevrolet Tracker models

1986 through 2001

Does not cover information specific to 1.8L four-cylinder or V6 models

(6Y9 - 90010)

Haynes Publishing Group
Sparkford Nr Yeovil
Somerset BA22 7JJ England

Haynes North America, Inc
859 Lawrence Drive
Newbury Park
California 91320 USA
www.haynes.com

Acknowledgements

Geo Tracker wiring diagrams originated exclusively for Haynes North America by Valley Forge Technical Information Services. Suzuki Samurai wiring diagrams originated by George Edward Brodd. Technical consultants who contributed to this project include Ken Freund, Jon LaCourse and Robert Maddox.

A book in the Haynes Automotive Repair Manual Series

ISBN-10: 1-56392-432-3

ISBN-13: 978-1-56392-432-3

Library of Congress Control Number 2001093181

While every attempt is made to ensure that the information in this manual is correct, no liability can be accepted by the authors or publishers for loss, damage or injury caused by any errors in, or omissions from, the information given.

Contents

Haynes photographer, mechanic and author with 1986 Suzuki Samurai

About this manual

Its purpose

The purpose of this manual is to help you get the best value from your vehicle. It can do so in several ways. It can help you decide what work must be done, even if you choose to have it done by a dealer service department or a repair shop; it provides information and procedures for routine maintenance and servicing; and it offers diagnostic and repair procedures to follow when trouble occurs.

We hope you use the manual to tackle the work yourself. For many simpler jobs, doing it yourself may be quicker than arranging an appointment to get the vehicle into a shop and making the trips to leave it and pick it up. More importantly, a lot of money can be saved by avoiding the expense the shop must pass on to you to cover its labor and overhead costs. An added benefit is the sense of satisfaction and accomplishment that you feel after doing the job yourself.

Using the manual

The manual is divided into Chapters. Each Chapter is divided into numbered Sections, which are headed in bold type between horizontal lines. Each Section consists of consecutively numbered paragraphs.

At the beginning of each numbered Section you will be referred to any illustrations which apply to the procedures in that Section. The reference numbers used in illustration captions pinpoint the pertinent Section and the Step within that Section. That is, illustration 3.2 means the illustration refers to Section 3 and Step (or paragraph) 2 within that Section.

Procedures, once described in the text, are not normally repeated. When it's necessary to refer to another Chapter, the reference will be given as Chapter and Section number. Cross references given without use of the word "Chapter" apply to Sections and/or paragraphs in the same Chapter. For example, "see Section 8" means in the same Chapter.

References to the left or right side of the vehicle assume you are sitting in the driver's seat, facing forward.

Even though we have prepared this manual with extreme care, neither the publisher nor the author can accept responsibility for any errors in, or omissions from, the information given.

NOTE

A **Note** provides information necessary to properly complete a procedure or information which will make the procedure easier to understand.

CAUTION

A **Caution** provides a special procedure or special steps which must be taken while completing the procedure where the Caution is found. Not heeding a Caution can result in damage to the assembly being worked on.

WARNING

A **Warning** provides a special procedure or special steps which must be taken while completing the procedure where the Warning is found. Not heeding a Warning can result in personal injury.

Introduction to the Suzuki Samurai/Sidekick/X-90/Vitara and Geo/Chevrolet Tracker

The vehicles covered by this manual are available in 2-and 4-door soft top or hard top styles.

The front mounted inline four-cylinder engine used on these vehicles is equipped with a carburetor or fuel injection, depending on model. The engine drives the rear wheels through either a five-speed manual or three- or four-speed automatic transmission via a driveshaft running between the transmission (2-wheel drive models) or transfer case (4-wheel drive) and solid rear axle. The transfer case and another driveshaft are used to drive the front axle.

The suspension features a solid axle at the rear on all models, supported by leaf springs (Samurai) or coil springs (Sidekick/X-90/Vitara/Tracker). The front axle on Samurai models is also solid, with leaf spring suspension. All other models use an independent suspension arrangement, with the wheels supported by control arms and MacPherson struts.

The steering box is on earlier models mounted to the left of the engine and is connected to the steering arms through a series of rods which incorporates a damper (Samurai). On later models with rack and pinion steering the steering gear is mounted in front of the engine and connected by rods to the steering arms. Power assist is optional on most models.

The brakes are disc at the front and drums at the rear, with power assist standard.

Vehicle identification numbers

Modifications are a continuing and unpublicized process in vehicle manufacturing. Since spare parts manuals and lists are compiled on a numerical basis, the individual vehicle numbers are essential to correctly identify the component required.

Vehicle Identification Number (VIN)

This very important identification number is stamped on a plate attached to the left side windshield pillar, the left end of the dash (visible when the door is opened) or on the left side of the dash, visible through the windshield. The VIN also appears on the Vehicle Certificate of Title and Registration. It contains information such as where and when the vehicle was manufactured, the model year and the body style.

Safety Certification label

The Safety Certification label is affixed to the left front door pillar. The plate contains the name of the manufacturer, the month and year of manufacture, the Gross Vehicle Weight Rating (GVWR), and the certification statement.

Engine identification number

The engine ID number is located on a machined surface on the left side of the block where the transmission bolts to the engine.

Transmission identification number

The ID number on manual transmissions is located on the top of the case. On automatic transmissions the ID numbers are located on the top of the bellhousing.

Buying parts

Replacement parts are available from many sources, which generally fall into one of two categories - authorized dealer parts departments and independent retail auto parts stores. Our advice concerning these parts is as follows:

Retail auto parts stores: Good auto parts stores will stock frequently needed components which wear out relatively fast, such as clutch components, exhaust systems, brake parts, tune-up parts, etc. These stores often supply new or reconditioned parts on an exchange basis, which can save a considerable amount of money. Discount auto parts stores are often very good places to buy materials and parts needed for general vehicle maintenance such as oil, grease, filters, spark plugs, belts, touch-up paint, bulbs, etc. They also usually sell tools and general accessories, have convenient hours, charge lower prices and can often be found not far from home.

Authorized dealer parts department: This is the best source for parts which are unique to the vehicle and not generally available elsewhere (such as major engine parts, transmission parts, trim pieces, etc.).

Warranty information: If the vehicle is still covered under warranty, be sure that any replacement parts purchased - regardless of the source - do not invalidate the warranty!

To be sure of obtaining the correct parts, have engine and chassis numbers available and, if possible, take the old parts along for positive identification.

Maintenance techniques, tools and working facilities

Maintenance techniques

There are a number of techniques involved in maintenance and repair that will be referred to throughout this manual. Application of these techniques will enable the home mechanic to be more efficient, better organized and capable of performing the various tasks properly, which will ensure that the repair job is thorough and complete.

Fasteners

Fasteners are nuts, bolts, studs and screws used to hold two or more parts together. There are a few things to keep in mind when working with fasteners. Almost all of them use a locking device of some type, either a lockwasher, locknut, locking tab or thread adhesive. All threaded fasteners should be clean and straight, with undamaged threads and undamaged corners on the hex head where the wrench fits. Develop the habit of replacing all damaged nuts and bolts with new ones. Special locknuts with nylon or fiber inserts can only be used once. If they are removed, they lose their locking ability and must be replaced with new ones.

Rusted nuts and bolts should be treated with a penetrating fluid to ease removal and prevent breakage. Some mechanics use turpentine in a spout-type oil can, which works quite well. After applying the rust penetrant, let it work for a few minutes before trying to loosen the nut or bolt. Badly rusted fasteners

may have to be chiseled or sawed off or removed with a special nut breaker, available at tool stores.

If a bolt or stud breaks off in an assembly, it can be drilled and removed with a special tool commonly available for this purpose. Most automotive machine shops can perform this task, as well as other repair procedures, such as the repair of threaded holes that have been stripped out.

Flat washers and lockwashers, when removed from an assembly, should always be replaced exactly as removed. Replace any damaged washers with new ones. Never use a lockwasher on any soft metal surface (such as aluminum), thin sheet metal or plastic.

Fastener sizes

For a number of reasons, automobile manufacturers are making wider and wider use of metric fasteners. Therefore, it is important to be able to tell the difference between standard (sometimes called U.S. or SAE) and metric hardware, since they cannot be interchanged.

All bolts, whether standard or metric, are sized according to diameter, thread pitch and length. For example, a standard 1/2 - 13 x 1 bolt is 1/2 inch in diameter, has 13 threads per inch and is 1 inch long. An M12 - 1.75 x 25 metric bolt is 12 mm in diameter, has a thread pitch of 1.75 mm (the distance between threads) and is 25 mm long. The two bolts are

nearly identical, and easily confused, but they are not interchangeable.

In addition to the differences in diameter, thread pitch and length, metric and standard bolts can also be distinguished by examining the bolt heads. To begin with, the distance across the flats on a standard bolt head is measured in inches, while the same dimension on a metric bolt is sized in millimeters (the same is true for nuts). As a result, a standard wrench should not be used on a metric bolt and a metric wrench should not be used on a standard bolt. Also, most standard bolts have slashes radiating out from the center of the head to denote the grade or strength of the bolt, which is an indication of the amount of torque that can be applied to it. The greater the number of slashes, the greater the strength of the bolt. Grades 0 through 5 are commonly used on automobiles. Metric bolts have a property class (grade) number, rather than a slash, molded into their heads to indicate bolt strength. In this case, the higher the number, the stronger the bolt. Property class numbers 8.8, 9.8 and 10.9 are commonly used on automobiles.

Strength markings can also be used to distinguish standard hex nuts from metric hex nuts. Many standard nuts have dots stamped into one side, while metric nuts are marked with a number. The greater the number of dots, or the higher the number, the greater the strength of the nut.

Metric studs are also marked on their ends according to property class (grade). Larger studs are numbered (the same as metric bolts), while smaller studs carry a geometric code to denote grade.

It should be noted that many fasteners, especially Grades 0 through 2, have no distinguishing marks on them. When such is the case, the only way to determine whether it is standard or metric is to measure the thread pitch or compare it to a known fastener of the same size.

Standard fasteners are often referred to as SAE, as opposed to metric. However, it should be noted that SAE technically refers to a non-metric fine thread fastener only. Coarse thread non-metric fasteners are referred to as USS sizes.

Since fasteners of the same size (both standard and metric) may have different strength ratings, be sure to reinstall any bolts, studs or nuts removed from your vehicle in their original locations. Also, when replacing a fastener with a new one, make sure that the new one has a strength rating equal to or greater than the original.

Tightening sequences and procedures

Most threaded fasteners should be tightened to a specific torque value (torque is the twisting force applied to a threaded component such as a nut or bolt). Overtightening the fastener can weaken it and cause it to break, while undertightening can cause it to eventually come loose. Bolts, screws and

studs, depending on the material they are made of and their thread diameters, have specific torque values, many of which are noted in the Specifications at the beginning of each Chapter. Be sure to follow the torque recommendations closely. For fasteners not assigned a specific torque, a general torque value chart is presented here as a guide. These torque values are for dry (unlubricated) fasteners threaded into steel or cast iron (not aluminum). As was previously mentioned, the size and grade of a fastener determine the amount of torque that can safely be applied to it. The figures listed here are approximate for Grade 2 and Grade 3 fasteners. Higher grades can tolerate higher torque values.

Fasteners laid out in a pattern, such as cylinder head bolts, oil pan bolts, differential

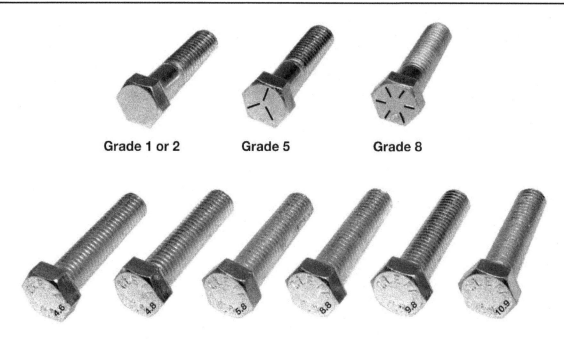

Grade 1 or 2   Grade 5   Grade 8

Bolt strength marking (standard/SAE/USS; bottom - metric)

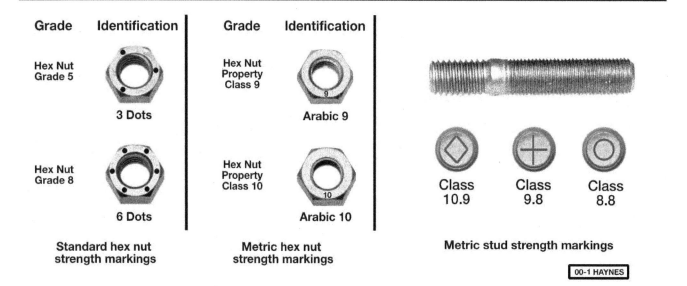

Grade	Identification	Grade	Identification
Hex Nut Grade 5	3 Dots	Hex Nut Property Class 9	Arabic 9
Hex Nut Grade 8	6 Dots	Hex Nut Property Class 10	Arabic 10

Class 10.9   Class 9.8   Class 8.8

Standard hex nut strength markings   **Metric hex nut strength markings**   **Metric stud strength markings**

00-1 HAYNES

cover bolts, etc., must be loosened or tightened in sequence to avoid warping the component. This sequence will normally be shown in the appropriate Chapter. If a specific pattern is not given, the following procedures can be used to prevent warping.

Initially, the bolts or nuts should be assembled finger-tight only. Next, they should be tightened one full turn each, in a criss-cross or diagonal pattern. After each one has been tightened one full turn, return to the first one and tighten them all one-half turn, following the same pattern. Finally, tighten each of them one-quarter turn at a time until each fastener has been tightened to the proper torque. To loosen and remove the fasteners, the procedure would be reversed.

Component disassembly

Component disassembly should be done with care and purpose to help ensure that the parts go back together properly. Always keep track of the sequence in which parts are removed. Make note of special characteristics or marks on parts that can be installed more than one way, such as a grooved thrust washer on a shaft. It is a good idea to lay the disassembled parts out on a clean surface in the order that they were removed. It may also be helpful to make sketches or take instant photos of components before removal.

When removing fasteners from a component, keep track of their locations. Sometimes threading a bolt back in a part, or putting the washers and nut back on a stud, can prevent mix-ups later. If nuts and bolts cannot be returned to their original locations, they should be kept in a compartmented box or a

Metric thread sizes	Ft-lbs	Nm
M-6	6 to 9	9 to 12
M-8	14 to 21	19 to 28
M-10	28 to 40	38 to 54
M-12	50 to 71	68 to 96
M-14	80 to 140	109 to 154
Pipe thread sizes		
1/8	5 to 8	7 to 10
1/4	12 to 18	17 to 24
3/8	22 to 33	30 to 44
1/2	25 to 35	34 to 47
U.S. thread sizes		
1/4 - 20	6 to 9	9 to 12
5/16 - 18	12 to 18	17 to 24
5/16 - 24	14 to 20	19 to 27
3/8 - 16	22 to 32	30 to 43
3/8 - 24	27 to 38	37 to 51
7/16 - 14	40 to 55	55 to 74
7/16 - 20	40 to 60	55 to 81
1/2 - 13	55 to 80	75 to 108

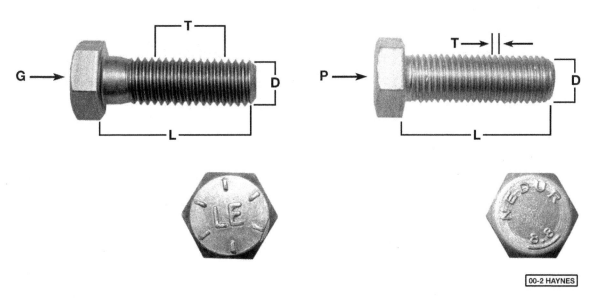

Standard (SAE and USS) bolt dimensions/grade marks

    G   Grade marks (bolt strength)
    L   Length (in inches)
    T   Thread pitch (number of threads per inch)
    D   Nominal diameter (in inches)

Metric bolt dimensions/grade marks

    P   Property class (bolt strength)
    L   Length (in millimeters)
    T   Thread pitch (distance between threads in millimeters)
    D   Diameter

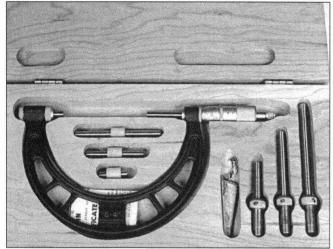

Micrometer set

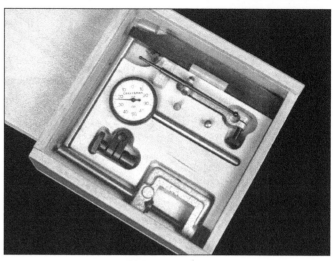

Dial indicator set

series of small boxes. A cupcake or muffin tin is ideal for this purpose, since each cavity can hold the bolts and nuts from a particular area (i.e. oil pan bolts, valve cover bolts, engine mount bolts, etc.). A pan of this type is especially helpful when working on assemblies with very small parts, such as the carburetor, alternator, valve train or interior dash and trim pieces. The cavities can be marked with paint or tape to identify the contents.

Whenever wiring looms, harnesses or connectors are separated, it is a good idea to identify the two halves with numbered pieces of masking tape so they can be easily reconnected.

Gasket sealing surfaces

Throughout any vehicle, gaskets are used to seal the mating surfaces between two parts and keep lubricants, fluids, vacuum or pressure contained in an assembly.

Many times these gaskets are coated with a liquid or paste-type gasket sealing compound before assembly. Age, heat and pressure can sometimes cause the two parts to stick together so tightly that they are very difficult to separate. Often, the assembly can be loosened by striking it with a soft-face hammer near the mating surfaces. A regular hammer can be used if a block of wood is placed between the hammer and the part. Do not hammer on cast parts or parts that could be easily damaged. With any particularly stubborn part, always recheck to make sure that every fastener has been removed.

Avoid using a screwdriver or bar to pry apart an assembly, as they can easily mar the gasket sealing surfaces of the parts, which must remain smooth. If prying is absolutely necessary, use an old broom handle, but keep in mind that extra clean up will be necessary if the wood splinters.

After the parts are separated, the old gasket must be carefully scraped off and the gasket surfaces cleaned. Stubborn gasket material can be soaked with rust penetrant or treated with a special chemical to soften it so

it can be easily scraped off. A scraper can be fashioned from a piece of copper tubing by flattening and sharpening one end. Copper is recommended because it is usually softer than the surfaces to be scraped, which reduces the chance of gouging the part. Some gaskets can be removed with a wire brush, but regardless of the method used, the mating surfaces must be left clean and smooth. If for some reason the gasket surface is gouged, then a gasket sealer thick enough to fill scratches will have to be used during reassembly of the components. For most applications, a non-drying (or semi-drying) gasket sealer should be used.

Hose removal tips

Warning: *If the vehicle is equipped with air conditioning, do not disconnect any of the A/C hoses without first having the system depressurized by a dealer service department or a service station.*

Hose removal precautions closely parallel gasket removal precautions. Avoid scratching or gouging the surface that the hose mates against or the connection may leak. This is especially true for radiator hoses. Because of various chemical reactions, the rubber in hoses can bond itself to the metal spigot that the hose fits over. To remove a hose, first loosen the hose clamps that secure it to the spigot. Then, with slip-joint pliers, grab the hose at the clamp and rotate it around the spigot. Work it back and forth until it is completely free, then pull it off. Silicone or other lubricants will ease removal if they can be applied between the hose and the outside of the spigot. Apply the same lubricant to the inside of the hose and the outside of the spigot to simplify installation.

As a last resort (and if the hose is to be replaced with a new one anyway), the rubber can be slit with a knife and the hose peeled from the spigot. If this must be done, be careful that the metal connection is not damaged.

If a hose clamp is broken or damaged, do not reuse it. Wire-type clamps usually

weaken with age, so it is a good idea to replace them with screw-type clamps whenever a hose is removed.

Tools

A selection of good tools is a basic requirement for anyone who plans to maintain and repair his or her own vehicle. For the owner who has few tools, the initial investment might seem high, but when compared to the spiraling costs of professional auto maintenance and repair, it is a wise one.

To help the owner decide which tools are needed to perform the tasks detailed in this manual, the following tool lists are offered: *Maintenance and minor repair, Repair/overhaul* and *Special.*

The newcomer to practical mechanics should start off with the *maintenance and minor repair* tool kit, which is adequate for the simpler jobs performed on a vehicle. Then, as confidence and experience grow, the owner can tackle more difficult tasks, buying additional tools as they are needed. Eventually the basic kit will be expanded into the *repair and overhaul* tool set. Over a period of time, the experienced do-it-yourselfer will assemble a tool set complete enough for most repair and overhaul procedures and will add tools from the special category when it is felt that the expense is justified by the frequency of use.

Maintenance and minor repair tool kit

The tools in this list should be considered the minimum required for performance of routine maintenance, servicing and minor repair work. We recommend the purchase of combination wrenches (box-end and open-end combined in one wrench). While more expensive than open end wrenches, they offer the advantages of both types of wrench.

Combination wrench set (1/4-inch to 1 inch or 6 mm to 19 mm)
Adjustable wrench, 8 inch
Spark plug wrench with rubber insert

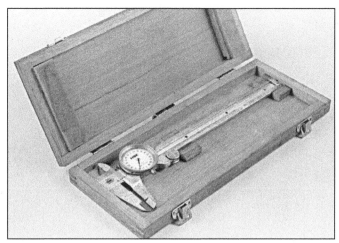

Dial caliper

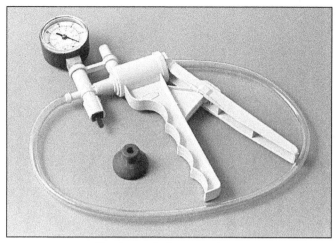

Hand-operated vacuum pump

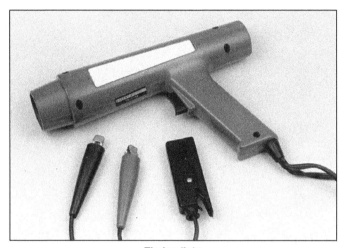

Timing light

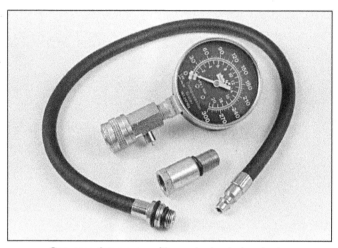

Compression gauge with spark plug hole adapter

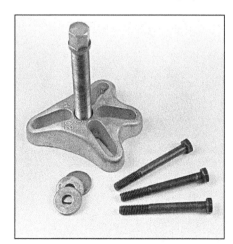

Damper/steering wheel puller

General purpose puller

Hydraulic lifter removal tool

Spark plug gap adjusting tool
Feeler gauge set
Brake bleeder wrench
Standard screwdriver (5/16-inch x
    6 inch)
Phillips screwdriver (No. 2 x 6 inch)
Combination pliers - 6 inch
Hacksaw and assortment of blades

Tire pressure gauge
Grease gun
Oil can
Fine emery cloth
Wire brush
Battery post and cable cleaning tool
Oil filter wrench
Funnel (medium size)

Safety goggles
Jackstands (2)
Drain pan

Note: *If basic tune-ups are going to be part of routine maintenance, it will be necessary to purchase a good quality stroboscopic timing light and combination tachometer/dwell*

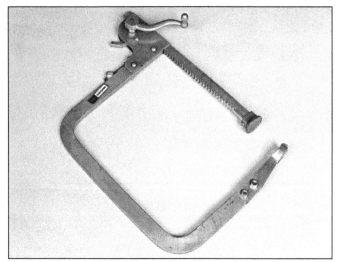

Valve spring compressor

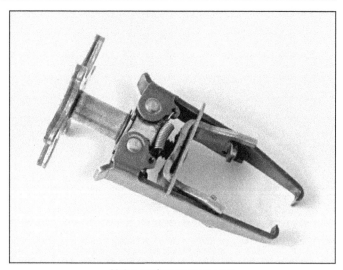

Valve spring compressor

Ridge reamer

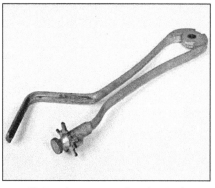

Piston ring groove cleaning tool

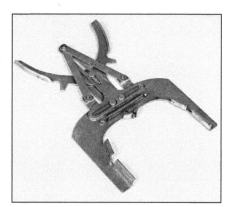

Ring removal/installation tool

meter. Although they are included in the list of special tools, it is mentioned here because they are absolutely necessary for tuning most vehicles properly.

Repair and overhaul tool set

These tools are essential for anyone who plans to perform major repairs and are in addition to those in the maintenance and minor repair tool kit. Included is a compre-

Ring compressor

hensive set of sockets which, though expensive, are invaluable because of their versatility, especially when various extensions and drives are available. We recommend the 1/2-inch drive over the 3/8-inch drive. Although the larger drive is bulky and more expensive, it has the capacity of accepting a very wide range of large sockets. Ideally, however, the mechanic should have a 3/8-inch drive set and a 1/2-inch drive set.

Socket set(s)
Reversible ratchet
Extension - 10 inch
Universal joint
Torque wrench (same size drive as
    sockets)
Ball peen hammer - 8 ounce
Soft-face hammer (plastic/rubber)
Standard screwdriver (1/4-inch x 6 inch)
Standard screwdriver (stubby -
    5/16-inch)
Phillips screwdriver (No. 3 x 8 inch)
Phillips screwdriver (stubby - No. 2)
Pliers - vise grip
Pliers - lineman's
Pliers - needle nose
Pliers - snap-ring (internal and external)
Cold chisel - 1/2-inch
Scribe

Scraper (made from flattened copper
    tubing)
Centerpunch
Pin punches (1/16, 1/8, 3/16-inch)
Steel rule/straightedge - 12 inch
Allen wrench set (1/8 to 3/8-inch or
    4 mm to 10 mm)
A selection of files
Wire brush (large)
Jackstands (second set)
Jack (scissor or hydraulic type)

Note: *Another tool which is often useful is an electric drill with a chuck capacity of 3/8-inch and a set of good quality drill bits.*

Special tools

The tools in this list include those which are not used regularly, are expensive to buy, or which need to be used in accordance with their manufacturer's instructions. Unless these tools will be used frequently, it is not very economical to purchase many of them. A consideration would be to split the cost and use between yourself and a friend or friends. In addition, most of these tools can be obtained from a tool rental shop on a temporary basis.

This list primarily contains only those tools and instruments widely available to the

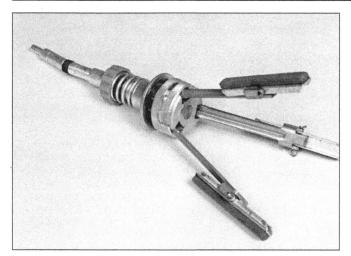

Cylinder hone

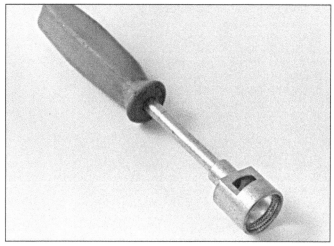

Brake hold-down spring tool

public, and not those special tools produced by the vehicle manufacturer for distribution to dealer service departments. Occasionally, references to the manufacturer's special tools are included in the text of this manual. Generally, an alternative method of doing the job without the special tool is offered. However, sometimes there is no alternative to their use. Where this is the case, and the tool cannot be purchased or borrowed, the work should be turned over to the dealer service department or an automotive repair shop.

Valve spring compressor
Piston ring groove cleaning tool
Piston ring compressor
Piston ring installation tool
Cylinder compression gauge
Cylinder ridge reamer
Cylinder surfacing hone
Cylinder bore gauge
Micrometers and/or dial calipers
Hydraulic lifter removal tool
Balljoint separator
Universal-type puller
Impact screwdriver
Dial indicator set
*Stroboscopic timing light (inductive
    pick-up)*

Hand operated vacuum/pressure pump
Tachometer/dwell meter
Universal electrical multimeter
Cable hoist
*Brake spring removal and installation
    tools*
Floor jack

Buying tools

For the do-it-yourselfer who is just starting to get involved in vehicle maintenance and repair, there are a number of options available when purchasing tools. If maintenance and minor repair is the extent of the work to be done, the purchase of individual tools is satisfactory. If, on the other hand, extensive work is planned, it would be a good idea to purchase a modest tool set from one of the large retail chain stores. A set can usually be bought at a substantial savings over the individual tool prices, and they often come with a tool box. As additional tools are needed, add-on sets, individual tools and a larger tool box can be purchased to expand the tool selection. Building a tool set gradually allows the cost of the tools to be spread over a longer period of time and gives the mechanic the freedom to choose only those tools that will actually be used.

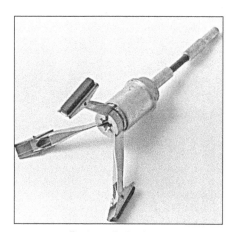

Brake cylinder hone

Tool stores will often be the only source of some of the special tools that are needed, but regardless of where tools are bought, try to avoid cheap ones, especially when buying screwdrivers and sockets, because they won't last very long. The expense involved in replacing cheap tools will eventually be greater than the initial cost of quality tools.

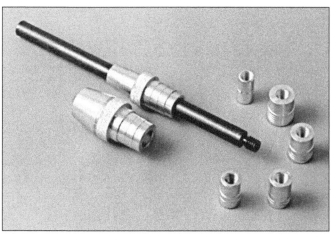

Clutch plate alignment tool

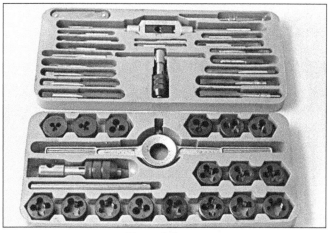

Tap and die set

Care and maintenance of tools

Good tools are expensive, so it makes sense to treat them with respect. Keep them clean and in usable condition and store them properly when not in use. Always wipe off any dirt, grease or metal chips before putting them away. Never leave tools lying around in the work area. Upon completion of a job, always check closely under the hood for tools that may have been left there so they won't get lost during a test drive.

Some tools, such as screwdrivers, pliers, wrenches and sockets, can be hung on a panel mounted on the garage or workshop wall, while others should be kept in a tool box or tray. Measuring instruments, gauges, meters, etc. must be carefully stored where they cannot be damaged by weather or impact from other tools.

When tools are used with care and stored properly, they will last a very long time. Even with the best of care, though, tools will wear out if used frequently. When a tool is damaged or worn out, replace it. Subsequent jobs will be safer and more enjoyable if you do.

How to repair damaged threads

Sometimes, the internal threads of a nut or bolt hole can become stripped, usually from overtightening. Stripping threads is an all-too-common occurrence, especially when working with aluminum parts, because aluminum is so soft that it easily strips out.

Usually, external or internal threads are only partially stripped. After they've been cleaned up with a tap or die, they'll still work. Sometimes, however, threads are badly damaged. When this happens, you've got three choices:

1) *Drill and tap the hole to the next suitable oversize and install a larger diameter bolt, screw or stud.*

2) *Drill and tap the hole to accept a threaded plug, then drill and tap the plug to the original screw size. You can also buy a plug already threaded to the original size. Then you simply drill a hole to the specified size, then run the threaded plug into the hole with a bolt and jam nut. Once the plug is fully seated, remove the jam nut and bolt.*

3) *The third method uses a patented thread repair kit like Heli-Coil or Slimsert. These easy-to-use kits are designed to repair damaged threads in straight-through holes and blind holes. Both are available as kits which can handle a variety of sizes and thread patterns. Drill the hole, then tap it with the special included tap. Install the Heli-Coil and the hole is back to its original diameter and thread pitch.*

Regardless of which method you use, be sure to proceed calmly and carefully. A little impatience or carelessness during one of these relatively simple procedures can ruin your whole day's work and cost you a bundle if you wreck an expensive part.

Working facilities

Not to be overlooked when discussing tools is the workshop. If anything more than routine maintenance is to be carried out, some sort of suitable work area is essential.

It is understood, and appreciated, that many home mechanics do not have a good workshop or garage available, and end up removing an engine or doing major repairs outside. It is recommended, however, that the overhaul or repair be completed under the cover of a roof.

A clean, flat workbench or table of comfortable working height is an absolute necessity. The workbench should be equipped with a vise that has a jaw opening of at least four inches.

As mentioned previously, some clean, dry storage space is also required for tools, as well as the lubricants, fluids, cleaning solvents, etc. which soon become necessary.

Sometimes waste oil and fluids, drained from the engine or cooling system during normal maintenance or repairs, present a disposal problem. To avoid pouring them on the ground or into a sewage system, pour the used fluids into large containers, seal them with caps and take them to an authorized disposal site or recycling center. Plastic jugs, such as old antifreeze containers, are ideal for this purpose.

Always keep a supply of old newspapers and clean rags available. Old towels are excellent for mopping up spills. Many mechanics use rolls of paper towels for most work because they are readily available and disposable. To help keep the area under the vehicle clean, a large cardboard box can be cut open and flattened to protect the garage or shop floor.

Whenever working over a painted surface, such as when leaning over a fender to service something under the hood, always cover it with an old blanket or bedspread to protect the finish. Vinyl covered pads, made especially for this purpose, are available at auto parts stores.

Booster battery (jump) starting

Observe the following precautions when using a booster battery to start a vehicle:

a) *Before connecting the booster battery, make sure the ignition switch is in the Off position.*
b) *Turn off the lights, heater and other electrical loads.*
c) *Your eyes should be shielded. Safety goggles are a good idea.*
d) *Make sure the booster battery is the same voltage as the dead one in the vehicle.*
e) *The two vehicles MUST NOT TOUCH each other.*
f) *Make sure the transmission is in Neutral (manual transaxle) or Park (automatic transaxle).*
g) *If the booster battery is not a maintenance-free type, remove the vent caps and lay a cloth over the vent holes.*

Connect the red jumper cable to the positive (+) terminals of each battery.

Connect one end of the black cable to the negative (-) terminal of the booster battery. The other end of this cable should be connected to a good ground on the engine block **(see illustration)**. Make sure the cable will not come into contact with the fan, drivebelts or other moving parts of the engine.

Start the engine using the booster battery, then, with the engine running at idle speed, disconnect the jumper cables in the reverse order of connection.

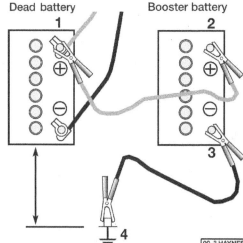

Make the booster battery cable connections in the numerical order shown (note that the negative cable of the booster battery is NOT attached to the negative terminal of the dead battery)

Jacking and towing

Jacking

The jack supplied with the vehicle should only be used for raising the vehicle when changing a tire or placing jackstands under the frame. **Warning:** *Never work under the vehicle or start the engine while this jack is being used as the only means of support.*

The vehicle should be on level ground with the wheels blocked and the transmission in Park (automatic) or Reverse (manual). If the wheel is being replaced, loosen the wheel nuts one-half turn and leave them in place until the wheel is raised off the ground.

Position the jack directly under the axle on the leaf spring seat on Samurai models. On Tracker/Sidekick models, position the jack under the rocker flange approximately 9 inches from the wheel opening (front) or 7 inches from wheel opening on the rear. Operate the jack with a slow, smooth motion until the wheel is raised off the ground.

Install the wheel and lug nuts, tightening the nuts as securely as possible. Lower the vehicle, remove the jack and tighten the nuts (if loosened or removed) in a criss-cross sequence to the torque listed in the Chapter 1 Specifications Section.

Towing

As a general rule, vehicles can be towed with all four wheels on the ground, provided that the driveshafts are removed (Chapter 8).

Equipment specifically designed for towing should be used and should be attached to the main structural members of the vehicle, not the bumper or brackets.

Safety is a major consideration when towing and all applicable state and local laws must be obeyed. A safety chain system must be used for all towing.

While towing, the parking brake should be released and the transmission and (if equipped) transfer case must be in Neutral. The steering must be unlocked (ignition switch in the Off position). Remember that power steering and power brakes will not work with the engine off.

Automotive chemicals and lubricants

A number of automotive chemicals and lubricants are available for use during vehicle maintenance and repair. They include a wide variety of products ranging from cleaning solvents and degreasers to lubricants and protective sprays for rubber, plastic and vinyl.

Cleaners

Carburetor cleaner and choke cleaner is a strong solvent for gum, varnish and carbon. Most carburetor cleaners leave a dry-type lubricant film which will not harden or gum up. Because of this film it is not recommended for use on electrical components.

Brake system cleaner is used to remove brake dust, grease and brake fluid from the brake system, where clean surfaces are absolutely necessary. It leaves no residue and often eliminates brake squeal caused by contaminants.

Electrical cleaner removes oxidation, corrosion and carbon deposits from electrical contacts, restoring full current flow. It can also be used to clean spark plugs, carburetor jets, voltage regulators and other parts where an oil-free surface is desired.

Demoisturants remove water and moisture from electrical components such as alternators, voltage regulators, electrical connectors and fuse blocks. They are non-conductive, non-corrosive and non-flammable.

Degreasers are heavy-duty solvents used to remove grease from the outside of the engine and from chassis components. They can be sprayed or brushed on and, depending on the type, are rinsed off either with water or solvent.

Lubricants

Motor oil is the lubricant formulated for use in engines. It normally contains a wide variety of additives to prevent corrosion and reduce foaming and wear. Motor oil comes in various weights (viscosity ratings) from 0 to 50. The recommended weight of the oil depends on the season, temperature and the demands on the engine. Light oil is used in cold climates and under light load conditions. Heavy oil is used in hot climates and where high loads are encountered. Multi-viscosity oils are designed to have characteristics of both light and heavy oils and are available in a number of weights from 5W-20 to 20W-50.

Gear oil is designed to be used in differentials, manual transmissions and other areas where high-temperature lubrication is required.

Chassis and wheel bearing grease is a heavy grease used where increased loads and friction are encountered, such as for wheel bearings, balljoints, tie-rod ends and universal joints.

High-temperature wheel bearing grease is designed to withstand the extreme temperatures encountered by wheel bearings in disc brake equipped vehicles. It usually contains molybdenum disulfide (moly), which is a dry-type lubricant.

White grease is a heavy grease for metal-to-metal applications where water is a problem. White grease stays soft under both low and high temperatures (usually from -100 to +190-degrees F), and will not wash off or dilute in the presence of water.

Assembly lube is a special extreme pressure lubricant, usually containing moly, used to lubricate high-load parts (such as main and rod bearings and cam lobes) for initial start-up of a new engine. The assembly lube lubricates the parts without being squeezed out or washed away until the engine oiling system begins to function.

Silicone lubricants are used to protect rubber, plastic, vinyl and nylon parts.

Graphite lubricants are used where oils cannot be used due to contamination problems, such as in locks. The dry graphite will lubricate metal parts while remaining uncontaminated by dirt, water, oil or acids. It is electrically conductive and will not foul electrical contacts in locks such as the ignition switch.

Moly penetrants loosen and lubricate frozen, rusted and corroded fasteners and prevent future rusting or freezing.

Heat-sink grease is a special electrically non-conductive grease that is used for mounting electronic ignition modules where it is essential that heat is transferred away from the module.

Sealants

RTV sealant is one of the most widely used gasket compounds. Made from silicone, RTV is air curing, it seals, bonds, waterproofs, fills surface irregularities, remains flexible, doesn't shrink, is relatively easy to remove, and is used as a supplementary sealer with almost all low and medium temperature gaskets.

Anaerobic sealant is much like RTV in that it can be used either to seal gaskets or to form gaskets by itself. It remains flexible, is solvent resistant and fills surface imperfections. The difference between an anaerobic sealant and an RTV-type sealant is in the curing. RTV cures when exposed to air, while an anaerobic sealant cures only in the absence of air. This means that an anaerobic sealant cures only after the assembly of parts, sealing them together.

Thread and pipe sealant is used for sealing hydraulic and pneumatic fittings and vacuum lines. It is usually made from a Teflon compound, and comes in a spray, a paint-on liquid and as a wrap-around tape.

Chemicals

Anti-seize compound prevents seizing, galling, cold welding, rust and corrosion in fasteners. High-temperature anti-seize, usually made with copper and graphite lubricants, is used for exhaust system and exhaust manifold bolts.

Anaerobic locking compounds are used to keep fasteners from vibrating or working loose and cure only after installation, in the absence of air. Medium strength locking compound is used for small nuts, bolts and screws that may be removed later. High-strength locking compound is for large nuts, bolts and studs which aren't removed on a regular basis.

Oil additives range from viscosity index improvers to chemical treatments that claim to reduce internal engine friction. It should be noted that most oil manufacturers caution against using additives with their oils.

Gas additives perform several functions, depending on their chemical makeup. They usually contain solvents that help dissolve gum and varnish that build up on carburetor, fuel injection and intake parts. They also serve to break down carbon deposits that form on the inside surfaces of the combustion chambers. Some additives contain upper cylinder lubricants for valves and piston rings, and others contain chemicals to remove condensation from the gas tank.

Miscellaneous

Brake fluid is specially formulated hydraulic fluid that can withstand the heat and pressure encountered in brake systems. Care must be taken so this fluid does not come in contact with painted surfaces or plastics. An opened container should always be resealed to prevent contamination by water or dirt.

Weatherstrip adhesive is used to bond weatherstripping around doors, windows and trunk lids. It is sometimes used to attach trim pieces.

Undercoating is a petroleum-based, tar-like substance that is designed to protect metal surfaces on the underside of the vehicle from corrosion. It also acts as a sound-deadening agent by insulating the bottom of the vehicle.

Waxes and polishes are used to help protect painted and plated surfaces from the weather. Different types of paint may require the use of different types of wax and polish. Some polishes utilize a chemical or abrasive cleaner to help remove the top layer of oxidized (dull) paint on older vehicles. In recent years many non-wax polishes that contain a wide variety of chemicals such as polymers and silicones have been introduced. These non-wax polishes are usually easier to apply and last longer than conventional waxes and polishes.

Conversion factors

Length (distance)

Inches (in)	X	25.4	= Millimeters (mm)	X	0.0394	= Inches (in)
Feet (ft)	X	0.305	= Meters (m)	X	3.281	= Feet (ft)
Miles	X	1.609	= Kilometers (km)	X	0.621	= Miles

Length (distance)
Inches (in) X 25.4 = Millimeters (mm)   X 0.0394 = Inches (in)
Feet (ft) X 0.305 = Meters (m)   X 3.281 = Feet (ft)
Miles X 1.609 = Kilometers (km)   X 0.621 = Miles

Volume (capacity)

Cubic inches (cu in; in^3) X 16.387 = Cubic centimeters (cc; cm^3)   X 0.061 = Cubic inches (cu in; in^3)
Imperial pints (Imp pt) X 0.568 = Liters (l)   X 1.76 = Imperial pints (Imp pt)
Imperial quarts (Imp qt) X 1.137 = Liters (l)   X 0.88 = Imperial quarts (Imp qt)
Imperial quarts (Imp qt) X 1.201 = US quarts (US qt)   X 0.833 = Imperial quarts (Imp qt)
US quarts (US qt) X 0.946 = Liters (l)   X 1.057 = US quarts (US qt)
Imperial gallons (Imp gal) X 4.546 = Liters (l)   X 0.22 = Imperial gallons (Imp gal)
Imperial gallons (Imp gal) X 1.201 = US gallons (US gal)   X 0.833 = Imperial gallons (Imp gal)
US gallons (US gal) X 3.785 = Liters (l)   X 0.264 = US gallons (US gal)

Mass (weight)

Ounces (oz) X 28.35 = Grams (g)   X 0.035 = Ounces (oz)
Pounds (lb) X 0.454 = Kilograms (kg)   X 2.205 = Pounds (lb)

Force

Ounces-force (ozf; oz) X 0.278 = Newtons (N)   X 3.6 = Ounces-force (ozf; oz)
Pounds-force (lbf; lb) X 4.448 = Newtons (N)   X 0.225 = Pounds-force (lbf; lb)
Newtons (N) X 0.1 = Kilograms-force (kgf; kg)   X 9.81 = Newtons (N)

Pressure

Pounds-force per square inch (psi; lbf/in^2; lb/in^2) X 0.070 = Kilograms-force per square centimeter (kgf/cm^2; kg/cm^2)   X 14.223 = Pounds-force per square inch (psi; lbf/in^2; lb/in^2)
Pounds-force per square inch (psi; lbf/in^2; lb/in^2) X 0.068 = Atmospheres (atm)   X 14.696 = Pounds-force per square inch (psi; lbf/in^2; lb/in^2)
Pounds-force per square inch (psi; lbf/in^2; lb/in^2) X 0.069 = Bars   X 14.5 = Pounds-force per square inch (psi; lbf/in^2; lb/in^2)
Pounds-force per square inch (psi; lbf/in^2; lb/in^2) X 6.895 = Kilopascals (kPa)   X 0.145 = Pounds-force per square inch (psi; lbf/in^2; lb/in^2)
Kilopascals (kPa) X 0.01 = Kilograms-force per square centimeter (kgf/cm^2; kg/cm^2)   X 98.1 = Kilopascals (kPa)

Torque (moment of force)

Pounds-force inches (lbf in; lb in) X 1.152 = Kilograms-force centimeter (kgf cm; kg cm)   X 0.868 = Pounds-force inches (lbf in; lb in)
Pounds-force inches (lbf in; lb in) X 0.113 = Newton meters (Nm)   X 8.85 = Pounds-force inches (lbf in; lb in)
Pounds-force inches (lbf in; lb in) X 0.083 = Pounds-force feet (lbf ft; lb ft)   X 12 = Pounds-force inches (lbf in; lb in)
Pounds-force feet (lbf ft; lb ft) X 0.138 = Kilograms-force meters (kgf m; kg m)   X 7.233 = Pounds-force feet (lbf ft; lb ft)
Pounds-force feet (lbf ft; lb ft) X 1.356 = Newton meters (Nm)   X 0.738 = Pounds-force feet (lbf ft; lb ft)
Newton meters (Nm) X 0.102 = Kilograms-force meters (kgf m; kg m)   X 9.804 = Newton meters (Nm)

Vacuum

Inches mercury (in. Hg) X 3.377 = Kilopascals (kPa)   X 0.2961 = Inches mercury
Inches mercury (in. Hg) X 25.4 = Millimeters mercury (mm Hg)   X 0.0394 = Inches mercury

Power

Horsepower (hp) X 745.7 = Watts (W)   X 0.0013 = Horsepower (hp)

Velocity (speed)

Miles per hour (miles/hr; mph) X 1.609 = Kilometers per hour (km/hr; kph)   X 0.621 = Miles per hour (miles/hr; mph)

Fuel consumption*

Miles per gallon, Imperial (mpg) X 0.354 = Kilometers per liter (km/l)   X 2.825 = Miles per gallon, Imperial (mpg)
Miles per gallon, US (mpg) X 0.425 = Kilometers per liter (km/l)   X 2.352 = Miles per gallon, US (mpg)

Temperature

Degrees Fahrenheit = (°C x 1.8) + 32   Degrees Celsius (Degrees Centigrade; °C) = (°F - 32) x 0.56

*It is common practice to convert from miles per gallon (mpg) to liters/100 kilometers (l/100km), where mpg (Imperial) x l/100 km = 282 and mpg (US) x l/100 km = 235

DECIMALS to MILLIMETERS

Decimal	mm	Decimal	mm
0.001	0.0254	0.500	12.7000
0.002	0.0508	0.510	12.9540
0.003	0.0762	0.520	13.2080
0.004	0.1016	0.530	13.4620
0.005	0.1270	0.540	13.7160
0.006	0.1524	0.550	13.9700
0.007	0.1778	0.560	14.2240
0.008	0.2032	0.570	14.4780
0.009	0.2286	0.580	14.7320
		0.590	14.9860
0.010	0.2540		
0.020	0.5080		
0.030	0.7620		
0.040	1.0160	0.600	15.2400
0.050	1.2700	0.610	15.4940
0.060	1.5240	0.620	15.7480
0.070	1.7780	0.630	16.0020
0.080	2.0320	0.640	16.2560
0.090	2.2860	0.650	16.5100
		0.660	16.7640
0.100	2.5400	0.670	17.0180
0.110	2.7940	0.680	17.2720
0.120	3.0480	0.690	17.5260
0.130	3.3020		
0.140	3.5560		
0.150	3.8100		
0.160	4.0640	0.700	17.7800
0.170	4.3180	0.710	18.0340
0.180	4.5720	0.720	18.2880
0.190	4.8260	0.730	18.5420
		0.740	18.7960
0.200	5.0800	0.750	19.0500
0.210	5.3340	0.760	19.3040
0.220	5.5880	0.770	19.5580
0.230	5.8420	0.780	19.8120
0.240	6.0960	0.790	20.0660
0.250	6.3500		
0.260	6.6040		
0.270	6.8580	0.800	20.3200
0.280	7.1120	0.810	20.5740
0.290	7.3660	0.820	21.8280
		0.830	21.0820
0.300	7.6200	0.840	21.3360
0.310	7.8740	0.850	21.5900
0.320	8.1280	0.860	21.8440
0.330	8.3820	0.870	22.0980
0.340	8.6360	0.880	22.3520
0.350	8.8900	0.890	22.6060
0.360	9.1440		
0.370	9.3980		
0.380	9.6520		
0.390	9.9060		
		0.900	22.8600
0.400	10.1600	0.910	23.1140
0.410	10.4140	0.920	23.3680
0.420	10.6680	0.930	23.6220
0.430	10.9220	0.940	23.8760
0.440	11.1760	0.950	24.1300
0.450	11.4300	0.960	24.3840
0.460	11.6840	0.970	24.6380
0.470	11.9380	0.980	24.8920
0.480	12.1920	0.990	25.1460
0.490	12.4460	1.000	25.4000

FRACTIONS to DECIMALS to MILLIMETERS

Fraction	Decimal	mm	Fraction	Decimal	mm
1/64	0.0156	0.3969	33/64	0.5156	13.0969
1/32	0.0312	0.7938	17/32	0.5312	13.4938
3/64	0.0469	1.1906	35/64	0.5469	13.8906
1/16	0.0625	1.5875	9/16	0.5625	14.2875
5/64	0.0781	1.9844	37/64	0.5781	14.6844
3/32	0.0938	2.3812	19/32	0.5938	15.0812
7/64	0.1094	2.7781	39/64	0.6094	15.4781
1/8	0.1250	3.1750	5/8	0.6250	15.8750
9/64	0.1406	3.5719	41/64	0.6406	16.2719
5/32	0.1562	3.9688	21/32	0.6562	16.6688
11/64	0.1719	4.3656	43/64	0.6719	17.0656
3/16	0.1875	4.7625	11/16	0.6875	17.4625
13/64	0.2031	5.1594	45/64	0.7031	17.8594
7/32	0.2188	5.5562	23/32	0.7188	18.2562
15/64	0.2344	5.9531	47/64	0.7344	18.6531
1/4	0.2500	6.3500	3/4	0.7500	19.0500
17/64	0.2656	6.7469	49/64	0.7656	19.4469
9/32	0.2812	7.1438	25/32	0.7812	19.8438
19/64	0.2969	7.5406	51/64	0.7969	20.2406
5/16	0.3125	7.9375	13/16	0.8125	20.6375
21/64	0.3281	8.3344	53/64	0.8281	21.0344
11/32	0.3438	8.7312	27/32	0.8438	21.4312
23/64	0.3594	9.1281	55/64	0.8594	21.8281
3/8	0.3750	9.5250	7/8	0.8750	22.2250
25/64	0.3906	9.9219	57/64	0.8906	22.6219
13/32	0.4062	10.3188	29/32	0.9062	23.0188
27/64	0.4219	10.7156	59/64	0.9219	23.4156
7/16	0.4375	11.1125	15/16	0.9375	23.8125
29/64	0.4531	11.5094	61/64	0.9531	24.2094
15/32	0.4688	11.9062	31/32	0.9688	24.6062
31/64	0.4844	12.3031	63/64	0.9844	25.0031
1/2	0.5000	12.7000	1	1.0000	25.4000

Safety first!

Regardless of how enthusiastic you may be about getting on with the job at hand, take the time to ensure that your safety is not jeopardized. A moment's lack of attention can result in an accident, as can failure to observe certain simple safety precautions. The possibility of an accident will always exist, and the following points should not be considered a comprehensive list of all dangers. Rather, they are intended to make you aware of the risks and to encourage a safety conscious approach to all work you carry out on your vehicle.

Essential DOs and DON'Ts

DON'T rely on a jack when working under the vehicle. Always use approved jackstands to support the weight of the vehicle and place them under the recommended lift or support points.

DON'T attempt to loosen extremely tight fasteners (i.e. wheel lug nuts) while the vehicle is on a jack - it may fall.

DON'T start the engine without first making sure that the transmission is in Neutral (or Park where applicable) and the parking brake is set.

DON'T remove the radiator cap from a hot cooling system - let it cool or cover it with a cloth and release the pressure gradually.

DON'T attempt to drain the engine oil until you are sure it has cooled to the point that it will not burn you.

DON'T touch any part of the engine or exhaust system until it has cooled sufficiently to avoid burns.

DON'T siphon toxic liquids such as gasoline, antifreeze and brake fluid by mouth, or allow them to remain on your skin.

DON'T inhale brake lining dust - it is potentially hazardous (see *Asbestos* below).

DON'T allow spilled oil or grease to remain on the floor - wipe it up before someone slips on it.

DON'T use loose fitting wrenches or other tools which may slip and cause injury.

DON'T push on wrenches when loosening or tightening nuts or bolts. Always try to pull the wrench toward you. If the situation calls for pushing the wrench away, push with an open hand to avoid scraped knuckles if the wrench should slip.

DON'T attempt to lift a heavy component alone - get someone to help you.

DON'T rush or take unsafe shortcuts to finish a job.

DON'T allow children or animals in or around the vehicle while you are working on it.

DO wear eye protection when using power tools such as a drill, sander, bench grinder, etc. and when working under a vehicle.

DO keep loose clothing and long hair well out of the way of moving parts.

DO make sure that any hoist used has a safe working load rating adequate for the job.

DO get someone to check on you periodically when working alone on a vehicle.

DO carry out work in a logical sequence and make sure that everything is correctly assembled and tightened.

DO keep chemicals and fluids tightly capped and out of the reach of children and pets.

DO remember that your vehicle's safety affects that of yourself and others. If in doubt on any point, get professional advice.

Asbestos

Certain friction, insulating, sealing, and other products - such as brake linings, brake bands, clutch linings, torque converters, gaskets, etc. - may contain asbestos. Extreme care must be taken to avoid inhalation of dust from such products, since it is hazardous to health. If in doubt, assume that they do contain asbestos.

Fire

Remember at all times that gasoline is highly flammable. Never smoke or have any kind of open flame around when working on a vehicle. But the risk does not end there. A spark caused by an electrical short circuit, by two metal surfaces contacting each other, or even by static electricity built up in your body under certain conditions, can ignite gasoline vapors, which in a confined space are highly explosive. Do not, under any circumstances, use gasoline for cleaning parts. Use an approved safety solvent.

Always disconnect the battery ground (-) cable at the battery before working on any part of the fuel system or electrical system. Never risk spilling fuel on a hot engine or exhaust component. It is strongly recommended that a fire extinguisher suitable for use on fuel and electrical fires be kept handy in the garage or workshop at all times. Never try to extinguish a fuel or electrical fire with water.

Fumes

Certain fumes are highly toxic and can quickly cause unconsciousness and even death if inhaled to any extent. Gasoline vapor falls into this category, as do the vapors from some cleaning solvents. Any draining or pouring of such volatile fluids should be done in a well ventilated area.

When using cleaning fluids and solvents, read the instructions on the container carefully. Never use materials from unmarked containers.

Never run the engine in an enclosed space, such as a garage. Exhaust fumes contain carbon monoxide, which is extremely poisonous. If you need to run the engine, always do so in the open air, or at least have the rear of the vehicle outside the work area.

If you are fortunate enough to have the use of an inspection pit, never drain or pour gasoline and never run the engine while the vehicle is over the pit. The fumes, being heavier than air, will concentrate in the pit with possibly lethal results.

The battery

Never create a spark or allow a bare light bulb near a battery. They normally give off a certain amount of hydrogen gas, which is highly explosive.

Always disconnect the battery ground (-) cable at the battery before working on the fuel or electrical systems.

If possible, loosen the filler caps or cover when charging the battery from an external source (this does not apply to sealed or maintenance-free batteries). Do not charge at an excessive rate or the battery may burst.

Take care when adding water to a non maintenance-free battery and when carrying a battery. The electrolyte, even when diluted, is very corrosive and should not be allowed to contact clothing or skin.

Always wear eye protection when cleaning the battery to prevent the caustic deposits from entering your eyes.

Household current

When using an electric power tool, inspection light, etc., which operates on household current, always make sure that the tool is correctly connected to its plug and that, where necessary, it is properly grounded. Do not use such items in damp conditions and, again, do not create a spark or apply excessive heat in the vicinity of fuel or fuel vapor.

Secondary ignition system voltage

A severe electric shock can result from touching certain parts of the ignition system (such as the spark plug wires) when the engine is running or being cranked, particularly if components are damp or the insulation is defective. In the case of an electronic ignition system, the secondary system voltage is much higher and could prove fatal.

Troubleshooting

Contents

This Section provides an easy reference guide to the more common problems that may occur during the operation of your vehicle. Various symptoms and their probable causes are grouped under headings denoting components or systems, such as Engine, Cooling system, etc. They also refer to the Chapter and/or Section that deals with the problem.

Remember that successful troubleshooting isn't a mysterious "black art" practiced only by professional mechanics, it's simply the result of knowledge combined with an intelligent, systematic approach to a problem. Always use a process of elimination starting with the simplest solution and working through to the most complex - and never overlook the obvious. Anyone can run the gas tank dry or leave the lights on overnight, so don't assume that you're exempt from such oversights.

Finally, always establish a clear idea why a problem has occurred and take steps to ensure that it doesn't happen again. If the electrical system fails because of a poor connection, check all other connections in the system to make sure they don't fail as well. If a particular fuse continues to blow, find out why - don't just go on replacing fuses. Remember, failure of a small component can often be indicative of potential failure or incorrect functioning of a more important component or system.

Engine and performance

1 Engine will not rotate when attempting to start

1 Battery terminal connections loose or corroded. Check the cable terminals at the battery; tighten cable clamp and/or clean off corrosion as necessary (see Chapter 1).
2 Battery discharged or faulty. If the cable ends are clean and tight on the battery posts, turn the key to the On position and switch on the headlights or windshield wipers. If they won't run, the battery is discharged.
3 Automatic transmission not engaged in park (P) or Neutral (N).
4 Broken, loose or disconnected wires in the starting circuit. Inspect all wires and connectors at the battery, starter solenoid and ignition switch (on steering column).
5 Starter motor pinion jammed in flywheel ring gear. If manual transmission, place transmission in gear and rock the vehicle to manually turn the engine. Remove starter (Chapter 5) and inspect pinion and flywheel (Chapter 2) at earliest convenience.
6 Starter solenoid faulty (Chapter 5).
7 Starter motor faulty (Chapter 5).
8 Ignition switch faulty (Chapter 13).
9 Engine seized. Try to turn the crankshaft with a large socket and breaker bar on the pulley bolt.

2 Engine rotates but will not start

1 Fuel tank empty.
2 Battery discharged (engine rotates slowly). Check the operation of electrical components as described in previous Section.
3 Battery terminal connections loose or corroded. See previous Section.
4 Fuel not reaching carburetor or fuel injector. Check for clogged fuel filter or lines and defective fuel pump. Also make sure the tank vent lines aren't clogged (Chapter 4).
5 Choke not operating properly (Chapter 1).
6 Faulty distributor components. Check the cap and rotor (Chapter 1).
7 Low cylinder compression. Check as described in Chapter 2.
8 Valve clearances not properly adjusted (Chapter 2).
9 Water in fuel. Drain tank and fill with new fuel.
10 Defective ignition coil (Chapter 5).
11 Dirty or clogged carburetor jets or fuel injector. Carburetor or fuel injection system out of adjustment. Check the float level (Chapter 4).
12 Wet or damaged ignition components (Chapters 1 and 5).
13 Worn, faulty or incorrectly gapped spark plugs (Chapter 1).
14 ·Broken, loose or disconnected wires in the starting circuit (see previous Section).
15 Loose distributor (changing ignition timing). Turn the distributor body as necessary to start the engine, then adjust the ignition timing as soon as possible (Chapter 1).
16 Broken, loose or disconnected wires at the ignition coil or faulty coil (Chapter 5).
17 Timing belt failure or wear affecting valve timing (Chapter 2).

3 Starter motor operates without turning engine

1 Starter pinion sticking. Remove the starter (Chapter 5) and inspect.
2 Starter pinion or flywheel/driveplate teeth worn or broken. Remove the inspection cover and inspect.

4 Engine hard to start when cold

1 Battery discharged or low. Check as described in Chapter 1.
2 Fuel not reaching the carburetor or fuel injectors. Check the fuel filter, lines and fuel pump (Chapters 1 and 4).
3 Choke inoperative (Chapters 1 and 4).
4 Defective spark plugs (Chapter 1).

5 Engine hard to start when hot

1 Air filter dirty (Chapter 1).

2 Fuel not reaching carburetor or fuel injectors (see Section 4). Check for a vapor lock situation, brought about by clogged fuel tank vent lines.
3 Bad engine ground connection.
4 Carburetor choke sticking (Chapter 1).
5 Defective pick-up coil in distributor (Chapter 5).
6 Carburetor float level too high (Chapter 4).

6 Starter motor noisy or engages roughly

1 Pinion or flywheel/driveplate teeth worn or broken. Remove the inspection cover on the left side of the engine and inspect.
2 Starter motor mounting bolts loose or missing.

7 Engine starts but stops immediately

1 Loose or damaged wire harness connections at distributor, coil or alternator.
2 Intake manifold vacuum leaks. Make sure all mounting bolts/nuts are tight and all vacuum hoses connected to the manifold are attached properly and in good condition.
3 Insufficient fuel flow (see Chapter 4).

8 Engine 'lopes' while idling or idles erratically

1 Vacuum leaks. Check mounting bolts at the intake manifold for tightness. Make sure that all vacuum hoses are connected and in good condition. Use a stethoscope or a length of fuel hose held against your ear to listen for vacuum leaks while the engine is running. A hissing sound will be heard. A soapy water solution will also detect leaks. Check the intake manifold gasket surfaces.
2 Leaking EGR valve or plugged PCV valve (see Chapters 1 and 6).
3 Air filter clogged (Chapter 1).
4 Fuel pump not delivering sufficient fuel (Chapter 4).
5 Leaking head gasket. Perform a cylinder compression check (Chapter 2).
6 Timing belt worn (Chapter 2).
7 Camshaft lobes worn (Chapter 2).
8 Valve clearances out of adjustment (Chapter 2). Valves burned or otherwise leaking (Chapter 2).
9 Ignition timing out of adjustment (Chapter 1).
10 Ignition system not operating properly (Chapters 1 and 5).
11 Thermostatic air cleaner not operating properly (carbureted models) (Chapter 1).
12 Carburetor choke not operating properly (Chapters 1 and 4).
13 Dirty or clogged injectors. Carburetor or fuel injection system dirty, clogged or out of

adjustment. Check the float level (Chapter 4).
14 Idle speed out of adjustment (Chapter 1).

9 Engine misses at idle speed

1 Spark plugs faulty or not gapped properly (Chapter 1).
2 Faulty spark plug wires (Chapter 1).
3 Wet or damaged distributor components (Chapter 1).
4 Short circuits in ignition, coil or spark plug wires.
5 Sticking or faulty emissions systems (see Chapter 6).
6 Clogged fuel filter and/or foreign matter in fuel. Remove the fuel filter (Chapter 1) and inspect.
7 Vacuum leaks at intake manifold or hose connections. Check as described in Section 8.
8 Incorrect idle speed (Chapter 1) or idle mixture (Chapter 4).
9 Incorrect ignition timing (Chapter 1).
10 Low or uneven cylinder compression. Check as described in Chapter 2.
11 Carburetor choke not operating properly (Chapter 1).
12 Clogged or dirty fuel injectors (Chapter 4).

10 Excessively high idle speed

1 Sticking throttle linkage (Chapter 4).
2 Carburetor choke opened excessively at idle (Chapter 4).
3 Idle speed incorrectly adjusted (Chapter 1).

11 Battery will not hold a charge

1 Alternator drivebelt defective or not adjusted properly (Chapter 1).
2 Battery cables loose or corroded (Chapter 1).
3 Alternator not charging properly (Chapter 5).
4 Loose, broken or faulty wires in the charging circuit (Chapter 5).
5 Short circuit causing a continuous drain on the battery.
6 Battery defective internally.

12 Alternator light stays on

1 Fault in alternator or charging circuit (Chapter 5).
2 Alternator drivebelt defective or not properly adjusted (Chapter 1).

13 Alternator light fails to come on when key is turned on

1 Faulty bulb (Chapter 12).

2 Defective alternator (Chapter 5).
3 Fault in the printed circuit, dash wiring or bulb holder (Chapter 12).

14 Engine misses throughout driving speed range

1 Fuel filter clogged and/or impurities in the fuel system. Check fuel filter (Chapter 1) or clean system (Chapter 4).
2 Faulty or incorrectly gapped spark plugs (Chapter 1).
3 Incorrect ignition timing (Chapter 1).
4 Cracked distributor cap, disconnected distributor wires or damaged distributor components (Chapter 1).
5 Defective spark plug wires (Chapter 1).
6 Emissions system components faulty (Chapter 6).
7 Low or uneven cylinder compression pressures. Check as described in Chapter 2.
8 Weak or faulty ignition coil (Chapter 5).
9 Weak or faulty ignition system (Chapter 5).
10 Vacuum leaks at intake manifold or vacuum hoses (see Section 8).
11 Dirty or clogged carburetor or fuel injector (Chapter 4).
12 Leaky EGR valve (Chapter 6).
13 Carburetor or fuel injection system out of adjustment (Chapter 4).
14 Idle speed out of adjustment (Chapter 1).

15 Hesitation or stumble during acceleration

1 Ignition timing incorrect (Chapter 1).
2 Ignition system not operating properly (Chapter 5).
3 Dirty or clogged carburetor or fuel injector (Chapter 4).
4 Low fuel pressure. Check for proper operation of the fuel pump and for restrictions in the fuel filter and lines (Chapter 4).
5 Carburetor or fuel injection system out of adjustment (Chapter 4).

16 Engine stalls

1 Idle speed incorrect (Chapter 1).
2 Fuel filter clogged and/or water and impurities in the fuel system (Chapter 1).
3 Choke not operating properly (Chapter 1).
4 Distributor cap and/or wires are damaged or wet.
5 Emissions system components faulty (Chapter 6).
6 Faulty or incorrectly gapped spark plugs (Chapter 1). Also check the spark plug wires (Chapter 1).
7 Vacuum leak at the carburetor intake manifold, vacuum hoses or fuel injection system. Check as described in Section 8.
8 Valve clearances incorrect (Chapter 2).

17 Engine lacks power

1 Incorrect ignition timing (Chapter 1).
2 Excessive play in distributor shaft. At the same time check for faulty distributor cap, wires, etc. (Chapter 1).
3 Faulty or incorrectly gapped spark plugs (Chapter 1).
4 Air filter dirty (Chapter 1).
5 Ignition coil is faulty (Chapter 5).
6 Brakes binding (Chapters 1 and 10).
7 Automatic transmission fluid level incorrect, causing slippage (Chapter 1).
8 Clutch slipping (Chapter 8).
9 Fuel system filter is clogged and/or there are impurities in the fuel (Chapters 1 and 4).
10 EGR system not functioning properly (Chapter 6).
11 The fuel octane is incorrect. Fill tank with proper octane fuel.
12 Low or uneven cylinder compression pressures. Check as described in Chapter 2.
13 The carburetor, intake manifold or fuel injection system has an air leak (check as described in Section 8).
14 Carburetor jets are dirty or clogged or there is a choke malfunction or a fuel injection system malfunction (Chapters 1 and 4).
15 The cylinder compression is too low. Check as described in Chapter 2.

18 Engine backfires

1 EGR system not functioning properly (Chapter 6).
2 Ignition timing is incorrect (Chapter 1).
3 Thermostatic air cleaner system not operating properly (Chapter 6).
4 Vacuum leak (refer to Section 8).
5 Valve clearances incorrect (Chapter 2).
6 Damaged valve springs or sticking valves (Chapter 2).
7 Carburetor float level or fuel injection system out of adjustment (Chapter 4).

19 Engine surges while holding accelerator steady

1 Vacuum leak (see Section 8).
2 Fuel pump not working properly (Chapter 4).

20 Pinging or knocking engine sounds when engine is under load

1 Incorrect grade of fuel. Fill tank with fuel of the proper octane rating.
2 Ignition timing incorrect (Chapter 1).
3 Carbon build-up in combustion chambers. Remove cylinder head and clean combustion chambers (Chapter 2).
4 Incorrect spark plugs (Chapter 1).

21 Engine diesels (continues to run) after being turned off

1   Idle speed too high (Chapter 1).
2   Ignition timing incorrect (Chapter 1).
3   Incorrect spark plug heat range (Chapter 1).
4   Intake air leak (see Section 8).
5   Carbon build-up in combustion chambers. Remove the cylinder head and clean the combustion chambers (Chapter 2).
6   Valves sticking (Chapter 2).
7   Valve clearances incorrect (Chapter 2).
8   EGR system not operating properly (Chapter 6).
9   Fuel shut-off system not operating properly (Chapter 6).
10   Check for causes of overheating (Section 27).

22 Low oil pressure

1   Improper grade of oil.
2   Oil pump worn or damaged (Chapter 2).
3   Engine overheating (see Section 27).
4   Clogged oil filter (Chapter 1).
5   Clogged oil strainer (Chapter 2).
6   Oil pressure gauge not working properly (Chapter 2).

23 Excessive oil consumption

1   Oil drain plug loose.
2   Oil pan gasket bolts loose or damaged (Chapter 2).
3   Loose bolts or damaged front cover gasket (Chapter 2).
4   Front or rear crankshaft oil seal leaking (Chapter 2).
5   Loose bolts or damaged camshaft cover gasket (Chapter 2).
6   Loose oil filter (Chapter 1).
7   Loose or damaged oil pressure switch (Chapter 2).
8   Pistons and cylinders excessively worn (Chapter 2).
9   Piston rings not installed correctly on pistons (Chapter 2).
10   Worn or damaged piston rings (Chapter 2).
11   Intake and/or exhaust valve oil seals worn or damaged (Chapter 2).
12   Worn valve stems.
13   Worn or damaged valves and/or guides (Chapter 2).

24 Excessive fuel consumption

1   Dirty or clogged air filter element (Chapter 1).
2   Ignition timing is incorrect (Chapter 1).
3   Incorrect idle speed (Chapter 1).
4   Low tire pressure or incorrect tire size (Chapter 11).
5   Check all connections, lines and components in the fuel system (Chapter 4) for fuel leakage.
6   Choke not operating properly (Chapter 1).
7   Dirty or clogged carburetor jets or fuel injectors (Chapter 4).

25 Fuel odor

1   Fuel leakage. Check all connections, lines and components in the fuel system (Chapter 4).
2   Fuel tank overfilled. Fill only to automatic shut-off.
3   Charcoal canister filter in Evaporative Emissions Control system clogged (Chapter 1).
4   Vapor leaks from Evaporative Emissions Control system lines (Chapter 6).

26 Miscellaneous engine noises

1   A strong dull noise that becomes more rapid as the engine accelerates indicates worn or damaged crankshaft bearings or an unevenly worn crankshaft. To pinpoint the trouble spot, remove the spark plug wire from one plug at a time and crank the engine over. If the noise stops, the cylinder with the removed plug wire indicates the problem area. Replace the bearing and/or service or replace the crankshaft (Chapter 2).
2   A similar (yet slightly higher pitched) noise to the crankshaft knocking described in the previous paragraph, that becomes more rapid as the engine accelerates, indicates worn or damaged connecting rod bearings (Chapter 2). The procedure for locating the problem cylinder is the same as described in Paragraph 1.
3   An overlapping metallic noise that increases in intensity as the engine speed increases, yet diminishes as the engine warms up indicates abnormal piston and cylinder wear (Chapter 2). To locate the problem cylinder, use the procedure described in Paragraph 1.
4   A rapid clicking noise that becomes faster as the engine accelerates indicates a worn piston pin or piston pin hole. This sound occurs each time the piston hits the highest and lowest points in the stroke (Chapter 2). The procedure for locating the problem piston is described in Paragraph 1.
5   A metallic clicking noise coming from the water pump indicates worn or damaged water pump bearings or pump. Replace the water pump with a new one (Chapter 3).
6   A rapid tapping sound or clicking sound that becomes faster as the engine speed increases indicates "valve tapping" or improperly adjusted valve clearances. This can be identified by holding one end of a section of hose to your ear and placing the other end at different spots along the rocker arm cover. The point where the sound is loudest indicates the problem valve. Adjust the valve clearances (Chapter 2). If the problem persists, you likely have a damaged valve train component. Changing the engine oil and adding a high viscosity oil treatment will sometimes cure a stuck lifter problem. If the problem still persists, the lifters and rocker arms must be removed for inspection (Chapter 2).

Cooling system

27 Overheating

1   Insufficient coolant in system (Chapter 1).
2   Drivebelt defective or not adjusted properly (Chapter 1).
3   Radiator core blocked or radiator grille dirty and restricted (Chapter 3).
4   Thermostat faulty (Chapter 3).
5   Fan not functioning properly (Chapter 3).
6   Radiator cap is not maintaining proper pressure. Have the cap pressure tested by gas station or repair shop.
7   The ignition timing setting is incorrect (Chapter 1).
8   Defective water pump (Chapter 3).
9   Engine oil viscosity improper.
10   Temperature gauge is inaccurate or faulty (Chapter 12).

28 Overcooling

1   Thermostat faulty (Chapter 3).
2   Inaccurate or faulty temperature gauge (Chapter 12).

29 External coolant leakage

1   Hoses are deteriorated or damaged. Loose clamps at hose connections (Chapter 1).
2   Water pump seals defective. If this is the case, water will drip from the weep hole in the water pump body (Chapter 3).
3   Leakage from radiator core or header tank. This will require the radiator to be professionally repaired (see Chapter 3 for removal procedures).
4   Engine drain plugs or water jacket freeze plugs leaking (see Chapters 1 and 2).
5   Leak from coolant temperature switch (Chapter 3).
6   Leak from damaged gaskets or small cracks (Chapter 2).
7   Damaged head gasket. This can sometimes be verified by checking the condition of the engine oil as noted in Section 30.

30   Internal coolant leakage

Note: *Internal coolant leaks can usually be detected by examining the oil. Check the dipstick and inside the rocker arm cover for water deposits and an oil consistency like that of a milkshake.*

1   Leaking cylinder head gasket. Have the system pressure tested or remove the cylinder head (Chapter 2) and inspect.
2   Cylinder bore and/or cylinder head are cracked. Dismantle engine and inspect (Chapter 2).
3   The cylinder head bolts are loose (tighten as described in Chapter 2).

31   Abnormal coolant loss

1   Overfilled system (Chapter 1).
2   Coolant boiling away due to overheating (see causes in Section 27).
3   Internal or/or external leakage (see Sections 29 and 30).
4   Faulty or leaking radiator cap. Have the cap pressure tested.
5   The cooling system being pressurized by engine compression. This could be due to a cracked head or block or leaking head gasket.

32   Poor coolant circulation

1   Inoperative water pump. A quick test is to pinch the top radiator hose closed with your hand while the engine is idling, then release it. You should feel a surge of coolant if the pump is working properly (Chapter 3).
2   Restriction in cooling system. Drain, flush and refill the system (Chapter 1). If necessary, remove the radiator (Chapter 3) and have it reverse flushed or professionally cleaned.
3   Water pump drivebelt is loose (Chapter 1).
4   The thermostat is sticking (Chapter 3).
5   Coolant level is too low (Chapter 1).

33   Corrosion

1   Excessive impurities in the water. Soft, clean water is recommended. Distilled or rainwater is satisfactory.
2   Insufficient antifreeze solution (refer to Chapter 1 for the proper ratio of water to antifreeze).
3   Infrequent flushing and draining of system. Regular flushing of the cooling system should be carried out at the specified intervals as described in (Chapter 1).

Clutch

Note: *All clutch related service information is located in Chapter 8, unless otherwise noted.*

34   Fails to release (pedal pressed to the floor - shift lever does not move freely in and out of Reverse)

1   Clutch contaminated with oil. Remove clutch plate and inspect. Determine source of oil leak.
2   Clutch plate warped, distorted or otherwise damaged.
3   Diaphragm spring fatigued. Remove clutch cover/pressure plate assembly and inspect.
4   Clutch cable broken or improperly adjusted.
5   Insufficient pedal stroke. Check and adjust as necessary.
6   Lack of grease on pilot bushing.
7   Leak in the clutch hydraulic system. Check the master cylinder, slave cylinder and lines (Chapters 1 and 8).

35   Clutch slips (engine speed increases with no increase in vehicle speed)

1   Worn or oil soaked clutch plate. Determine source of oil leak.
2   Clutch plate not broken in. It may require 30 or 40 normal starts for a new clutch to seat.
3   Diaphragm spring weak or damaged. Remove clutch cover/pressure plate assembly and inspect.
4   Flywheel warped (Chapter 2).
5   Clutch cable sticking.

36   Grabbing (chattering) as clutch is engaged

1   Oil on clutch plate. Remove and inspect. Repair any leaks.
2   Worn or loose engine or transmission mounts. They may move slightly when clutch is released. Inspect mounts and bolts.
3   Worn splines on transmission input shaft. Remove and inspect the clutch components.
4   Warped pressure plate or flywheel. Remove and inspect the clutch components.
5   Diaphragm spring fatigued. Remove clutch cover/pressure plate assembly and inspect.
6   Clutch linings hardened or warped.
7   Clutch lining rivets loose.

37   Squeal or rumble with clutch engaged (pedal released)

1   Improper pedal adjustment. Adjust pedal free play.
2   Release bearing binding on transmission shaft. Remove clutch components and

check bearing. Remove any burrs or nicks, clean and relubricate before reinstallation.
3   Clutch rivets loose.
4   Clutch plate cracked.
5   Fatigued clutch plate torsion springs. Replace clutch plate.

38   Squeal or rumble with clutch disengaged (pedal depressed)

1   Worn or damaged release bearing.
2   Worn or broken pressure plate diaphragm fingers.
3   Pilot bearing worn or damaged.

39   Clutch pedal stays on floor when disengaged

Binding cable or release bearing. Inspect cable or remove clutch components as necessary.

Manual transmission

Note: *All manual transmission service information is located in Chapter 7A, unless otherwise noted.*

40   Noisy in Neutral with engine running

1   Input shaft bearing worn.
2   Damaged main drive gear bearing.
3   Insufficient transmission lubricant (Chapter 1).
4   Transmission lubricant in poor condition. Drain and fill with proper grade lubricant. Check old lubricant for water and debris (Chapter 1).
5   Noise can be caused by variations in engine torque. Change the idle speed and see if noise disappears.

41   Noisy in all gears

1   Any of the above causes, and/or:
2   Worn or damaged output gear bearings or shaft.

42   Noisy in one particular gear

1   Worn, damaged or chipped gear teeth.
2   Worn or damaged synchronizer.

43   Slips out of gear

1   Transmission loose on clutch housing.
2   Stiff shift lever seal.
3   Shift linkage binding.
4   Broken or loose input gear bearing retainer.

5 Dirt between clutch lever and engine housing.
6 Worn linkage.
7 Damaged or worn check balls, fork rod ball grooves or check springs.
8 Worn mainshaft or countershaft bearings.
9 Loose engine mounts (Chapter 2).
10 Excessive gear end play.
11 Worn synchronizers.

44 Oil leaks

1 Excessive amount of lubricant in transmission (see Chapter 1 for correct checking procedures). Drain lubricant as required.
2 Rear oil seal or speedometer oil seal damaged.
3 To pinpoint a leak, first remove all built-up dirt and grime from the transmission. Degreasing agents and/or steam cleaning will achieve this. With the underside clean, drive the vehicle at low speeds so the air flow will not blow the leak far from its source. Raise the vehicle and determine where the leak is located.

45 Difficulty engaging gears

1 Clutch not releasing completely.
2 Loose or damaged shift linkage. Make a thorough inspection, replacing parts as necessary.
3 Insufficient transmission lubricant (Chapter 1).
4 Transmission lubricant in poor condition. Drain and fill with proper grade lubricant. Check lubricant for water and debris (Chapter 1).
5 Worn or damaged striking rod.
6 Sticking or jamming gears.

46 Noise occurs while shifting gears

1 Check for proper operation of the clutch (Chapter 8).
2 Faulty synchronizer assemblies. Check for wear or damage to baulk rings or any parts of the synchromesh assemblies.

Automatic transmission

Note: *Due to the complexity of the automatic transmission, it's difficult for the home mechanic to properly diagnose and service. For problems other than the following, the vehicle should be taken to a reputable mechanic.*

47 Fluid leakage

1 Automatic transmission fluid is a deep red color, and fluid leaks should not be confused with engine oil which can easily be blown by air flow to the transmission.
2 To pinpoint a leak, first remove all built-up dirt and grime from the transmission. Degreasing agents and/or steam cleaning will achieve this. With the underside clean, drive the vehicle at low speeds so the air flow will not blow the leak far from its source. Raise the vehicle and determine where the leak is located. Common areas of leakage are:

a) *Fluid pan: tighten mounting bolts and/or replace pan gasket as necessary (Chapter 1).*
b) *Rear extension: tighten bolts and/or replace oil seal as necessary.*
c) *Filler pipe: replace the rubber oil seal where pipe enters transmission case.*
d) *Transmission oil lines: tighten fittings where lines enter transmission case and/or replace lines.*
e) *Vent pipe: transmission overfilled and/or water in fluid (see checking procedures, Chapter 1).*
f) *Speedometer connector: replace the O-ring where speedometer cable or speed sensor sender enters transmission case.*

48 General shift mechanism problems

Chapter 7B deals with checking and adjusting the shift linkage on automatic transmissions. Common problems that could be caused by out of adjustment linkage are:

a) *Engine starting in gears other than P (park) or N (Neutral).*
b) *Indicator pointing to a gear other than the one actually engaged.*
c) *Vehicle moves with transmission in P (Park) position.*

49 Transmission will not downshift with the accelerator pedal pressed to the floor

Chapter 7B deals with adjusting the TV linkage to enable the transmission to downshift properly.

50 Engine will start in gears other than Park or Neutral

Chapter 7B deals with adjusting the Neutral start switch installed on automatic transmissions.

51 Transmission slips, shifts rough, is noisy or has no drive in forward or Reverse gears

1 There are many probable causes for the above problems, but the home mechanic should concern himself only with one possibility: fluid level.
2 Before taking the vehicle to a shop, check the fluid level and condition as described in Chapter 1. Add fluid, if necessary, or change the fluid and filter if needed. If problems persist, have a professional diagnose the transmission.

Driveshaft

52 Leaks at front of driveshaft

Defective transmission rear seal. See Chapter 7 for replacement procedure. As this is done, check the splined yoke for burrs or roughness that could damage the new seal. Remove burrs with a fine file or whetstone.

53 Knock or clunk when transmission is under initial load (just after transmission is put into gear)

1 Loose or disconnected rear suspension components. Check all mounting bolts and bushings (Chapters 7 and 10).
2 Loose driveshaft bolts. Inspect all bolts and nuts and tighten them securely.
3 Worn or damaged universal joint bearings. Replace the universal joint(s) (Chapter 8).
4 Worn sleeve yoke and mainshaft spline.

54 Metallic grating sound consistent with vehicle speed

Pronounced wear in the universal joint bearings. Replace U-joints as necessary.

55 Vibration

Note: *Before blaming the driveshaft, make sure the tires are perfectly balanced and perform the following test.*
1 Install a tachometer inside the vehicle to monitor engine speed as the vehicle is driven. Drive the vehicle and note the engine speed at which the vibration (roughness) is most pronounced. Now shift the transmission to a different gear and bring the engine speed to the same point.
2 If the vibration occurs at the same engine speed (rpm) regardless of which gear the transmission is in, the driveshaft is NOT at fault since the driveshaft speed varies.
3 If the vibration decreases or is eliminated when the transmission is in a different gear at the same engine speed, refer to the following probable causes.
4 Bent or dented driveshaft. Inspect and replace as necessary.
5 Undercoating or built-up dirt, etc. on the driveshaft. Clean the shaft thoroughly.

6   Worn universal joint bearings. Replace the U-joints as necessary.
7   Driveshaft and/or companion flange out of balance. Check for missing weights on the shaft. Remove driveshaft and reinstall 180-degrees from original position, then recheck. Have the driveshaft balanced if problem persists.
8   Loose driveshaft mounting bolts/nuts.
9   Defective center bearing, if so equipped.
10   Worn transmission rear bushing (Chapter 7).

56   Scraping noise

Make sure the dust cover on the sleeve yoke isn't rubbing on the transmission extension housing.

57   Whining or whistling noise

Defective center bearing, if so equipped.

Rear axle and differential

58   Noise - same when in drive as when vehicle is coasting

1   Road noise. No corrective action available.
2   Tire noise. Inspect tires and check tire pressures (Chapter 1).
3   Front wheel bearings loose, worn or damaged (Chapter 1).
4   Insufficient differential oil (Chapter 1).
5   Defective differential.

59   Knocking sound when starting or shifting gears

Defective or incorrectly adjusted differential.

60   Noise when turning

Defective differential.

61   Vibration

See probable causes under Driveshaft. Proceed under the guidelines listed for the driveshaft. If the problem persists, check the rear wheel bearings by raising the rear of the vehicle and spinning the wheels by hand. Listen for evidence of rough (noisy) bearings. Remove and inspect (Chapter 8).

62   Oil leaks

1   Pinion oil seal damaged (Chapter 8).

2   Axleshaft oil seals damaged (Chapter 8).
3   Differential cover leaking. Tighten mounting bolts or replace the gasket as required.
4   Loose filler or drain plug on differential (Chapter 1).
5   Clogged or damaged breather on differential.

Transfer case

63   Gear jumping out of mesh

1   Interference between the control lever and the console.
2   Play or fatigue in the transfer case mounts.
3   Internal wear or incorrect adjustments.

64   Difficult shifting

1   Lack of oil.
2   Internal wear, damage or incorrect adjustment.

65   Noise

1   Lack of oil in transfer case.
2   Noise in 4H and 4L, but not in 2H indicates cause is in the front differential or front axle.
3   Noise in 2H, 4H and 4L indicates cause is in rear differential or rear axle.
4   Noise in 2H and 4H but not in 4L, or in 4L only, indicates internal wear or damage in transfer case.

Brakes

Note: *Before assuming a brake problem exists, make sure the tires are in good condition and inflated properly, the front end alignment is correct and the vehicle is not loaded with weight in an unequal manner. All service procedures for the brakes are included in Chapter 9, unless otherwise noted.*

66   Vehicle pulls to one side during braking

1   Defective, damaged or oil contaminated brake pad on one side. Inspect as described in Chapter 1. Refer to Chapter 9 if replacement is required.
2   Excessive wear of brake pad material or disc on one side. Inspect and repair as necessary.
3   Loose or disconnected front suspension components. Inspect and tighten all bolts securely (Chapters 1 and 10).
4   The caliper assembly may be defective. Remove the caliper and inspect for stuck pis-

ton or damage.
5   Scored brake disc.
6   Caliper mounting bolts may be loose.
7   Wheel bearing adjustment may be incorrect.

67   Noise (high-pitched squeal)

1   Front brake pads worn out. This noise comes from the wear sensor rubbing against the disc. Replace pads with new ones immediately!
2   Glazed or contaminated pads.
3   Dirty or scored disc.
4   Bent support plate.

68   Excessive brake pedal travel

1   Partial brake system failure. Inspect entire system (Chapter 1) and correct as required.
2   Insufficient fluid in master cylinder. Check (Chapter 1) and add fluid - bleed system if necessary.
3   Air in system. Bleed system.
4   Excessive lateral disc play.
5   Brakes out of adjustment. Check the operation of the automatic adjusters.

69   Brake pedal feels spongy when depressed

1   Air in brake lines. Bleed the brake system.
2   Deteriorated rubber brake hoses. Inspect all system hoses and lines. Replace parts as necessary.
3   Master cylinder-mounting nuts loose. Inspect master cylinder nuts and tighten them securely.
4   Master cylinder faulty.
5   Incorrect shoe or pad clearance.
6   Clogged reservoir cap vent hole.
7   Deformed rubber brake lines.
8   Soft or swollen caliper seals.
9   Poor quality brake fluid. Bleed entire system and fill with new approved fluid.

70   Excessive effort required to stop vehicle

1   Power brake booster not operating properly.
2   Excessively worn linings or pads. Check and replace if necessary.
3   One or more caliper pistons seized or sticking. Inspect and rebuild as required.
4   Brake pads or linings contaminated with oil or grease. Inspect and replace as required.
5   New pads or linings installed and not yet seated. It'll take a while for the new material to seat against the rotor or drum.
6   Worn or damaged master cylinder or

caliper assemblies. Check particularly for frozen pistons.
7   Refer to the causes listed under Section 69.

71 Pedal travels to the floor with little resistance

1   Little or no fluid in the master cylinder reservoir caused by leaking caliper piston(s) or loose, damaged or disconnected brake lines. Inspect entire system and repair as necessary.
2   Defective master cylinder.

72 Brake pedal pulsates during brake application

1   Wheel bearings damaged, worn or out of adjustment (Chapter 1).
2   Caliper not sliding properly due to improper installation or obstructions. Remove and inspect.
3   Disc not within specifications. Check the disc for excessive lateral runout and parallelism. Have the discs resurfaced or replace them with new ones. Also make sure that all discs are the same thickness.
4   Out of round rear brake drums. Remove the drums and have them machined or replace them with new ones.

73 Brakes drag (indicated by sluggish engine performance or wheels being very hot after driving)

1   Output rod adjustment incorrect at the brake pedal.
2   Obstructed master cylinder compensator. Disassemble master cylinder and clean.
3   Master cylinder piston seized in bore. Overhaul master cylinder.
4   Brake caliper assembly in need of overhaul.
5   Brake pads or shoes worn out.
6   Piston cups in master cylinder or caliper assembly deformed. Overhaul master cylinder.
7   Parking brake assembly will not release.
8   Clogged brake lines.
9   Out of adjustment wheel bearings (Chapter 1).
10   Brake pedal height improperly adjusted.
11   Wheel cylinder needs overhaul.
12   Shoe to drum clearance improper. Adjust as necessary.

74 Rear brakes lock up under light brake application

1   Tire pressures too high.

2   Tires excessively worn (Chapter 1).
3   The Load Sensing Proportioning Valve is defective or out of adjustment.

75 Rear brakes lock up under heavy brake application

1   Tire pressures too high.
2   Tires excessively worn (Chapter 1).
3   Front brake pads contaminated with oil, mud or water. Clean or replace the pads.
4   Front brake pads excessively worn.
5   The master cylinder and/or caliper assembly is defective.

Suspension and steering

Note: *All service procedures for the suspension and steering systems are included in Chapter 10, unless otherwise noted.*

76 Vehicle pulls to one side

1   Tire pressures uneven (Chapter 1).
2   Defective tire (Chapter 1).
3   Excessive wear in suspension or steering components (Chapter 1).
4   Front wheel alignment is incorrect.
5   Front brakes dragging. Inspect as described in Section 73.
6   Wheel bearings improperly adjusted (Chapter 1).
7   Wheel lug nuts loose.

77 Shimmy, shake or vibration

1   Tires and/or wheels out of balance or out of round. Have them balanced on the vehicle.
2   Loose, worn or out of adjustment wheel bearings (Chapter 1).
3   Shock absorbers and/or suspension components are worn or damaged. Check for worn bushings.
4   Wheel lug nuts loose.
5   Incorrectly set tire pressures.
6   Excessively worn or damaged tire.
7   Loosely mounted steering gear housing.
8   Steering gear improperly adjusted.
9   Loose, worn or damaged steering components.
10   Damaged idler arm.
11   Worn balljoint.

78 Excessive pitching and/or rolling around corners or during braking

1   Defective shock absorbers. Replace as a set.
2   Broken or weak springs and/or suspension components.
3   Stabilizer bar or bushings are worn or damaged.

79 Wandering or general instability

1   Improper tire pressures.
2   Worn or damaged tension rod bushings.
3   Incorrect front end alignment.
4   Worn or damaged steering linkage or suspension components.
5   Improperly adjusted steering gear.
6   Out of balance wheels.
7   Loose wheel lug nuts.
8   Worn rear shock absorbers.
9   Fatigued or damaged springs.

80 Excessively stiff steering

1   Lack of lubricant in power steering fluid reservoir, where appropriate (Chapter 1).
2   Tire pressures are incorrect (Chapter 1).
3   Balljoints lack lubrication (Chapter 1).
4   Front end out of alignment.
5   Steering gear is out of adjustment or lacks lubrication.
6   Improperly adjusted wheel bearings.
7   Worn or damaged steering gear.
8   Interference of steering column with turn signal switch.
9   Tire pressures low.
10   Balljoints worn or damaged.

81 Excessive play in steering

1   Loose wheel bearings (Chapter 1).
2   Excessive wear in suspension bushings (Chapter 1).
3   Steering gear improperly adjusted.
4   Front wheel alignment is incorrect.
5   Steering gear mounting bolts loose.
6   Steering linkage is worn.

82 Lack of power assistance

1   Power steering pump drivebelt faulty or not adjusted properly (Chapter 1).
2   Fluid level low (Chapter 1).
3   Hoses or pipes restrict the flow. Inspect and replace parts as necessary.
4   The power steering system has air in it. Bleed system.
5   Defective power steering pump.

83 Steering wheel fails to return to straight-ahead position

1   Incorrect front end alignment.
2   Tire pressures low.
3   Steering gears improperly engaged.
4   Steering column out of alignment.
5   Balljoint is worn or damaged.
6   Steering linkage is worn or damaged.
7   Improperly lubricated idler arm.
8   Lack of oil in steering gear.
9   Lack of fluid in the pump for the power steering.

84 Steering effort not the same in both directions (power system)

1 Leaks in steering gear.
2 Clogged fluid passage in steering gear.

85 Noisy power steering pump

1 Insufficient oil in pump.
2 Clogged hoses or oil filter in pump.
3 Loose pulley.
4 Improperly adjusted drivebelt (Chapter 1).
5 Defective pump.

86 Miscellaneous noises

1 Improper tire pressures.
2 Insufficiently lubricated balljoint or steering linkage.
3 Steering gear assembly and/or steering linkage and suspension components loose or worn.
4 Defective shock absorber.
5 Wheel bearing is defective.
6 Suspension bushings worn or damaged.
7 Damaged leaf spring.
8 Loose wheel lug nuts.
9 Worn or damaged rear axleshaft spline.
10 Worn or damaged rear shock absorber mounting bushing.
11 Rear axle end play is incorrect.
12 See also causes of noises at the rear axle and driveshaft.

87 Excessive tire wear (not specific to one area)

1 Incorrect tire pressures.
2 Tires out of balance. Have them balanced on the vehicle.
3 Wheels damaged. Inspect and replace as necessary.
4 Suspension or steering components worn (Chapter 1).

88 Excessive tire wear on outside edge

1 Incorrect tire pressure.
2 Excessive speed in turns.
3 Front end alignment incorrect (excessive toe-in).

89 Excessive tire wear on inside edge

1 Incorrect tire pressure.
2 Front end alignment incorrect (toe-out).
3 Loose or damaged steering components (Chapter 1).

90 Tire tread worn in one place

1 Tires out of balance. Have them balanced on the vehicle.
2 Damaged or buckled wheel. Inspect and replace if necessary.
3 Defective tire.

Chapter 1   Tune-up and routine maintenance

Contents

Specifications

Recommended lubricants and fluids

Note: *Listed here are manufacturer recommendations at the time this manual was written. Manufacturers occasionally upgrade their lubricant specifications, so check with your local auto parts store for current recommendations.*

Engine oil	API rated multigrade and fuel-efficient oil
Viscosity	See accompanying chart
Automatic transmission fluid	Dexron II or III automatic transmission fluid
Manual transmission lubricant	API GL-4 SAE 75W-90 gear lubricant
Transfer case lubricant	API GL-4 SAE 75W-90 gear lubricant
Rear axle differential lubricant	API GL-5 SAE 75W-90 or 80W-90 gear lubricant

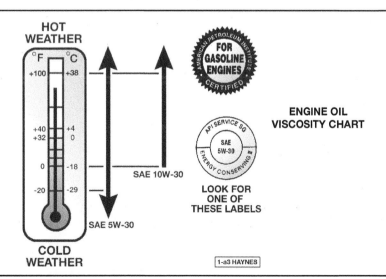

ENGINE OIL VISCOSITY CHART

Recommended lubricants and fluids (continued)

Front axle differential lubricant .... API GL-5 SAE 75W-90 or 80W-90 gear lubricant
Brake fluid .... DOT 3 brake fluid
Power steering system fluid .... Dexron II or III automatic transmission fluid

Capacities*

Engine oil (including filter)
    1.3L engine .... 4.0 qts
    1.6L engine .... 4.5 qts
    2.0L engine .... 5.5 qts
Cooling system.. Approximately 5.3 qts
Fuel tank... 10.6 gal
Differential lubricant
    Samurai
        Front .... 2.1 qts
        Rear .... 1.6 qts
    Sidekick/X-90/Vitara/Tracker
        Front .... 1.1 qts
        Rear .... 2.3 qts
Automatic transmission fluid .... 3.0 qts
Manual transmission lubricant
    Samurai .... 2.7 pts
    Sidekick/X-90/Vitara/Tracker .... 3.2 pts
Transfer case lubricant
    Samurai .... 1.7 pts
    Sidekick/X-90/Vitara/Tracker .... 3.6 pts

All capacities are approximate. Add as necessary to bring to appropriate level.

Ignition system

Spark plug
    Type
        1990 and earlier.. NGK BPR-5ES
        1991 through 1997 .... NGK BKR-6E
        1998 and later... NGK BKR-6E-11
    Gap
        1997 and earlier.. 0.028 to 0.031 inch (0.7 to 0.8 mm)
        1998 and later... 0.039 to 0.043 inch (1.0 to 1.1mm)
Ignition timing (not adjustable on later models)
    Samurai .... 10-degrees BTDC at 800 RPM
    Sidekick/X-90/Vitara/Tracker .... 8-degrees BTDC at 800 RPM
Engine firing order 1-3-4-2

8-VALVE   **16-VALVE**

0791H   0791H

FIRING ORDER   **FIRING ORDER**
1-3-4-2   **1-3-4-2**

Cylinder location and distributor rotation

The blackened terminal shown on the distributor cap indicates the Number One spark plug wire position

Valve clearances (1.3L and 1.6L engines only)

8-valve engine
    Intake
        Cold .... 0.005 to 0.007 inch (0.13 to 0.17 mm)
        Hot .... 0.009 to 0.011 inch (0.23 to 0.27 mm)
    Exhaust
        Cold .... 0.006 to 0.008 inch (0.16 to 0.20 mm)
        Hot .... 0.010 to 0.012 inch (0.26 to 0.30 mm)
16-valve engine (intake and exhaust)
    Cold.. 0.005 to 0.007 inch (0.13 to 0.17 mm)
    Hot .... 0.007 to 0.008 inch (0.17 to 0.21 mm)

Thermostat rating

Starts to open... 179-degrees F (82-degrees C)
Fully open .... 212-degrees F (100-degrees C)

Drivebelt deflection (under moderate pressure)

Alternator 1/4-inch
Air conditioning compressor 1/4-inch
Power steering pump 1/2-inch

Clutch

Clutch pedal
    Freeplay .... 0.6 to 1.0 in (15 to 25 mm)
    Height
        Samurai .... Level with brake pedal
        Sidekick/X-90/Vitara/Tracker.. 0.2 in (5.0 mm) above brake pedal
Clutch release arm freeplay .... 0.02 to 0.06 in (0.5 to 1.5 mm)

Brakes

Disc brake pad lining thickness (minimum)	0.04 in (1 mm)
Drum brake shoe lining thickness (minimum)	0.04 in (1 mm)
Brake pedal freeplay	0.04 to 0.32 in (1 to 8 mm)
Brake pedal height	
1998 and earlier	5.12 in (130 mm)
1999 and later	7.0 in (185 mm)
Parking brake adjustment	7 to 9 clicks

Steering wheel freeplay limit

0.40 to 1.2 in (10 to 30 mm)

Torque specifications

Ft-lbs (unless otherwise indicated)

Automatic transmission pan bolts	108 in-lbs
Manual transmission drain and check/fill plugs	20
Engine mounting center member bolt	43
Wheel lug nuts	
Tracker	
1991 and earlier	50
1992	60
1993 and later	70
Vitara, X-90	70
Samurai	50
Sidekick	
1994 and earlier	50
1995 and later	70
Spark plugs	21
Samurai steering knuckle seal retainer bolts	96 in-lbs
Carburetor/throttle body mounting nuts	13.5 to 20
Oxygen sensor	40
Differential drain plug	
Front	32
Rear	36

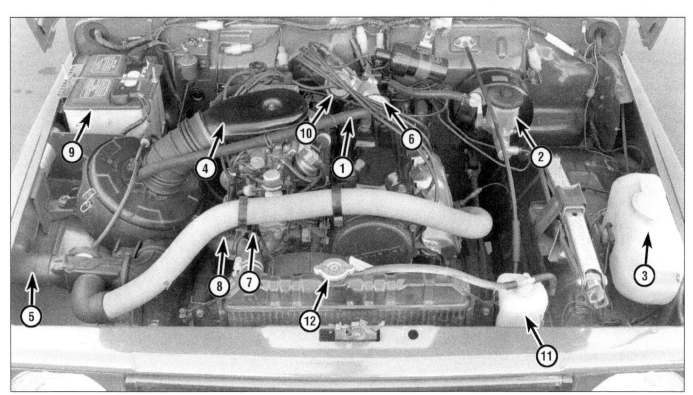

Engine compartment components (Samurai)

1	Crankcase Ventilation (CCV) system hose and fitting	3	Windshield washer reservoir	6	Oil filler cap	10	Distributor
				7	Engine oil dipstick	11	Coolant reservoir
		4	Air intake case	8	Radiator hose	12	Radiator cap
2	Brake fluid reservoir	5	Fresh air intake hose	9	Battery		

Front underside components (Samurai)

1	Steering knuckle	5	Brake hose	8	Front axle housing
2	Transmission	6	Steering linkage	9	Transfer case
3	Engine oil drain plug	7	Steering damper	10	Front differential
4	Exhaust pipe				

1 Introduction

This Chapter is designed to help the home mechanic maintain the Samurai/Sidekick/X-90/Vitara/Tracker with the goals of maximum performance, economy, safety and reliability in mind.

Included is a master maintenance schedule, followed by procedures dealing specifically with each item on the schedule. Visual checks, adjustments, component replacement and other helpful items are included. Refer to the accompanying illustrations of the engine compartment and the underside of the vehicle for the locations of various components.

Servicing your vehicle in accordance with the mileage/time maintenance schedule and the step-by-step procedures will result in a planned maintenance program that should produce a long and reliable service life. Keep in mind that it is a comprehensive plan, so maintaining some items but not others at the specified intervals will not produce the same results.

As you service your vehicle, you will dis-

Rear underside components (Samurai)

1	Rear leaf spring	5	Fuel filter	8	Fuel tank
2	Muffler	6	Parking brake cable	9	Exhaust pipe
3	Driveshaft	7	Rear differential	10	Shock absorber
4	Universal joint				

cover that many of the procedures can - and should - be grouped together because of the nature of the particular procedure you're performing or because of the close proximity of two otherwise unrelated components to one another.

For example, if the vehicle is raised for chassis lubrication, you should inspect the exhaust; suspension, steering and fuel sys-

tems while you're under the vehicle. When you're rotating the tires, it makes good sense to check the brakes since the wheels are already removed. Finally, let's suppose you have to borrow or rent a torque wrench. Even if you only need it to tighten the spark plugs, you might as well check the torque of as many critical fasteners as time allows.

The first step in this maintenance pro-

gram is to prepare yourself before the actual work begins. Read through all the procedures you're planning to do, then gather up all the parts and tools needed. If it looks like you might run into problems during a particular job, seek advice from a mechanic or an experienced do-it-yourselfer.

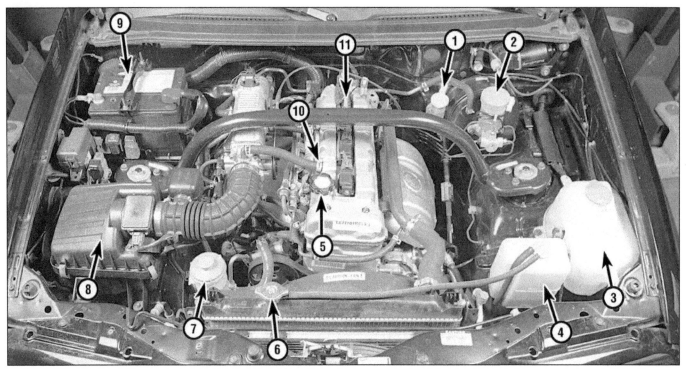

Engine compartment components (Sidekick/X-90/Vitara/Tracker)

1	Brake fluid reservoir	5	Oil filler cap	9	Battery
2	Clutch fluid reservoir	6	Radiator cap	10	Positive Crankcase Ventilation (PCV)
3	Windshield washer reservoir	7	Power steering fluid reservoir		valve
4	Coolant reservoir	8	Air cleaner housing	11	Spark plug ignition coil

Front underside components (Sidekick/X-90/Vitara/Tracker 2WD model)

1	Lower control arm bushing	4	Engine oil filter	7	Engine oil drain plug
2	Rack and pinion steering housing	5	Catalytic converter and exhaust pipe	8	Stabilizer bar
3	Tie rod end	6	Transmission drain plug	9	Balljoint

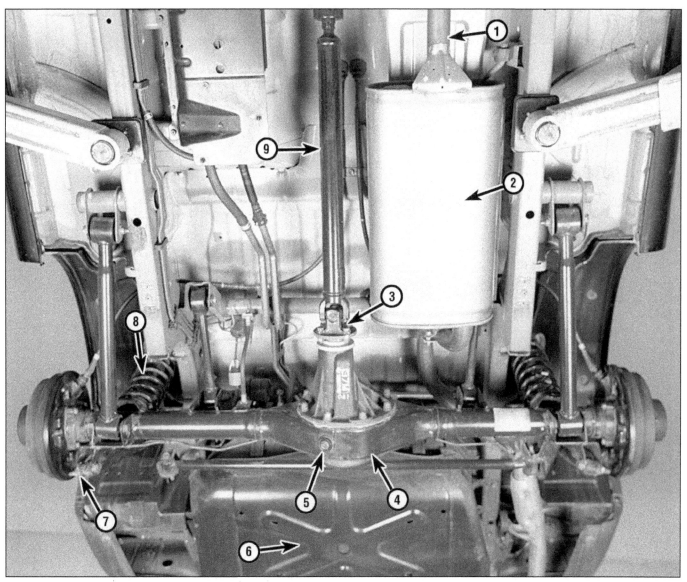

Rear underside components (Sidekick/X-90/Vitara/Tracker)

1	Exhaust pipe	4	Rear differential	7	Shock absorber
2	Muffler	5	Differential drain plug	8	Rear spring
3	Universal joint	6	Fuel tank	9	Driveshaft

2 Samurai/Sidekick/X-90/Vitara/Tracker Maintenance schedule

The following maintenance intervals are based on the assumption that the vehicle owner will be doing the maintenance or service work, as opposed to having a dealer service department do the work. Although the time/mileage intervals are loosely based on factory recommendations, most have been shortened to ensure, for example, that such items as lubricants and fluids are checked/changed at intervals that promote maximum engine/driveline service life. Also, subject to the preference of the individual owner interested in keeping his or her vehicle in peak condition at all times, and with the vehicle's ultimate resale in mind, many of the maintenance procedures may be performed more often than recommended in the following schedule. We encourage such owner initiative.

When the vehicle is new it should be serviced initially by a factory authorized dealer service department to protect the factory warranty. In many cases the initial maintenance check is done at no cost to the owner (check with your dealer service department for more information).

Every 250 miles or weekly, whichever comes first

Check the engine oil level (Section 4)
Check the engine coolant level (Section 4)
Check the windshield washer fluid level (Section 4)
Check the brake and clutch fluid levels (Section 4)
Check the tires and tire pressures (Section 5)

Every 3000 miles or 3 months, whichever comes first

All items listed above plus . . .
Check the automatic transmission fluid level (Section 6)
Check the power steering fluid level (Section 7)
Check and service the battery (Section 8)
Check the cooling system (Section 9)
Inspect and replace if necessary all underhood hoses (Section 10)
Inspect and replace if necessary the windshield wiper blades (Section 11)
Change the engine oil and oil filter (Section 12)

Every 7500 miles or 6 months, whichever comes first

All items listed above plus . . .
Check and adjust if necessary, the engine drivebelts (Section 13)
Rotate the tires (Section 14)
Check the clutch pedal for proper height and freeplay (Section 15)
Check and adjust if necessary, the brake pedal height (Section 17)

Inspect the brake system (Section 18)
Check the manual transmission lubricant (Section 19)*
Check the transfer case lubricant (Section 20)*
Check the differential lubricant (Section 21)*

Every 15,000 miles or 12 months, whichever comes first

All items listed above plus . . .
Inspect the suspension and steering components (Section 22)**
Check the driveaxle boots (Sidekick/X-90/Vitara/Tracker) (Section 23)**
Check the driveshafts and lubricate the driveaxles (Section 24)
Inspect the fuel system (Section 25)
Check the carburetor choke (Section 26)
Check and adjust if necessary, the engine valve clearances (1.3L and 1.6L engines) (Section 16)
Check, and adjust if necessary, the engine idle speed (Section 33)
Check and repack the front wheel bearings (see Chapter 8)

Every 22,500 miles or 18 months, whichever comes first

Replace the steering knuckle oil seals (Samurai only) (Section 27)

Every 30,000 miles or 24 months, whichever comes first

Replace the spark plugs (Section 28)**
Check and replace, if necessary the spark plug wires (Section 29)
Inspect, and replace if necessary, distributor cap and rotor (Section 30)**
Replace the air filter (Section 31)**
Check the operation of the thermostatic air cleaner (Samurai) (Section 32)
Service the cooling system (drain, flush and refill) (Section 34)
Inspect the exhaust system (Section 35)
Change the automatic transmission fluid and filter (Sidekick/X-90/Vitara/Tracker) (Section 36)
Change the manual transmission lubricant (Section 37)
Change the transfer case lubricant (Section 38)
Change the differential lubricant (Section 39)
Replace the fuel filter (Section 40)

Every 50,000 miles or 40 months, whichever comes first

Check the EGR system (Section 41)
Replace the PCV valve (Section 42)

Every 60,000 miles or 48 months, whichever comes first

Replace the fuel tank cap (Section 43)
Replace the brake fluid (Section 44)***
Replace the charcoal canister and inspect the evaporative emissions control system (Section 45)
Check the ignition timing (Section 46)
Replace the timing belt (Chapter 2A)
Check the distributor advance operation (Chapter 5)
Have the fuel/air mixture checked by a dealer service department
Check the carburetor mounting nut torque (Section 47)
Replace the oxygen sensor (1986 Samurai models only) (Section 48)

Every 80,000 miles or 80 months, whichever comes first

Replace the oxygen sensor (1987 through 1995 models only) (Section 48)

Every 100,000 miles or 100 months, whichever comes first

Have the ECM and associated sensors checked by a dealer service department (1988 and later models)
Check the operation of the fuel injection system (Sidekick/X-90/Vitara/Tracker) (Chapter 4)
Have the operation of the catalytic converter checked by a dealer or other properly equipped repair facility

 * *Transmission, transfer case, and differential oil changes performed initially at the 7500 mile/6 month service and at 30,000 mile intervals thereafter (also see "severe" conditions below).*

 ** *This item is affected by "severe" operating conditions as described below. If your vehicle is operated under "severe" conditions, perform all maintenance indicated with a (*) at 7500 mile/6 month intervals.*

Consider the conditions "severe" if most driving is done . . .
In dusty areas
Idling for extended periods and/or low speed operation
When outside temperatures remain below freezing and most trips are less than four miles

 *** *If operated under one or more of the following conditions, change the brake fluid every 15,000 miles:*

Continuous hard driving
Operation in extremely humid climate
After extensive use of brakes in hilly or mountainous terrain

---

3   Tune-up general information

The term tune-up is used in this manual to represent a combination of individual operations rather than one specific procedure.

If, from the time the vehicle is new, the routine maintenance schedule is followed closely and frequent checks are made of fluid levels and high wear items, as suggested throughout this manual, the engine will be kept in relatively good running condition and the need for additional work will be minimized.

More likely than not, however, there will be times when the engine is running poorly due to lack of regular maintenance. This is even more likely if a used vehicle, which has not received regular and frequent maintenance checks, is purchased. In such cases, an engine tune-up will be needed outside of the regular routine maintenance intervals.

The first step in any tune-up or diagnostic procedure to help correct a poor running engine is a cylinder compression check. Checking the compression (see Chapter 2 Part C) will help determine the condition of internal engine components and should be used as a guide for tune-up and repair procedures. If, for instance, a compression check indicates serious internal engine wear, a conventional tune-up will not improve the performance of the engine and would be a waste of

time and money. Because of its importance, someone should do the compression check with the right equipment and the knowledge to use it properly.

The following procedures are those most often needed to bring a generally poor running engine back into a proper state of tune.

Minor tune-up

Check all engine related fluids (Section 4)
Clean, inspect and test the battery (Section 8)
Check and adjust the drivebelts (Section 13)
Replace the spark plugs (Section 28)
Inspect the distributor cap and rotor (Section 30)
Inspect the spark plug and coil wires (Section 29)
Check and adjust the ignition timing (Section 46)
Check the PCV valve (Section 42)
Check the air and PCV filters (Section 31)
Check the cooling system (Section 9)
Check all underhood hoses (Section 10)

Major tune-up

All items listed under Minor tune-up plus . . .
Check the EGR system (Section 41)
Check the ignition system (Chapter 5)

Check the charging system (Chapter 5)
Check the fuel system (Section 25)
Replace the air and PCV filters (Section 31)
Replace the distributor cap and rotor (Section 30)
Replace the spark plug wires (Section 29)

4   Fluid level checks

Refer to illustrations 4.2, 4.4, 4.6, 4.8, 4.14 and 4.19

Note: *The following are fluid level checks to be done on a 250 mile or weekly basis. Additional fluid level checks can be found in specific maintenance procedures which follow. Regardless of intervals, be alert to fluid leaks under the vehicle which would indicate a fault to be corrected immediately.*

1   Fluids are an essential part of the lubrication, cooling, brake, clutch and windshield washer systems. Because the fluids gradually become depleted and/or contaminated during normal operation of the vehicle, they must be periodically replenished. See *Recommended lubricants and fluids* at the beginning of this Chapter before adding fluid to any of the following components. **Note:** *The vehicle must be on level ground when fluid levels are checked.*

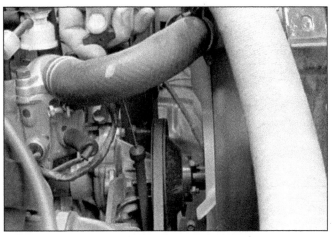

4.2 The dipstick is located on the passenger side of the engine near the water pump

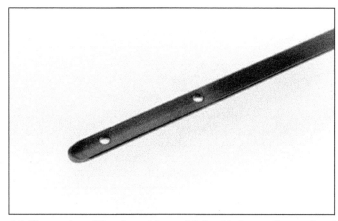

4.4 The engine oil level must be maintained between the marks at all times - it takes one full quart of oil to raise the level from the Add mark (lower hole) to the Full mark (upper hole)

Engine oil

2   Engine oil level is checked with a dip-stick that extends through a tube and into the oil pan at the bottom of the engine (see illustration).

3   The oil level should be checked before the vehicle has been driven, or about 5 minutes after the engine has been shut off. If the oil is checked immediately after driving the vehicle, some of the oil will remain in the upper engine components, resulting in an inaccurate reading on the dipstick.

4   Pull the dipstick from the tube and wipe all the oil from the end with a clean rag or paper towel. Insert the clean dipstick all the way back into the tube, then pull it out again. Note the oil at the end of the dipstick. Add oil as necessary to keep the level between the ADD mark and the FULL mark on the dipstick (see illustration).

5   Do not overfill the engine by adding too much oil since this may result in oil-fouled spark plugs, oil leaks or oil seal failures.

6   Oil is added to the engine after removing the cap (see illustration). A funnel may help to reduce spills.

7   Checking the oil level is an important preventive maintenance step. A consistently low oil level indicates oil leakage through damaged seals, defective gaskets or past worn rings or valve guides. If the oil looks milky in color or has water droplets in it, the cylinder head gasket may be blown or the head or block may be cracked. The engine should be checked immediately. The condition of the oil should also be checked. Whenever you check the oil level, slide your thumb and index finger up the dipstick before wiping off the oil. If you see small dirt or metal particles clinging to the dipstick, the oil should be changed (see Section 12).

Engine coolant

Warning: *Do not allow antifreeze to come in contact with your skin or painted surfaces of the vehicle. Flush contaminated areas immediately with plenty of water. Don't store new coolant or leave old coolant lying around*

4.6 The twist-off oil filler cap is located on the camshaft cover - always make sure the area around this opening is clean before unscrewing the cap to prevent dirt from contaminating the engine

where it's accessible to children or pets - they're attracted by its sweet taste. Ingestion of even a small amount of coolant can be fatal! Wipe up garage floor and drip pan coolant spills immediately. Keep antifreeze containers covered and repair leaks in your cooling system immediately.

8   All vehicles covered by this manual are equipped with a pressurized coolant recovery system. A white plastic coolant reservoir located in the engine compartment is connected by a hose to the radiator filler neck (see illustration). If the engine overheats, coolant escapes through a valve in the radiator cap and travels through the hose into the reservoir. As the engine cools, the coolant is automatically drawn back into the cooling system to maintain the correct level.

9   The coolant level in the reservoir should be checked regularly. **Warning:** *Do not remove the radiator cap to check the coolant level when the engine is warm.* The level in the reservoir varies with the temperature of

4.8 Remember to add coolant to the reservoir only when the engine is off and cold in order to obtain an accurate level

the engine. When the engine is cold, the coolant level should be at or slightly above the FULL COLD mark on the reservoir. Once the engine has warmed up, the level should be at or near the FULL HOT mark. If it isn't, allow the engine to cool, then remove the cap from the reservoir and add a 50/50 mixture of ethylene glycol-based antifreeze and water.

10   Drive the vehicle and recheck the coolant level. If only a small amount of coolant is required to bring the system up to the proper level, water can be used. However, repeated additions of water will dilute the antifreeze and water solution. In order to maintain the proper ratio of antifreeze and water, always top up the coolant level with the correct mixture. An empty plastic milk jug or bleach bottle makes an excellent container for mixing coolant. Do not use rust inhibitors or additives.

11   If the coolant level drops consistently, there may be a leak in the system. Inspect the radiator, hoses, filler cap, drain plugs and water pump (see Section 9). If no leaks are noted, have the radiator cap pressure tested

4.14 The windshield washer reservoir is located on the left fender panel - special windshield washer solvent and/or water can be added after opening the cap

4.18 The clutch cylinder fluid level should remain at the reservoir top mark

4.19 The brake fluid level is easily checked by looking through the clear brake fluid reservoir

by a service station.

12   If you have to remove the radiator cap, wait until the engine has cooled, then wrap a thick cloth around the cap and turn it to the first stop. If coolant or steam escapes, let the engine cool down longer, then remove the cap.

13   Check the condition of the coolant as well. It should be relatively clear. If it's brown or rust colored, the system should be drained, flushed and refilled. Even if the coolant appears to be normal, the corrosion inhibitors wear out, so it must be replaced at the specified intervals.

Windshield washer fluid

14   Fluid for the windshield washer system is located in a plastic reservoir in the engine compartment **(see illustration)**.

15   In milder climates, plain water can be used in the reservoir, but it should be kept no more than 2/3 full to allow for expansion if the water freezes. In colder climates, use windshield washer system antifreeze, available at any auto parts store, to lower the freezing point of the fluid. Mix the antifreeze with water in accordance with the manufacturer's directions on the container. **Caution:** *Don't use cooling system antifreeze - it will damage the vehicle's paint.*

16   To help prevent icing in cold weather, warm the windshield with the defroster before using the washer.

Battery electrolyte

17   All vehicles with which this manual is concerned are equipped with a battery which is permanently sealed (except for vent holes) and has no filler caps. Water doesn't have to be added to these batteries at any time. If a maintenance-type battery is installed, the caps on the top of the battery should be removed periodically to check for a low water level. This check is most critical during the warm summer months.

Brake and clutch fluid

Refer to illustration 4.18

18   The brake master cylinder is mounted on the front of the power booster unit in the engine compartment. The clutch cylinder used on later models with manual transmissions is mounted adjacent to it on the firewall **(see illustration)**.

19   The fluid inside is readily visible. The level should be above the MIN marks on the reservoirs **(see illustration)**. If a low level is indicated, be sure to wipe the top of the reservoir cover with a clean rag to prevent contamination of the brake and/or clutch system before removing the cover.

20   When adding fluid, pour it carefully into the reservoir to avoid spilling it onto surrounding painted surfaces. Be sure the specified fluid is used, since mixing different types of brake fluid can cause damage to the system. See Recommended lubricants and fluids at the front of this Chapter or your owner's manual. **Warning:** *Brake fluid can harm your eyes and damage painted surfaces, so use extreme caution when handling or pouring it. Do not use brake fluid that has been standing open or is more than one year old. Brake fluid absorbs moisture from the air. Excess moisture can cause a dangerous loss of braking effectiveness.*

21   At this time the fluid and master cylinder can be inspected for contamination. The system should be drained and refilled if deposits, dirt particles or water droplets are seen in the fluid.

22   After filling the reservoir to the proper level, make sure the cover is on tight to prevent fluid leakage.

23   Brake fluid level in the master cylinder will drop slightly as the pads and the brake shoes at each wheel wear down during normal operation. If the master cylinder requires repeated additions to keep it at the proper level, it's an indication of leakage in the brake system, which should be corrected immediately. Check all brake lines and connections (see Section 18 for more information).

24   If upon checking the master cylinder fluid level, you discover one or both reservoirs empty or nearly empty, the brake system should be bled (see Chapter 9).

5   Tire and tire pressure checks

Refer to illustrations 5.2, 5.3, 5.4a, 5.4b and 5.8

1   Periodic inspection of the tires may spare you the inconvenience of being stranded with a flat tire. It can also provide you with vital information regarding possible problems in the steering and suspension systems before major damage occurs.

2   The original tires on this vehicle are equipped with 1/2-inch wide bands that will appear when tread depth reaches 1/16-inch, at which point the tires can be considered worn out. Tread wear can be monitored with a simple, inexpensive device known as a tread depth indicator **(see illustration)**.

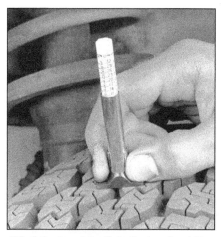

5.2 A tire tread depth indicator should be used to monitor tire wear - they are available at auto parts stores and service stations and cost very little

UNDERINFLATION

CUPPING

Cupping may be caused by:
- Underinflation and/or mechanical irregularities such as out-of-balance condition of wheel and/or tire, and bent or damaged wheel.
- Loose or worn steering tie-rod or steering idler arm.
- Loose, damaged or worn front suspension parts.

OVERINFLATION

INCORRECT TOE-IN OR EXTREME CAMBER

FEATHERING DUE TO MISALIGNMENT

5.3 This chart will help you determine the condition of the tires, the probable cause(s) of abnormal wear and the corrective action necessary

3   Note any abnormal tread wear **(see illustration)**. Tread pattern irregularities such as cupping, flat spots and more wear on one side than the other are indications of front end alignment and/or balance problems. If any of these conditions are noted, take the vehicle to a tire shop or service station to correct the problem.

4   Look closely for cuts, punctures and embedded nails or tacks. Sometimes a tire will hold air pressure for a short time or leak down very slowly after a nail has embedded itself in the tread. If a slow leak persists, check the valve stem core to make sure it's tight **(see illustration)**. Examine the tread for an object that may have embedded itself in the tire or for a "plug" that may have begun to leak (radial tire punctures are repaired with a plug that's installed in a puncture). If a puncture is suspected, it can be easily verified by spraying a solution of soapy water onto the puncture area **(see illustration)**. The soapy solution will bubble if there's a leak. Unless the puncture is unusually large, a tire shop or service station can usually repair the tire.

5   Carefully inspect the inner sidewall of each tire for evidence of brake fluid leakage. If you see any, inspect the brakes immediately.

6   Correct air pressure adds miles to the lifespan of the tires, improves mileage and

enhances overall ride quality. Tire pressure cannot be accurately estimated by looking at a tire, especially if it's a radial. A tire pressure gauge is essential. Keep an accurate gauge in the vehicle. The pressure gauges attached to the nozzles of air hoses at gas stations are often inaccurate.

7   Always check tire pressure when the tires are cold. Cold, in this case, means the vehicle has not been driven over a mile in the three hours preceding a tire pressure check. A pressure rise of four to eight pounds is not uncommon once the tires are warm.

8   Unscrew the valve cap protruding from

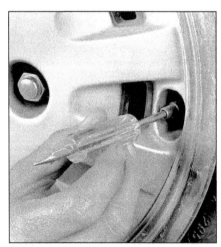

5.4a If a tire loses air on a steady basis, check the valve core first to make sure it's snug (special inexpensive wrenches are commonly available at auto parts stores)

5.4b If the valve core is tight, jack up the low tire and spray a soapy water solution onto the tread as the tire is turned slowly - leaks will cause small bubbles to appear

5.8 To extend the life of the tires, check the air pressure at least once a week with an accurate gauge (don't forget the spare!)

the wheel or hubcap and push the gauge firmly onto the valve stem **(see illustration)**. Note the reading on the gauge and compare the figure to the recommended tire pressure shown on the placard on the driver's side door pillar. Be sure to reinstall the valve cap to keep dirt and moisture out of the valve stem mechanism. Check all four tires and, if necessary, add enough air to bring them up to the recommended pressure.

9   Don't forget to keep the spare tire inflated to the specified pressure (refer to your owner's manual or the tire sidewall). Note that the pressure recommended for the compact spare is higher than for the tires on the vehicle.

6   Automatic transmission fluid level check

Refer to illustration 6.6

1   The automatic transmission fluid level should be carefully maintained. Low fluid level can lead to slipping or loss of drive, while overfilling can cause foaming and loss of fluid.

2   The automatic transmission fluid level is normally checked with the engine at normal operating temperature. This temperature is reached after approximately 15 miles of highway driving. With the parking brake set, start the engine, then move the shift lever through all the gear ranges, ending in Park. The fluid level must be checked with the vehicle level and the engine running at idle. **Note:** *Incorrect fluid level readings will result if the vehicle has just been driven at high speeds for an extended period, in hot weather in city traffic, or if it has been pulling a trailer. If any of these conditions apply, wait until the fluid has cooled (about 30 minutes).*

3   With the transmission at normal operating temperature, remove the dipstick from the filler tube.

4   Wipe the fluid from the dipstick with a clean rag and push it back into the filler tube until the cap seats.

5   Pull the dipstick out again and note the fluid level.

6   The level should at or near the upper HOT mark. If additional fluid is required, add it directly into the tube using a funnel. It takes about one pint to raise the level from the LOW HOT notch to the FULL HOT notch with a hot transmission, so add the fluid a little at a time and keep checking the level until it's correct.

7   The condition of the fluid should also be checked along with the level. If the fluid at the end of the dipstick is a dark reddish-brown color, or if it smells burned, it should be changed. If you are in doubt about the condition of the fluid, purchase some new fluid and compare the two for color and smell.

7   Power steering fluid level check

Refer to illustration 7.6

1   Unlike manual steering, the power steering system relies on fluid which may, over a period of time, require replenishing.

2   On earlier models the fluid reservoir for the power steering pump is located on the inner fender on the driver's side and is checked with a dipstick. On later models the reservoir is mounted adjacent to the power steering pump and is translucent plastic so it can be checked visually.

3   For the check, the front wheels should be pointed straight ahead and the engine should be off.

4   On earlier dipstick-type reservoirs, use a clean rag to wipe off the reservoir cap and the area around the cap. This will help prevent any foreign matter from entering the reservoir during the check.

5   Twist off the cap and check the temperature of the fluid at the end of the dipstick with your finger.

6   Wipe off the fluid with a clean rag, reinsert the dipstick, then withdraw it and read the fluid level. The level should be at the HOT mark if the fluid was hot to the touch **(see illustration)**. It should be at the COLD mark if the fluid was cool to the touch. Never allow the fluid level to drop below the ADD mark.

7   On later models with a translucent plastic fluid reservoir mounted remotely from the power steering pump housing allows the fluid level to be checked without removing the cap.

8   Should additional fluid be required, pour the specified type directly into the reservoir, using a funnel to prevent spills.

9   If the reservoir requires frequent fluid additions, all power steering hoses, hose connections and the power steering pump should be carefully checked for leaks.

8   Battery check and maintenance

Refer to illustrations 8.1 and 8.6
Warning: *Certain precautions must be followed when checking and servicing the battery. Hydrogen gas, which is highly flammable, is always present in the battery cells, so keep lighted tobacco and all other open flames and sparks away from the battery. The electrolyte inside the battery is actually dilute sulfuric acid, which will cause injury if splashed on your skin or in your eyes. It will also ruin clothes and painted surfaces. When removing the battery cables, always detach the negative cable first and hook it up last!*

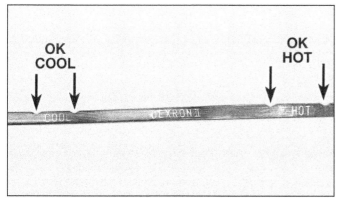

6.6 The automatic transmission fluid level must be maintained between the ADD mark and the FULL mark (arrows) at the indicated operating temperature

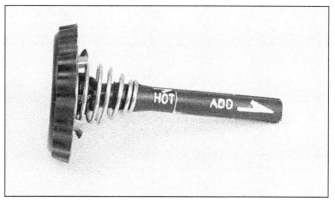

7.6 The marks on the power steering fluid dipstick indicate the safe range

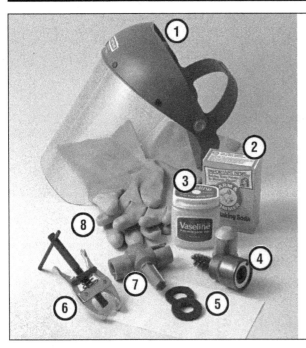

8.1 Regardless of the type of tool used on the battery posts, a clean, shiny surface should be the result

1 *Face shield/safety goggles – When removing corrosion with a brush, the acidic particles can easily fly up into your eyes*
2 *Baking soda – A solution of baking soda and water can be used to neutralize corrosion*
3 *Petroleum jelly – A layer of this on the battery posts will help prevent corrosion*
4 *Battery post/cable cleaner – This wire brush cleaning tool will remove all traces of corrosion from the battery posts and cable clamps*
5 *Treated felt washers – Placing one of these on each port, directly under the cable clamps, will help prevent corrosion*
6 *Puller – Sometime the cable clamps are very difficult to pull off the posts, even after the nut/bolt has been completely loosened. This tool pulls the clamp straight up and off the post without damage.*
7 *Battery post/cable cleaner – Here is another cleaning tool which is a slightly different version of number 4 above, but it does the same thing*
8 *Rubber gloves – Another safety item to consider when servicing the battery; remember that's acid inside the battery!*

1   Battery maintenance is an important procedure which will help ensure that you are not stranded because of a dead battery. Several tools are required for this procedure **(see illustration)**.
2   When checking/servicing the battery, always turn the engine and all accessories off.
3   A sealed (sometimes called maintenance-free), side-terminal battery is standard equipment on these vehicles. The cell caps cannot be removed, no electrolyte checks are required and water cannot be added to the cells. However, if a standard top-terminal aftermarket battery has been installed, the following maintenance procedure can be used.

8.6 Removing the cable from a battery post with a wrench - sometimes special battery pliers are required for this procedure if corrosion has caused deterioration of the nut hex (always remove the ground cable first and hook it up last!)

4   Remove the caps and check the electrolyte level in each of the battery cells. It must be above the plates. There's usually a split-ring indicator in each cell to indicate the correct level. If the level is low, add distilled water only, then reinstall the cell caps. **Caution:** *Overfilling the cells may cause electrolyte to spill over during periods of heavy charging, causing corrosion and damage to nearby components.*
5   The external condition of the battery should be checked periodically. Look for damage such as a cracked case.
6   Check the tightness of the battery cable bolts **(see illustration)** to ensure good electrical connections. Inspect the entire length of each cable, looking for cracked or abraded insulation and frayed conductors.
7   If corrosion (visible as white, fluffy deposits) is evident, remove the cables from the terminals, clean them with a battery brush and reinstall them. Corrosion can be kept to a minimum by applying a layer of petroleum jelly or grease to the bolt threads.
8   Make sure the battery carrier is in good condition and the hold-down clamp is tight. If the battery is removed (see Chapter 5 for the removal and installation procedure), make sure that no parts remain in the bottom of the carrier when it's reinstalled. When reinstalling the hold-down clamp, don't overtighten the bolt.
9   Corrosion on the carrier, battery case and surrounding areas can be removed with a solution of water and baking soda. Apply the mixture with a small brush, let it work, then rinse it off with plenty of clean water.
10   Any metal parts of the vehicle damaged by corrosion should be coated with a zinc-based primer, then painted.
11   Additional information on the battery, charging and jump starting can be found in the front of this manual and in Chapter 5.

9   Cooling system check

Refer to illustration 9.4
1   Many major engine failures can be attributed to a faulty cooling system. If the vehicle is equipped with an automatic transmission, the cooling system also cools the transmission fluid and thus plays an important role in prolonging transmission life.
2   The cooling system should be checked with the engine cold. Do this before the vehicle is driven for the day or after it has been shut off for at least three hours.
3   Remove the radiator cap by turning it to the left until it reaches a stop. If you hear a hissing sound (indicating there is still pressure in the system), wait until this stops. Now press down on the cap with the palm of your hand and continue turning to the left until the cap can be removed. Thoroughly clean the cap, inside and out, with clean water. Also clean the filler neck on the radiator. All traces of corrosion should be removed. The coolant inside the radiator should be relatively transparent. If it is rust colored, the system should be drained and refilled (see Section 37). If the coolant level is not up to the top, add additional antifreeze/coolant mixture (see Section 4).
4   Carefully check the large upper and lower radiator hoses along with the smaller diameter heater hoses which run from the engine to the firewall. On some models the heater return hose runs directly to the radiator. Inspect each hose along its entire length, replacing any hose which is cracked, swollen or shows signs of deterioration. Cracks may become more apparent if the hose is squeezed **(see illustration)**. Regardless of condition, it's a good idea to replace hoses with new ones every two years.
5   Make sure that all hose connections are tight. A leak in the cooling system will usually

Check for a chafed area that could fail prematurely.

Check for a soft area indicating the hose has deteriorated inside.

Overtightening the clamp on a hardened hose will damage the hose and cause a leak.

Check each hose for swelling and oil-soaked ends. Cracks and breaks can be located by squeezing the hose.

9.4 Hoses, like drivebelts, have a habit of failing at the worst possible time - to prevent the inconvenience of a blown radiator or heater hose, inspect them carefully as shown here

show up as white or rust colored deposits on the areas adjoining the leak. If wire-type clamps are used at the ends of the hoses, it may be a good idea to replace them with more secure screw-type clamps.

6   Use compressed air or a soft brush to remove bugs, leaves, etc. from the front of the radiator or air conditioning condenser. Be careful not to damage the delicate cooling fins or cut yourself on them.

7   Every other inspection, or at the first indication of cooling system problems, have the cap and system pressure tested. If you don't have a pressure tester, most gas stations and repair shops will do this for a minimal charge.

10   Underhood hose check and replacement

Refer to illustration 10.1

General

1   **Caution:** *Replacement of air conditioning hoses must be left to a dealer service department or air conditioning shop that has the equipment to depressurize the system safely. Never remove air conditioning components or hoses* **(see illustration)** *until the system has been depressurized.*

2   High temperatures in the engine compartment can cause the deterioration of the rubber and plastic hoses used for engine, accessory and emission systems operation. Periodic inspection should be made for cracks, loose clamps, material hardening and leaks. Information specific to the cooling system hoses can be found in Section 9.

3   Some, but not all, hoses are secured to the fittings with clamps. Where clamps are used, check to be sure they haven't lost their tension, allowing the hose to leak. If clamps aren't used, make sure the hose has not

expanded and/or hardened where it slips over the fitting, allowing it to leak.

Vacuum hoses

4   It's quite common for vacuum hoses, especially those in the emissions system, to be color coded or identified by colored stripes molded into them. Various systems require hoses with different wall thicknesses, collapse resistance and temperature resistance. When replacing hoses, be sure the new ones are made of the same material.

5   Often the only effective way to check a hose is to remove it completely from the vehicle. If more than one hose is removed, be sure to label the hoses and fittings to ensure correct installation.

6   When checking vacuum hoses, be sure to include any plastic T-fittings in the check. Inspect the fittings for cracks and the hose where it fits over the fitting for distortion,

10.1 Air conditioning hoses are best identified by the metal tubes used at all connection points (arrows) - DO NOT disconnect or accidentally damage the air conditioning hoses as the system is under high pressure

which could cause leakage.

7   A small piece of vacuum hose (1/4-inch inside diameter) can be used as a stethoscope to detect vacuum leaks. Hold one end of the hose to your ear and probe around vacuum hoses and fittings, listening for the "hissing" sound characteristic of a vacuum leak. **Warning:** *When probing with the vacuum hose stethoscope, be very careful not to come into contact with moving engine components such as the drivebelt, cooling fan, etc.*

Fuel hose

Warning: *There are certain precautions which must be taken when inspecting or servicing fuel system components. Work in a well ventilated area and do not allow open flames (cigarettes, appliance pilot lights, etc.) or bare light bulbs near the work area. Mop up any spills immediately and do not store fuel soaked rags where they could ignite. On vehicles equipped with fuel injection, the fuel system is under pressure, so if any fuel lines are to be disconnected, the pressure in the system must be relieved first (see Chapter 4 for more information).*

8   Check all rubber fuel lines for deterioration and chafing. Check especially for cracks in areas where the hose bends and just before fittings, such as where a hose attaches to the fuel filter.

9   High quality fuel line must be used for fuel line replacement. Never, under any circumstances, use unreinforced vacuum line, clear plastic tubing or water hose for fuel lines.

10   Spring-type clamps are commonly used on fuel lines. These clamps often lose their tension over a period of time, and can be "sprung" during removal. Replace all spring-type clamps with screw clamps whenever a hose is replaced.

Metal lines

11   Sections of metal line are often used for fuel line between the fuel pump and carburetor or fuel injection unit. Check carefully to be sure the line has not been bent or crimped and that cracks have not started in the line.

12   If a section of metal fuel line must be replaced, only seamless steel tubing should be used, since copper and aluminum tubing don't have the strength necessary to withstand normal engine vibration.

13   Check the metal brake lines where they enter the master cylinder and brake proportioning unit (if used) for cracks in the lines or loose fittings. Any sign of brake fluid leakage calls for an immediate thorough inspection of the brake system.

11   Wiper blade inspection and replacement

Refer to illustrations 11.6a and 11.6b

1   The windshield wiper and blade assembly should be inspected periodically for damage, loose components and cracked or worn

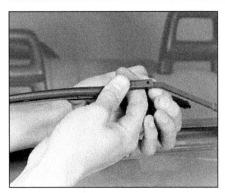

11.6a Remove the blade assembly by angling it slightly away from the retaining stud

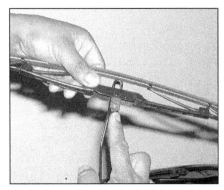

11.6b Depress the release lever (finger is on it here) and slide the wiper assembly down the wiper arm and out of the hook at the end of the arm

blade elements.

2   Road film can build up on the wiper blades and affect their efficiency, so they should be washed regularly with a mild detergent solution.

3   The action of the wiping mechanism can loosen the bolts, nuts and fasteners, so they should be checked and tightened, as necessary, at the same time the wiper blades are checked.

4   If the wiper blade elements (sometimes called inserts) are cracked, worn or warped, they should be replaced with new ones.

5   Pull the wiper blade/arm assembly away from the glass.

6   On earlier models, depress the blade-to-arm connector and slide the blade assembly off the wiper arm and over the retaining stud. Angle the blade slightly **(see illustration)**. On later models, rotate the blade assembly for access to the release lever, press the lever and slide the assembly out the hook in the wiper arm **(see illustration)**.

7   Pinch the tabs at the end, then slide the element out of the blade assembly.

8   Compare the new element with the old for length, design, etc.

9   Slide the new element into place. It will automatically lock at the correct location.

10   Reinstall the blade assembly on the arm, wet the windshield and check for proper operation.

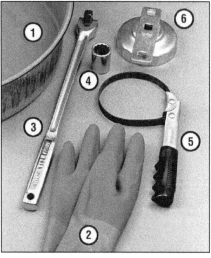

12.3 These tools are required when changing the engine oil and filter

1   **Drain pan** - It should be fairly shallow in depth, but wide to prevent spills
2   **Rubber gloves** - When removing the drain plug and filter, you will get oil on your hands (the gloves will prevent burns)
3   **Breaker bar** - Sometimes the oil drain plug is tight, and a long breaker bar is needed to loosen it
4   **Socket** - To be used with the breaker bar or a ratchet (must be the correct size to fit the drain plug - six-point preferred)
5   **Filter wrench** - This is a metal band-type wrench, which requires clearance around the filter to be effective
6   **Filter wrench** - This type fits on the bottom of the filter and can be turned with a ratchet or breaker bar (different-size wrenches are available for different types of filters)

---

12 Engine oil and filter change

Refer to illustrations 12.3, 12.9, 12.14 and 12.18

1   Frequent oil changes are the most important preventive maintenance procedures that can be done by the home mechanic. As engine oil ages, it becomes diluted and contaminated, which leads to premature engine wear.

2   Although some sources recommend oil filter changes every other oil change, we feel that the minimal cost of an oil filter and the relative ease with which it is installed dictate that a new filter be installed every time the oil is changed.

3   Gather together all necessary tools and materials before beginning this procedure **(see illustration)**.

4   You should have plenty of clean rags and newspapers handy to mop up any spills. Access to the underside of the vehicle is greatly improved if the vehicle can be lifted on a hoist, driven onto ramps or supported

12.9 The oil drain plug is located at the bottom of the pan and should be removed with a socket or box-end wrench, as the corners on the bolt can be easily rounded off

by jackstands. **Warning:** *Do not work under a vehicle which is supported only by a bumper, hydraulic or scissors-type jack.*

5   If this is your first oil change, get under the vehicle and familiarize yourself with the locations of the oil drain plug and the oil filter. The engine and exhaust components will be warm during the actual work, so note how they are situated to avoid touching them when working under the vehicle.

6   Warm the engine to normal operating temperature. If the new oil or any tools are needed, use this warm-up time to gather everything necessary for the job. The correct type of oil for your application can be found in Recommended lubricants and fluids at the beginning of this Chapter.

7   With the engine oil warm (warm engine oil will drain better and more built-up sludge will be removed with it), raise and support the vehicle. Make sure it's safely supported!

8   Move all necessary tools, rags and newspapers under the vehicle. Set the drain pan under the drain plug. Keep in mind that the oil will initially flow from the pan with some force; position the pan accordingly.

9   Being careful not to touch any of the hot exhaust components, use a wrench to remove the drain plug near the bottom of the oil pan **(see illustration)**. Depending on how hot the oil is, you may want to wear gloves while unscrewing the plug the final few turns.

10   Allow the old oil to drain into the pan. It may be necessary to move the pan as the oil flow slows to a trickle.

11   After all the oil has drained, wipe off the drain plug with a clean rag. Small metal particles may cling to the plug and would immediately contaminate the new oil.

12   Clean the area around the drain plug opening and reinstall the plug. Tighten the plug securely with the wrench. If a torque wrench is available, use it to tighten the plug.

13   Move the drain pan into position under the oil filter.

14   Use the filter wrench to loosen the oil filter **(see illustration)**. Chain or metal band fil-

12.14 Use a strap-type oil filter wrench to loosen the filter - if access makes removal difficult, other types of filter wrenches are available

12.18 Lubricate the oil filter gasket with clean engine oil before installing the filter on the engine

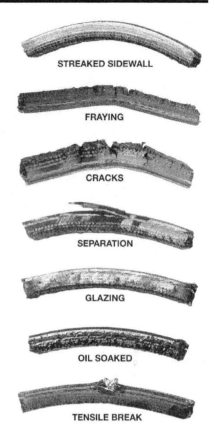

13.3 Here are some of the more common problems associated with drivebelts (check the belts very carefully to prevent an untimely breakdown)

ter wrenches may distort the filter canister, but it doesn't matter since the filter will be discarded anyway.

15   Completely unscrew the old filter. Be careful; it's full of oil. Empty the oil inside the filter into the drain pan.

16   Compare the old filter with the new one to make sure they're the same type.

17   Use a clean rag to remove all oil, dirt and sludge from the area where the oil filter mounts to the engine. Check the old filter to make sure the rubber gasket isn't stuck to the engine. If the gasket is stuck to the engine (use a flashlight if necessary), remove it.

18   Apply a light coat of clean oil to the rubber gasket on the new oil filter **(see illustration)**.

19   Attach the new filter to the engine, following the tightening directions printed on the filter canister or packing box. Most filter manufacturers recommend against using a filter wrench due to the possibility of over-tightening and damage to the seal.

20   Remove all tools, rags, etc. from under the vehicle, being careful not to spill the oil in the drain pan, then lower the vehicle.

21   Move to the engine compartment and locate the oil filler cap.

22   If an oil can spout is used, push the spout into the top of the oil can and pour the fresh oil through the filler opening. A funnel may also be used.

23   Pour four quarts of fresh oil into the engine. Wait a few minutes to allow the oil to drain into the pan, then check the level on the oil dipstick (see Section 4 if necessary). If the oil level is above the ADD mark, start the engine and allow the new oil to circulate.

24   Run the engine for only about a minute and then shut it off. Immediately look under the vehicle and check for leaks at the oil pan drain plug and around the oil filter. If either is leaking, tighten with a bit more force.

25   With the new oil circulated and the filter now completely full, recheck the level on the dipstick and add more oil as necessary.

26   During the first few trips after an oil change, make it a point to check frequently

for leaks and proper oil level.

27   The old oil drained from the engine cannot be reused in its present state and should be disposed of. Check with your local auto parts store, disposal facility or environmental agency to see if they will accept the oil for recycling. After the oil has cooled it can be drained into a container (capped plastic jugs, topped bottles, milk cartons, etc.) for transport to one of these disposal sites. Don't dispose of the oil by pouring it on the ground or down a drain!

13 Drivebelt check, adjustment and replacement

Refer to illustrations 13.3, 13.4, 13.5a and 13.5b

1   The drivebelts, or V-belts as they are often called, are located at the front of the engine and play an important role in the overall operation of the engine and accessories. Due to their function and material makeup, the belts are prone to failure after a period of time and should be inspected and adjusted periodically to prevent major engine damage.

2   The number of belts used on a particular vehicle depends on the accessories installed. Drivebelts are used to turn the alternator, power steering pump, water pump and air conditioning compressor. Depending on the

pulley arrangement, more than one of the components may be driven by a single belt.

3   With the engine off, locate the drivebelts at the front of the engine. Using your fingers (and a flashlight, if necessary), move along the belts checking for cracks and separation of the belt plies. Also check for fraying and glazing, which gives the belt a shiny appearance **(see illustration)**. Both sides of each belt should be inspected, which means you'll have to twist each belt to check the underside. Check the pulleys for nicks, cracks, distortion and corrosion.

4   The tension of each belt is checked by pushing on it at a distance halfway between

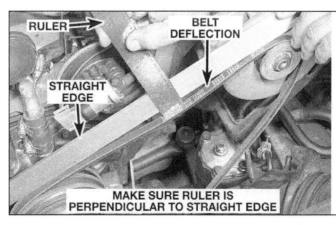

13.4 Measuring drivebelt deflection with a straightedge and ruler

13.5a Loosen the bracket bolt (arrow) to allow the alternator to slide within the pulley adjustment bracket

13.5b Loosen the two lower mounting bolts (viewed from under the vehicle) to allow the alternator to pivot

the pulleys. Push firmly with your thumb and see how much the belt moves (deflects) **(see illustration)**. A rule of thumb is that if the distance from pulley center-to-pulley center is between 7 and 11 inches, the belt should deflect 1/4-inch. If the belt travels between pulleys spaced 12-to-16 inches apart, the belt should deflect 1/2-inch.

5   If adjustment is needed, either to make the belt tighter or looser, it's done by moving the belt-driven accessory on the bracket **(see illustrations)**.

6   Each component usually has an adjusting bolt and a pivot bolt. Both bolts must be loosened slightly to enable you to move the component. Some components have an adjusting bolt that can be turned to change the belt tension after the mounting bolt is loosened. Others are equipped with an idler pulley that must be moved to change the belt tension.

7   After the two bolts have been loosened, move the component away from the engine to tighten the belt or toward the engine to

loosen the belt. Hold the accessory in position and check the belt tension. If it's correct, tighten the two bolts until just snug, then recheck the tension. If the tension is correct, tighten the bolts.

8   You may have to use some sort of pry bar to move the accessory while the belt is adjusted. If this must be done to gain the proper leverage, be very careful not to damage the component being moved or the part being pried against.

9   To replace a belt, follow the above procedures for drivebelt adjustment but slip the belt off the pulleys and remove it. Since belts tend to wear out more or less at the same time, it's a good idea to replace all of them at the same time. Mark each belt and the corresponding pulley grooves so the replacement belts can be installed properly.

10   Take the old belts with you when purchasing new ones in order to make a direct comparison for length, width and design.

11   Adjust the belts as described earlier in this Section.

14 Tire rotation

Refer to illustration 14.2

1   The tires should be rotated at the specified intervals and whenever uneven wear is noticed.

2   Refer to the **accompanying illustration** for the preferred tire rotation pattern.

3   Refer to the information in *Jacking and Towing* at the front of this manual for the proper procedures to follow when raising the vehicle and changing a tire. If the brakes are to be checked, don't apply the parking brake as stated. Make sure the tires are blocked to prevent the vehicle from rolling as it's raised.

4   Preferably, the entire vehicle should be raised at the same time. This can be done on a hoist or by jacking up each corner and then lowering the vehicle onto jackstands placed under the frame rails. Always use four jackstands and make sure the vehicle is safely supported.

5   After rotation, check and adjust the tire pressures as necessary and be sure to check the lug nut tightness.

6   For additional information on the wheels and tires, refer to Chapter 10.

15 Clutch pedal height and free play check and adjustment

Refer to illustrations 15.2 and 15.3

1   On vehicles equipped with a manual transmission, the clutch pedal height and free play must be correctly adjusted.

2   The height of the clutch pedal is the distance the pedal sits off the floor **(see illustration)**. The distance should be as specified. If the pedal height is not within the specified range, loosen the locknut on the pedal stopper or switch located to the rear of the clutch pedal and turn the stopper or switch in or out until the pedal height is correct. Retighten the locknut.

3   The free play is the pedal slack, or the

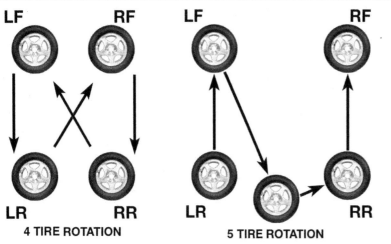

LF   RF
LR   RR
4 TIRE ROTATION

LF   RF
LR   RR
5 TIRE ROTATION

90010-1-14.2 HAYNES

14.2 Tire rotation diagram

15.2 Clutch pedal height is adjusted by loosening the lock nut behind the pedal bracket and turning the adjusting bolt

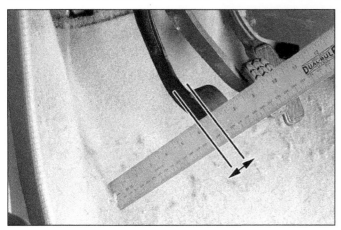

15.3 To determine clutch freeplay, depress the clutch pedal and stop the moment the clutch resistance is felt

16.5a On 8-valve engines, with the number one piston at TDC on the compression stroke, adjust the valves indicated by the arrows

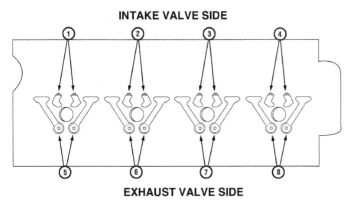

INTAKE VALVE SIDE

① ② ③ ④

⑤ ⑥ ⑦ ⑧

EXHAUST VALVE SIDE

16.5b Valve numbering on 16-valve engines - with the number one piston at TDC on the compression stroke, adjust valves 1, 2, 5 and 7; with the number four piston at TDC on the compression stroke, adjust valves 3, 4, 6 and 8

16.6 On 8-valve engines, insert the feeler gauge between the valve stem and the adjusting screw

distance the pedal can be depressed before it begins to have any effect on the clutch **(see illustration)**. The distance should be as specified. If it isn't, loosen the locknuts on the clutch cable bracket (located on the transmission housing) and adjust until freeplay is correct. Then tighten the nuts.

4   Adjust the clutch release arm free play by turning the joint nut at the end of the clutch cable.

16 Valve clearance check and adjustment (every 15,000 miles or 12 months)

Refer to illustrations 16.5a, 16.5b, 16.6, 16.7 and 16.10

Note: *On 16-valve engines, a special tool (see Step 7) is required for this procedure.*

1   Start the engine and allow it to reach normal operating temperature, then shut it off.

2   Remove the camshaft cover (see Chapter 2).

3   Position the number one piston at TDC on the compression stroke (see Chapter 2).

4   Make sure the rocker arms for the number one cylinder valves are loose and number four are tight. If they aren't, the number one

piston is not at TDC on the compression stroke.

5   On 8-valve engines, check and adjust only the valves shown **(see illustration)**. On 16-valve engines, check and adjust the valves numbered 1, 2, 5 and 7 in the accompanying illustration **(see illustration)**. The valve clearances can be found in the Specifications at the beginning of this Chapter.

6   On 8-valve engines, the clearance is measured by inserting the specified size feeler gauge between the end of the valve stem and the adjusting screw **(see illustration)**. On 16-valve engines, measure between the camshaft and the face of the rocker arm where it rides on the camshaft. On either engine type, you should feel a slight amount of drag when the feeler gauge is moved back-and-forth.

7   If the gap is too large or too small, loosen the locknut and turn the adjusting screw to obtain the correct gap **(see illustration)**. On 16-valve engines, you will need a special tool, available at most auto parts stores, to loosen and tighten the locknuts and screws because they are difficult to reach.

8   Once the gap has been set, hold the screw in position with a screwdriver (8-valve) or special tool (16-valve) and retighten the locknut. Recheck the valve clearance - some-

times it will change slightly when the locknut is tightened. If so, readjust it until it's correct.

9   Repeat the procedure for the remaining valves **(see illustrations 16.5a and 16.5b)**, then turn the crankshaft one complete revolution (360-degrees) and realign the notch in

16.7 To adjust the valve clearance on 8-valve engines, use a box end wrench to slightly loosen the lock nut and a screwdriver to change the adjustment

16.10 On 8-valve engines, turn the crankshaft one complete revolution (360-degrees), then adjust the remaining valves (arrows)

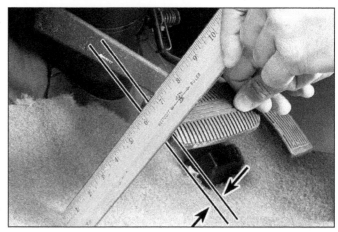

17.2 To check brake pedal freeplay, measure the distance between the natural resting place of the pedal and the point at which you first encounter resistance

the pulley with the zero on the engine.

10   Adjust the remaining valves. On 8-valve engines, adjust the valves shown **(see illustration)**. On 16-valve engines, adjust valves 3, 4, 6 and 8 in **illustration 16.5b**.

11   Reinstall the camshaft cover.

17   Brake pedal check and adjustment

Refer to illustration 17.2

1   Brake pedal height is the distance the pedal sits away from the floor. The distance should be as listed in this Chapter's Specifications). If the pedal height is not within the specified range, loosen the locknut on the brake light switch located in the bracket to the rear of the brake pedal and turn the switch in or out until the pedal height is correct. Retighten the locknut.

2   The free play is the pedal slack, or the distance the pedal can be depressed before it begins to have any effect on the brakes **(see illustration)**. It should be as specified. If it isn't, loosen the locknut on the brake booster input rod, to which the brake pedal is attached. Turn the input rod until the free play is correct, then retighten the locknut.

18   Brake check

Refer to illustrations 18.6, 18.11, 18.13a, 18.13b and 18.15

Note: *For detailed photographs of the brake system, refer to Chapter 9.*

Warning: *Dust created by the brake system may contain asbestos, which is harmful to your health. Never blow it out with compressed air and don't inhale any of it. An approved filtering mask should be worn when working on the brakes. Do not, under any circumstances, use petroleum-based solvents to clean brake parts. Use brake cleaner or denatured alcohol only!*

1   In addition to the specified intervals, the

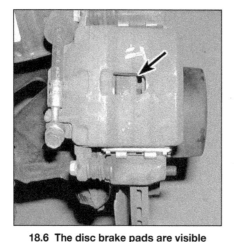

18.6 The disc brake pads are visible through the opening in the caliper (arrow) - be sure to measure only the lining material and not the metal backing to which the lining is attached

brakes should be inspected every time the wheels are removed or whenever a defect is suspected.

2   To check the brakes, the vehicle must be raised and supported securely on jackstands.

Disc brakes

3   Disc brakes are used on the front wheels. Extensive disc damage can occur if the pads are allowed to wear beyond the limit listed in this Chapter's Specifications.

4   Raise the vehicle and support it securely on jackstands, then remove all four wheels (see *Jacking and Towing* at the front of the manual if necessary).

5   The disc brake calipers, which contain the pads, are visible with the wheels removed. There's an outer pad and an inner pad in each caliper. All four pads should be inspected.

6   Each caliper has openings, which will allow you to inspect the pads **(see illustration)**. If the pad material has worn down near the minimum thickness listed in this Chap-

18.11 Use a slide hammer to remove stubborn brake drums

ter's Specifications, the pads should be replaced.

7   If you're unsure about the exact thickness of the remaining lining material, remove the pads for further inspection or replacement (see Chapter 9).

8   Before installing the wheels, check for leakage and/or damage (cracks, splitting, etc.) around the brake hose connections. Replace the hose or fittings as necessary, referring to Chapter 9.

9   Check the condition of the disc. Look for score marks, deep scratches and burned spots. If these conditions exist, the hub/disc assembly should be removed for servicing.

Drum brakes

10   On rear brakes, remove the drum by pulling it off the axle and brake assembly. If it's stuck, make sure the parking brake is released, then squirt penetrating oil into the joint between the hub and drum. Allow the oil to soak in and try to pull the drum off again (see Chapter 9).

11   As a last resort, use a slide hammer on the wheel bolts to force the drum off **(see illustration)**.

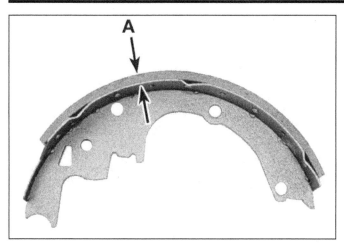

18.13a If the lining is bonded to the brake shoe, measure the lining thickness from the outer surface to the metal shoe, as shown here; if the lining is riveted to the shoe, measure from the lining outer surface to the rivet head

18.13b Check the thickness of the lining at several points on the shoes (arrows)

18.15 Carefully peel back the rubber boots on both sides of the wheel cylinder - if there's any brake fluid behind the boots, the wheel cylinders must be rebuilt or replaced

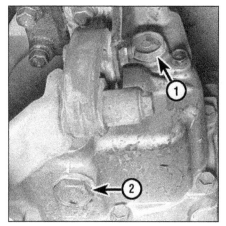

19.1a The manual transmission check/fill plugs are accessible from under the vehicle (Samurai)

1 Oil level check plug
2 Oil drain plug

19.1b Manual transmission fill (upper) - drain plug located in bottom section of transmission

12   With the drum removed, be careful not to touch any brake dust (see the Warning at the beginning of this Section).
13   Note the thickness of the lining material on both the leading and trailing brake shoes. If the material has worn down near the minimum thickness listed in this Chapter's Specifications, the shoes should be replaced **(see illustration)**. The shoes should also be replaced if they're cracked, glazed (shiny surface) or contaminated with brake fluid or oil.
14   Make sure that all the brake assembly springs are connected and in good condition.
15   Check the brake components for signs of fluid leakage. Carefully pry back the rubber cups on the wheel cylinders located at the top of the brake shoes with your finger **(see illustration)**. Any leakage is an indication that the wheel cylinders should be overhauled immediately (see Chapter 9). Also check brake hoses and connections for leakage.
16   Wipe the inside of the drum with a clean rag and brake cleaner or denatured alcohol.

Again, be careful not to breath the asbestos dust.
17   Check the inside of the drum for cracks, score marks, deep scratches and hard spots, which will appear as small discolorations. If imperfections cannot be removed with fine emery cloth, the drum must be taken to a machine shop equipped to resurface the drums.
18   If after the inspection process all parts are in good working condition, reinstall the brake drum.
19   Install the wheels and lower the vehicle.

Parking brake

20   The parking brake operates from a hand lever and locks the rear brake system. The easiest, and perhaps most obvious method of periodically checking the operation of the parking brake assembly is to park the vehicle on a steep hill with the parking brake set and the transmission in Neutral. If the parking brake cannot prevent the vehicle from rolling

within 6 to 10 clicks, it's in need of adjustment (see Chapter 9).

19 Manual transmission lubricant level check

Refer to illustrations 19.1a and 19.1b
1   Manual transmissions don't have a dipstick. The oil level is checked by removing a plug from the side of the transmission case **(see illustrations)**. Locate the plug and use a rag to clean the plug and the area around it. If the vehicle is raised to gain access to the plug, be sure to support it safely on jackstands - DO NOT crawl under the vehicle when it's supported only by a jack!
2   With the engine and transmission cold, remove the plug. If lubricant immediately starts leaking out, thread the plug back into the transmission - the level is correct. If it doesn't, completely remove the plug and reach inside the hole with your little finger. The level should be even with the bottom of the plug hole.

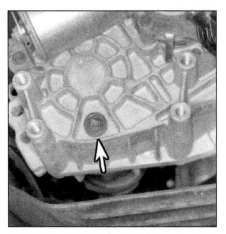

20.1a On Samurai models, locate the two transfer case plugs - the upper one is used for checking and refilling and the lower one is used for draining

20.1b Location of the transfer case drain plug. The check/fill plug is located in the upper section of the transfer case.

21.2a Front differential check/fill plug and drain plug locations

3   If the transmission needs more lubricant, use a syringe or small pump to add it through the plug hole.
4   Thread the plug back into the transmission and tighten it securely. Drive the vehicle, then check for leaks around the plug.

20   Transfer case lubricant level check

Refer to illustrations 20.1a and 20.1b
1   The transfer case oil level is checked by removing a plug from the side of the case **(see illustrations)**. Remove the rock guard, then locate the plug and use a rag to clean the plug and the area around it. If the vehicle is raised to gain access to the plug, be sure to support it safely on jackstands - DO NOT crawl under the vehicle w hen it's supported only by a jack!
2   With the engine and transfer case cold, remove the plug. If lubricant immediately starts leaking out, thread the plug back into the case - the level is correct. If it doesn't, completely remove the plug and reach inside the hole with your little finger. The level should be even with the bottom of the plug hole.
3   If more oil is needed, use a syringe or small pump to add it through the opening.
4   Thread the plug back into the case and tighten it securely. Drive the vehicle, then check for leaks around the plug. Install the rock guard.

21   Differential oil level check

Refer to illustrations 21.2a, 21.2b and 21.3
1   The differential has a check/fill plug which must be removed to check the oil level. If the vehicle is raised to gain access to the plug, be sure to support it safely on jackstands - DO NOT crawl under the vehicle when it's supported only by a jack.

21.2b Fill/check (upper) and drain (lower) bolt locations on the front differential of the Sidekick/X-90/Vitara/Tracker

2   Remove the oil check/fill plug from the differential **(see illustrations)**.
3   The oil level should be at the bottom of the plug opening **(see illustration)**. If not, use a syringe to add the recommended lubricant until it just starts to run out of the opening.
4   Install the plug and tighten it securely.

22   Suspension and steering check

1   Whenever the front of the vehicle is raised for any reason, it's a good idea to visually check the suspension and steering components for wear.
2   Indications of steering or suspension problems include excessive play in the steering wheel before the front wheels react, excessive swaying around corners or body movement over rough roads and binding at some point as the steering wheel is turned.
3   Before the vehicle is raised for inspection, test the shock absorbers by pushing down aggressively at each corner. If the vehicle doesn't come back to a level position within one or two bounces, the shocks are worn and should be replaced. As this is done

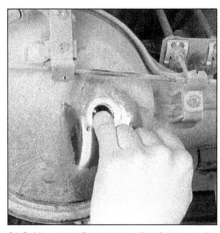

21.3 Use your finger as a dipstick to make sure the lubricant level is even with the bottom of the hole

listen for squeaks and other noises from the suspension components. Information on shock absorber and suspension components can be found in Chapter 10.
4   Raise the front end of the vehicle and support it on jackstands. Make sure it's safely supported!
5   Crawl under the vehicle and check for loose bolts, broken or disconnected parts and deteriorated rubber bushings on all suspension and steering components. Look for grease or fluid leaking from around the steering gear assembly and shock absorbers. If equipped, check the power steering hoses and connections for leaks.
6   The balljoint seals should be checked at this time (Sidekick/X-90/Vitara/Tracker). This includes not only the upper and lower suspension balljoints, but those connecting the steering linkage parts as well. After cleaning around the balljoints, inspect the seals for cracks and damage.
7   Grip the top and bottom of each wheel and try to move it in and out. It won't take a lot of effort to be able to feel any play in the wheel bearings. If the play is noticeable it would be a good idea to adjust it right away

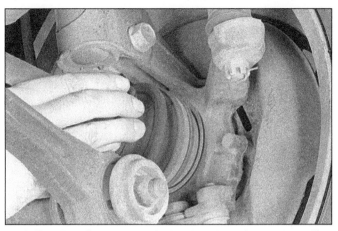

23.2 Flex the driveaxle boots by hand to check for cracks and/or leaking grease

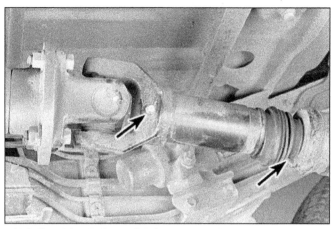

24.9 The driveshafts are lubricated at the fitting and also at the boot (arrows)

or it could confuse further inspections.

8   Grip each side of the wheel and try rocking it laterally. Steady pressure will, of course, turn the steering, but back and forth pressure will reveal a loose steering joint. If some play is felt it would be easier to get assistance from someone so while one person rocks the wheel from side to side, the other can look at the joints, bushings and connections in the steering linkage. On models equipped with a steering gearbox there are eight places where the play may occur. The two outer balljoints on the tie-rods are the most likely, followed by the two inner joints on the same rods, where they join to the center rod. Any play in them means replacement of the tie-rod end. Next are two swivel bushings, one at each end of the center gear rod. Finally, check the steering gear arm balljoint and the one on the idler arm that supports the center rod on the side opposite the steering box. This unit is bolted to the side of the frame member and any play calls for replacement of the bushings.

9   To check the steering box, first make sure the bolts holding the steering box to the frame are tight. Then get another person to help examine the mechanism. One should look at, or hold onto, the arm at the bottom of the steering box while the other turns the steering wheel a little from side to side. The amount of lost motion between the steering wheel and the gear arm indicates the degree of wear in the steering box mechanism. This check should be carried out with the wheels first in the straight ahead position and then at nearly full lock on each side. If the play only occurs noticeably in the straight ahead position then the wear is most likely in the worm and/or nut. If it occurs at all positions, then the wear is probably in the sector shaft bearing. Oil leaks from the unit are another indication of such wear. In either case the steering box will need removal for closer examination and repair.

10   Moving to the vehicle interior, check the play in the steering wheel by turning it slowly in both directions until the wheels can just be felt turning. The steering wheel free play

should be less than 1 3/8-inch (35 mm). Excessive play is another indication of wear in the steering gear or linkage.

11   Following the inspection of the front, a similar inspection should be made of the rear suspension components, again checking for loose bolts, damaged or disconnected parts and deteriorated rubber bushings.

23   Driveaxle boot check

Refer to illustration 23.2

1   The driveaxle boots are very important because they prevent dirt, water and foreign material from entering and damaging the constant velocity (CV) joints. Oil and grease can cause the boot material to deteriorate prematurely, so it's a good idea to wash the boots with soap and water.

2   Inspect the boots for tears and cracks as well as loose clamps **(see illustration)**. If there is any evidence of cracks and leaking lubricant, they must be replaced as described in Chapter 8.

24   Driveshaft check and lubrication

Refer to illustration 24.9

1   Raise the rear of the vehicle and support it securely on jackstands. Block the front wheels. The transmission should be in Neutral.

2   Crawl under the vehicle and visually inspect the driveshaft. Look for dents and cracks in the tube. If any are found, the driveshaft must be replaced (see Chapter 8).

3   Check for oil leakage at the front and rear of the driveshaft. Leakage where the driveshaft enters the transmission indicates a defective rear transmission seal. Leakage where the driveshaft enters the differential indicates a defective pinion seal. For these repair operations refer to Chapters 7 and 8 respectively.

4   While still under the vehicle, have an assistant turn the rear wheel so the driveshaft

will rotate. As it does, check for binding, noise and excessive play in the U-joints.

5   The universal joints can also be checked with the driveshaft motionless, by gripping both sides of the joint and attempting to twist it. Any movement at all in the joint is a sign of considerable wear. Lifting up on the shaft will also indicate movement in the universal joints.

6   Check the driveshaft mounting bolts at both ends to make sure they're tight.

7   The above driveshaft checks should be repeated on all driveshafts. In addition check for grease leakage around the sleeve yoke, indicating failure of the yoke seal.

8   Check for leakage where the driveshafts connect to the transfer case.

9   The driveshafts must be lubricated periodically as indicated in the maintenance schedule. Use the grease gun to insert grease into the fitting **(see illustration)**. Also pull back the boot at the rear of the axle and insert grease.

25   Fuel system check

Warning: *There are certain precautions to take when inspecting or servicing the fuel system components. Work in a well ventilated area and don't allow open flames (cigarettes, natural gas appliances, etc.) in the work area. Mop up spills immediately and don't store fuel soaked rags where they could ignite. On fuel injection equipped models the fuel system is under pressure. No components should be disconnected until the pressure has been relieved (see Chapter 4).*

1   On most models the main fuel tank is located under the left side of the vehicle.

2   The fuel system is most easily checked with the vehicle raised on a hoist so the components underneath the vehicle are readily visible and accessible.

3   If the smell of gasoline is noticed while driving or after the vehicle has been in the sun, the system should be thoroughly inspected immediately.

4   Remove the gas tank cap and check for damage, corrosion and an unbroken sealing

26.3 With the air intake case removed, the choke plate is visible at the top of the carburetor

27.2 Pry the seal retainer away from the steering knuckle, being careful not to damage the felt gasket on the back of the retainer

27.3 Pull the old seal out with a pair of needle-nose pliers

imprint on the gasket. Replace the cap with a new one if necessary.

5   With the vehicle raised, check the gas tank and filler neck for punctures, cracks and other damage. The connection between the filler neck and the tank is especially critical. Sometimes a rubber filler neck will leak due to loose clamps or deteriorated rubber, problems a home mechanic can usually rectify. **Warning:** *Do not, under any circumstances, try to repair a fuel tank yourself (except rubber components). A welding torch or any open flame can easily cause the fuel vapors to explode if the proper precautions are not taken!*

6   Carefully check all rubber hoses and metal lines leading away from the fuel tank. Look for loose connections, deteriorated hoses, crimped lines and other damage. Follow the lines to the front of the vehicle, carefully inspecting them all the way. Repair or replace damaged sections as necessary.

7   If a fuel odor is still evident after the inspection, refer to Chapter 4.

26   Carburetor choke check

Refer to illustration 26.3

1   The choke operates only when the engine is cold, so this check should be performed before the engine has been started for the day.

2   Open the hood and remove the air intake case from the carburetor. It's held in place by a nut at the center. If any vacuum hoses must be disconnected, tag them to ensure reinstallation in their original positions.

3   Look at the center of the carburetor. You'll notice a flat plate at the carburetor opening **(see illustration).**

4   Have an assistant press the throttle pedal to the floor. The plate should close completely. Start the engine while you watch the plate at the carburetor. Don't position your face near the carburetor, as the engine could backfire, causing serious burns! When the engine starts, the choke plate should

open slightly.

5   Allow the engine to continue running at an idle speed. As the engine warms up to operating temperature, the plate should slowly open, allowing more air to enter through the top of the carburetor.

6   After a few minutes, the choke plate should be completely open to the vertical position. Blip the throttle to make sure the fast idle cam disengages.

7   You'll notice that engine speed corresponds to the plate opening. With the plate closed, the engine should run at a fast idle speed. As the plate opens and the throttle is moved to disengage the fast idle cam, the engine speed will decrease.

8   With the engine off and the throttle held half-way open, open and close the choke several times. Check the linkage to see if it's hooked up correctly and make sure it doesn't bind.

9   If the choke or linkage binds, sticks or works sluggishly, clean it with choke cleaner (an aerosol spray available at auto parts stores). If the condition persists after cleaning, replace the troublesome parts.

10   Visually inspect all vacuum hoses to be

sure they're securely connected and look for cracks and deterioration. Replace as necessary.

11   Refer to Chapter 4 for more information on the choke.

27   Steering knuckle oil seal replacement (Samurai)

Refer to illustrations 27.2, 27.3, 27.4, 27.5 and 27.6

1   Raise the front of the vehicle and support it securely on jackstands.

2   Remove the bolts that secure the seal retainer to the steering knuckle. Using a screwdriver or pry bar, pry the retainer from the knuckle and slide it inward onto the axle housing **(see illustration).**

3   Grasp the old seal with a pair of pliers and pull it out **(see illustration).** Cut the seal with a knife or pair of scissors and remove it from the vehicle.

4   Cut the new seal in one place, lubricate its inner circumference with grease and position it in the steering knuckle, with the cut portion at the top **(see illustration).**

5   Apply a bead of RTV sealant to the

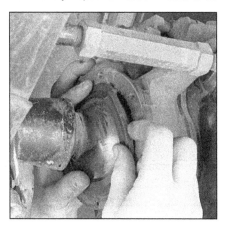

27.4 Cut the new seal, apply grease to the inside surface, and install it with the cut portion on top

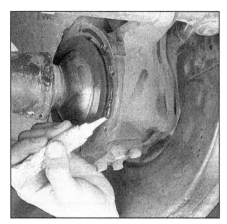

27.5 Apply a bead of RTV sealant to the steering knuckle, inside of the bolt holes

27.6 Apply some sealant to the ends of the seal retainers when installing them

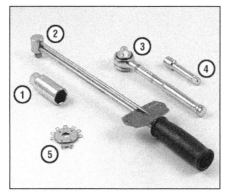

28.2a Tools required for changing spark plugs

1   **Spark plug socket** - This will have special padding inside to protect the spark plug's porcelain insulator
2   **Torque wrench** - Although not mandatory, using this tool is the best way to ensure the plugs are tightened properly
3   **Ratchet** - Standard hand tool to fit the spark plug socket
4   **Extension** - Depending on model and accessories, you may need special extensions and universal joints to reach one or more of the plugs
5   **Spark plug gap gauge** - This gauge for checking the gap comes in a variety of styles. Make sure the gap for your engine is included

28.2b On later models it will be necessary to unplug the electrical connector (A), remove the coil pack bolt (B) and remove the assembly for access to the spark plugs

retainer mating surface on the steering knuckle **(see illustration)**.

6   Install the seal retainers up to the steering knuckle. Align the holes and install the bolts. Install one half loosely before tightening both of them to the torque listed in this Chapter's Specifications **(see illustration)**.

28 Spark plug replacement

Refer to illustrations 28.2a, 28.2b, 28.5a, 28.5b, 28.6 and 28.10

1   Replace the spark plugs with new ones at the intervals recommended in the *Routine maintenance* schedule.
2   In most cases, the tools necessary for spark plug replacement include a spark plug socket which fits onto a ratchet (spark plug sockets are padded inside to prevent damage to the porcelain insulators on the new plugs), various extensions and a gap gauge to check and adjust the gaps on the new plugs **(see illustration)**. A special plug wire removal tool is available for separating the

wire boots from the spark plugs, but it isn't absolutely necessary. A torque wrench should be used to tighten the new plugs.
3   The best approach when replacing the spark plugs is to purchase the new ones in advance, adjust them to the proper gap and replace them one at a time. When buying the new spark plugs, be sure to obtain the correct plug type for your particular engine. This information can be found on the Emission Control Information label located under the

hood and in the factory owner's manual. If differences exist between the plug specified on the emissions label and in the owner's manual, assume the emissions label is correct.
4   Allow the engine to cool completely before attempting to remove any of the plugs. While you're waiting for the engine to cool, check the new plugs for defects and adjust the gaps.
5   The gap is checked by inserting the proper thickness gauge between the electrodes at the tip of the plug **(see illustration)**. The gap between the electrodes should be the same as the one specified on the Emissions Control Information label. The wire should just slide between the electrodes with a slight amount of drag. If the gap is incorrect, use the adjuster on the gauge body to bend the curved side electrode slightly until the proper gap is obtained **(see illustration)**. If the side electrode is not exactly over the center electrode, bend it with the adjuster until it is. Check for cracks in the porcelain insulator (if any are found, the plug shouldn't be used).
6   With the engine cool, remove the spark plug wire from one spark plug. Pull only on the boot at the end of the wire - don't pull on the wire. A plug wire removal tool should be used if available **(see illustration)**.

28.5a Spark plug manufacturers recommend using a wire-type gauge when checking the gap - if the wire does not slide between the electrodes with a slight drag, adjustment is required

28.5b To change the gap, bend the side electrode only, as indicated by the arrows, and be very careful not to crack or chip the porcelain insulator surrounding the center electrode

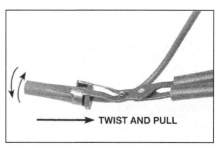

TWIST AND PULL

28.6 When removing the spark plug wires, pull only on the boot and twist it back-and-forth

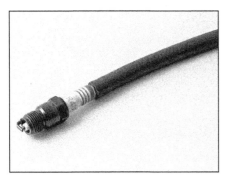

28.10 A length of rubber hose will save time and prevent damaged threads when installing the spark plugs

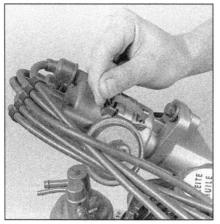

30.2 Release the two spring clips, then remove the distributor cap - the distributor cap will only fit on the distributor body one way, so make sure the new cap is in the same position as the old one when transferring the wires

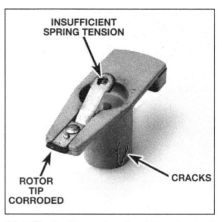

30.4 The ignition rotor should be checked for wear and corrosion as indicated here (if in doubt about its condition, buy a new one)

7   If compressed air is available, use it to blow any dirt or foreign material away from the spark plug hole. The idea here is to eliminate the possibility of debris falling into the cylinder as the spark plug is removed.

8   Place the spark plug socket over the plug and remove it from the engine by turning it in a counterclockwise direction.

9   Compare the spark plug to those shown on the inside back cover to get an indication of the general running condition of the engine.

10   Thread one of the new plugs into the hole until you can no longer turn it with your fingers, then tighten it with a torque wrench (if available) or the ratchet. It might be a good idea to slip a short length of rubber hose over the end of the plug to use as a tool to thread it into place **(see illustration)**. The hose will grip the plug well enough to turn it, but will start to slip if the plug begins to cross-thread in the hole - this will prevent damaged threads and the accompanying repair costs.

11   Before pushing the spark plug wire onto the end of the plug, inspect it following the procedures outlined in Section 29.

12   Attach the plug wire to the new spark plug, again using a twisting motion on the boot until it's seated on the spark plug.

13   Repeat the procedure for the remaining spark plugs, replacing them one at a time to prevent mixing up the spark plug wires.

29 Spark plug wire check and replacement

1   The spark plug wires should be checked at the recommended intervals and whenever new spark plugs are installed in the engine.

2   The wires should be inspected one at a time to prevent mixing up the order, which is essential for proper engine operation.

3   Disconnect the plug wire from one spark plug. To do this, grab the rubber boot, twist slightly and pull the wire free. Do not pull on the wire itself, only on the rubber boot **(see illustration 28.6)**.

4   Check inside the boot for corrosion, which will look like a white crusty powder. Push the wire and boot back onto the end of the spark plug. It should be a tight fit on the

plug. If it isn't, remove the wire and use a pair of pliers to carefully crimp the metal connector inside the boot until it fits securely on the end of the spark plug.

5   Using a clean rag, wipe the entire length of the wire to remove any built-up dirt and grease. Once the wire is clean, check for holes, burned areas, cracks and other damage. Don't bend the wire excessively or the conductor inside might break.

6   Disconnect the wire from the distributor cap. Again, pull only on the rubber boot. Check for corrosion and a tight fit in the same manner as the spark plug end. Reattach the wire to the distributor cap.

7   Check the remaining spark plug wires one at a time, making sure they are securely fastened at the distributor and the spark plug when the check is complete.

8   If new spark plug wires are required, purchase a new set for your specific engine model. Wire sets are available pre-cut, with the rubber boots already installed. Remove and replace the wires one at a time to avoid mix-ups in the firing order. The wire routing is extremely important, so be sure to note exactly how each wire is situated before removing it.

30 Distributor cap and rotor check and replacement

Refer to illustrations 30.2, 30.4 and 30.7
Note: *It's common practice to install a new distributor cap and rotor whenever new spark plug wires are installed.*

1   Although the breakerless distributor used on these vehicles requires much less maintenance than a conventional distributor, periodic inspections should be performed at the intervals specified in the routine maintenance schedule and whenever any work is performed on the distributor.

2   Disconnect the ignition coil wire from the coil, then unsnap the spring clips or loosen the screws that hold the cap to the distributor body **(see illustration)**. Detach the distributor cap and wires.

3   Place the cap, with the spark plug and coil wires still attached, out of the way. Use a length of wire or rope to secure it, if necessary.

4   The rotor is now visible on the end of the distributor shaft. Check it carefully for cracks and carbon tracks. Make sure the center terminal spring tension is adequate (not all models) and look for corrosion and wear on the rotor tip **(see illustration)**. If in doubt about its condition, replace it with a new one.

30.7 Shown here are some of the common defects to look for when inspecting the distributor cap (if in doubt about its condition, install a new one)

31.3a Disengage the clips on the air cleaner housing, remove the nut securing the air intake case to the carburetor . . .

31.3b . . . then lift the assembly up to gain access to the air filter (Samurai)

31.3c On Sidekick/X-90/Vitara/Tracker models, the air filter housing is located in the engine compartment on the inner fender wall - remove the screws (arrows) or detach the clips

31.3d Hold the cover up out of the way and lift out the filter element

5   If replacement is required, detach the rotor from the shaft and install a new one.
6   While the distributor cap is off, check the air gap as described in Chapter 5.
7   Check the distributor cap for carbon tracks, cracks and other damage. Closely examine the terminals on the inside of the cap for excessive corrosion and damage **(see illustration)**. Slight deposits are normal. Again, if in doubt about the condition of the cap, replace it with a new one.
8   When replacing the cap, simply transfer the spark plug and coil wires, one at a time, from the old cap to the new cap. Be very careful not to mix up the wires!
9   Reattach the cap to the distributor, then tighten the screws or reposition the spring clips to hold it in place.

31 Air filter replacement

Refer to illustrations 31.3a, 31.3b, 31.3c and 31.3d

1   At the specified intervals, the air filter and (if equipped) PCV filter should be replaced with new ones. A thorough program of preventative maintenance would also call for the filter to be inspected periodically between changes, especially if the vehicle is often driven in dusty conditions.
2   The air filter is located inside the air cleaner housing which is mounted on the inner fender panel on the passenger side (Samurai) or on the drivers side (Sidekick/Tracker).
3   Release the clips or remove the screws securing the top plate to the air cleaner body. On Samurai models, remove the air intake case-to-carburetor nut. Disconnect any electrical connections or vacuum lines and carefully mark them for reinstallation purposes **(see illustrations)**. Lift the air filter out of the housing. If it's covered with dirt, it should be replaced **(see illustration)**.
4   Wipe the inside of the air cleaner housing with a clean rag.
5   Place the old filter (if in good condition) or the new filter in the air cleaner housing.

6   Reinstall the top plate and any hoses which were disconnected. Secure the clips or tighten the screws.

32 Thermostatically Controlled Air Cleaner (TCAC) check

The Thermostatically Controlled Air Cleaner and related components are covered in detail in Chapter 6.

33 Idle speed check and adjustment

Note: *On later models idle speed is controlled by the ECM and is not adjustable. If the information in this Section differs from the Vehicle Emission Control Information label in the engine compartment of your vehicle, the label should be considered correct.*
Refer to illustration 33.8

1   Engine idle speed is the speed at which the engine operates when no throttle pedal pressure is applied. The idle speed is critical to the performance of the engine itself, as well as many engine sub-systems.
2   A hand-held tachometer must be used when adjusting idle speed to get an accurate reading. The exact hook-up for these meters varies with the manufacturer, so follow the particular directions included.
3   Set the parking brake and block the wheels. Be sure the transmission is in Neutral (manual transmission) or Park (automatic transmission).
4   Turn off the air conditioner (if equipped), the headlights and any other accessories during this procedure.
5   Start the engine and allow it to reach normal operating temperature.
6   Open the hood and run the engine at about 2000 rpm for approximately three minutes, then allow it to idle again for about one minute.
7   Check the engine idle speed with the tachometer and compare it to the VECI label on the hood.
8   If the idle speed is not correct, turn the

idle speed adjusting screw (clockwise for faster, counterclockwise for slower) until the idle speed is correct **(see illustration)**.

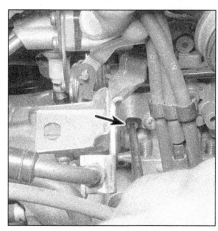

33.8 Set the idle to the speed listed on the Vehicle Emission Control Information label under the hood - turning the adjustment screw (arrow) clockwise will cause the idle speed to increase, turning it counterclockwise will cause it to decrease

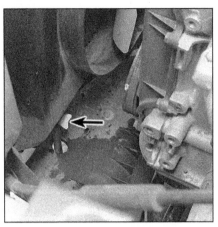

34.4 Open the drain fitting (arrow) but be careful not to completely remove it

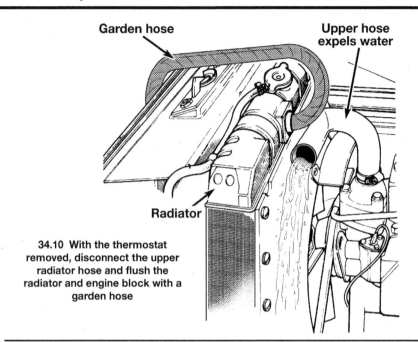

34.10 With the thermostat removed, disconnect the upper radiator hose and flush the radiator and engine block with a garden hose

34   Cooling system servicing (draining, flushing and refilling)

Warning: *Do not allow engine coolant (antifreeze) to come in contact with your skin or painted surfaces of the vehicle. Rinse off spills immediately with plenty of water. Antifreeze is highly toxic if ingested. Never leave antifreeze laying around in an open container or in puddles on the floor; children and pets are attracted by it's sweet smell and may drink it. Check with local authorities about disposing of used antifreeze. Many communities have collection centers which will see that antifreeze is disposed of safely.*

1   Periodically, the cooling system should be drained, flushed and refilled to replenish the antifreeze mixture and prevent formation of rust and corrosion, which can impair the performance of the cooling system and cause engine damage. When the cooling system is serviced, all hoses and the radiator cap should be checked and replaced if necessary.

Draining

Refer to illustration 34.4

2   Apply the parking brake and block the wheels. If the vehicle has just been driven, wait several hours to allow the engine to cool down before beginning this procedure.

3   Once the engine is completely cool, remove the radiator cap.

4   Move a large container under the radiator drain to catch the coolant. Attach a 3/8-inch inner diameter hose to the drain fitting to direct the coolant into the container (some models are already equipped with a hose), then open the drain fitting (a pair of pliers may be required to turn it) **(see illustration)**.

5   After the coolant stops flowing out of the radiator, move the container under the engine block drain plug. Loosen the plug and allow the coolant in the block to drain.

6   While the coolant is draining, check the condition of the radiator hoses, heater hoses and clamps (refer to Section 13 if necessary).

7   Replace any damaged clamps or hoses (see Chapter 3).

Flushing

Refer to illustration 34.10

8   Once the system is completely drained, remove the thermostat from the engine (see Chapter 3). Then reinstall the thermostat housing without the thermostat. This will allow the system to be flushed.

9   Reinstall the engine block drain plugs and tighten the radiator drain plug. Turn your heating system controls to Hot, so that the heater core will be flushed at the same time as the rest of the cooling system.

10   Disconnect the upper radiator hose from the radiator. Place a garden hose in the upper radiator inlet, turn the water on and flush the system until the water runs clear out of the upper radiator hose **(see illustration)**.

11   In severe cases of contamination or clogging of the radiator, remove the radiator (see Chapter 3) and have a radiator repair facility clean and repair it if necessary. Many deposits can be removed by the chemical action of a cleaner available at auto parts stores. Follow the procedure outlined in the manufacturer's instructions. **Note:** *When the coolant is regularly drained and the system refilled with the correct antifreeze/water mixture, there should be no need to use chemical cleaners or descalers.*

12   After flushing, drain the radiator and remove the block drain plugs once again to drain the water from the system.

Refilling

13   Close and tighten the radiator drain. Install and tighten the block drain plug.

14   Place the heater temperature control in the maximum heat position.

15   Slowly add new coolant (a 50/50 mixture of water and antifreeze) to the radiator until it's full. Add coolant to the reservoir up to the lower mark.

16   Leave the radiator cap off and run the engine in a well-ventilated area until the thermostat opens (coolant will begin flowing through the radiator and the upper radiator hose will become hot).

17   Turn the engine off and let it cool. Add more coolant mixture to bring the level back up to the lip on the radiator filler neck.

18   Squeeze the upper radiator hose to expel air, then add more coolant mixture if necessary. Replace the radiator cap.

19   Start the engine, allow it to reach normal operating temperature and check for leaks.

35   Exhaust system check

1   With the engine cold (at least three hours after the vehicle has been driven), check the complete exhaust system from the manifold to the end of the tailpipe. Be careful around the catalytic converter, which may be hot even after three hours. The inspection should be done with the vehicle on a hoist to permit unrestricted access. If a hoist isn't available, raise the vehicle and support it securely on jackstands.

2   Check the exhaust pipes and connections for signs of leakage and/or corrosion indicating a potential failure. Make sure that all brackets and hangers are in good condition and tight.

3   Inspect the underside of the body for holes, corrosion, open seams, etc. which may allow exhaust gases to enter the passenger compartment. Seal all body openings with silicone or body putty.

4   Rattles and other noises can often be traced to the exhaust system, especially the hangers, mounts and heat shields. Try to move the pipes, mufflers and catalytic converter. If the components can come in contact with the body or suspension parts, secure the exhaust system with new brackets and hangers.

36   Automatic transmission fluid and filter change

1   At the specified time intervals, the transmission fluid should be drained and replaced. Since the fluid will remain hot long after driving, perform this procedure only after the engine has cooled down completely.

2   Before beginning work, purchase the specified transmission fluid (see Recommended lubricants and fluids at the front of this Chapter) and a new filter.

3   Other tools necessary for this job include jackstands to support the vehicle in a raised position, a drain pan capable of holding at least eight pints, newspapers and clean rags.

4   Raise the vehicle and support it securely on jackstands.

5   To provide adequate clearance, unbolt the universal joint flange of the lower driveshaft and pull it aside to the right.

6   With a drain pan in place, remove the pan front and side mounting bolts.

7   Loosen the rear bolts approximately four turns.

8   Carefully pry the transmission pan loose with a screwdriver, allowing the fluid to drain.

9   Remove the remaining bolts, pan and gasket. Carefully clean the gasket surface of the transmission to remove all traces of the old gasket and sealant.

10   Drain the fluid from the transmission pan, clean it with solvent and dry it with compressed air.

11   Remove the filter from the mount inside the transmission.

12   Install a new filter and O-ring.

13   Make sure the gasket surface on the transmission pan is clean, then install a new gasket. Put the pan in place against the transmission and, working around the pan, tighten each bolt a little at a time until the torque figure listed in this Chapter's Specifications is reached.

14   Install the universal joint flange.

15   Lower the vehicle and add the specified amount of automatic transmission fluid through the filler tube (see Section 6).

16   With the transmission in Park and the parking brake set, run the engine at a fast idle, but don't race it.

17   Move the gear selector through each range and back to Park. Check the fluid level.

18   Check under the vehicle for leaks during the first few trips.

19   Finalize your automatic transmission fluid level check by driving the vehicle to reach normal operating temperature and then filling to the "FULL HOT" position on the dipstick (see Section 6).

37   Manual transmission lubricant change

1   Drive the vehicle for a few miles to thoroughly warm up the transmission oil.

2   Raise the vehicle and support it securely on jackstands.

3   Move a drain pan, rags, newspapers and a wrench under the vehicle. With the drain pan and newspapers in position under the transmission, use the wrench to loosen the drain plug located in the bottom of the transmission case (see illustrations 19.1a and 19.1b).

4   Once loosened, carefully unscrew it with your fingers until you can remove it from the transmission. Allow all of the oil to drain into the pan. If the plug is too hot to touch, use the wrench to remove it.

5   If the transmission is equipped with a magnetic drain plug, see if there are bits of metal clinging to it. If there are, it's a sign of excessive internal wear, indicating that the transmission should be carefully inspected in the near future. If the transmission isn't equipped with a magnetic drain plug, allow the oil in the pan to cool, then feel with your hands along the bottom of the drain pan for debris.

6   Clean the drain plug, then reinstall it in the transmission and tighten it to the torque listed in this Chapter's Specifications.

7   Remove the transmission oil check/fill plug (see Section 19). Using a hand pump or syringe, fill the transmission with the correct amount and grade of oil (see the Specifications), until the level is just at the bottom of the plug hole.

8   Reinstall the check/fill plug and tighten it securely.

38   Transfer case lubricant change

1   Drive the vehicle for at least 15 minutes in 4WD to warm up the oil in the case.

2   Raise the vehicle and support it securely on jackstands. Remove the rock guard.

3   Move a drain pan, rags, newspapers and a breaker bar or ratchet (to fit the square drive hole in the transfer case plugs) under the vehicle.

4   Remove the check/fill plug (see illustrations 20.1a and 20.1b).

5   Remove the drain plug from the lower part of the case and allow the old oil to drain completely.

6   Carefully clean and install the drain plug after the case is completely drained. Tighten the plug to the torque listed in this Chapter's Specifications.

7   Fill the case with the specified lubricant until it's level with the lower edge of the filler hole.

8   Install the check/fill plug and tighten it securely.

9   Lower the vehicle.

10   Check carefully for leaks around the drain plug after the first few miles of driving.

39   Differential lubricant change

Note: *The following procedure can be used for both differential.*

1   Drive the vehicle for several miles to warm up the differential oil, then raise the vehicle and support it securely on jackstands.

2   Move a drain pan, rags, newspapers and a wrench under the vehicle.

3   With the drain pan under the differential, use the wrench to loosen the drain plug. It's the lower of the two plugs (see illustrations 21.2a and 21.2b).

4   Once loosened, carefully unscrew the plug with your fingers until you can remove it from the case.

5   Allow all of the oil to drain into the pan, then replace the drain plug and tighten it to the torque listed in this Chapter's Specifications.

6   Feel with your hands along the bottom of the drain pan for any metal bits that may have come out with the oil. If there are any, it's a sign of excessive wear, indicating that the internal components should be carefully inspected in the near future.

7   Remove the differential check/fill plug located above the drain plug. Using a hand pump, syringe or funnel, fill the differential with the correct amount and grade of oil (see this Chapter's Specifications) until the level is just at the bottom of the plug hole.

8   Reinstall the plug and tighten it securely.

9   Lower the vehicle. Check for leaks at the drain plug after the first few miles of driving.

40   Fuel filter replacement

Refer to illustrations 40.2 and 40.7

Warning: *Gasoline is extremely flammable, so extra safety precautions must be observed when working on the fuel system. DO NOT smoke or allow open flames or bare light bulbs near the vehicle. Also, don't perform fuel system maintenance procedures in a garage where a natural gas type appliance, such as a water heater or clothes dryer, is present.*

1   This job should be done with the engine cold (after sitting at least three hours). Place a metal container, rags or newspapers under the filter to catch spilled fuel. **Warning:** Before attempting to remove the fuel filter, disconnect the negative cable from the battery and position it out of the way so it can't accidentally contact the battery post.

Carburetor equipped models

2   The fuel filter is located under the vehicle near the rear differential **(see illustration)**.

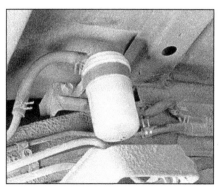

40.2 The fuel filter is located under the vehicle on the frame, near the rear axle (Samurai)

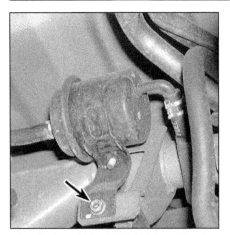

40.7 On later Sidekick/X-90/Vitara/Tracker models the fuel filter is held in place by a bolt

41.2 On earlier models it is possible to check the EGR valve by reaching under it and pushing on the diaphragm (arrow) with a finger - it should move smoothly within the housing

42.1 Locate the positive crankcase ventilation valve (PCV valve) on the intake manifold (arrow)

3   To replace the filter, loosen the clamps and slide them down the hoses, past the fittings on the filter.
4   Carefully twist and pull on the hoses to separate them from the filter. If the hoses are in bad shape, now would be a good time to replace them with new ones.
5   Pull the filter out of the clip and install the new one, then hook up the hoses and tighten the clamps securely. Start the engine and check carefully for leaks at the filter hose connections.

Fuel injected models

6   Depressurize the fuel system (see Chapter 4).
7   The fuel filter can be found in the fuel feed line attached on the chassis frame above the upper right hand side of the rear differential. To replace the filter, loosen the flange fittings (use two wrenches) or remove the clips and detach the hoses **(see illustration)**. Replace the old seals on flange-type fittings.
8   Note how the filter is installed (which end is facing up) so the new filter doesn't get installed backwards. If the hoses are in bad shape, now would be a good time to replace them with new ones.
9   Remove the filter.
10   Install the new filter in the clamp and tighten the bolts. Make sure the filter is properly oriented - fuel filters usually have an arrow on the canister that indicates the direction of fuel flow.
11   Connect the hoses to the new filter and tighten the fittings securely.
12   Start the engine and check carefully for leaks at the filter hose connections.

41   Exhaust Gas Recirculation (EGR) system check

Refer to illustration 41.2

1   The EGR valve is usually located on the intake manifold, adjacent to the carburetor or TBI unit. Most of the time when a problem

develops in this emissions system, it's due to a stuck or corroded EGR valve. On earlier models its possible to check the valve operation as follows.
2   With the engine cold to prevent burns, push on the EGR valve diaphragm. Using moderate pressure, you should be able to press the diaphragm in-and-out within the housing **(see illustration)**.
3   If the diaphragm doesn't move or moves only with much effort, replace the EGR valve with a new one. If in doubt about the condition of the valve, compare the free movement of your EGR valve with a new valve.
4   Refer to Chapter 6 for more information on the EGR system.

42   Positive Crankcase Ventilation (PCV) valve check and replacement

Refer to illustrations 42.1 and 42.2

1   On carbureted models and models with throttle body injection, the PCV valve threads into the intake manifold and is connected by a rubber hose to the camshaft cover **(see illustration)**. On models with sequential multiport fuel injection, the PCV hose connects the camshaft cover to the air intake plenum. The PCV valve is installed in the camshaft cover.
2   To check a PCV valve, unscrew it from the manifold or pull it out of the camshaft cover, start the engine and, with the engine idling, plug the open end of the valve with your finger **(see illustration)** and verify that there is vacuum. If there isn't, replace the PCV valve.
3   Loosen the clamps securing the hose to the PCV valve and to the camshaft cover and then disconnect the hose from the valve.
4   Unscrew the valve from the intake manifold (carbureted and throttle body injection models) or pull it out of the camshaft cover (sequential multiport fuel injection models)..

42.2 To test a PCV valve, unscrew it from the manifold or pull it out of the camshaft cover, start the engine and, with the engine idling, plug the open end of the valve with your finger and verify that there is vacuum

2   When purchasing a replacement PCV valve, make sure it's the correct one for your vehicle. Compare the old valve with the new one to make sure they're the same.
6   On carbureted models and models with throttle body injection, screw the new valve into the manifold and connect the hose to it. On models with sequential multiport fuel injection, attach the valve to the hose, secure it with the hose clamp and then push the valve into its grommet in the camshaft cover.
7   More information on the PCV system can be found in Chapter 6.

43   Fuel tank cap gasket replacement

1   Obtain a new gasket.
2   Remove the tank cap and carefully pry the old gasket out of the recess. Be very careful not to damage the sealing surface inside the cap.

45.1a The charcoal canister functions as a holding tank for fuel vapors - inspect it periodically as indicated in the maintenance schedule (early Samurai shown)

3   Work the new gasket into the cap recess.
4   Install the cap, then remove it and make sure the gasket seals all the way around.

44 Brake fluid replacement

1   Because brake fluid absorbs moisture which could ultimately cause corrosion of the brake components, and air which could make the braking system less effective, the fluid should be replaced at the specified intervals. This job can be accomplished for a nominal fee by a properly equipped brake shop using a pressure bleeder. The task can also be done by the home mechanic with the help of an assistant. To bleed the air and old fluid and replace it with fresh fluid from sealed containers, refer to the brake bleeding procedure in Chapter 9.
2   If there is any possibility that incorrect fluid has been used in the system, drain all the fluid and flush the system with brake cleaner. Replace all piston seals and cups, as they will be affected and could possibly fail under pressure.

45 Evaporative emissions control system check and canister replacement

Refer to illustrations 45.1a and 45.1b
1   The function of the evaporative emissions control system is to draw fuel vapors from the gas tank and fuel system, store them in a charcoal canister **(see illustrations)** and route them to the intake manifold during normal engine operation.
2   The most common symptom of a fault in the evaporative emissions system is a strong fuel odor in the engine compartment. If a fuel odor is detected, inspect the charcoal canister and all hoses for damage and deterioration.

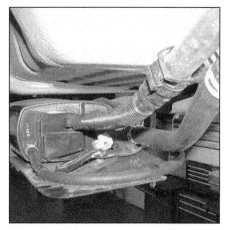

45.1b On late-model fuel-injected vehicles equipped with an On-Board Refueling Vapor Recovery (ORVR) system, the charcoal canister is located under the vehicle, on the right side, in front of the fuel tank

3   At the specified intervals, the charcoal canister must be replaced with a new one. Disconnect the hoses, and then unbolt the bracket from the firewall or from the underside of the vehicle. Installation is the reverse of removal.
4   The evaporative emissions control system is explained in more detail in Chapter 6.

46 Ignition timing check and adjustment

Refer to illustrations 46.1 and 46.3
Note: *If the information in this Section differs from the Vehicle Emission Control Information label in the engine compartment of your vehicle, the label should be considered correct.*
1   Some special tools are required for this procedure **(see illustration)**. The engine must be at normal operating temperature and the air conditioner must be Off.
2   Apply the parking brake and block the wheels to prevent movement of the vehicle. The transmission must be in Park (automatic) or Neutral (manual).

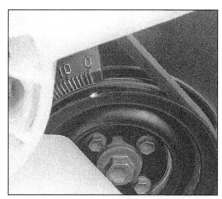

46.3 The ignition timing marks are located at the front of the engine

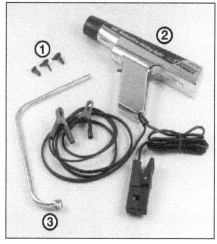

46.1 Tools needed to check and adjust the ignition timing

1   Vacuum plugs - Vacuum hoses will, in most cases, have to be disconnected and plugged. Molded plugs in various shapes and sizes are available for this
2   Inductive pick-up timing light - Flashes a bright, concentrated beam of light when the number one spark plug fires. Connect the leads according to the instructions supplied with the light
3   Distributor wrench - On some models, the hold-down bolt for the distributor is difficult to reach and turn with conventional wrenches or sockets. A special wrench like this must be used

3   Locate the timing marks at the front of the engine (they should be visible from above after the hood is opened) **(see illustration)**. The crankshaft pulley or vibration damper has a notch in it and a plate with raised numbers is attached to the timing cover. Clean the plate with solvent so the numbers are visible.
4   Use chalk or white paint to mark the notch in the pulley/vibration damper.
5   Highlight the point on the timing plate that corresponds to the ignition timing specification on the Vehicle Emission Control Information label.
6   Hook up the timing light by following the manufacturer's instructions (an inductive pick-up timing light is preferred). Generally, the power leads are attached to the battery terminals and the pick-up lead is attached to the number one spark plug wire. The number one spark plug is the one closest to the drive-belt end of the engine. **Caution:** *If an inductive pick-up timing light isn't available, don't puncture the spark plug wire to attach the timing light pick-up lead. Instead, use an adapter between the spark plug and plug wire. If the insulation on the plug wire is damaged, the secondary voltage will jump to ground at the damaged point and the engine will misfire.*
7   Make sure the timing light wires are

routed away from the drivebelts and fan, then start the engine.

8   Allow the idle speed to stabilize, then point the flashing timing light at the timing marks - be very careful of moving engine components!

9   The mark on the pulley/vibration damper will appear stationary. If it's aligned with the specified point on the timing plate, the ignition timing is correct.

10   If the marks aren't aligned, adjustment is required. Loosen the distributor mounting nut and turn the distributor very slowly until the marks are aligned.

11   Tighten the nut and recheck the timing.

12   Turn off the engine and remove the timing light (and adapter, if used).

47   Carburetor/throttle body mounting nut torque check

1   The carburetor or TBI unit used on earlier models is attached to the top of the intake manifold by several bolts or nuts. The fasteners can sometimes work loose from vibration and temperature changes during normal engine operation and cause a vacuum leak.

2   If you suspect that a vacuum leak exists at the bottom of the carburetor or throttle body, obtain a two-foot length of fuel line hose. Start the engine and place one end of the hose next to your ear as you probe around the base with the other end. You'll hear a hissing sound if a leak exists (be careful of hot or moving engine components).

3   Remove the air cleaner assembly, tagging each hose that's disconnected with a piece of numbered tape to make reassembly easier.

4   Locate the mounting nuts or bolts at the base of the carburetor or throttle body. Decide what special tools or adapters will be necessary, if any, to tighten the fasteners.

5   Tighten the nuts or bolts to the torque listed in this Chapter's specifications. Don't overtighten them, as the threads could strip.

48   Oxygen sensor replacement

Refer to illustration 48.2

1   The oxygen sensor used on 1986 Samurai models, should be replaced at the intervals specified in the Maintenance Schedule at the beginning of this Chapter.

2   The sensor is threaded into the exhaust manifold and can be identified by the wires attached to it **(see illustration)**. Replacement consists of disconnecting the wire harness

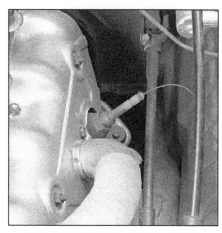

48.2 To remove the oxygen sensor, disconnect the wire and unscrew the sensor from the exhaust manifold

and unthreading the sensor from the manifold. Before installing it in the manifold, coat the threads with an anti-seize compound. Tighten the sensor to the torque listed in this Chapter's Specifications, then reconnect the wire harness. A special oxygen sensor socket, which can be purchased at most auto parts stores, should be used.

Chapter 2 Part A
SOHC engines

Contents

Specifications

General
Firing order	1-3-4-2
Cylinder numbers (drivebelt end-to-transmission end)	1-2-3-4

Cylinder head
Warpage limit	0.002 in (0.5 mm)

Oil pump
Outer gear-to-oil pump housing clearance limit	0.0122 in (0.310 mm)
Gear end play limit	0.0059 in (0.15 mm)
Pressure relief spring free length	1.77 in (45 mm)

The blackened terminal shown on the distributor cap indicates the number one spark plug wire position

Cylinder location and distributor rotation

Rocker arms and shafts
Rocker shaft diameter
8-valve engine	0.628 to 0.629 in (15.951 to 15.977 mm)
16-valve engine	0.6287 to 0.6293 in (15.969 to 15.984 mm)

Rocker arm inside diameter
8-valve engine	0.629 to 0.630 in (15.977 to 16.002 mm)

16-valve engine
Through 1998	0.629 to 0.630 in (15.977 to 16.002 mm)
1999 on	0.6298 to 0.6305 in (15.996 to 16.014 mm

Shaft-to-arm clearance
Standard
8-valve engine	0.0005 to 0.0017 in (0.012 to 0.045 mm)

16-valve engine
Through 1998	0.0001 to 0.0014 in (0.003 to 0.004 mm)
1999 on	0.0005 to 0.0018 in (0.013 to 0.046 mm)
Service limit	0.0035 in (0.09 mm)

Torque specifications

	Ft-lbs (unless otherwise indicated)
Camshaft cover bolts	36 to 42 in-lbs
Intake/exhaust manifold nuts/bolts	13.5 to 20
Camshaft bearing cap bolts (16-valve engines)	80 to 106 in-lbs
Camshaft sprocket bolt	41 to 46
Cylinder head bolts	
8-valve engine	46 to 50
16-valve engine	
Step 1	25
Step 2	40
Step 3	48 to 50
Crankshaft pulley bolts	90 to 108 in-lbs
Crankshaft pulley center bolt	52
8-valve engine	52
16-valve engine	76 to 83
Oil pump gear plate bolts	84 to 102 in-lbs
Oil pump-to-block bolts	84 to 102 in-lbs
Oil pan bolts/nuts	84 to 102 in-lbs
Flywheel/driveplate bolts	58
Timing belt cover	84 to 102 in-lbs
Rear main oil seal housing bolts	96 in-lbs
Rocker arm shaft retaining screws	84 to 102 in-lbs
Timing belt tensioner bolt	17.5 to 21.5
Timing belt tensioner nut	84 to 102 in-lbs

1   General information

This Part of Chapter 2 is devoted to in-vehicle repair procedures for the 1.3L and 1.6L engine. All information concerning engine removal and installation and engine block and cylinder head overhaul can be found in Part C of this Chapter.

All 1986 through 1991 models are equipped with a SOHC 8-valve engine (two valves per cylinder). Some 1992 through 1998 models are equipped with the 8-valve engine and some are equipped with a SOHC 16-valve engine (four valves per cylinder). Other than the valve components, the two engines are virtually identical. The 16-valve engine is also used on some 1999 and later models; the rest are equipped with a 2.0L DOHC engine (see Chapter 2B).

The following repair procedures are based on the assumption that the engine is installed in the vehicle. If the engine has been removed from the vehicle and mounted on a stand, many of the steps outlined in this Part of Chapter 2 will not apply.

The Specifications included in this Part of Chapter 2 apply only to the procedures contained in this Part. Part C of Chapter 2 contains the Specifications necessary for cylinder head and engine block rebuilding.

2   Repair operations possible with the engine in the vehicle

Many major repair operations can be accomplished without removing the engine from the vehicle.

Clean the engine compartment and the exterior of the engine with some type of degreaser before any work is done. It will make the job easier and help keep dirt out of the internal areas of the engine.

Depending on the components involved, it may be helpful to remove the hood to improve access to the engine as repairs are performed (see Chapter 11, if necessary). Cover the fenders to prevent damage to the paint. Special pads are available, but an old bedspread or blanket will also work.

If vacuum, exhaust, oil or coolant leaks develop, indicating a need for gasket or seal replacement, the repairs can generally be made with the engine in the vehicle. The intake and exhaust manifold gaskets, oil pan gasket, crankshaft oil seals and cylinder head gasket are all accessible with the engine in place.

Exterior engine components, such as the intake and exhaust manifolds, the oil pan (and the oil pump), the water pump, the starter motor, the alternator, the distributor and the fuel system components can be removed for repair with the engine in place.

Since the cylinder head can be removed without pulling the engine, valve component servicing can also be accomplished with the engine in the vehicle. Replacement of the camshaft, timing belt and sprockets is also possible with the engine in the vehicle.

In extreme cases caused by a lack of necessary equipment, repair or replacement of piston rings, pistons, connecting rods and rod bearings is possible with the engine in the vehicle. However, this practice is not recommended because of the cleaning and preparation work that must be done to the components involved.

3   Top Dead Center (TDC) for number one piston - locating

Refer to illustrations 3.6, 3.8 and 3.9

Note: *The following procedure is based on the assumption that the spark plug wires and distributor are correctly installed. If you are trying to locate TDC to install the distributor correctly, piston position must be determined by feeling for compression at the number one spark plug hole, then aligning the ignition timing marks as described in Step 8.*

1   Top Dead Center (TDC) is the highest point in the cylinder that each piston reaches as it travels up-and-down when the crankshaft turns. Each piston reaches TDC on the compression stroke and again on the exhaust stroke, but TDC generally refers to piston position on the compression stroke.

2   Positioning the piston(s) at TDC is an essential part of many procedures such as rocker arm removal, camshaft and timing belt/sprocket removal and distributor removal.

3   Before beginning this procedure, be sure to place the transmission in Neutral and apply the parking brake or block the rear wheels. Also, on vehicles equipped with a conventional distributor-type ignition system, disable the ignition system by detaching the coil wire from the center terminal of the distributor cap and grounding it on the block with a jumper wire. On vehicles with a distributorless system, unplug the electrical connectors to the ignition coils (see Chapter 5). On all models, remove the spark plugs (see Chapter 1).

4   In order to bring any piston to TDC, the crankshaft must be turned using one of the

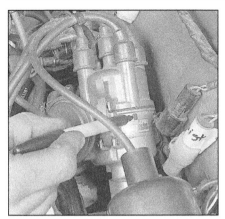

3.6 Make a mark on the distributor housing below the number one terminal on the cap

3.8 Turn the crankshaft until the notch in the pulley is aligned with the O on the timing belt cover (arrows)

3.9 When the number one piston is at Top Dead Center on the compression stroke, the rotor should point toward the mark you made on the distributor

methods outlined below. When looking at the front of the engine, normal crankshaft rotation is clockwise.

a) *The preferred method is to turn the crankshaft with a socket and ratchet attached to the bolt threaded into the front of the crankshaft.*

b) *A remote starter switch, which may save some time, can also be used. Follow the instructions included with the switch. Once the piston is close to TDC, use a socket and ratchet as described in the previous paragraph.*

c) *If an assistant is available to turn the ignition switch to the Start position in short bursts, you can get the piston close to TDC without a remote starter switch. Make sure your assistant is out of the vehicle, away from the ignition switch, then use a socket and ratchet as described in Paragraph a) to complete the procedure.*

Vehicles with a conventional distributor-type ignition system

5   Note the position of the terminal for the number one spark plug wire on the distributor cap. If the wire isn't marked, follow the plug wire from the number one cylinder spark plug to the cap.

6   Use a felt-tip pen or chalk to make a mark on the distributor body directly under the terminal **(see illustration)**.

7   On vehicles with a conventional distributor-type ignition system, detach the cap from the distributor and set it aside (see Chapter 1 if necessary).

8   Turn the crankshaft clockwise (see Paragraph 3 above) until the notch in the crankshaft pulley is aligned with the 0 on the timing plate (located at the front of the engine) **(see illustration)**.

9   Look at the distributor rotor - it should be pointing directly at the mark you made on the distributor body **(see illustration)**. If it is, go to Step 12.

10   If the rotor is 180-degrees off, the num-

ber one piston is at TDC on the exhaust stroke. Go to Step 11.

11   To get the piston to TDC on the compression stroke, turn the crankshaft one complete turn (360-degrees) clockwise. The rotor should now be pointing at the mark on the distributor. When the rotor is pointing at the number one spark plug wire terminal in the distributor cap and the ignition timing marks are aligned, the number one piston is at TDC on the compression stroke.

12   After the number one piston has been positioned at TDC on the compression stroke, TDC for any of the remaining pistons can be located by turning the crankshaft and following the firing order. Mark the remaining spark plug wire terminal locations on the distributor body just like you did for the number one terminal, then number the marks to correspond with the cylinder numbers. As you turn the crankshaft, the rotor will also turn. When it's pointing directly at one of the marks on the distributor, the piston for that particular cylinder is at TDC on the compression stroke.

Vehicles with a distributorless ignition system

13   Have an assistant turn the crankshaft with a socket and wrench while you hold a finger over the number one spark plug hole. **Note:** *See the Specifications for the number one cylinder location.*

14   When the piston approaches TDC, you will fee pressure at the spark plug hole. Have your assistant stop turning the crankshaft when the timing marks are aligned **(see illustration 3.8)**.

15   If the notch in the pulley is already past the timing mark, turn the crankshaft two complete revolutions in a clockwise direction until the notch is aligned with the timing mark.

16   After the number one piston has been positioned at TDC on the compression stroke, TDC for any of the remaining pistons can be located by turning the crankshaft one-half turn (180 degrees) to get to TDC for the next cylinder in the firing order.

4   Camshaft cover - removal and installation

Refer to illustrations 4.5a and 4.5b

1   Disconnect the negative cable from the battery.

2   On throttle body fuel-injected models, remove the air intake case from the throttle body (see Chapter 4). On multi-port fuel-injected models, remove the air intake pipe.

3   On multi-port fuel-injected models, remove the PCV hose from the camshaft cover. Also on multi-port fuel-injected models, disconnect the accelerator and (if equipped) cruise control cables from the throttle body.

4   Detach the spark plug wires from the plugs, unclip the wire loom from the top of the cover, then set the wires aside, leaving them attached to the loom. Disconnect the breather hose from the camshaft cover.

5   Remove the camshaft cover bolts **(see illustrations)** and lift off the cover. If the cover sticks to the cylinder head, tap on it with a soft-face hammer or place a block of wood against the cover and tap on the wood with a hammer.

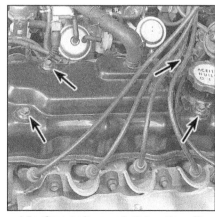

4.5a Camshaft cover bolt locations (8-valve engine)

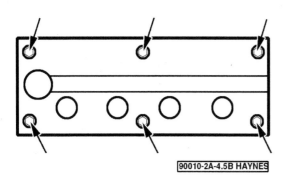

4.5b Camshaft cover bolt locations (16-valve engine)

5.4 Label the connections before detaching them

6   Thoroughly clean the camshaft cover and remove all traces of old gasket material. On 16-valve engines, be sure to remove the spark plug opening O-rings.

7   Install a new gasket and, if equipped, O-rings on the cover, using RTV to hold them in place. Place new grommets, if equipped, in the bolt holes in the cover, install the cover, then install the bolts.

8   Working from the center out, tighten the bolts to the torque listed in this Chapter's Specifications.

9   The remaining steps are the reverse of removal. When finished, run the engine and check for oil leaks.

5   Intake manifold - removal and installation

Refer to illustrations 5.4, 5.9 and 5.10
Note: *The intake manifold can be removed with the throttle body and plenum assembly still attached. If, however, you plan to remove the plenum and throttle body anyway (for example, if you're replacing the manifold), it may be easier to remove the throttle body and intake plenum before you begin removing the manifold (see Chapter 4 for the throttle body and plenum removal procedures).*

1   Disconnect the negative cable from the battery and relieve the fuel injection system pressure (see Chapter 4).

2   Drain the cooling system (see Chapter 1). If the coolant is in good condition it can be reused.

3   Remove the air cleaner (carbureted models), air intake case (throttle body fuel-injected models) or disconnect the air intake pipe (multi-port fuel-injected models) (see Chapter 4).

4   Clearly label, then disconnect all hoses, wires, brackets and emission lines which run to the carburetor/throttle body and intake manifold **(see illustration)**.

5   Disconnect the fuel lines and cap the fittings to prevent leakage (see Chapter 4). On carbureted models, disconnect the line at the carburetor. On throttle body fuel-injected models, disconnect the lines at the throttle body assembly. On multi-port fuel-injected models, it may be necessary to raise the vehicle and support it securely on jackstands, since the line connections are usually at the lower rear of the engine compartment. Also, on multi-port fuel-injected models, use a back-up wrench when disconnecting the fuel feed line.

6   Disconnect the throttle cable from the carburetor/throttle body (see Chapter 4).

7   Detach the cable which runs from the carburetor/throttle body to the transmission (automatic transmission only) and the cruise control cable, on vehicles so equipped.

8   On multi-port fuel-injected models, unbolt and remove the three brackets that attach the intake manifold/intake plenum assembly to the engine.

9   Using a socket, ratchet and long extension, unscrew the bolts and nuts and remove the intake manifold from the engine **(see illustration)**. If it sticks, tap the manifold with a soft-face hammer. **Caution:** *Do not pry between gasket sealing surfaces or tap on the carburetor/throttle body. If there are any clamps under the bolts or nuts, note their locations before you remove them so they can be returned to their original locations.*

10   Thoroughly clean the manifold and cylinder head mating surfaces, removing all traces of gasket material **(see illustration)**. Be very careful not to scratch or gouge the delicate aluminum gasket surfaces on the manifold and cylinder head. Gasket removal solvents are available from auto parts stores and may prove helpful.

11   Install the manifold, using a new gasket and tighten the bolts and nuts in several stages, working from the center out, until you reach the torque listed in this Chapter's Specifications.

12   Reinstall the remaining parts in the reverse order of removal.

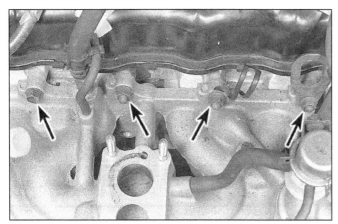

5.9 The intake manifold has eight mounting bolts/nuts - four upper (arrows) and four lower (hidden by the manifold in this photo)

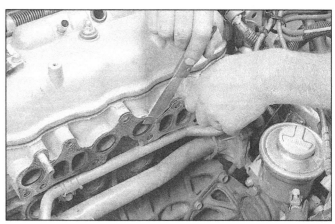

5.10 Remove all traces of old gasket material - scrape gently to avoid gouging the aluminum

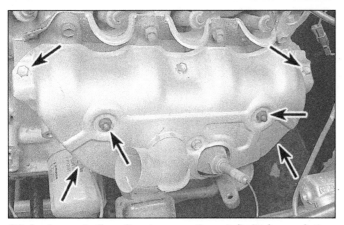

6.6 Apply penetrating oil and remove the nuts/bolts (arrows), then remove the heat shields from the manifold

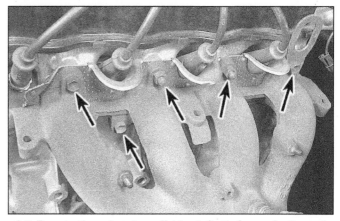

6.7 Remove the exhaust manifold nuts/bolts (arrows) - three of the lower ones are hidden from view in this photo

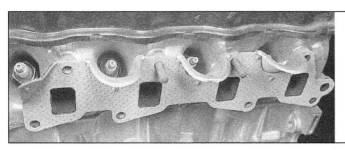

6.11 Slip the gasket over the studs with the spark plug wells facing outward

13   Add coolant, run the engine and check for leaks and proper operation.

6   Exhaust manifold - removal and installation

Refer to illustrations 6.6, 6.7 and 6.11
Warning: *Allow the engine to cool completely before following this procedure.*
1   Disconnect the negative cable from the battery.
2   Set the parking brake and block the rear wheels. Raise the front of the vehicle and support it securely on jackstands.
3   Working from under the vehicle, remove the nuts that secure the exhaust system to the bottom of the exhaust manifold. Apply penetrating oil to the threads to make removal easier.
4   On fuel injected models, remove the air cleaner assembly (see Chapter 4).
5   Unplug the oxygen sensor wire (see Chapter 6).
6   Remove the fasteners that secure the heat shields to the exhaust manifold and to each other **(see illustration)**. Remove the heat shields.
7   Apply penetrating oil to the threads, then remove the exhaust manifold mounting nuts/bolts **(see illustration)**.
8   Slip the manifold off the studs and remove it from the engine compartment.
9   Clean and inspect all threaded fasteners and repair as necessary.
10   Remove all traces of gasket material from the mating surfaces and inspect them for wear and cracks.

11   Install a new gasket **(see illustration)**, install the manifold and tighten the nuts in several stages, working from the center out, to the torque listed in this Chapter's Specifications.
12   Reinstall the remaining parts in the reverse order of removal.
13   Run the engine and check for exhaust leaks.

7   Timing belt and sprockets - removal, inspection and installation

Removal

> ## ** CAUTION **
>
> The timing system is complex. Severe engine damage will occur if you make any mistakes. Do not attempt this procedure unless you are highly experienced with this type of repair. If you are at all unsure of your abilities, consult an expert. Double-check all your work and be sure everything is correct before you attempt to start the engine.

Refer to illustrations 7.4, 7.8a, 7.8b, 7.9, 7.10, 7.11, 7.14, 7.15a, and 7.15b
Warning: *The air conditioning system is under high pressure. Do not loosen any fittings or remove any components until after the system has been discharged by an air conditioning technician. Always wear eye protection when disconnecting refrigerant fittings.*
Caution 1: *Do not try to turn the crankshaft*

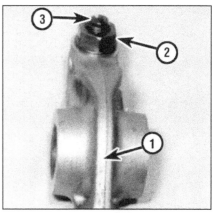

7.4 On 8-valve engines, loosen the locknut and back off the valve adjustment screw

1   *Rocker arm*
2   *Locknut*
3   *Adjustment screw*

with the camshaft sprocket bolt and do not rotate the crankshaft counterclockwise.
Caution 2: *Do not bend, twist or turn the timing belt inside out. Do not allow it to come in contact with oil, coolant or fuel. Do not utilize timing belt tension to keep the camshaft or crankshaft from turning when installing the pulley bolt(s). Do not turn the crankshaft or camshaft more than a few degrees (necessary for tooth alignment) while the timing belt is removed.*
1   On air conditioned models, if necessary for clearance, have the refrigerant discharged by an air conditioning technician (see **Warning** above), then detach the hose from the suction pipe on the compressor (the suction pipe is the larger diameter pipe).
2   Disconnect the negative cable from the battery.
3   Raise the vehicle and support it securely on jackstands (Vitara models only).
4   On 8-valve engines, remove the camshaft cover (see Section 4). Loosen the locknuts and back off the valve adjustment screws until they're not in contact with the valves **(see illustration)**.

7.8a Samurai crankshaft pulley mounting details

1	Crankshaft pulley bolts	3	Center bolt
2	Indexing notch		

7.8b Wrap an old belt or a rag around the pulley and grip it with a chain wrench

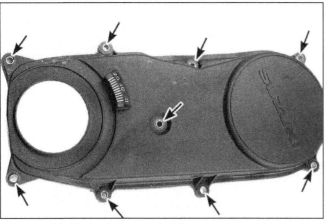

7.9 Timing belt cover bolt locations (arrows) - cover removed for clarity

7.10 If the timing belt doesn't have arrows like these to indicate direction of rotation, paint one on

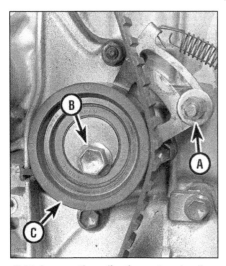

7.11 Loosen the adjusting nut and pulley bolt and move the tensioner pulley as far as possible towards the water pump

A	Adjusting nut
B	Pulley bolt
C	Tensioner pulley

7.14 The belt guide has a notch which allows it to fit over the crankshaft key (arrows)

5   Set the parking brake and block the rear wheels. Raise the front of the vehicle and support it securely on jackstands.

6   Loosen the four water pump pulley nuts, then remove the drivebelts (see Chapter 1).

7   Unbolt and remove the fan, fan shroud and water pump pulley (see Chapter 3).

8   Remove the crankshaft pulley bolts **(see illustration)**. Note: *If you're only replacing the timing belt, it's not necessary to remove the crankshaft center bolt; however, if you will be removing the crankshaft sprocket to replace the oil pump or oil seal, you must remove the center bolt. Do this before you remove the pulley bolts. The center bolt is very tight, so, to break it loose, remove the splash pan from beneath the front of the engine, wrap a rag or an old belt around the pulley and attach a chain wrench to hold the pulley in place* **(see illustration)**. Use a breaker bar and socket to loosen the bolt.

9   Remove the bolts that secure the timing belt cover and lift the cover off **(see illustration)**.

10   If you plan to reuse the timing belt, and it doesn't already have arrows painted on it, paint one on to indicate the direction of rotation (clockwise) **(see illustration)**.

11   Loosen the adjusting nut and pulley bolt. Move the tensioner pulley towards the water pump as far as possible **(see illustration)**.

7.15a Hold the camshaft sprocket from turning with a large screwdriver

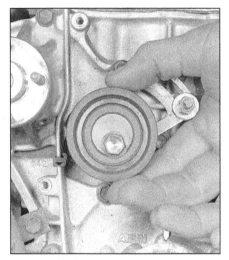

7.16 Check the tensioner pulley for roughness and excess play

12   Temporarily secure the tensioner pulley by tightening the adjusting nut.
13   Slip the timing belt off the sprockets and set it aside.
14   If you intend to replace the oil pump or crankshaft front oil seal, slide off the crankshaft sprocket and the belt guide located behind the crankshaft sprocket (see illustration). When removing the guide, note the way it's installed (the chamfered side faces out).
15   If you intend to replace the camshaft oil seal, unscrew the camshaft sprocket securing bolt and slide the sprocket off - a large screwdriver inserted through a hole in the sprocket will keep it from turning while you remove the bolt (see illustration). Unbolt the cover (see illustration) to access the seal.

Inspection

Refer to illustrations 7.16 and 7.17

16   Rotate the tensioner pulley by hand and move it side-to-side to detect roughness and excess play (see illustration). Visually

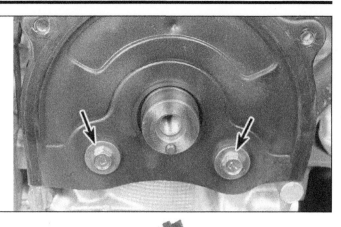

7.15b To get to the camshaft seal, remove the two mounting bolts and detach the cover (arrows)

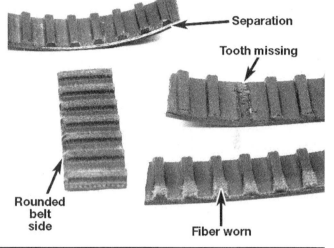

7.17 Carefully inspect the timing belt for the conditions shown here

Separation

Tooth missing

Rounded belt side

Fiber worn

inspect the sprockets for any signs of damage and wear. Replace parts as necessary.
17   Inspect the timing belt for cracks, separation, wear, missing teeth and oil contamination (see illustration). Replace the belt if it's worn or damaged. **Note:** *Unless the engine has very low mileage, it's common practice to replace the timing belt with a new one every time it's removed. Don't reinstall the original belt unless it's in like-new condition. Never reinstall a belt in questionable condition.*

Installation

```
** CAUTION **

Before starting the engine, carefully rotate
the crankshaft by hand through at least two
full revolutions (use a socket and breaker bar
on the crankshaft pulley center bolt). If you
feel any resistance, STOP! There is some-
thing wrong - most likely, valves are contact-
ing the pistons. You must find the problem
before proceeding. Check your work and
see if any updated repair
information is available.
```

Refer to illustrations 7.18a, 7.18b, 7.19a and 7.19b
Note: *Earlier models use timing belts with squared-off teeth, while later models have rounded teeth on the belt. Be sure to use the*

correct belt type, since it must match the pulleys.
18   Reinstall the camshaft seal cover and timing belt sprockets, if they were removed. Note that the camshaft sprocket is indexed by a dowel (see illustration). Slip the belt guide onto the crankshaft before installing the crankshaft sprocket - the chamfered side of the guide faces away from the belt. The crankshaft sprocket has a keyway which matches the key on the crankshaft (see illustration).

7.18a The camshaft sprocket is indexed by a dowel (arrow)

7.18b The crankshaft sprocket has a slot (keyway) which must align with the key in the crankshaft

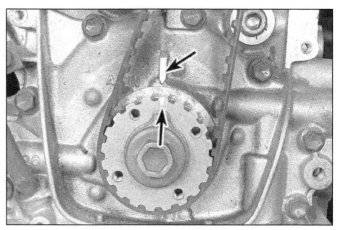

7.19a The mark on the lower timing belt sprocket must align with the mark on the engine front cover (arrows)

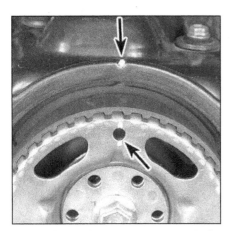

7.19b The timing mark on the camshaft sprocket must align with the V mark on the camshaft seal cover (arrows)

8.2 Carefully pry the seal out with a small screwdriver - wrap the tip with tape to prevent damaging the seal bore and crankshaft sealing surface

8.4 Install the new seal by gently tapping it into place with a socket and a small hammer

19   Align the valve timing marks located on the camshaft and crankshaft sprockets **(see illustrations)**. **Note:** *On 16-valve engines, there's an "E" adjacent to the timing mark.*
20   Slip the timing belt onto the crankshaft sprocket. While maintaining tension on the side of the belt opposite to the tensioner, slip the belt onto the camshaft sprocket.
21   Release the tensioner adjusting nut to allow spring tension against the belt. Rotate the crankshaft clockwise two complete revolutions (720-degrees). Retighten the nut.
22   Temporarily install the crankshaft pulley, taking care to align the notch in the pulley with the raised area on the sprocket. Install the crankshaft pulley bolts and the center bolt, if removed. Tighten the center bolt to the torque listed in this Chapter's Specifications. When tightening the bolts, hold the crankshaft in place using the method discussed in Step 8. Remove the pulley, leaving the center bolt in place.
23   Using the bolt in the center of the crankshaft sprocket, turn the crankshaft clockwise through two complete revolutions (720-degrees). Recheck the alignment of the valve timing marks. If the marks do not align properly, loosen the tensioner, slip the belt

off the camshaft sprocket, align the marks, reinstall the belt, and check the alignment again.
24   Tighten the tensioner bolt and nut to the torque listed in this Chapter's Specifications. Start with the nut, then tighten the bolt.
25   Reinstall the remaining parts in the reverse order of removal.
26   On eight-valve engines, set the valve clearances with the engine cold (see Chapter 1).
27   Start the engine, allow it to reach normal operating temperature, set the ignition timing and, on eight-valve engines, check the valve clearance (see Chapter 1). Road test the vehicle.

8 Crankshaft front oil seal - replacement

Refer to illustrations 8.2 and 8.4

1   Remove the timing belt, crankshaft sprocket and inner belt guide (see Section 7).
2   Wrap the tip of a small screwdriver with tape. Working from below, use the screwdriver to pry the seal out of its bore **(see illus-**

tration). Take care to prevent damaging the crankshaft and the seal bore.
3   Thoroughly clean and inspect the seal bore and sealing surface on the crankshaft. Minor imperfections can be removed with emery cloth. If there is a groove worn in the crankshaft sealing surface (from contact with the seal), installing a new seal will probably not stop the leak. Try installing a repair sleeve which fits over the crankshaft sealing surface. These are normally available at larger auto parts stores.
4   Lubricate the new seal with engine oil and drive the seal into place with a hammer and socket **(see illustration)**.
5   Reinstall the timing belt and related components as described in Section 7.
6   Run the engine, checking for oil leaks.

9 Camshaft oil seal - replacement

Refer to illustrations 9.2 and 9.4

1   Remove the timing belt, camshaft sprocket and camshaft seal cover (see Section 7).
2   Note how far the seal is seated in the bore, then carefully pry it out with a small

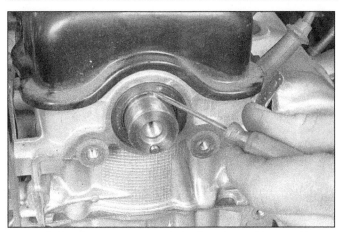

9.2 Carefully pry the camshaft oil seal out with a small screwdriver - wrap the tip with tape to prevent damaging the seal bore and camshaft sealing surface

9.4 Gently tap the new oil seal into place with a socket and a small hammer

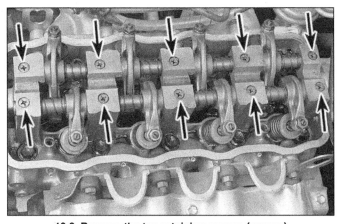

10.6 Remove the ten retaining screws (arrows)

10.8 Push the rear of the rocker arm shaft until it protrudes enough at the front to grip it

screwdriver **(see illustration)**. Wrap the tip of the screwdriver with tape so you don't scratch the bore or damage the camshaft in the process (if the camshaft is damaged, the new seal will end up leaking).

3   Clean the bore and coat the outer edge of the new seal with engine oil or multi-purpose grease. Apply moly-base grease to the seal lip.

4   Using a socket with an outside diameter slightly smaller than the outside diameter of the seal, carefully drive the new seal into place with a hammer **(see illustration)**. Make sure it's installed squarely and driven in to the same depth as the original. If a socket isn't available, a short section of pipe will also work.

5   Reinstall the seal cover, camshaft sprocket and timing belt (see Section 7).

6   Run the engine and check for oil leaks at the camshaft seal.

10   Rocker arms and shafts - removal, inspection and installation

Note: *This procedure also includes camshaft removal and installation on 16-valve engines.*

Removal

1   Disconnect the negative cable from the battery.

2   Remove the radiator (see Chapter 3).

3   Set the number one piston at top dead center and remove the camshaft cover (see Sections 3 and 4).

4   Loosen the locknuts and back off the valve adjustment screws until they're not in contact with the valves **(see illustration 7.4)**.

5   Remove the timing belt and camshaft sprocket (see Section 7).

Eight-valve engines

Refer to illustrations 10.6, 10.8 and 10.9

6   Remove the ten rocker arm shaft retaining screws **(see illustration)**.

7   Number the rocker arms with a scribe. Start with number one at the front and criss-cross until all are marked. When you are done, the rocker arms on the intake manifold side should be numbered 1, 3, 5, 7 and the ones on the exhaust manifold side should be marked 2, 4, 6, 8.

8   Push the rear of one rocker arm shaft until it protrudes enough at the front to grip **(see illustration)**.

9   Slowly pull each rocker arm shaft out the front of the engine **(see illustration)**. Lift

10.9 Slowly pull the shaft out the front of the cylinder head

the rocker arms and springs out of the head as they are released from the shaft.

Sixteen-valve engines

10   Remove the radiator grille (see Chapter 11) and remove the hood latch and front upper member (the long, narrow panel the hood latch is attached to).

11   Remove the distributor (see Chapter 5)

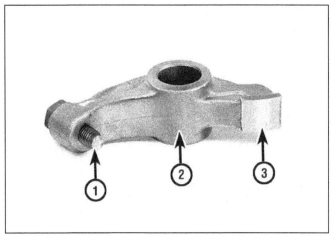

10.21 Inspect the tips of the valve adjusting screws and the cam-riding faces (8-valve engine shown, 16-valve engine similar)

1	Valve adjusting screw	3	Cam-riding face
2	Rocker arm		

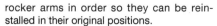

10.23 On 8-valve engines, the shaft on the intake manifold side must be installed with the stepped side toward the camshaft sprocket - the shaft on the exhaust manifold side must be installed with the stepped side toward the distributor

1	Intake rocker arm shaft	4	0.59-inch (15 mm)
2	0.55-inch (14 mm)	5	Camshaft sprocket side
3	Exhaust rocker arm shaft	6	Distributor side

and housing (see *Cylinder head - disassembly* in Chapter 2C). **Note:** *Position a drain pan under the housing before removing it, since a small amount of oil usually flows out when the housing is removed.*

12   Remove the camshaft cover, timing belt and camshaft sprocket (see Sections 4 and 7).

13   Reverse the order shown in **illustration 10.24** to loosen the camshaft bearing cap bolts, until they are all loose, then remove them and lift the camshaft bearing caps and camshaft off the cylinder head. **Note:** *The camshaft bearing caps must be installed in their original locations, with the same ends facing forward. The caps should be numbered and an arrow on each cap should point toward the front of the engine. If the caps are not marked, mark them before disassembly.*

14   Remove the timing belt inner cover and, using a hex-drive tool, unscrew the rocker arm shaft plug from the front of the cylinder head.

15   Mark the intake and exhaust rocker arms so they can be returned to their original locations on reassembly.

16   Taking care not to bend the clips, remove the intake rocker arms. Keep the

rocker arms in order so they can be reinstalled in their original positions.

17   Remove the rocker arm shaft bolts, which are at the top of the cylinder head, just above the rocker arm shaft.

18   Push the rocker arm shaft out of the cylinder head, toward the distributor opening, and detach the O-ring on the end of the shaft. Install a new O-ring when reinstalling the shaft.

19   Slowly pull the rocker arm shaft out the front of the engine. Lift the exhaust valve rocker arms and wave washers out of the head as they are released from the shaft. Keep the rocker arms and washers in order.

Inspection

Refer to illustrations 10.21

Note: *Camshaft inspection procedures and specifications are in Part B of this Chapter.*

20   Thoroughly clean the parts in solvent and wipe them off with a lint-free cloth. On 16-valve engines, be sure to clean the camshaft bearing cap mating surfaces and camshaft saddles in the cylinder head.

21   Measure the inside diameter of each rocker arm and the outside diameter of the shaft where the rocker arm rides. Compare

the measurements to this Chapter's Specifications. Subtract the rocker arm shaft diameter from the rocker arm inside diameter to calculate the shaft-to-arm clearance. Visually inspect the rocker arms for wear **(see illustration)** and the springs (if equipped) for wear and damage. Replace any components that are not within specifications, damaged or excessively worn.

Installation

Refer to illustrations 10.23 and 10.24

22   Thoroughly lubricate the shafts, rocker arms and camshaft journals and lobes (16-valve engines) with engine assembly lube.

23   Slowly push a rocker arm shaft into the cylinder head while guiding it into the rocker arms and springs or washers. Note the rocker arm markings and be sure to install them in the same positions they were in originally. On eight-valve engines, note that the shafts are different and must be installed with the stepped sides facing the correct direction **(see illustration)**. On 16-valve engines, make sure the shaft is rotated so the flattened area on the front end of the shaft is facing down and is parallel with the head gasket surface.

24   On 16-valve engines, install the intake rocker arms, then the camshaft and bearing caps. Apply a thin layer of anaerobic sealant to the rear bearing cap before installation. Tighten the camshaft bearing cap bolts, in the sequence shown **(see illustration)** to the torque listed in this Chapter's Specifications.

25   When all the rocker arm components are positioned correctly, install the retaining screws and tighten them to the torque listed in this Chapter's Specifications. You may have to rotate the shafts to get the bolt holes to line up.

26   Adjust the valves as described in Chapter 1.

27   Install the remaining components in the reverse order of removal.

10.24 Camshaft bearing cap bolt TIGHTENING sequence (16-valve engines)

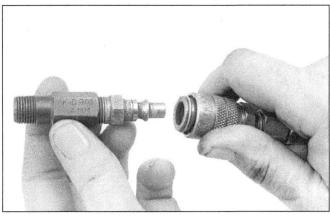

11.4 This is what the air hose adapter that threads into the spark plug hole looks like - they're commonly available from auto parts stores

11.9 Compress the spring enough to release the keepers (arrows)

11 Valve spring, retainer and seals - replacement

Refer to illustrations 11.4, 11.9 and 11.17
Note: *Broken valve springs and defective valve stem seals can be replaced without removing the cylinder head. Two special tools and a compressed air source are normally required to perform this operation, so read through this Section carefully and rent or buy the tools before beginning the job. If compressed air isn't available, a length of nylon rope can be used to keep the valves from falling into the cylinder during this procedure.*

1　Refer to Section 10 and remove the rocker arms and shafts from the cylinder head.

2　Remove the spark plug from the cylinder which has the defective component. If all of the valve stem seals are being replaced, all of the spark plugs should be removed.

3　Turn the crankshaft until the piston in the affected cylinder is at top dead center on the compression stroke (refer to Section 3 for instructions). If you're replacing all of the valve stem seals, begin with cylinder number one and work on the valves for one cylinder at a time. Move from cylinder-to-cylinder following the firing order sequence (see this Chapter's Specifications).

4　Thread an adapter into the spark plug hole **(see illustration)** and connect an air hose from a compressed air source to it. Most auto parts stores can supply the air hose adapter. **Note:** *Many cylinder compression gauges utilize a screw-in fitting that may work with your air hose quick-disconnect fitting.*

5　Apply compressed air to the cylinder. **Warning:** *The piston may be forced down by compressed air, causing the crankshaft to turn suddenly. If the wrench used when positioning the number one piston at TDC is still attached to the bolt in the crankshaft nose, it could cause damage or injury when the crankshaft moves. Keep your hands clear of the drivebelts and rotating engine components.*

6　The valves should be held in place by the air pressure. If the valve faces or seats are in poor condition, leaks may prevent air pressure from retaining the valves - refer to the alternative procedure below.

7　If you don't have access to compressed air, an alternative method can be used. Position the piston at a point just before TDC on the compression stroke.

8　Feed a long piece of nylon rope through the spark plug hole until it fills the combustion chamber. Be sure to leave the end of the rope hanging out of the engine so it can be removed easily. Use a large ratchet and socket to rotate the crankshaft in the normal direction of rotation (clockwise) until slight resistance is felt.

9　Stuff shop rags into the cylinder head holes above and below the valves to prevent parts and tools from falling into the engine, then use a valve spring compressor to compress the spring **(see illustration)**. Remove the keepers with a small needle-nose pliers or a magnet. **Note:** *A couple of different types of tools are available for compressing the valve springs with the head in place. One type, shown here, grips the lower spring coils and presses on the retainer as the handle is turned, while the other type utilizes the rocker arm shaft for leverage.*

10　Remove the spring retainer and valve spring, then remove the guide seal. **Note:** *If air pressure fails to hold the valve in the closed position during this operation, the valve face and/or seat is probably damaged. If so, the cylinder head will have to be removed for additional repair operations.*

11　Wrap a rubber band or tape around the top of the valve stem so the valve won't fall into the combustion chamber, then release the air pressure. **Note:** *If a rope was used instead of air pressure, turn the crankshaft slightly in the direction opposite normal rotation.*

12　Inspect the valve stem for damage. Rotate the valve in the guide and check the end for eccentric movement, which would indicate that the valve is bent.

13　Move the valve up-and-down in the guide and make sure it doesn't bind. If the valve stem binds, either the valve is bent or the guide is damaged. In either case, the head will have to be removed for repair.

14　Reapply air pressure to the cylinder to retain the valve in the closed position, then remove the tape or rubber band from the valve stem. If a rope was used instead of air pressure, rotate the crankshaft clockwise until slight resistance is felt.

15　Lubricate the valve stem with engine oil and install a new guide seal.

16　Install the spring in position over the valve.

17　Install the valve spring retainer. Compress the valve spring and carefully position the keepers in the groove. Apply a small dab of grease to the inside of each keeper to hold it in place if necessary **(see illustration)**.

18　Remove the pressure from the spring tool and make sure the keepers are seated.

19　Disconnect the air hose and remove the adapter from the spark plug hole. If a rope was used in place of air pressure, pull it out of the cylinder.

20　Refer to Section 10 and install the rocker arm(s) and shaft(s).

21　Install the spark plugs(s) and hook up the wire(s).

11.17 Apply a small dab of grease to each keeper as shown here before installation - it'll hold them in place on the valve stem as the spring is released

12.11 Use casting protrusions to pry against - don't pry between the gasket surfaces

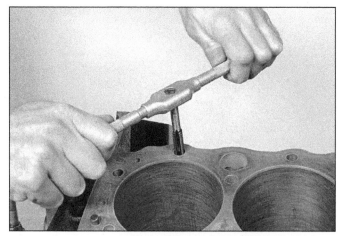

12.12 A tap should be used to remove sealant and corrosion from the head bolt threads prior to installation

22   Refer to Section 4 and install the camshaft cover.
23   Start and run the engine, then check for oil leaks and unusual sounds coming from the rocker arm cover area.

12   Cylinder head - removal and installation

Refer to illustrations 12.11, 12.12, 12.13, 12.14a and 12.14b
Caution: *Allow the engine to cool completely before following this procedure.*
1   Position the number one piston at Top Dead Center (see Section 3).
2   Disconnect the negative cable from the battery.
3   Drain the cooling system and remove the spark plugs (see Chapter 1).
4   Remove the intake manifold (see Section 5).
5   Remove the exhaust manifold (see Section 6).
6   Remove the distributor (see Chapter 5), including the cap and wires.
7   On carbureted models, remove the fuel pump (see Chapter 4).
8   Remove the timing belt (see Section 7).

9   Remove the camshaft cover (see Section 4). On 16-valve engines, also remove the camshaft and intake rocker arms (see Section 10).
10   Using a 14 mm socket, loosen the cylinder head bolts, 1/4-turn at a time, in the reverse of the sequence shown **(see illustrations 12.14a and 12.14b)** until they can be removed by hand.
11   Carefully lift the cylinder head straight up and place the head on wooden blocks to prevent damage to the sealing surfaces. If the head sticks to the engine block, dislodge it by prying against a protrusion on the head casting **(see illustration)**. **Note:** *Cylinder head disassembly and inspection procedures are covered in Chapter 2, Part B. It's a good idea to inspect the camshaft and have the head checked for warpage, even if you're just replacing the gasket.*
12   Remove all traces of old gasket material from the block and head. Do not allow anything to fall into the engine. Clean and inspect all threaded fasteners and be sure the threaded holes in the block are clean and dry.
13   Place a new gasket **(see illustration)** and the cylinder head in position.
14   The cylinder head bolts should be tightened in several stages following the proper sequence **(see illustrations)** to the torque

12.13 The head gasket must be installed with the word Top facing up

listed in this Chapter's Specifications.
15   Reinstall the timing belt (see Section 7).
16   Reinstall the remaining parts in the reverse order of removal.
17   Be sure to refill the cooling system and check all fluid levels. Rotate the crankshaft clockwise slowly by hand through two complete revolutions. Recheck the camshaft tim-

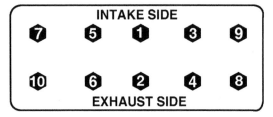

INTAKE SIDE

❼ ❺ ❶ ❸ ❾

❿ ❻ ❷ ❹ ❽

EXHAUST SIDE

90010-2a-12.10a HAYNES

12.14a Cylinder head bolt TIGHTENING sequence (8-valve engine) - to loosen the bolts, reverse this sequence

INTAKE SIDE

❼ ❸ ❶ ❺ ❾

❽ ❹ ❷ ❻ ❿

EXHAUST SIDE

90010-2a-12.10b HAYNES

12.14b Cylinder head bolt TIGHTENING sequence (16-valve engines) - to loosen the bolts, reverse this sequence

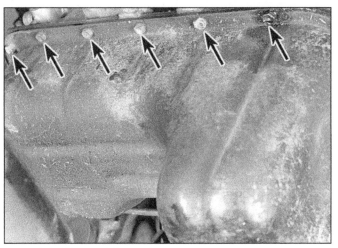

13.7a The bolts (arrows) are spaced evenly around the perimeter of the oil pan

13.7b Use a soft-face hammer to dislodge the oil pan

13.8a On Samurai models, lower the oil pan and unbolt the oil pickup tube at the front (arrow) . . .

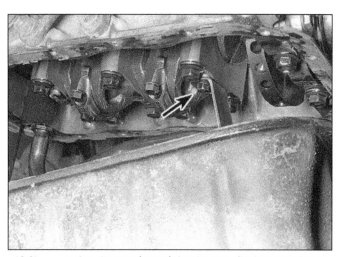

13.8b . . . and at the rear (arrow); let the tube fall into the oil pan and remove the pan and tube as a unit

ing marks (see Section 7).

18   Start the engine and set the ignition timing (see Chapter 1). Run the engine until normal operating temperature is reached. Check for leaks and proper operation. Shut off the engine, remove the camshaft cover and retorque the cylinder head bolts, unless the gasket manufacturer states otherwise. Recheck the valve adjustment.

13 Oil pan - removal and installation

Refer to illustrations 13.7a, 13.7b, 13.8a and 13.8b

1   Raise the front of the vehicle and support it securely on jackstands placed under the frame.
2   Drain the engine oil (see Chapter 1).
3   Remove the splash shield from under the engine.
4   On Tracker and Sidekick models, remove the front differential carrier assembly as described in Chapter 8.

5   Remove the bellhousing lower plate.
6   On 1996 models, disconnect the electrical connector from the crankshaft position sensor. Remove the bolt and detach the sensor from the oil pan.
7   Remove the bolts and detach the oil pan **(see illustration)**. Don't pry between the block and pan or damage to the sealing surfaces may result and oil leaks could develop. Use a soft-face hammer to dislodge the pan if it's stuck **(see illustration)**.
8   On Samurai models, the oil pickup tube must be unbolted and dropped into the oil pan before the oil pan can be removed **(see illustrations)**.
9   Use a scraper to remove all traces of old gasket material and sealant from the block and oil pan. Clean the gasket sealing surfaces with lacquer thinner or acetone and make sure the bolt holes in the block are clean.
10   Check the oil pan flange for distortion, particularly around the bolt holes. If necessary, place the pan on a block of wood and use a hammer to flatten and restore the gasket surface.

11   Before installing the oil pan, apply a thin coat of RTV sealant to the flange. Attach the new gasket to the pan (make sure the bolt holes are aligned).
12   Position the oil pan against the engine block and install the mounting bolts. Tighten them to the torque listed in this Chapter's Specifications in a criss-cross pattern.
13   Wait at least 30 minutes before filling the engine with oil, then start the engine and check the pan for leaks.

14 Oil pump - removal, inspection and installation

Removal

Refer to illustrations 14.3, 14.4a, 14.4b, 14.5 and 14.7

1   Remove the timing belt, tensioner, crankshaft sprocket and belt guide (see Section 7).
2   Remove the oil pan and oil pickup tube (see Section 13).

14.3 Oil pump mounting details

1   Alternator bracket nut   2   Dipstick tube mounting nut

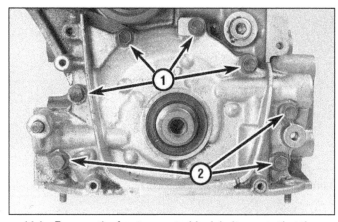

14.4a Remove the front cover-to-block bolts - note that they come in different lengths

1   Short bolts   2   Long bolts

14.4b Gently separate the front cover from the engine with a prybar (arrow)

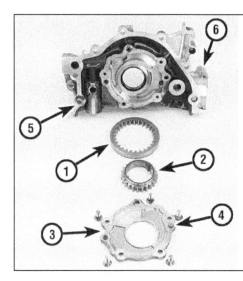

14.5 Oil pump components - exploded view

1   Oil pump outer gear
2   Oil pump inner gear
3   Gear plate
4   Gear plate pin
5   Dowel
6   Oil pump housing

3   Remove the alternator (see Chapter 5) and bracket **(see illustration)**. Unbolt the dipstick tube and pull the tube out of the engine front case.

4   Remove the front cover-to-block bolts and carefully separate the front cover from the engine **(see illustrations)**.

5   Detach the cover from the rear of the case and remove the inner and outer oil pump gears **(see illustration)**.

6   Remove the crankshaft oil seal from the front case (see Section 8).

7   Remove the pressure relief valve retainer, spring and plunger **(see illustration)**.

Inspection

Refer to illustrations 14.9a, 14.9b and 14.9c

8   Clean all parts thoroughly and remove all traces of old gasket material from the sealing surfaces. Visually inspect all parts for wear, cracks and other damage. Replace parts as necessary.

9   Install the oil pump outer and inner gears and measure the clearances **(see illustrations)**. Compare the clearances to this Chapter's Specifications. Measure the free

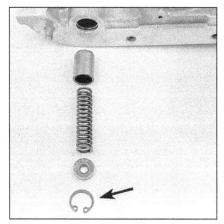

14.7 After you remove the snap-ring (arrow) with a pair of snap-ring pliers, the retainer, spring and plunger can be removed - if the plunger sticks in the bore, tap the front cover on a block of wood to dislodge it

height of the relief valve spring and compare the measurement to this Chapter's Specifications. Replace parts as necessary. Pack the

14.9a Install the outer gear with the dot (arrow) visible - the inner gear will only fit one way: with the raised center collar facing out

pump cavity with petroleum jelly and install the cover. Tighten the bolts to the torque listed in this Chapter's Specifications.

14.9b Measure the outer gear-to-oil pump housing clearance with a feeler gauge

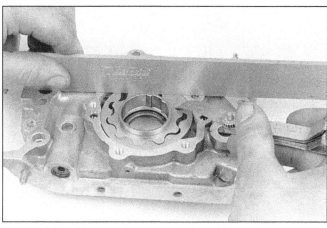

14.9c Measure the gear end play with a precision straightedge and feeler gauge

14.12a The oil pump has two tangs (arrows) . . .

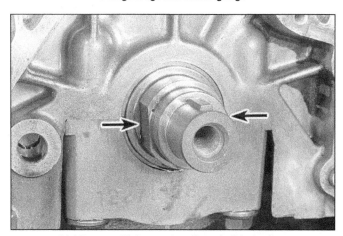

14.12b . . . which must be aligned with the flats on the crankshaft (arrows)

Installation

Refer to illustrations 14.12a and 14.12b

10   Replace the crankshaft front oil seal with a new seal (see Section 8).

11   Install the pressure relief valve components.

12   Using a new gasket, position the front cover on the engine **(see illustrations)**. Install the bolts in their proper locations. Tighten the bolts to the torque listed in this Chapter's Specifications.

13   Reinstall the remaining parts in the reverse order of removal.

14   Add oil, start the engine and check for oil pressure and leaks.

15   Flywheel/driveplate - removal and installation

Refer to illustrations 15.8a and 15.8b

1   Raise the vehicle and support it securely on jackstands, then refer to Chapter 7 and remove the transmission. If it's leaking, now would be a very good time to replace the front pump seal/O-ring (automatic transmission only).

2   Remove the pressure plate and clutch disc (see Chapter 8) (manual transmission equipped models). Now is a good time to check/replace the clutch components and pilot bearing.

3   On driveplate-equipped models, mark the relationship between the driveplate and crankshaft to ensure correct alignment during reinstallation.

4   Remove the bolts that secure the flywheel/driveplate to the crankshaft. If the crankshaft turns on flywheel-equipped models, immobilize it by reinstalling two of the pressure plate bolts halfway and holding a prybar between them. On driveplate-equipped models, wedge a screwdriver through the starter opening and into the ring gear teeth to immobilize the crankshaft.

5   Remove the flywheel/driveplate from the crankshaft. Since the flywheel is fairly heavy, be sure to support it while removing the last bolt.

6   Using brake system cleaner, clean the flywheel to remove grease and oil. Inspect the surface for cracks, rivet grooves, burned areas and score marks. Light scoring can be removed with emery cloth. Check for cracked and broken ring gear teeth. Lay the flywheel on a flat surface and use a straightedge to check for warpage.

7   Clean and inspect the mating surfaces of the flywheel/driveplate and the crankshaft. If the crankshaft rear seal is leaking, replace it before reinstalling the flywheel/driveplate (see Section 16).

8   Position the flywheel/driveplate against the crankshaft. On flywheel-equipped models, be sure to align the hole in the flywheel

15.8a An alignment dowel (arrow) ensures the flywheel can only be installed one way . . .

15.8b . . . it must be aligned with the dowel hole
in the flywheel (arrow)

16.2a Gently pry out the old seal with a thin screwdriver - wrap
the tip with tape to prevent damaging the seal bore and
crankshaft sealing surface

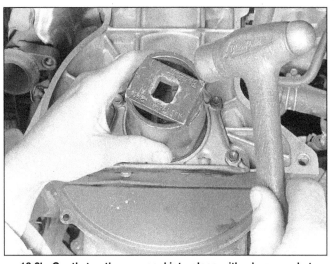

16.2b Gently tap the new seal into place with a large socket

16.5 After removing the retainer from the block, support it on two
wood blocks and drive out the old seal with a punch and hammer

with the alignment dowel in the crankshaft **(see illustrations)**. On driveplate-equipped models, align the marks you made during removal. Before installing the bolts, apply thread locking compound to the threads.

9   Tighten the bolts to the torque listed in this Chapter's Specifications. Hold the crankshaft from turning as described previously.

10   The remainder of installation is the reverse of the removal procedure.

16   Rear crankshaft oil seal - replacement

Refer to illustrations 16.2a, 16.2b, 16.5 and 16.6

1   The transmission must be removed from the vehicle for this procedure (see Chapter 7).

2   The seal can be replaced without drop-ping the oil pan or removing the seal retainer. However, this method is not recommended because the lip of the seal is quite stiff and it's possible to cock the seal in the retainer bore or damage it during installation. If you want to take the chance, pry out the old seal **(see illustration)**. Apply multi-purpose grease to the crankshaft seal journal and the lip of the new seal and carefully tap the new seal into place **(see illustration)**. The lip is stiff so carefully work it onto the seal journal of the crankshaft with a smooth object like the end of an extension as you tap the seal into place. Don't rush it or you may damage the seal.

3   The following method is recommended but requires removal of the oil pan (see Section 13) and the seal retainer.

4   After the oil pan has been removed, remove the bolts, detach the seal retainer and peel off all the old gasket material.

5   Position the seal and retainer assembly on a couple of wood blocks on a workbench and drive the old seal out from the back side with a punch and hammer **(see illustration)**.

6   Drive the new seal into the retainer with a block of wood **(see illustration)** or a section of pipe slightly smaller in diameter than the outside diameter of the seal.

7   Lubricate the crankshaft seal journal and the lip of the new seal with multi-purpose grease. Position a new gasket on the engine block.

8   Slowly and carefully push the seal onto the crankshaft. The seal lip is stiff, so work it onto the crankshaft with a smooth object such as the end of an extension as you push the retainer against the block.

9   Install and tighten the retainer bolts to the torque listed in this Chapter's Specifications.

10   The remaining steps are the reverse of removal.

11   Run the engine and check for oil leaks.

16.6 Drive the new seal into the retainer with a wood block or a
section of pipe, if you have one large enough - make sure you
don't cock the seal in the retainer bore

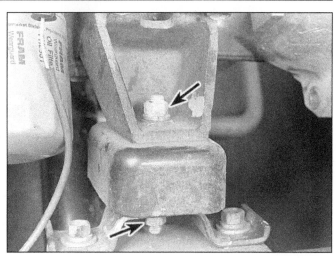

17.8 The engine mounts are secured to the frame and engine
brackets with nuts (arrows)

17 Engine mounts - check and replacement

Refer to illustration 17.8

1   Engine mounts seldom require attention, but broken or deteriorated mounts should be replaced immediately or the added strain placed on the driveline components may cause damage or wear.

Check

2   During the check, the engine must be raised slightly to remove the weight from the mounts.

3   Raise the vehicle and support it securely on jackstands, then position a jack under the engine bellhousing. Place a large block of wood between the jack head and the bellhousing, then carefully raise the engine just enough to take the weight off the mounts. **Warning:** *DO NOT place any part of your body under the engine when it's supported only by a jack!*

4   Check the mounts to see if the rubber is cracked, hardened or separated from the metal plates. Sometimes the rubber will split right down the center.

5   Check for relative movement between the mount plates and the engine or frame (use a large screwdriver or pry bar to attempt to move the mounts). If movement is noted, lower the engine and tighten the mount fasteners.

6   Rubber preservative should be applied to the mounts to slow deterioration.

Replacement

7   Disconnect the negative battery cable from the battery, then raise the vehicle and support it securely on jackstands (if not already done).

8   Remove the nut and detach the mount from the frame bracket **(see illustration)**.

9   Raise the engine slightly with a jack or hoist (make sure the fan doesn't hit the radiator or shroud). Remove the mount-to-engine bracket nut and detach the mount.

10   Installation is the reverse of removal. Use thread locking compound on the mount bolts and be sure to tighten them securely.

Notes

Chapter 2 Part B
DOHC engine

Contents

Specifications

General

Firing order	1-3-4-2
Cylinder numbers (timing chain end-to-transmission end)	1-2-3-4

Cylinder head

Warpage limit (gasket surface)	0.002 in (0.5 mm)
Warpage limit (manifold seating surfaces)	0.004 in (0.10 mm)

Oil pump

Radial clearance (outer rotor-to-oil pump housing clearance) limit	0.0059 in (0.15 mm)
Axial clearance (endplay) limit	0.0043 in (0.11 mm)
Pressure relief spring	
Free length	2.55 in (63.5 mm)
Preload	62.2 lbs at 2.05 in (8.6 kg at 52.0 mm)

Torque specifications

	Ft-lbs (unless otherwise indicated)
Alternator drivebelt idler pulley nut	33
Alternator drivebelt tensioner mounting bolt	18.5
Camshaft bearing cap bolts/studs	106 in-lbs
Camshaft cover nuts	96 in-lbs
Camshaft timing sprocket bolt (intake and exhaust)	57.5
Crankshaft pulley bolt	109
Cylinder head bolts	
Step 1	39
Step 2	61
Step 3	Loosen all bolts
Step 4	39
Step 5	61
Step 6	76

Torque specifications

Ft-lbs (unless otherwise indicated)

Cylinder head hex hole bolt (M6 bolt).. 96 in-lbs
Exhaust manifold bolts/nuts 37
Flywheel/driveplate bolts 51
Intake manifold bolts and nuts 19
Oil pan bolts/nuts .... 96 in-lbs
Oil pump
    Pressure relief spring retainer 21
    Pump housing Allen bolts .... 108 in-lbs
    Pump mounting bolts... 19.5
    Pump sprocket cover bolts .... 96 in-lbs
Oil pump pick-up tube mounting bolts.. 96 in-lbs
Timing chain cover bolts and nut .... 96 in-lbs
Timing chain tensioner (small timing chain)
    Tensioner bolts.. 96 in-lbs
    Tensioner nut... 33
Timing chain (big timing chain)
    Tensioner nut... 18.5
    Tensioner bolts.. 96 in-lbs
    Timing chain guide bolts .... 78 in-lbs

1   General information

This Part of Chapter 2 is devoted to in-vehicle repair procedures for the 2.0L DOHC engine, which was introduced in 1999. All information concerning engine removal and installation and engine block and cylinder head overhaul can be found in Part B of this Chapter.

The following repair procedures are based on the assumption that the engine is installed in the vehicle. If the engine has been removed from the vehicle and mounted on a stand, many of the steps outlined in this Part of Chapter 2 will not apply.

The Specifications included in this Part of Chapter 2 apply only to the procedures contained in this Part. Part C of Chapter 2 contains the Specifications necessary for cylinder head and engine block rebuilding.

2   Repair operations possible with the engine in the vehicle

Many major repair operations can be accomplished without removing the engine from the vehicle.

Clean the engine compartment and the exterior of the engine with some type of degreaser before any work is done. It will make the job easier and help keep dirt out of the internal areas of the engine.

Depending on the components involved, it may be helpful to remove the hood to improve access to the engine as repairs are performed (see Chapter 11, if necessary). Cover the fenders to prevent damage to the paint. Special pads are available, but an old bedspread or blanket will also work.

If vacuum, exhaust, oil or coolant leaks develop, indicating a need for gasket or seal replacement, the repairs can generally be made with the engine in the vehicle. The intake and exhaust manifold gaskets, oil pan gasket, crankshaft oil seals and cylinder head gasket are all accessible with the engine in place.

Exterior engine components, such as the intake and exhaust manifolds, the oil pan (and the oil pump), the water pump, the starter motor, the alternator, the distributor and the fuel system components can be removed for repair with the engine in place.

Since the cylinder head can be removed without pulling the engine, valve component servicing can also be accomplished with the engine in the vehicle. Replacement of the camshaft, timing belt and sprockets is also possible with the engine in the vehicle.

In extreme cases caused by a lack of necessary equipment, repair or replacement of piston rings, pistons, connecting rods and rod bearings is possible with the engine in the vehicle. However, this practice is not recommended because of the cleaning and preparation work that must be done to the components involved.

3   Top Dead Center (TDC) for number one piston - locating

Refer to illustration 3.6

Note: *The following procedure is based on the assumption that the spark plug wires and distributor are correctly installed. If you are trying to locate TDC to install the distributor correctly, piston position must be determined by feeling for compression at the number one spark plug hole, then aligning the ignition timing marks as described in Step 8.*

1   Top Dead Center (TDC) is the highest point in the cylinder that each piston reaches as it travels up-and-down when the crankshaft turns. Each piston reaches TDC on the compression stroke and again on the exhaust stroke, but TDC generally refers to piston position on the compression stroke.

2   Positioning the piston(s) at TDC is an essential part of many procedures such as rocker arm removal, camshaft and timing belt/sprocket removal.

3   Before beginning this procedure, be sure to place the transmission in Neutral and apply the parking brake or block the rear wheels. Also, disable the ignition system by unplugging the electrical connectors from the ignition coils (see Chapter 5). Remove the spark plugs (see Chapter 1).

4   In order to bring any piston to TDC, the crankshaft must be turned using one of the methods outlined below. When looking at the front of the engine, normal crankshaft rotation is clockwise.

  a) *The preferred method is to turn the crankshaft with a socket and ratchet attached to the bolt threaded into the front of the crankshaft.*

  b) *A remote starter switch, which may save some time, can also be used. Follow the instructions included with the switch. Once the piston is close to TDC, use a socket and ratchet as described in the previous paragraph.*

  c) *If an assistant is available to turn the ignition switch to the Start position in short bursts, you can get the piston close to TDC without a remote starter switch. Make sure your assistant is out of the vehicle, away from the ignition switch, then use a socket and ratchet as described in Paragraph a) to complete the procedure.*

5   Have an assistant turn the crankshaft with a socket and wrench while you hold a finger over the number one spark plug hole.
Note: *See the Specifications for the number one cylinder location.*

6   When the piston approaches TDC, you

3.6 To bring the No. 1 piston up to Top Dead Center (TDC), align the notch in the crankshaft pulley (lower arrow) with the stationary timing mark on the timing cover (upper arrow)

will feel pressure at the spark plug hole. Have your assistant stop turning the crankshaft when the timing marks are aligned (see illustration).

7   If the notch in the pulley is already past the timing mark, turn the crankshaft two complete revolutions in a clockwise direction until the notch is aligned with the timing mark.

8   After the number one piston has been positioned at TDC on the compression stroke, TDC for any of the remaining pistons can be located by turning the crankshaft one-half turn (180 degrees) to get to TDC for the next cylinder in the firing order.

4   Camshaft cover - removal and installation

Refer to illustrations 4.2, 4.6, 4.7a, 4.7b, 4.7c and 4.10

1   Disconnect the negative cable from the battery.

2   Disconnect the breather hose and PCV hose from the camshaft cover (see illustration).

3   Remove the ignition coils (see Chapter 5).

4   Detach the throttle cable from its bracket on the camshaft cover (see Chapter 4).

5   Remove the dipstick.

6   Remove the six camshaft cover retaining nuts (see illustration), remove and discard the old rubberized washers and lift off the cover. If the cover sticks to the cylinder head, tap on it with a soft-face hammer or place a block of wood against the cover and tap on the wood.

7   If necessary for camshaft removal or if it's leaking, remove the camshaft cover side seal (the rubber half-circle located at the rear end of the intake camshaft). Remove the camshaft cover gasket and the spark plug hole O-rings (see illustrations). Thoroughly clean the camshaft cover and remove all traces of old gasket material from the cylinder head and from the cover.

8   Inspect the spark plug hole O-rings and the cam cover gasket for deterioration. If any of these rubber parts are worn or damaged, replace them.

4.2 Before removing the camshaft cover, remove the breather hose (1) and the PCV hose (2)

4.6 To remove the camshaft cover, remove these six nuts (arrows)

4.7a If necessary, remove and discard the old cylinder head side seal (the rubber half-circle piece behind the rear end of the intake camshaft that seals the hole between the camshaft cover and the cylinder head)

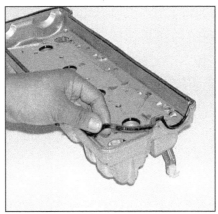

4.7b Remove the old camshaft cover gasket and then clean it thoroughly and inspect it for cracks, tears and deterioration; if the gasket is damaged or worn, replace it

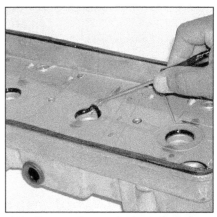

4.7c Remove the spark plug hole O-rings and inspect them for cracks, tears and deterioration; if any of the spark plug hole O-rings are damaged, replace all four as a set

9   Install the gasket and spark plug hole O-rings. Use RTV to hold them in place.
10   Apply RTV sealant to the indicated areas of the cylinder head **(see illustration)**, install a new camshaft cover side seal and then install the camshaft cover. Install new rubberized washers, install the nuts and tighten them by hand until they're all snug.
11   Working from the center out, tighten the cam cover retaining nuts to the torque listed in this Chapter's Specifications.
12   The remaining steps are the reverse of removal. When finished, run the engine and check for oil leaks.

5   Intake manifold - removal and installation

4.10 Before installing the camshaft cover, apply RTV sealant to the four indicated areas on the gasket mating surface of the head

Refer to illustrations 5.5, 5.11a, 5.11b and 5.13
1   Relieve the fuel injection system pressure (see Chapter 4).
2   Disconnect the negative cable from the battery.
3   Drain the cooling system (see Chapter 1). If the coolant is in good condition it can be reused.
4   Remove the air intake duct between the air cleaner housing and the throttle body (see Chapter 4).
5   Clearly label, then disconnect, the following electrical connectors **(see illustration)**:
a) *Exhaust Gas Recirculation (EGR) valve connector (see Chapter 6).*
b) *Idle Air Control (IAC) valve connector (see Chapter 6).*
c) *Throttle Position (TP) sensor connector (see Chapter 6).*
d) *Evaporative Emissions (EVAP) canister purge valve connector (see Chapter 6).*
e) *Ground terminal on intake manifold.*
f) *Manifold   Differential   Pressure (MDP)/Manifold   Absolute   Pressure (MAP) sensor connector (see Chapter 6).*
6   Clearly label, then disconnect, the following hoses from the indicated components:

a) *Brake booster hose from the intake manifold (see Chapter 9)*
b) *PCV hose from the intake manifold (see Chapter 6.*
c) *Fuel pressure regulator vacuum hose from the intake manifold (see Chapter 4).*
d) *Canister purge hose from the EVAP canister purge valve (see Chapter 6).*
e) *Vacuum hose from the intake manifold.*
f) *Coolant hoses from the throttle body (see Chapter 4).*
g) *Coolant hose from the coolant bypass pipe.*
h) *Breather hose from the throttle body.*
7   Disconnect the throttle cable from its bracket on the air intake plenum (see Chapter 4).
8   On vehicles equipped with an automatic transmission, disconnect the throttle valve cable (see Chapter 7B).
9   If you're going to replace the intake manifold, remove the throttle body (see Chapter 4). **Note:** *If you're only removing the intake manifold to service something else, such as the cylinder head, the intake manifold and throttle body can remain attached and be removed as a single assembly. If, however, you plan to replace the intake manifold, you'll have to remove the throttle body. It's easier to unbolt the throttle body from the intake mani-*

fold while the manifold is still bolted to the cylinder head, so do it now.
10   Remove the fuel rail and the fuel injectors (see Chapter 4).
11   Remove the two intake manifold stiffener plates **(see illustrations)**.
12   Detach the coolant pipe from the intake manifold.
13   Remove the intake manifold mounting bolts and nuts **(see illustration)** and remove the intake manifold from the engine. If it sticks, tap the manifold with a soft-face hammer. **Caution:** *Do not pry between gasket sealing surfaces or tap on the throttle body (if still bolted to the manifold).* Remove and discard the old intake manifold gasket.
14   Thoroughly clean the intake manifold and cylinder head mating surfaces, removing all traces of gasket material. Be very careful not to scratch or gouge the delicate aluminum gasket surfaces on the manifold and cylinder head. Gasket removal solvents are available from auto parts stores and may prove helpful.
15   Install the manifold, using a new gasket and tighten the bolts and nuts in several stages, working from the center out, until you reach the torque listed in this Chapter's Specifications.
16   Reinstall the remaining parts in the reverse order of removal.

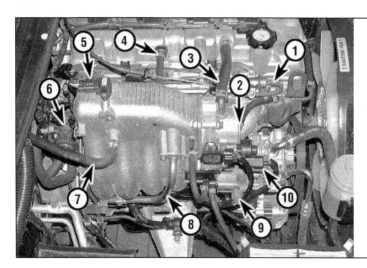

5.5 To remove the intake manifold, disconnect the following parts:

1   *Fuel pressure regulator vacuum hose*
2   *Throttle Position (TP) sensor electrical connector*
3   *Breather hose*
4   *PCV hose*
5   *Manifold Differential Pressure (MDP)/Manifold Absolute Pressure (MAP) sensor electrical connector*
6   *Exhaust Gas Recirculation (EGR) valve electrical connector*
7   *Power brake booster hose*
8   *EGR pipe*
9   *Evaporative canister purge valve electrical connector*
10   *Idle Air Control (IAC) valve electrical connector*

17   Add coolant, run the engine and check for leaks and proper operation.

6   Exhaust manifold - removal and installation

Refer to illustrations 6.5, 6.6 and 6.7
Warning: *Allow the engine to cool completely before following this procedure.*
1   Disconnect the negative cable from the battery.
2   Set the parking brake and block the rear wheels.
3   Raise the front of the vehicle and support it securely on jackstands..
4   Unplug the oxygen sensor connector (see Chapter 6)
5   Remove the exhaust manifold heat shield bolts or nuts **(see illustration)** and then remove the heat shield.
6   Remove the exhaust manifold flange nuts/bolt **(see illustration)**. Apply penetrating oil to the threads of the exhaust pipe studs to make removal easier.

5.11a To remove the upper intake manifold stiffener bracket, remove these two bolts (arrows)

5.11b To remove the lower intake manifold stiffener bracket, remove these two bolts (arrows)

7   Apply penetrating oil to the threads of the exhaust manifold mounting studs, and then remove the exhaust manifold mounting nuts **(see illustration)**.

8   Slip the manifold off the studs and remove it from the engine compartment. Remove and discard the old gasket.
9   Clean and inspect all threaded fasteners

5.13 To remove the intake manifold, remove these nuts and bolts (arrows) (lower fasteners not visible in this photo)

6.5 To detach the heat shield from the exhaust manifold, remove these four bolts (arrows)

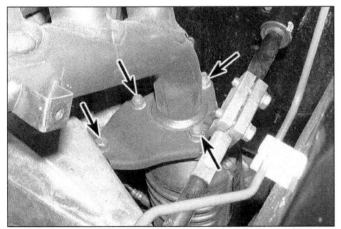

6.6 To separate the exhaust pipe from the exhaust manifold, remove these three nuts (arrows). The arrow at the lower left indicates the bolt that secures the flange to the left motor mount; this bolt must also be removed

6.7 To detach the exhaust manifold from the cylinder head, remove these bolts and nuts (arrows)

7.5 Before removing the timing chain cover, remove the coolant bypass hose (arrow) and pipe

7.9 To remove the alternator drivebelt tensioner (left arrow) remove the retaining bolt; to remove the alternator drivebelt idler pulley (right arrow), remove the retaining nut

7.12a To remove the crankshaft pulley, immobilize the pulley with a chain wrench or some other suitable holding device, break loose the bolt . . .

and repair as necessary.
10   Remove all traces of gasket material from the mating surfaces and inspect them for wear and cracks.
11   Install a new gasket, install the manifold and tighten the nuts in several stages, working from the center out, to the torque listed in this Chapter's Specifications.
12   Reinstall the remaining parts in the reverse order of removal.
13   Run the engine and check for exhaust leaks.

7   Crankshaft front oil seal - replacement

Note: *You can usually do this job without removing the timing chain cover - remove the radiator for access to the seal and proceed to Step 14. If clearance is too tight to remove the seal in-vehicle, remove the cover, as described below.*

Timing chain cover removal

Refer to illustrations 7.5, 7.9, 7.12a, 7.12b, 7.13a and 7.13b

Warning: *The air conditioning system is under high pressure. Do not loosen any fittings or remove any components until after the system has been discharged by an air conditioning technician. Always wear eye protection when disconnecting refrigerant fittings.*

1   Disconnect the negative battery cable.
2   Drain the engine oil and the coolant (see Chapter 1).
3   Remove the oil pan (see Section 12).
Note: *In some cases it may be possible to remove the front cover without removing the oil pan. Remove just the front pan-to-cover bolts. On reassembly, clean the mating surfaces thoroughly and seal the pan-to-cover joint with RTV sealant.*
4   Remove the camshaft cover (see Section 4).
5   Remove the coolant bypass pipe and

the bypass hose **(see illustration)**.
6   Remove the fan and cooling shroud (see Chapter 3), the fan belt (see Chapter 1) and the fan pulley (see Chapter 3).
7   Remove the alternator belt (see Chapter 1).
8   Remove the water pump pulley (see Chapter 3).
9   Remove the alternator drivebelt tensioner and idler pulley **(see illustration)**.
10   Disconnect the radiator outlet hose from the thermostat housing (see Chapter 3).
11   Detach the air conditioning compressor from its mounting bracket and set it aside (see Chapter 3). Do NOT disconnect either of the air conditioning refrigerant hoses. Remove the compressor mounting bracket.
12   Using a suitable pulley holding tool, remove the crankshaft pulley bolt **(see illustrations)** and then remove the pulley. If the pulley is hard to remove, use a suitable puller (such as a steering wheel puller) to get it off.
13   Remove the timing chain cover **(see illustrations)**. Note the locations of the two dowel pins.

7.12b . . . and then pull the pulley straight off; if the pulley is difficult to remove, use a small puller (like a steering wheel puller) to remove it

7.13a Timing chain cover bolts (arrows), left side; there's also a nut (left arrow) in the middle of cover, right below the stud for the alternator drivebelt idler pulley

7.13b Timing chain cover bolts (arrows), right side

7.15a Pry out the old crankshaft front oil seal with a seal removal tool

Seal replacement

Refer to illustrations 7.15a and 7.15b

Note: *The accompanying photos depict the crankshaft front seal being replaced with the timing cover installed, and the radiator removed, just to show that it can be done this way. However, be advised that installing a new seal in such a tight space is tricky. It's sometimes easier to do this job with the timing cover removed.*

14　Wipe off the timing chain cover with a shop rag. Remove all old oil, sealant and dirt.

15　Inspect the front crankshaft oil seal for cracks and any other damage. If the seal is damaged or worn, replace it. Use a seal removal tool **(see illustration)** to remove the old seal. To install the new seal, lay the timing chain cover on a clean flat surface and, using a large socket with an outside diameter (O.D.) slightly smaller than the O.D. of the seal, tap the new seal into place **(see illustration)**.

Timing chain cover installation

16　Apply RTV sealant to the mating surface of the timing cover.

17　Coat the lip of the front crankshaft oil seal with clean engine oil, make sure that the dowel pins are installed, and then install the

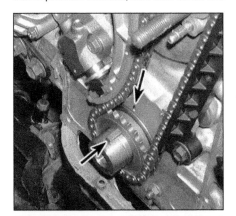

8.2a To bring the No. 1 piston to Top Dead Center (TDC), rotate the crankshaft so that the key for the crankshaft timing chain sprocket is aligned with the stationary mark on the front of the block . . .

7.15b Drive the new crankshaft front oil seal into the timing cover with a hammer and a large socket with an outside diameter slightly smaller than the outside diameter of the seal

timing cover. Install the timing chain cover bolts and nut and then tighten them to the torque listed in this Chapter's Specifications.

18　Install the crankshaft pulley and the pulley retaining bolt. Using a suitable pulley holding tool **(see illustration 7.12a)**, tighten the pulley bolt to the torque listed in this Chapter's Specifications.

19　On vehicles equipped with air conditioning, install the air conditioning compressor bracket and tighten the bracket bolts securely.

20　Install the alternator drivebelt idler pulley and tighten the idler pulley retaining nut to the torque listed in this Chapter's Specifications.

21　Reconnect the radiator outlet hose to the thermostat cap and tighten the hose clamp securely.

22　Install the alternator drivebelt tensioner and tighten the tensioner mounting bolt to the torque listed in this Chapter's Specifications.

23　Install the water pump pulley and tighten the pulley bolts to the torque listed in the Chapter 3 Specifications.

24　Install the alternator drivebelt (see Chapter 1).

8.2b . . . verify that the marks (arrows) on the camshaft sprockets are aligned with the index marks on the cylinder head

25　Install the fan pulley (see Chapter 3), the fan belt (see Chapter 1) and the cooling shroud and fan (see Chapter 3).

26　Install the coolant bypass hose and the bypass pipe.

27　Install the camshaft cover (see Section 4).

28　Install the oil pan (see Section 12).

29　Adjust the cooling fan drivebelt tension (see Chapter 1).

30　Refill the cooling system with coolant and refill the engine with clean engine oil (see Chapter 1). Start the engine and verify that there are no coolant or oil leaks.

8　Timing chains and sprockets - removal, inspection and installation

Removal

✱✱ CAUTION ✱✱

The timing system is complex. Severe engine damage will occur if you make any mistakes. Do not attempt this procedure unless you are highly experienced with this type of repair. If you are at all unsure of your abilities, consult an expert. Double-check all your work and be sure everything is correct before you attempt to start the engine.

Caution: *Do not try to turn the crankshaft with the camshaft sprocket bolt and do not rotate the crankshaft counterclockwise.*

Small timing chain and camshaft sprockets

Refer to illustrations 8.2a, 8.2b, 8.2c, 8.3 and 8.4

1　Remove the camshaft cover (see Section 4) and the timing chain cover (see Section 7).

2　Rotate the crankshaft so that the key for the crankshaft timing chain sprocket is aligned with the stationary mark on the front of the block **(see illustration)**. Then verify that the marks on the camshaft sprockets are

8.2c . . . and then verify that the arrow mark (arrow) at the 6 o'clock position on the idler sprocket is pointing straight up

8.3 Remove the timing chain tensioner adjuster bolts and nut (arrows), put some slack in the chain by rotating the camshaft sprocket counterclockwise while pushing against the "shoe" (the surface of the tensioner on which the chain rides) and then remove the tensioner

8.4 Using a wrench to immobilize each camshaft, remove the sprocket retaining bolts from the intake and exhaust camshafts (put the wrench on the "hex" cast into each camshaft to hold the cam)

aligned with the stationary marks on the cylinder head and that the arrow mark (at the 6 o'clock position) on the idler sprocket is

8.7 Remove the timing chain tensioner adjuster bolt and nut (arrows) and then remove the tensioner

8.8 Remove the timing chain tensioner nut (left arrow) and the timing chain guide bolts (right arrows) and then remove the tensioner and the guide

pointing straight up **(see illustrations)**. The No. 1 piston is now at Top Dead Center (TDC). **Caution:** *Do NOT rotate the crankshaft or the camshafts again until the timing chain has been re-installed.*

3   Remove the timing chain tensioner adjuster bolts and nut **(see illustration)** and then remove the tensioner adjuster. To disengage the tensioner from the small chain, put some slack in the chain by rotating the camshaft sprocket counterclockwise while pushing against the "shoe" (the surface of the tensioner on which the chain rides).

4   Using a wrench to immobilize each camshaft, remove the sprocket retaining bolts from the intake and exhaust camshafts **(see illustration)**.

5   Remove the camshaft timing chain and the camshaft sprockets. **Caution:** *Once the small timing chain has been removed, do NOT turn the intake and exhaust camshafts independently of the crankshaft, or you risk causing serious damage to the valves and/or*

8.10 Remove the idler sprocket and the timing chain; also remove the idler sprocket bushing

the pistons, as well as other parts related to those components.

Big timing chain, idler sprocket and crankshaft sprocket

Refer to illustrations 8.7, 8.8 and 8.10

6   Remove camshaft cover (see Section 4) and the timing chain cover (see Section 7). Bring the No. 1 piston to TDC (see Step 2) and then remove the small timing chain and camshaft sprockets (see Steps 3 through 5).

7   Remove the timing chain tensioner **(see illustration)**.

8   Remove the timing chain guide **(see illustration)**.

9   Remove the timing chain tensioner **(see illustration 8.8)**

10   Remove the idler sprocket and the timing chain **(see illustration)**. Also remove the idler sprocket bushing.

11   Remove the crankshaft sprocket.

Inspection

Small timing chain and camshaft sprockets

Refer to illustration 8.13

12   Wipe off the timing chain guide, the tensioner adjuster, the camshaft sprockets and the timing chain. Remove all old oil and dirt.

13   Inspect the "shoes" (the contact surfaces) of the timing chain guide and the tensioner adjuster for excessive wear or damage **(see illustration)**. If either part is damaged or worn out, replace it.

14   Inspect the timing chain and the teeth on the camshaft sprockets for excessive wear and damage. If either part is damaged or worn excessively, replace it.

Big timing chain, idler sprocket and crankshaft sprocket

Refer to illustrations 8.16 and 8.17

15   Wipe off the timing chain guide, the tensioner, the tensioner adjuster, the idler and crankshaft sprockets and the timing chain.

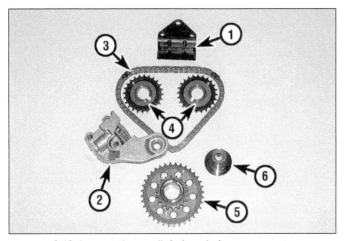

8.13 Inspect the small timing chain components

1 **Timing chain guide** - *inspect the "shoe" on the guide for excessive wear*
2 **Timing chain tensioner** - *inspect the "shoe" on the tensioner for excessive wear*
3 **Small timing chain** - *inspect the rollers and side plates for excessive wear and damage*
4 **Camshaft sprockets** - *inspect the teeth on the sprockets for excessive wear and damage*
5 **Idler sprocket** *(removed from big timing chain for clarity) - inspect the teeth on the idler sprocket for excessive wear*
6 **Idler sprocket bushing** *(removed from big timing chain for clarity) - inspect the bushing for excessive wear*

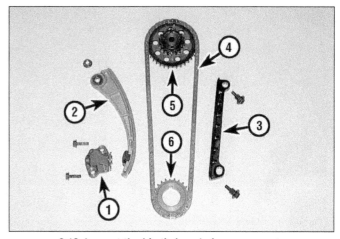

8.16 Inspect the big timing chain components

1 **Timing chain tensioner** - *see close-up in next illustration*
2 **Timing chain tensioner guide** - *inspect the "shoe" on the tensioner for excessive wear*
3 **Timing chain guide** - *inspect the "shoe" on the guide for excessive wear*
4 **Big timing chain** *(inspect the rollers and side plates for excessive wear and damage)*
5 **Idler sprocket** - *inspect the sprocket teeth for excessive wear (also inspect the idler sprocket bushing - shown in previous illustration - for wear)*
6 **Crankshaft sprocket** - *inspect the sprocket teeth for excessive wear*

sioner adjuster, the camshaft sprockets and the timing chain. Remove all old oil and dirt.
13 Inspect the "shoes" (the contact surfaces) of the timing chain guide and the tensioner adjuster for excessive wear or damage **(see illustration)**. If either part is damaged or worn out, replace it.
14 Inspect the timing chain and the teeth on the camshaft sprockets for excessive wear and damage. If either part is damaged or worn excessively, replace it.

Big timing chain, idler sprocket and crankshaft sprocket

Refer to illustrations 8.16 and 8.17
15 Wipe off the timing chain guide, the tensioner, the tensioner adjuster, the idler and crankshaft sprockets and the timing chain.

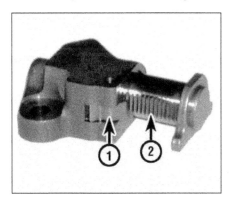

8.17 Verify that the latch (1) on the adjuster for the big timing chain functions correctly and that the toothed rack (2) of the adjuster is undamaged

Remove all old oil and dirt.

** CAUTION **

Before starting the engine, carefully rotate the crankshaft by hand through at least two full revolutions (use a socket and breaker bar on the crankshaft pulley center bolt). If you feel any resistance, STOP! There is something wrong - most likely, valves are contacting the pistons. You must find the problem before proceeding. Check your work and see if any updated repair information is available.

8.19a Verify that the crankshaft sprocket key is aligned with the stationary timing mark (arrow) on the block . . .

16 Inspect the shoes of the timing chain guide and the tensioner for excessive wear and damage **(see illustration)**. If either part is damaged or worn out, replace it.
17 Inspect the timing chain tensioner adjuster **(see illustration)**. Verify that the latch and the toothed rack are undamaged and that the latch functions correctly. If the tensioner adjuster is damaged, excessively worn or dysfunctional, replace it.
18 Inspect the timing chain and the teeth on the idler and crankshaft sprockets for excessive wear and damage. If either part is damaged or worn excessively, replace it. Also inspect the idler sprocket bushing for

8.19b . . . and then install the crankshaft sprocket with the match mark (arrow) facing out, toward the timing chain cover

8.21a When installing the big timing chain, make sure that the dark blue side plate (upper arrow) is aligned with the match mark (lower arrow) on the idler sprocket . . .

8.21b . . . and that the yellow side plate (lower arrow) is aligned with the match mark (upper arrow) on the crankshaft sprocket

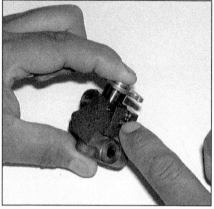

8.23a To depress the tensioner plunger, release the latch on the tensioner, push the plunger back into the tensioner adjuster body until it stops . . .

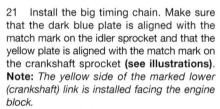

21   Install the big timing chain. Make sure that the dark blue plate is aligned with the match mark on the idler sprocket and that the yellow plate is aligned with the match mark on the crankshaft sprocket **(see illustrations)**. **Note:** *The yellow side of the marked lower (crankshaft) link is installed facing the engine block.*

22   Install the timing chain tensioner, secure it with the tensioner nut and tighten the nut to the torque listed in this Chapter's Specifications.

23   Release the latch on the tensioner and push the plunger back into the body **(see illustration)**, and then hold it in position by inserting a push pin or a piece of stiff wire between the latch and the tensioner body **(see illustration)**.

24   Install the timing chain tensioner, secure it with the tensioner mounting bolts and then tighten the bolts to the torque listed

in this Chapter's Specifications. Pull out the push-pin or piece of wire that you used to hold the plunger.

25   Install the timing chain guide, install the guide mounting bolts and tighten them to the torque listed in this Chapter's Specifications.

26   Verify that the dark blue and yellow plates on the timing chain are still aligned with their respective match marks on the idler sprocket and on the crankshaft sprocket and that the crankshaft sprocket key is still aligned with the index mark on the block **(see illustration)**.

Small timing chain and camshaft sprockets

Refer to illustrations 8.28, 8.29 and 8.32

27   Verify that the crankshaft sprocket key is aligned with the stationary match mark on the block **(see illustration 8.26)**.

28   Verify that the arrow mark on the idler sprocket faces upward and that the knock

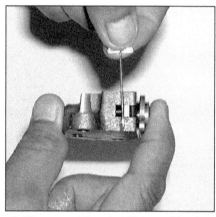

8.23b . . . and then hold it in position by inserting a push pin or a piece of stiff wire between the latch and the tensioner body

8.26 After installing the guide, tensioner guide and tensioner for the big timing chain, verify that the dark blue and yellow plates on the timing chain are still aligned with their respective match marks on the idler sprocket and on the crankshaft sprocket and that the crankshaft sprocket key is still aligned with the index mark on the block.

8.28 Verify that the arrow mark at 6 o'clock on the idler sprocket (lower arrow) faces upward and that the knock pins (upper arrows) on the intake and exhaust camshafts are aligned with the match marks on the cylinder head

8.29 When installing the small timing chain and the intake and exhaust camshaft sprockets, make sure that the yellow plate on the small timing chain is aligned with the arrow mark on the idler sprocket (lower arrows) and make sure that the dark blue plates on the small timing chain are aligned with the arrow marks on the intake and exhaust camshaft sprockets (upper arrows)

pins on the intake and exhaust camshafts are aligned with the match marks on the cylinder head **(see illustration)**.

29   Install the small timing chain and the intake and exhaust camshaft sprockets. Make sure that the yellow plate on the small timing chain is aligned with the arrow mark on the idler sprocket **(see illustration)**. Also make sure that the dark blue plates on the small timing chain are aligned with the arrow marks on the intake and exhaust camshaft sprockets. **Note:** *There are arrow marks on both sides of each camshaft sprocket, so the sprockets can be installed facing in either direction.*

30   Install the intake and exhaust camshaft sprocket bolts and hand tighten them until they're snug.

31   Using a wrench to immobilize each camshaft **(see illustration 8.4)**, tighten the bolts to the torque listed in this Chapter's Specifications.

32   Push the plunger into the tensioner body and hold it in place with a suitable stopper, such as a tack or push-pin **(see illustration)**.

33   Install the timing chain tensioner and gasket. Install the tensioner bolts and nut and

tighten them to the torque listed in this Chapter's Specifications. Remove the stopper from the tensioner.

34   Rotate the crankshaft in a clockwise direction for two complete revolutions, and then verify that the timing marks on the crankshaft and camshaft sprockets are aligned with their respective stationary timing marks on the block and on the cylinder head **(see illustrations 8.2a and 8.2b)**. **Caution:** *If you feel resistance, stop and re-check the timing components.*

35   Lubricate the timing chains, tensioners, tensioner adjusters, sprockets and chain guides with clean engine oil.

---

9   Camshafts and lifters - removal and installation

Refer to illustration 9.9

1   Disconnect the negative battery cable.

2   Drain the engine oil and coolant (see Chapter 1).

3   Remove the oil pan and the oil pump strainer (see Sections 12 and 13).

4   Remove the camshaft cover (see Section 4).

5   Remove the timing chain cover (see Section 7).

6   Remove the small timing chain, the chain tensioner and the camshaft sprockets (see Section 8).

7   Remove the Camshaft Position (CMP) sensor (see Chapter 6).

8   Rotate the crankshaft 90 degrees in a clockwise direction to prevent interference between the pistons and the valves.

9   Loosen the camshaft bearing cap bolts in the indicated order **(see illustration)**. Note the location of the bearing cap *studs*; these studs must be re-installed at the same locations. Remove the camshaft bearing caps and store them in order. Remove the dowel pin for the big camshaft bearing cap and store it in a plastic bag.

10   Remove the camshafts. Note the offset slot in the end of the exhaust camshaft; this slot engages the dogs on the CMP sensor.

11   Wipe off the top of each lifter and clearly label it with a laundry marker. Remove the lifters and store them, upside down, in a pan of clean engine oil. **Caution:** *Do NOT store the lifters dry. Make sure that they're immersed in oil (or they might bleed down while they're out of the engine). Finally, don't try to disassemble the lifters, and don't apply force to the hydraulic valve lash adjuster mechanism on the underside of each lifter.*

Inspection

12   Refer to Section 9 in Chapter 2C.

Installation

Refer to illustrations 9.16, 9.17 and 9.18

13   Before installing the lifters, squirt clean engine oil into the cylinder head oil passages. Squirt oil into the oil holes of the lifters and verify that the oil comes out the holes in the friction surfaces of the lifters. Install the lifters in the cylinder head.

14   Rotate the crankshaft sprocket counter-clockwise 90 degrees so that the sprocket key is realigned with the timing mark.

15   Lubricate the friction and bearing sur-

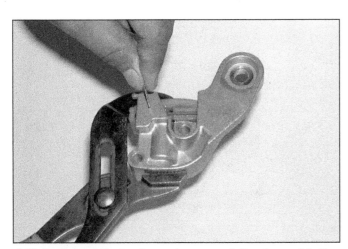

8.32 To depress the tensioner plunger, push the plunger into the tensioner body with a pair of water pump pliers and hold it in place with a suitable stopper, such as a tack or push-pin

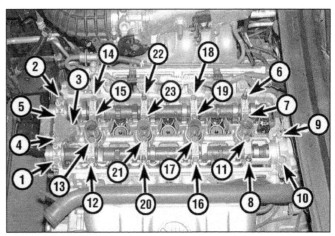

9.9 Camshaft bearing cap bolt/stud loosening sequence

faces of the camshafts with clean engine oil and then install the cams. Make sure that the pins on the cams are aligned with the match marks on the cylinder head **(see illustration 8.28)**. **Note:** *If you've mixed up the camshafts, the exhaust cam has a slot in its rear end for the CMP sensor.*

16 Install the dowel pin for the big camshaft bearing cap and apply sealant to the mating surface of that cap **(see illustration)**.

17 Install the cam bearing caps in their correct locations. A letter, a number and an arrow are embossed on each cap **(see illustration)**:

   a) *The letter "I" indicates the intake side; the letter "E" indicates the exhaust side.*

   b) *The number ("1," "2," "3" or "4") indicates the position of the cap in relation to the timing chain end of the engine (numbered in ascending order, starting at the timing chain end of the engine).*

   c) *The arrow points toward the timing chain end of the engine.*

18 Lubricate the camshaft bearing cap bolts and studs with clean engine oil and then install them in their correct locations and tighten them by hand. Then, following the tightening sequence **(see illustration)**, tighten the camshaft bearing cap bolts, gradually and evenly, to the torque listed in this Chapter's Specifications.

19 Install the CMP sensor (see Chapter 6).

20 Install the camshaft sprockets, the small timing chain and the tensioner (see Section 7).

21 Install the timing chain cover (see Section 7).

22 Install the camshaft cover (see Section 4).

23 Install the oil pump strainer (see Section 13) and the oil pan (see Section 12).

24 Refill the cooling system with coolant

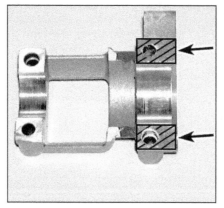

9.16 Apply sealant to the indicated surfaces (arrows) of the big camshaft bearing cap at the rear of the exhaust camshaft

and the engine with oil (see Chapter 1).

25 Start the engine and verify that there are no leaks. Check ignition timing (see Chapter 5) and adjust as necessary.

10 Cylinder head - removal and installation

Removal

Refer to illustration 10.15

Caution: *Allow the engine to cool completely before following this procedure.*

1 Relieve the system fuel pressure (see Chapter 4).

2 Disconnect the negative cable from the battery.

3 Drain the engine oil and the coolant (see Chapter 1).

4 Remove the strut tower bar (see Chap-

ter 10).

5 Remove the air intake duct between the air cleaner housing and the throttle body (see Chapter 4).

6 Disconnect the following electrical harnesses and/or connectors **(see illustration 5.5)**:

   a) *EGR valve connector*

   b) *IAC valve connector*

   c) *TP sensor connector*

   d) *EVAP canister purge valve connector*

   e) *Oxygen sensor connector*

   f) *CMP sensor connector (see Chapter 6)*

   g) *ECT sensor connector (see Chapter 6)*

   h) *Fuel injector wire harness connectors (see Chapter 4)*

   i) *Ignition coil connectors (see Chapter 5)*

   j) *All ground wires*

   k) *All wiring harness brackets, clamps and clips*

7 Disconnect the throttle cable (see Chapter 4).

8 On vehicles with an automatic transmission, disconnect the TV cable (see Chapter 7B).

9 Disconnect the following hoses:

   a) *Brake booster hose, from the intake manifold* **(see illustration 5.5)**

   b) *All vacuum hoses, from the intake manifold*

   c) *Canister purge hoses from the EVAP canister purge valve* **(see illustration 5.5)**

   d) *Coolant hose from the bypass pipe* **(see illustration 7.5)**

   e) *Fuel supply hose and return hose from their respective pipes (see Chapter 4)*

   f) *Heater hose from the heater outlet pipe (see Chapter 3)*

   g) *Radiator inlet hose from coolant outlet pipe (see Chapter 3)*

10 Remove the intake manifold rear sup-

9.17 A letter, a number and an arrow are embossed on each camshaft bearing cap:

A   *The letter "I" indicates the intake side; the letter "E" indicates the exhaust side.*

B   *The number ("1," "2," "3" or "4") indicates the position of the cap in relation to the timing chain end of the engine (numbered in ascending order, starting at the timing chain end of the engine).*

C   *The arrow points toward the timing chain end of the engine.*

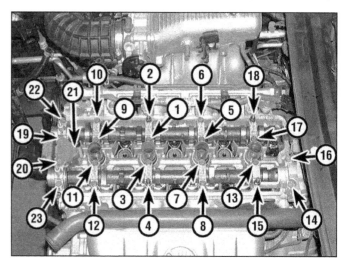

9.18 Camshaft bearing cap bolt/stud tightening sequence

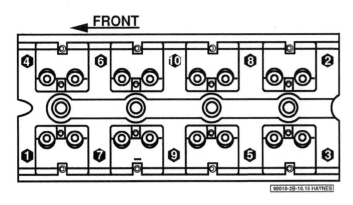

10.15 Cylinder head bolt loosening sequence

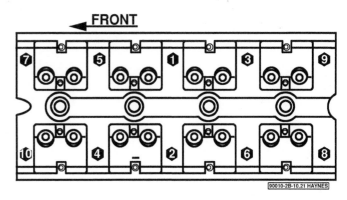

10.21 Cylinder head bolt tightening sequence

port bracket (see Section 5).

11   Detach the coolant pipe from the intake manifold.

12   Disconnect the exhaust pipe from the exhaust manifold (see Section 6 in this Chapter. Remove the exhaust manifold support bracket, if equipped **(see illustration 6.6)** and then remove the exhaust manifold (see Section 6).

13   Remove the camshaft cover (see Section 4), the intake manifold (see Section 5), the oil pan (see Section 12), the timing chain cover (see Section 7), the timing chains (see Section 8), the camshafts and the lifters (see Section 9).

14   Remove the coolant outlet pipe (the long pipe that runs along the top of the exhaust manifold heat shield) and the coolant outlet elbow from the cylinder head.

15   Loosen the cylinder head bolts in the indicated sequence **(see illustration)**. Don't forget to remove the M6 bolt from the left front corner of the head.

16   Make a final inspection of the cylinder head for any electrical connectors, hoses, brackets or anything else that might impede removal of the head, and then remove or detach such items as necessary.

17   Remove the cylinder head assembly. **Warning:** *The cylinder head is fairly heavy. Get help if necessary.* Carefully lift the cylinder head straight up and then place the head on wooden blocks to prevent damage to the sealing surfaces. If the head sticks to the engine block, dislodge it by prying against a protrusion on the head casting. **Note:** *Cylinder head disassembly and inspection procedures are covered in Chapter 2, Part C. It's a good idea to inspect the camshaft and have the head checked for warpage, even if you're just replacing the head gasket.*

Installation

Refer to illustration 10.21

18   Remove all traces of old gasket material from the block and head. Do not allow anything to fall into the engine. Clean and inspect all threaded fasteners and be sure the threaded holes in the block are clean and dry.

19   Install the two knock pins in the front left and right rear corners of the upper mating surface of the block.

20   Place a new gasket in position. Make sure that the TOP mark on the gasket faces up (toward the cylinder head), and that it's located near the left front end of the engine.

21   Install the cylinder head on the block. Apply engine oil to the cylinder head bolts and hand tighten them. Then, working from the center of the head to the end ends, in a criss-cross fashion, gradually and evenly tighten the bolts in the indicated sequence and in the Steps listed in this Chapter's Specifications **(see illustration)**. After the large bolts are torqued, tighten the small (M6) bolt to the torque listed in this Chapter's Specifications.

22   Install the coolant outlet elbow and the coolant outlet pipe on the cylinder head. Be sure to coat the threads of the forward coolant outlet pipe bolt with thread sealant. Tighten all fasteners securely.

23   Install the camshafts and lifters (see Section 9), the timing chains (see Section 8), the timing chain cover (see Section 7), the oil pan (see Section 11), the intake manifold (see Section 5) and the camshaft cover (see Section 4).

24   Install the exhaust manifold and, if equipped, the exhaust manifold support bracket (see Section 6). Reattach the exhaust pipe to the exhaust manifold (see Section 6 in this Chapter and Section 19 in Chapter 4).

25   Reattach the coolant pipe to the intake manifold.

26   Install the intake manifold rear support bracket (see Section 5).

27   Reconnect the following hoses:

a) *Brake booster hose from the intake manifold*

b) *Vacuum hose from the intake manifold*

c) *Canister purge hose from the EVAP canister*

d) *Coolant hose from the bypass pipe*

e) *Fuel supply hose and return hose from their respective pipes*

f) *Heater hose from the heater outlet pipe*

g) *Radiator inlet hose from coolant outlet pipe*

28   On vehicles with an automatic transmis-

sion, reconnect the TV cable (see Chapter 7B).

29   Reconnect the throttle cable (see Chapter 4).

30   Reconnect the following electrical harnesses and/or connectors:

a) *EGR valve connector*

b) *IAC valve connector*

c) *TP sensor connector*

d) *EVAP canister purge valve connector*

e) *Oxygen sensor connector*

f) *CMP sensor connector*

g) *ECT sensor connector*

h) *Fuel injector wire harness connector*

i) *Ground wire*

j) *Ignition coil connectors*

k) *All wiring harness brackets, clamps and clips*

31   Install the air intake duct between the air cleaner housing and the throttle body (see Chapter 4).

32   Install the strut tower bar (see Chapter 10).

33   Refill the engine with oil and the cooling system with coolant (see Chapter 1).

34   Reconnect the negative battery cable.

35   Start the engine and readjust the ignition timing, if necessary (see Chapter 5). Run the engine until normal operating temperature is reached. Check for leaks and proper operation.

11   Oil pan - removal and installation

Removal

Refer to illustrations 11.7a and 11.7b

1   Loosen the front wheel lug nuts. Raise the front of the vehicle and support it securely on jackstands placed under the frame. Remove the front wheels.

2   Drain the engine oil (see Chapter 1) and remove the dipstick.

3   On 4WD models, remove the front differential assembly (see Chapter 8).

4   Remove the transmission support bracket, if equipped (see Chapter 7).

5   Remove the clutch/torque converter access plate (see Chapter 7).

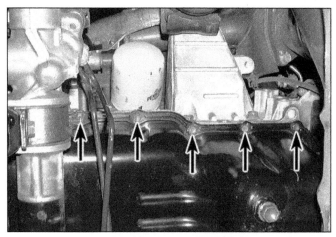

11.7a Oil pan bolts (arrows), left side (front and rear bolts not visible in this photo)

11.7b Oil pan bolts (arrows), right side (front and rear bolts not visible in this photo)

6   Disconnect the engine from the motor mounts (see Section 15) and raise it an inch or two with an engine hoist.

7   Remove the oil pan retaining bolts **(see illustrations)** and detach the oil pan. If the pan is stuck, use a soft-face hammer to dislodge the pan. Don't try to pry off the pan. Inserting a pry bar between the block and pan will damage the sealing surface. **Note:** *If the oil pan will not slide out after it's detached, try to raise the engine higher. If it still won't come out, reach inside and remove the pick-up tube bolts, dropping the tube into the pan.*

Installation

8   Use a scraper to remove all traces of old gasket material and sealant from the block and oil pan. Clean the gasket sealing surfaces with lacquer thinner or acetone and make sure the bolt holes in the block are clean.

9   Check the oil pan flange for distortion, particularly around the bolt holes. If necessary, place the pan on a block of wood and use a hammer to flatten and restore the gasket surface.

10   Using two new O-rings, install the oil pump pick-up tube (if removed) and tighten

the pick-up tube retaining bolts to the torque listed in this Chapter's Specifications.

11   To facilitate installation of the oil pan, remove the strut tower bar (see Chapter 10), disconnect the exhaust pipe from the exhaust manifold (see Chapter 4), remove the engine mounts (see Section 15) and then, using an engine hoist, raise the engine about one inch. **Caution:** *Do NOT raise the engine more than one inch or you might break wiring or hoses or damage something on the engine or transmission.*

12   Before installing the oil pan, apply a thin coat of RTV sealant to the flange. Attach the new gasket to the pan (make sure the bolt holes are aligned).

13   Position the oil pan against the engine block and install the mounting bolts. Tighten them to the torque listed in this Chapter's Specifications in a criss-cross pattern.

14   Lower the engine and reattach it to the motor mounts (see Section 15).

15   Install the clutch/torque converter access plate (see Chapter 7).

16   Install the transmission support bracket, if equipped (see Chapter 7).

17   On 4WD models, install the front differential assembly (see Chapter 8).

18   Refill the engine with oil (see Chapter 1) and insert the dipstick into the dipstick tube. **Note:** *Wait at least 30 minutes after the pan has been installed before filling the engine with oil.*

19   Install the front wheels and then lower the vehicle to the ground.

20   Start the engine and check the pan for leaks.

12 Oil pump - removal, inspection and installation

Removal

Refer to illustrations 12.2, 12.3a, 12.3b and 12.4

1   Remove the oil pan (see Section 11).

2   Remove the oil pump pickup tube **(see illustration)**.

3   Remove the oil pump sprocket cover bolts **(see illustration)** and then remove the sprocket cover. Loosen the oil pump drive chain guide nuts **(see illustration)** and slide the guide away from the chain to put some slack in the chain.

4   Remove the oil pump mounting bolts

12.2 Remove the bolts (arrows) and remove the oil pick-up tube

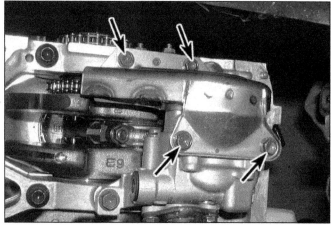

12.3a To detach the oil pump sprocket cover, remove these four bolts (arrows)

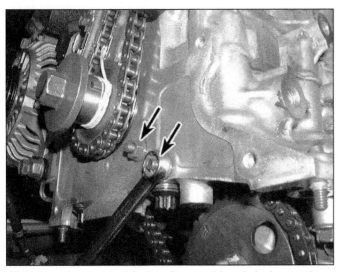

12.3b To put some slack in the oil pump drive chain, loosen these two nuts (arrows) that secure the chain guide

12.4 To detach the oil pump, remove these three bolts (arrows)

12.5a To separate the pump housing halves, remove five bolts (arrows) (fifth bolt not visible in this photo)

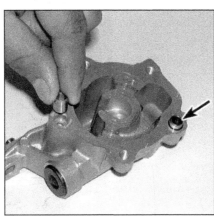

12.5b Remove the two dowel pins

12.5c Remove the outer rotor (the inner rotor is keyed onto the shaft and need not be removed)

(see illustration) and then remove the oil pump. **Caution:** *Do NOT try to remove the sprocket from the oil pump. Doing so might damage the oil pump center shaft.*

Inspection

Refer to illustrations 12.5a, 12.5b, 12.5c, 12.5d, 12.5e, 12.5f, 12.5g and 12.7

5   Disassemble the pump **(see illustrations)**.

6   Clean all parts thoroughly and remove all traces of old gasket material from the sealing surfaces. Visually inspect all parts for wear, cracks and other damage. Replace parts as necessary.

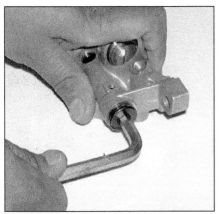

12.5d Using an Allen key, remove the relief spring retainer

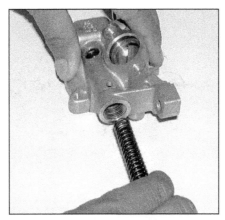

12.5e Remove the relief spring

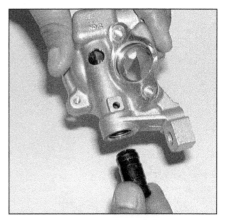

12.5f Remove the relief valve

7   Install the outer oil pump rotor and, using a feeler gauge, measure the clearance between the outer rotor and the case (see Chapter 2A), and between the rotor faces and a straightedge **(see illustration)**. Compare the indicated clearances to the clearances listed in this Chapter's Specifications.

8   Measure the free length of the relief valve spring and compare your measurement to this Chapter's Specifications. If the spring is too short, replace it.

9   Pack the pump cavity with petroleum jelly, install the rotors and reassemble the pump housing halves. Tighten the pump housing Allen bolts to the torque listed in this Chapter's Specifications.

10   Install the relief valve, relief spring and retainer. Tighten the retainer to the torque listed in this Chapter's Specifications.

11   Inspect the oil pump drive chain rollers and plates for excessive wear and damage. If the chain is worn or damaged, replace it. To remove the chain, you'll have to remove the crankshaft (see Chapter 2C).

Installation

Refer to illustration 12.13

12   With the oil pump drive chain installed, but not tensioned, install the oil pump on the block, install the pump mounting bolts and tighten them to the torque listed in this Chapter's Specifications.

13   To set the tension of the oil pump drive chain, use a fish scale or some other suitable device that can measure pulling force. Loosen the two chain guide nuts **(see illustration 12.3b)**, pull on the guide with the scale **(see illustration)** until the indicated pulling force on the scale is within the range of 4.4 to 6.6 pounds of force, and then tighten the guide nuts securely.

14   Install the pump sprocket cover and tighten the cover bolts to the torque listed in this Chapter's Specifications.

15   Install the oil pick-up tube and the oil

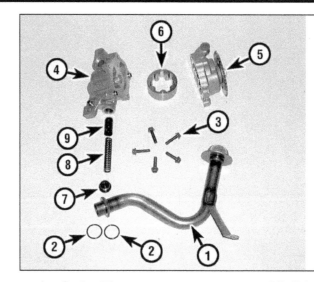

12.5g An exploded view of the oil pump assembly:

1   *Oil pump pickup tube (clean thoroughly and then inspect the strainer and tube for obstructions)*
2   *Pickup tube O-rings (always replace)*
3   *Oil pump housing bolts*
4   *Relief valve half of oil pump housing*
5   *Sprocket half of oil pump housing*
6   *Outer rotor*
7   *Retainer*
8   *Relief spring*
9   *Relief valve*

pan (see Section 11).

16   Add oil, start the engine and check for oil pressure and leaks.

13 Flywheel/driveplate - removal and installation

1   Raise the vehicle and support it securely on jackstands, then refer to Chapter 7 and remove the transmission.

2   On vehicles with a manual transmission, remove the pressure plate and clutch disc (see Chapter 8). Now is a good time to check/replace the clutch components and pilot bearing. If an automatic transmission is leaking, this would be a very good time to replace the front pump seal/O-ring.

3   On driveplate-equipped models, mark the relationship between the driveplate and crankshaft to ensure correct alignment during reinstallation.

4   Remove the bolts that secure the flywheel/driveplate to the crankshaft (see Chap-

ter 2A). If the crankshaft turns on flywheel-equipped models, immobilize it by reinstalling two of the pressure plate bolts halfway and holding a prybar between them. On driveplate-equipped models, wedge a screwdriver through the starter opening and into the ring gear teeth to immobilize the crankshaft.

5   Remove the flywheel/driveplate from the crankshaft. Since the flywheel is fairly heavy, be sure to support it while removing the last bolt.

6   Using brake system cleaner, clean the flywheel to remove grease and oil. Inspect the surface for cracks, rivet grooves, burned areas and score marks. Light scoring can be removed with emery cloth. Check for cracked and broken ring gear teeth. Lay the flywheel on a flat surface and use a straightedge to check for warpage.

7   Clean and inspect the mating surfaces of the flywheel/driveplate and the crankshaft. If the crankshaft rear seal is leaking, replace it before reinstalling the flywheel/driveplate (see Section 15).

12.7 To measure the axial play (rotor endplay), lay a straightedge across the pump housing half as shown and measure the clearance between the rotor faces and the straightedge with a feeler gauge

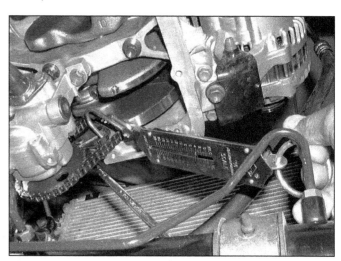

12.13 To correctly tension the oil pump drive chain, pull on the chain guide with a fish scale or some other suitable tool that can measure pulling force, and then tighten the two chain guide retaining nuts

8   Position the flywheel/driveplate against the crankshaft. On flywheel-equipped models, be sure to align the hole in the flywheel with the alignment dowel in the crankshaft **(see illustrations 15.8a and 15.8b in Chapter 2A)**. On driveplate-equipped models, align the marks you made during removal. Before installing the bolts, apply thread locking compound to the threads.

9   Tighten the bolts to the torque listed in this Chapter's Specifications. Hold the crankshaft from turning as described previously.

10   The remainder of installation is the reverse of the removal procedure.

14 Crankshaft rear oil seal - replacement

1   The transmission must be removed from the vehicle for this procedure (see Chapter 7).

2   The seal can be replaced without dropping the oil pan or removing the seal retainer. However, this method is not recommended because the lip of the seal is quite stiff and it's possible to cock the seal in the retainer bore or damage it during installation. If you want to take the chance, pry out the old seal (see Chapter 2A). Apply multi-purpose grease to the crankshaft seal journal and the lip of the new seal and carefully tap the new seal into place (see Chapter 2A). The lip is stiff so carefully work it onto the seal journal of the crankshaft with a smooth object like the end of an extension as you tap the seal into place. Don't rush it or you may damage the seal.

3   The following method is recommended but requires removal of the oil pan (see Section 13) and the seal retainer.

4   After the oil pan has been removed, remove the bolts, detach the seal retainer and peel off all the old gasket material.

5   Position the seal and retainer assembly on a couple of wood blocks on a workbench and drive the old seal out from the back side with a punch and hammer (see Chapter 2A).

6   Drive the new seal into the retainer with a block of wood (see Chapter 2A) or a section of pipe slightly smaller in diameter than the outside diameter of the seal.

7   Lubricate the crankshaft seal journal and the lip of the new seal with multi-purpose grease. Position a new gasket on the engine block.

8   Slowly and carefully push the seal onto the crankshaft. The seal lip is stiff, so work it onto the crankshaft with a smooth object such as the end of an extension as you push the retainer against the block.

9   Install and tighten the retainer bolts securely.

10   The remaining steps are the reverse of removal.

11   Run the engine and check for oil leaks.

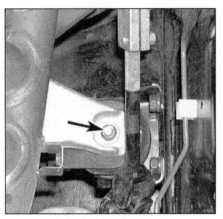

15.8a Left engine mount upper nut (arrow)

15.8a Right engine mount upper nut (arrow)

15.8b Left engine mount lower nuts (arrows)

15.8b Right engine mount lower nuts (arrows)

15 Engine mounts - check and replacement

1   Engine mounts seldom require attention, but broken or deteriorated mounts should be replaced immediately or the added strain placed on the driveline components may cause damage or wear.

Check

2   During the check, the engine must be raised slightly to remove the weight from the mounts.

3   Raise the vehicle and support it securely on jackstands, then position a jack under the engine bellhousing. Place a large block of wood between the jack head and the bellhousing, then carefully raise the engine just enough to take the weight off the mounts. **Warning:** *DO NOT place any part of your body under the engine when it's supported only by a jack!*

4   Check the mounts to see if the rubber is cracked, hardened or separated from the metal plates. Sometimes the rubber will split

right down the center.

5   Check for relative movement between the mount plates and the engine or frame (use a large screwdriver or pry bar to attempt to move the mounts). If movement is noted, lower the engine and tighten the mount fasteners.

6   Rubber preservative should be applied to the mounts to slow deterioration.

Replacement

Refer to illustrations 15.8a, 15.8b, 15.8c and 15.8d

7   Disconnect the negative battery cable from the battery, then raise the vehicle and support it securely on jackstands (if not already done).

8   Remove the upper and lower nuts **(see illustrations)** that attach the mount to the bracket.

9   Raise the engine slightly with a jack or hoist (make sure the fan doesn't hit the radiator or shroud). Carefully remove the mount(s).

10   Installation is the reverse of removal. Use thread locking compound on the mount bolts and be sure to tighten them securely.

Notes

Chapter 2 Part C
General engine overhaul procedures

Contents

Specifications

1.3L and 1.6L SOHC engines

General

Cylinder compression pressure (at 400 rpm)	
Standard	199 psi
Minimum	170 psi
Maximum allowable variation between cylinders	14.2 psi
Oil pressure (engine warm)	42 to 60 psi at 3000 rpm

Cylinder head

Warpage service limit	0.002 in (0.05 mm)

Camshaft

8-valve engine

Camshaft endplay (service limit)	0.0295 in (0.75 mm)
Cam lobe height (intake and exhaust)	
Standard	1.4763 in (37.5 mm)
Service limit	1.4724 in (37.4 mm)
Camshaft journal oil clearance	
Standard	0.002 to 0.0035 in (0.05 to 0.09 mm)
Service limit	0.006 in (0.15 mm)
Camshaft journal diameter (front to rear)	
1	1.7372 to 1.7381 in (44.125 to 44.150 mm)
2	1.7451 to 1.7460 in (44.325 to 44.350 mm)
3	1.7530 to 1.7539 in (44.525 to 44.550 mm)
4	1.7609 to 1.7618 in (44.725 to 44.750 mm)
5	1.7687 to 1.7697 in (44.925 to 44.950 mm)
Camshaft journal bore (inside) diameter	
1	1.7402 to 1.7407 in (44.200 to 44.216 mm)
2	1.7480 to 1.7486 in (44.400 to 44.416 mm)
3	1.7560 to 1.7565 in (44.600 to 44.616 mm)
4	1.7638 to 1.7644 in (44.800 to 44.816 mm)
5	1.7716 to 1.7723 in (45.000 to 45.016 mm)
Camshaft runout service limit	0.0039 inch

1.3L and 1.6L SOHC engines (continued)

16-valve engine

Camshaft lobe height
    1993 and earlier
        Intake
            Standard .... 1.4130 to 1.4192 in (35.888 to 36.048 mm)
            Service limit .... 1.4090 in (35.788 mm)
        Exhaust
            Standard .... 1.4300 to 1.4362 in (36.322 to 36.482 mm)
            Service limit .... 1.4261 in (36.222 mm)
    1994 through 1998
        Intake
            Standard .... 1.4241 to 1.4303 in (36.151 to 36.331 mm)
            Service limit .... 1.4202 in (36.071 mm)
        Exhaust
             Standard .... 1.4314 to 1.4376 in (36.356 to 36.516 mm)
            Service limit .... 1.4275 in (36.256 mm)
    1999 on
        Intake
            Standard .... 1.4246 to 1.4308 in (36.184 to 36.344 mm)
            Service limit .... 1.4206 in (36.084 mm)
        Exhaust
             Standard .... 1.4134 to 1.4197 in (35.900 to 36.060 mm)
            Service limit .... 1.4094 in (35.800 mm)
Camshaft journal oil clearance
    Standard.. 0.0016 to 0.0032 in (0.040 to 0.082 mm)
    Service limit .... 0.0047 in (0.12 mm)
Camshaft journal diameter .... 1.1000 to 1.1008 in (27.939 to 27.960 mm)
Camshaft journal bore (inside) diameter...................................... 1.1024 to 1.1031 in (28.000 to 28.021 mm)
Camshaft runout service limit... 0.0039 in (0.10 mm)

Valves and related components

8-valve engine

Valve margin width
    Intake
        Standard .... 0.039 in (1.0 mm)
        Service limit... 0.0236 in (0.6 mm)
    Exhaust
        Standard .... 0.039 in (1.0 mm)
        Service limit... 0.0275 in (0.7 mm)
Valve stem diameter
    Intake.. 0.2742 to 0.2748 in (6.965 to 6.980 mm)
    Exhaust... 0.2737 to 0.2742 in (6.950 to 6.965mm)
Valve stem-to-guide clearance
    Intake
        Standard .... 0.0008 to 0.0019 in (0.02 to 0.05 mm)
        Service limit... 0.0027 in (0.07 mm)
    Exhaust
        Standard .... 0.0014 to 0.0025 in (0.035 to 0.65 mm)
        Service limit... 0.0035 in (0.09 mm)
Valve spring
    Out-of-square limit .... 0.079 in (2.0 mm)
    Installed height .... 1.63 in (41.5 mm)
    Free length
        1986 and 1987
            Standard .... 1.9409 in (49.3 mm)
            Service limit .... 1.8937 in (48.1 mm)
        1988 on
            Standard .... 1.9866 in (50.46 mm)
            Service limit .... 1.9094 in (48.50 mm)
    Pressure
        Standard .... 54.7 to 64.3 lbs at 1.63 in (24.8 to 29.2 Kg at 41.5 mm)
        Service limit... 50.2 lbs at 1.63 in (22.8 Kg at 41.5 mm)

16-valve engine

Valve margin width
    Intake
        Standard .... 0.03 to 0.047 in (0.8 to 1.2 mm)
        Service limit... 0.024 in (0.6 mm)

Exhaust
    Standard .... 0.03 to 0.047 in (0.8 to 1.2 mm)
    Service limit .... 0.027 in (0.7 mm)
Valve stem diameter
    Intake .... 0.2152 to 0.2157 in (5.465 to 5.480 mm)
    Exhaust .... 0.2142 to 0.2148 in (5.440 to 5.455 mm)
Valve guide inner diameter (intake and exhaust) .... 0.2166 to 0.2170 in (5.500 to 5.512 mm)
Valve stem-to-guide clearance
    Intake
        Standard .... 0.0008 to 0.0018 in (0.02 to 0.05 mm)
        Service limit.. 0.0027 in (0.07 mm)
    Exhaust
        Standard .... 0.0018 to 0.0028 in (0.045 to 0.72 mm)
        Service limit.. 0.0035 in (0.09 mm)
Valve spring
    Out-of-square limit .... 0.079 in (2.0 mm)
    Free length
        Standard .... 1.4500 in (36.67 mm)
        Service limit.. 1.4043 in (35.67 mm)
    Pressure
        Standard .... 23.6 to 27.5 lbs at 1.24 in (10.7 to 12.5 kg at 31.5 mm)
        Service limit.. 20.5 lbs at 1.24 in (9.3 kg at 31.5 mm)

Crankshaft and connecting rods

Crankshaft endplay
    Standard.. 0.0044 to 0.0122 in (0.11 to 0.31 mm)
    Service limit .... 0.0149 in (0.38 mm)
Crankshaft runout limit (at middle).. 0.0023 in (0.06 mm)
Connecting rod journal
    Diameter
        1.3L engine .... 1.6529 to 1.6535 in (41.982 to 42.000 mm)
        1.6L engine .... 1.7316 to 1.7322 in (43.982 to 44.000 mm)
    Out-of-round/taper limits .... 0.0004 in (0.01 mm)
    Bearing oil clearance
        1.3L engine .... 0.0012 to 0.0019 in (0.030 to 0.050 mm)
        1.6L engine .... 0.0008 to 0.0019 in (0.020 to 0.050 mm)
        Service limit.. 0.0031 in (0.08 mm)
Connecting rod endplay (side clearance)
    Standard
        1992 through 1998.. 0.0039 to 0.0078 in (0.10 to 0.20 mm)
        1999 on... 0.0039 to 0.0098 in (0.10 to 0.25 mm)
    Service limit.. 0.0137 in (0.35 mm)
Main bearing journal
    Out-of-round/taper limits .... 0.0004 in (0.01 mm)
    Bearing oil clearance
        Standard
            1992 through 1998 .... 0.0008 to 0.0016 in (0.02 to 0.04 mm)
            1999 on.. 0.0006 to 0.0014 in (0.016 to 0.0014 mm)
        Service limit.. 0.0023 in (0.06 mm)
    Main bearing journal diameters (numbers stamped on crankshaft webs No. 2 and No. 3 (adjacent to No. 3 main bearing cap)
        Sidekick and Tracker models
            No. 1 .... 2.0470 to 2.0472 in (51.994 to 52.000 mm)
            No. 2 .... 2.0468 to 2.0470 in (51.988 to 51.994 mm)
            No. 3 .... 2.0465 to 2.0468 in (51.982 to 51.988 mm)
        Samurai models
            No. 1 .... 1.7714 to 1.7716 in (44.994 to 45.000 mm)
            No. 2 .... 1.7712 to 1.7714 in (44.988 to 44.994 mm)
            No. 3 .... 1.7710 to 1.7712 in (44.982 to 44.988 mm)
    Main bearing bore sizes (markings stamped on oil pan mating surface of block)
        Sidekick and Tracker models
            Marking "A" .... 2.2047 to 2.2050 in (56.000 to 56.006 mm)
            Marking "B" .... 2.2050 to 2.2052 in (56.006 to 56.012 mm)
            Marking "C" .... 2.2052 to 2.2054 in (56.012 to 56.018 mm)
        Samurai models
            Marking "A" .... 1.9292 to 1.9294 in (49.000 to 49.006 mm)
            Marking "B" .... 1.9294 to 1.9296 in (49.006 to 49.012 mm)
            Marking "C" .... 1.9296 to 1.9298 in (49.012 to 49.018 mm)

1.3L and 1.6L SOHC engines (continued)

Main bearing thicknesses (standard size bearing color codes)

1988 through 1998

Green	0.0786 to 0.0787 in (1.996 to 2.000 mm)
Black	0.0787 to 0.0788 in (1.999 to 2.003 mm)
Colorless	0.0788 to 0.0789 in (2.002 to 2.006 mm)
Yellow	0.0789 to 0.0790 in (2.005 to 2.009 mm)
Blue	0.0790 to 0.0791 in (2.008 to 2.012 mm)

1999 on

Green	0.0787 to 0.0788 in (1.998 to 2.002 mm)
Black	0.0788 to 0.0789 in (2.001 to 2.005 mm)
Colorless	0.0789 to 0.0790 in (2.004 to 2.008 mm)
Yellow	0.0790 to 0.0791 in (2.007 to 2.011 mm)
Blue	0.0791 to 0.0792 in (2.010 to 2.014 mm)

Main bearing thicknesses (undersize bearing color codes)

1988 through 1998

Green and red	0.0835 to 0.0836 in (2.121 to 2.125 mm)
Black and red	0.0836 to 0.0837 in (2.124 to 2.128 mm)
Red only	0.0837 to 0.0838 in (2.127 to 2.131 mm)
Yellow and red	0.0838 to 0.0839 in (2.130 to 2.134 mm)
Blue and red	0.0839 to 0.0840 in (2.133 to 2.137 mm)

1999 on

Green and red	0.0836 to 0.0837 in (2.123 to 2.127 mm)
Black and red	0.0837 to 0.0838 in (2.126 to 2.130 mm)
Red only	0.0838 to 0.0839 in (2.129 to 2.133 mm)
Yellow and red	0.0839 to 0.0840 in (2.132 to 2.136 mm)
Blue and red	0.0840 to 0.0841 in (2.135 to 2.139 mm)

Cylinder bore

Diameter

1300 cc engine

Marking "1"	2.9138 to 2.9142 in (74.01 to 74.02 mm)
Marking "2"	2.9134 to 2.9138 in (74.01 to 74.02 mm)

1600 cc engine

Marking "1"	2.9531 to 2.9535 in (75.01 to 75.02 mm)
Marking "2"	2.9528 to 2.9531 in (75.00 to 75.01 mm)
Out-of-round/taper limits	0.0039 in (0.10 mm)

Pistons and rings

Piston diameter*

1300 cc engine

Marking "1"	2.9126 to 2.9130 in (73.98 to 73.99 mm)
Marking "2"	2.9122 to 2.9126 in (73.97 to 73.98 mm)

1600 cc engine

Marking "1"	2.9520 to 2.9524 in (74.98 to 74.99 mm)
Marking "2"	2.9516 to 2.9520 in (74.97 to 74.98 mm)
Piston-to-bore clearance	0.0008 to 0.0015 in (0.02 to 0.04 mm)

Piston ring end gap

Standard

1992 through 1998

Top and 2nd rings	0.0079 to 0.0137 in (0.20 to 0.35 mm)
Oil ring	0.0079 to 0.0275 in (0.2 to 0.7 mm)

1999 on

Top ring	0.0079 to 0.0137 in (0.20 to 0.35 mm)
2nd ring	0.0138 to 0.0197 in (0.35 to 0.50 mm)
Oil ring	0.0039 to 0.0157 in (0.10 to 0.40 mm)

Service limit

Top and 2nd rings	0.0275 in (0.7 mm)
Oil ring	0.0669 in (1.7 mm)

Piston ring side clearance

No. 1 (top) compression ring	0.0012 to 0.0027 in (0.03 to 0.07 mm)
No. 2 compression ring	0.0008 to 0.0023 in (0.02 to 0.06 mm)

Measured 0.63-inch (16 mm) up from bottom of skirt

Torque specifications**

Ft-lbs (unless otherwise indicated)

Distributor gear case bolts	70 to 86 in-lbs
Main bearing cap bolts	36.5 to 41
Connecting rod cap nuts	24 to 26.5
Rear oil seal housing bolts	84 to 102 in-lbs

** **Note:** *Refer to Part A for additional torque specifications.*

2.0L DOHC engine

Cylinder head

Warpage service limit	0.002 in (0.05 mm)

Camshafts

Camshaft lobe height
  Intake
    Standard .... 1.5906 to 1.5969 in (40.402 to 40.562 mm)
    Service limit .... 1.5827 in (40.202 mm)
  Exhaust
    Standard .... 1.5717 to 1.5780 in (39.921 mm)
    Service limit... 1.5638 in (39.721 mm)
Camshaft journal oil clearance
  Standard.. 0.0039 to 0.0018 in (0.045 to 0.099 mm)
  Service limit .... 0.0047 in (0.12 mm)
Camshaft journal diameter .... 1.0210 to 1.0218 in (25.934 to 25.955 mm)
Camshaft journal bore (inside) diameter................................... 1.0236 to 1.0249 in (26.000 to 26.033 mm)
Camshaft runout service limit .... 0.0039 in (0.10 mm)
Lifters
  Lifter outside diameter .... 1.2189 to 1.2194 in (30.959 to 30.975 mm)
  Lifter bore inside diameter .... 1.2205 to 1.2214 in (31.000 to 31.025 mm)
  Lifter-to-bore clearance
    Standard .... 0.0010 to 0.0025 in (0.025 to 0.066 mm)
    Limit .... 0.0059 in (0.15 mm)

Valves and related components

Valve margin width
  Intake
    Standard .... 0.039 in (1.0 mm)
    Service limit .... 0.024 in (0.6 mm)
  Exhaust
    Standard .... 0.047 in (1.2 mm)
    Service limit... 0.027 in (0.7 mm)
Valve stem diameter
  Intake.. 0.2348 to 0.2354 in (5.965 to 5.980 mm)
  Exhaust.. 0.2339 to 0.2344 in (5.940 to 5.955 mm)
Valve guide inner diameter (intake and exhaust)..................... 0.2362 to 0.2366 in (6.000 to 6.012 mm)
Valve stem-to-guide clearance
  Intake
    Standard .... 0.0008 to 0.0018 in (0.020 to 0.047 mm)
    Service limit... 0.0027 in (0.07 mm)
  Exhaust
    Standard .... 0.0018 to 0.0028 in (0.045 to 0.72 mm)
    Service limit... 0.0035 in (0.09 mm)
Valve spring
  Out-of-square limit .... 0.079 in (2.0 mm)
  Free length
    Inner
      Standard .... 1.4204 in (36.08 mm)
      Limit .... 1.3780 in (35.00 mm)
    Outer
      Standard .... 1.5921 in (40.44 mm)
      Limit .... 1.5441 in (39.22 mm)
  Pressure
    Inner
      Standard .... 15.2 to 17.4 lbs at 1.08 in (6.9 to 7.9 kg at 27.5 mm)
      Limit .... 13.6 lbs at 1.08 in (6.2 kg at 27.5 mm)
    Outer
      Standard .... 33.9 to 39.2 lbs at 1.25 in (15.4 to 17.8 kg at 31.7 mm)
      Limit .... 30.4 lbs at 1.25 in (13.8 kg at 31.7 mm)

Crankshaft and connecting rods

Crankshaft endplay
  Standard.. 0.0039 to 0.0138 in (0.10 to 0.35 mm)
  Service limit .... 0.0165 in (0.42 mm)
Crankshaft runout limit (at middle).. 0.0023 in (0.06 mm)
Connecting rod journal
  Diameter .... 1.9678 to 1.9685 in (49.982 to 50.000 mm)
  Out-of-round/taper limits .... 0.0004 in (0.01 mm)

2.0L DOHC engine (continued)

Crankshaft and connecting rods (continued)

Bearing oil clearance
    Standard .... 0.0016 to 0.0022 in (0.039 to 0.057 mm)
    Service limit... 0.0031 in (0.08 mm)
Connecting rod endplay (side clearance)
    Standard... 0.0099 to 0.0157 in (0.25 to 0.40 mm)
    Service limit .... 0.0177 in (0.45 mm)
Main bearing journal
    Out-of-round/taper limits .... 0.0004 in (0.01 mm)
    Bearing oil clearance
        Standard .... 0.0013 to 0.0020 in (0.013 to 0.0020 mm)
        Service limit.. 0.0023 in (0.060 mm)
    Main bearing journal diameters (numbers stamped on crankshaft webs adjacent to No. 2 main bearing cap)
        No. 1 .... 2.2832 to 2.2834 in (57.994 to 58.000 mm)
        No. 2 .... 2.2830 to 2.2832 in (57.988 to 57.994 mm)
        No. 3 .... 2.2828 to 2.2829 in (57.982 to 57.988 mm)
    Main bearing bore sizes
        For bearing cap Nos. 1, 2, 3 and 5 (stamped on caps)
            Marking "A" .... 2.4409 to 2.4411 in (62.000 to 62.006 mm)
            Marking "B" .... 2.4412 to 2.4414 in (62.006 to 62.012 mm)
            Marking "C" .... 2.4414 to 2.4416 in (62.012 to 62.018 mm)
        For bearing cap No. 4 (stamped on cap)
            Marking "A" .... 2.4411 to 2.4414 in (62.006 to 62.012 mm)
            Marking "B" .... 2.4414 to 2.4416 in (62.012 to 62.018 mm)
            Marking "C" .... 2.4416 to 2.4419 in (62.018 to 62.024 mm)
    Main bearing thicknesses (standard size bearing color codes)
        Green .... 0.0783 to 0.0785 in (1.990 to 1.994 mm)
        Black.. 0.0785 to 0.0786 in (1.993 to 1.997 mm)
        Colorless.. 0.0786 to 0.0787 in (1.996 to 2.000 mm)
        Yellow .... 0.0787 to 0.0788 in (1.999 to 2.003 mm)
        Blue... 0.0786 to 0.0789 in (2.002 to 2.006 mm)
    Main bearing thicknesses (undersize bearing color codes)
        Green and red .... 0.0833 to 0.0834 in (2.115 to 2.119 mm)
        Black and red.. 0.0834 to 0.0835 in (2.118 to 2.122 mm)
        Red only... 0.0835 to 0.0837 in (2.121 to 2.125 mm)
        Yellow and red .... 0.0837 to 0.0838 in (2.124 to 2.128 mm)
        Blue and red.. 0.0838 to 0.0839 in (2.127 to 2.131 mm)

Cylinder bore

Diameter
    Red paint mark next to cylinder .... 3.3075 to 3.3078 in (84.01 to 84.02 mm)
    Blue paint mark next to cylinder .... 3.3071 to 3.3074 in (84.00 to 884.01 mm)
    Service limit .... 3.3090 in (84.050 mm)
Out-of-round/taper limits... 0.0039 in (0.10 mm)

Pistons and rings

Piston diameter*
    Number "1" on top of piston .... 3.3063 to 3.3066 in (83.98 to 83.99 mm)
    Number "2" on top of piston .... 3.3059 to 3.3062 in (83.97 to 83.98 mm)
Piston-to-bore clearance.. 0.0008 to 0.0015 in (0.02 to 0.04 mm)
Piston ring end gap
    Standard
        Top ring.. 0.0079 to 0.0137 in (0.20 to 0.35 mm)
        2nd ring.. 0.0138 to 0.0196 in (0.35 to 0.50 mm)
        Oil ring... 0.0079 to 0.0275 in (0.20 to 0.70 mm)
    Service limit
        Top and 2nd rings... 0.0276 in (0.7 mm)
        Oil ring... 0.0709 in (1.8 mm)
Piston ring side clearance
    No. 1 (top) compression ring.. 0.0012 to 0.0027 in (0.03 to 0.07 mm)
    No. 2 compression ring .... 0.0008 to 0.0023 in (0.02 to 0.06 mm)

* Measured 1.04-inch (26.5 mm) up from bottom of skirt

Torque specifications**   Ft-lbs

Connecting rod cap nuts 33
Main bearing cap bolts
    Step 1 31
    Step 2 44

1   General information

Included in this portion of Chapter 2 are the general overhaul procedures for the cylinder head(s) and internal engine components.

The information ranges from advice concerning preparation for an overhaul and the purchase of replacement parts to detailed, step-by-step procedures covering removal and installation of internal engine components and the inspection of parts.

The following Sections have been written based on the assumption that the engine has been removed from the vehicle. For information concerning in-vehicle engine repair, as well as removal and installation of the external components necessary for the overhaul, see Part A or B of this Chapter and Section 7 of this Part.

The Specifications included in this Part are only those necessary for the inspection and overhaul procedures which follow. Refer to Part A or B for additional Specifications.

2   Engine overhaul - general information

Refer to illustrations 2.4a and 2.4b

It's not always easy to determine when, or if, an engine should be completely overhauled, as a number of factors must be considered.

High mileage is not necessarily an indication that an overhaul is needed, while low mileage doesn't preclude the need for an overhaul. Frequency of servicing is probably the most important consideration. An engine that's had regular and frequent oil and filter changes, as well as other required maintenance, will most likely give many thousands of miles of reliable service. Conversely, a neglected engine may require an overhaul very early in its life.

Excessive oil consumption is an indication that piston rings, valve seals and/or valve guides are in need of attention. Make sure that oil leaks aren't responsible before deciding that the rings and/or guides are bad. Perform a cylinder compression check to determine the extent of the work required (see Section 3).

Check the oil pressure with a gauge installed in place of the oil pressure sending unit **(see illustrations)** and compare it to this Chapter's Specifications. If it's extremely low, the bearings and/or oil pump are probably worn out.

Loss of power, rough running, knocking or metallic engine noises, excessive valve train noise and high fuel consumption rates may also point to the need for an overhaul, especially if they're all present at the same time. If a complete tune-up doesn't remedy the situation, major mechanical work is the only solution.

An engine overhaul involves restoring

2.4a On SOHC engines, the oil pressure sending unit is located between the oil filter and exhaust manifold (arrow)

the internal parts to the specifications of a new engine. During an overhaul, the piston rings are replaced and the cylinder walls are reconditioned (rebored and/or honed). If a rebore is done by an automotive machine shop, new oversize pistons will also be installed. The main bearings, connecting rod bearings and camshaft bearings are generally replaced with new ones and, if necessary, the crankshaft may be reground to restore the journals. Generally, the valves are serviced as well, since they're usually in less-than-perfect condition at this point.

While the engine is being overhauled, other components, such as the distributor, starter and alternator, can be rebuilt as well. The end result should be a like new engine that will give many trouble free miles. **Note:** *Critical cooling system components such as the hoses, drivebelts, thermostat and water pump MUST be replaced with new parts when an engine is overhauled. The radiator should be checked carefully to ensure that it isn't clogged or leaking (see Chapter 3). Also, check the oil pump very carefully as described in Chapter 2, Part A.*

Before beginning the engine overhaul, read through the entire procedure to familiarize yourself with the scope and requirements of the job. Overhauling an engine isn't difficult if you follow all of the instructions carefully, have the necessary tools and equipment and pay close attention to all specifications; however, it can be time consuming. Plan on the vehicle being tied up for a minimum of two weeks, especially if parts must be taken to an automotive machine shop for repair or reconditioning. Check on availability of parts and make sure that any necessary special tools and equipment are obtained in advance. Most work can be done with typical hand tools, although a number of precision measuring tools are required for inspecting parts to determine if they must be replaced. Often an automotive machine shop will handle the inspection of parts and offer advice concerning reconditioning and replacement. **Note:** *Always wait until the engine has been completely disassembled and all components,*

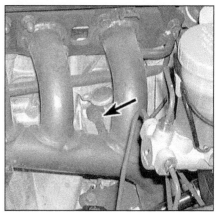

2.4b On DOHC engines, remove the exhaust manifold heat shield (see Chapter 2B) to access the oil pressure sending unit (arrow)

especially the engine block, have been inspected before deciding what service and repair operations must be performed by an automotive machine shop. Since the block's condition will be the major factor to consider when determining whether to overhaul the original engine or buy a rebuilt one, never purchase parts or have machine work done on other components until the block has been thoroughly inspected. As a general rule, time is the primary cost of an overhaul, so it doesn't pay to install worn or substandard parts.

As a final note, to ensure maximum life and minimum trouble from a rebuilt engine, everything must be assembled with care in a spotlessly clean environment.

3   Cylinder compression check

Refer to illustration 3.6

1   A compression check will tell you what mechanical condition the upper end (pistons, rings, valves, head gaskets) of your engine is in. Specifically, it can tell you if the compression is down due to leakage caused by worn piston rings, defective valves and seats or a blown head gasket. **Note:** *The engine must be at normal operating temperature and the battery must be fully charged for this check. Also, if the engine is equipped with a carburetor, the choke valve must be all the way open to get an accurate compression reading (if the engine's warm, the choke should be open).*

2   On DOHC engines, remove the ignition coils (see Chapter 5).

3   Clean the area around the spark plugs before you remove them (compressed air should be used, if available, otherwise a small brush or even a bicycle tire pump will work). The idea is to prevent dirt from getting into the cylinders as the compression check is being done. Then remove all of the spark plugs from the engine (see Chapter 1).

4   Block the throttle wide open.

5   On models with distributors, detach the coil wire from the center of the distributor cap

and ground it on the engine block. Use a jumper wire with alligator clips on each end to ensure a good ground. On EFI equipped vehicles, the fuel pump circuit should also be disabled (see Chapter 4).

6   Install the compression gauge in the number one spark plug hole **(see illustration)**.

7   Crank the engine over at least seven compression strokes and watch the gauge. The compression should build up quickly in a healthy engine. Low compression on the first stroke, followed by gradually increasing pressure on successive strokes, indicates worn piston rings. A low compression reading on the first stroke, which doesn't build up during successive strokes, indicates leaking valves or a blown head gasket (a cracked head could also be the cause). Deposits on the valve seats can also cause low compression. Record the highest gauge reading obtained.

8   Repeat the procedure for the remaining cylinders and compare the results to this Chapter's Specifications.

9   Add some engine oil (about three squirts from a plunger-type oil can) to each cylinder, through the spark plug hole, and repeat the test.

10   If the compression increases after the oil is added, the piston rings are definitely worn. If the compression doesn't increase significantly, the leakage is occurring at the valves or head gasket. Leakage past the valves may be caused by burned valve seats and/or faces or warped, cracked or bent valves.

11   If two adjacent cylinders have equally low compression, there's a strong possibility that the head gasket between them is blown. The appearance of coolant in the combustion chambers or the crankcase would verify this condition.

12   If one cylinder is 20 percent lower than the others, and the engine has a slightly rough idle, a worn exhaust lobe on the camshaft could be the cause.

13   If the compression is unusually high, the combustion chambers are probably coated with carbon deposits. If that's the case, the cylinder head(s) should be removed and decarbonized.

14   If compression is way down or varies greatly between cylinders, it would be a good idea to have a leak-down test performed by an automotive repair shop. This test will pinpoint exactly where the leakage is occurring and how severe it is.

---

4   Engine removal - methods and precautions

If you've decided that an engine must be removed for overhaul or major repair work, several preliminary steps should be taken.

Locating a suitable place to work is extremely important. Adequate workspace, along with storage space for the vehicle, will be needed. If a shop or garage isn't available, at the very least a flat, level, clean work sur-

3.6 A compression gauge with a threaded fitting for the spark plug hole is preferred over the type that requires hand pressure to maintain the seal - be sure to open the throttle and choke valves as far as possible during the compression check!

face made of concrete or asphalt is required.

Cleaning the engine compartment and engine before beginning the removal procedure will help keep tools clean and organized.

An engine hoist or A-frame will also be necessary. Make sure the equipment is rated in excess of the combined weight of the engine and accessories. Safety is of primary importance, considering the potential hazards involved in lifting the engine out of the vehicle.

If the engine is being removed by a novice, a helper should be available. Advice and aid from someone more experienced would also be helpful. There are many instances when one person cannot simultaneously perform all of the operations required when lifting the engine out of the vehicle.

Plan the operation ahead of time. Arrange for or obtain all of the tools and equipment you'll need prior to beginning the job. Some of the equipment necessary to remove and install an engine safely and with relative ease includes: an engine hoist, a heavy duty floor jack, suitable wrenches and sockets (described in the front of this manual), wooden blocks, and plenty of rags and cleaning solvent for mopping up spilled oil, coolant and gasoline. If the hoist must be rented, make sure that you arrange for it in advance and perform all of the operations possible without it beforehand. This will save you money and time.

Plan for the vehicle to be out of use for quite a while. A machine shop will be required to perform some of the work that the do-it-yourselfer can't accomplish without special equipment. These shops often have a busy schedule, so it would be a good idea to consult them before removing the engine in order to accurately estimate the amount of time required to rebuild or repair components that may need work.

Always be extremely careful when removing and installing the engine. Serious injury can result from careless actions. Plan ahead, take your time and a job of this

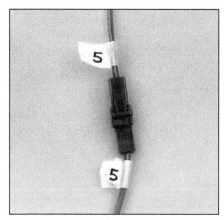

5.5 Label both ends of each wire before disconnecting them

nature, although major, can be accomplished successfully.

---

5   Engine - removal and installation

Removal

Refer to illustrations 5.5, 5.24a and 5.24b

Warning: *The air conditioning system is under high pressure! If it is necessary to disconnect air conditioning hoses for engine removal, first have a dealer service department or service station discharge the system. Carefully inspect the routing of air conditioning system refrigerant lines before beginning engine removal to see if line disconnection, and therefore professional discharging, is necessary.*

1   Refer to Chapter 4 and relieve the fuel system pressure (TBI and MPFI vehicles), then disconnect the negative cable from the battery.

2   Cover the fenders and cowl and remove the hood (see Chapter 11). Special pads are available to protect the fenders, but an old bedspread or blanket will also work.

3   Remove the air cleaner assembly (see Chapter 4).

4   Drain the cooling system (see Chapter 1).

5   Label the vacuum lines, emissions system hoses, wiring connectors, ground straps and fuel lines, to ensure correct reinstallation, then detach them. Pieces of masking tape with numbers or letters written on them work well **(see illustration)**. If there's any possibility of confusion, make a sketch of the engine compartment and clearly label the lines, hoses and wires.

6   Label and detach all coolant hoses from the engine (see Chapter 3).

7   Remove the cooling fan, shroud and radiator (see Chapter 3).

8   Remove the drivebelts (see Chapter 1).

9   Disconnect the fuel lines running from the engine to the chassis (see Chapter 4). Plug or cap all open fittings/lines. **Warning:** *Gasoline is extremely flammable, so extra precautions must be taken when working on*

5.24a Pull the engine forward as far as possible
to clear the transmission . . .

5.24b . . . then slowly raise the engine until it clears the body

any part of the fuel system. DO NOT smoke or allow open flames or bare light bulbs near the vehicle. Also, don't work in a garage if a natural gas appliance with a pilot light is present.

10 Disconnect the throttle linkage (see Chapter 4), the cruise control cable, if equipped, and, on automatics, the TV linkage (see Chapter 7B).

11 On power steering equipped vehicles, unbolt the power steering pump (see Chapter 10). Leave the lines/hoses attached and make sure the pump is kept in an upright position in the engine compartment (use wire or rope to restrain it out of the way).

12 On A/C equipped vehicles, unbolt the compressor (see Chapter 3) and set it aside. Do not disconnect the hoses unless it is necessary.

13 Drain the engine oil (see Chapter 1) and remove the filter.

14 Remove the starter motor (see Chapter 5).

15 Remove the alternator (see Chapter 5).

16 Unbolt the exhaust system from the engine (see Chapter 4).

17 If the vehicle is equipped with an automatic transmission, remove the torque converter-to-driveplate fasteners (see Chapter 7B).

18 Support the transmission with a jack. Position a block of wood between the transmission and the jack head to prevent damage to the transmission. Special transmission jacks with safety chains are available - use one if possible.

19 Attach an engine sling or a length of chain to the lifting brackets on the engine.

20 Roll the hoist into position and connect the sling to it. Take up the slack in the sling or chain, but don't lift the engine. **Warning:** *DO NOT place any part of your body under the engine when it's supported only by a hoist or other lifting device.*

21 Remove the transmission-to-engine block bolts.

22 Remove the engine mount-to-frame bolts.

23 Recheck to be sure nothing is still connecting the engine to the transmission or vehicle. Disconnect anything still remaining.

24 Raise the engine slightly. Carefully work it forward to separate it from the transmission **(see illustration)**. If you're working on a vehicle with an automatic transmission, be sure the torque converter stays in the transmission (clamp a pair of vise-grips to the housing to keep the converter from sliding out). If you're working on a vehicle with a manual transmission, the input shaft must be completely disengaged from the clutch. Slowly raise the engine out of the engine compartment **(see illustration)**. Check carefully to make sure nothing is hanging up.

25 Remove the flywheel/driveplate and mount the engine on an engine stand.

Installation

26 Check the engine and transmission mounts. If they're worn or damaged, replace them.

27 If you're working on a manual transmission equipped vehicle, install the clutch and pressure plate (see Chapter 8). Now is a good time to install a new clutch.

28 Carefully lower the engine into the engine compartment - make sure the engine mounts line up.

29 If you're working on an automatic transmission equipped vehicle, guide the torque converter into the crankshaft following the procedure outlined in Chapter 7.

30 If you're working on a manual transmission equipped vehicle, apply a dab of high-temperature grease to the input shaft and guide it into the crankshaft pilot bearing until the bellhousing is flush with the engine block.

31 Install the transmission-to-engine bolts and tighten them securely. **Caution:** *DO NOT use the bolts to force the transmission and engine together! If the engine and transmission will not mate correctly, make sure the clutch is correctly aligned (manual transmission, Chapter 8) or the torque converter is fully seated (automatic transmission, Chapter 7B)*

32 Reinstall the remaining components in the reverse order of removal.

33 Add coolant, oil, power steering and transmission fluid as needed.

34 Run the engine and check for leaks and proper operation of all accessories, then install the hood and test drive the vehicle.

35 Have the A/C system recharged and leak tested, if it was discharged.

6 Engine rebuilding alternatives

The do-it-yourselfer is faced with a number of options when performing an engine overhaul. The decision to replace the engine block, piston/connecting rod assemblies and crankshaft depends on a number of factors, with the number one consideration being the condition of the block. Other considerations are cost, access to machine shop facilities, parts availability, time required to complete the project and the extent of prior mechanical experience on the part of the do-it-yourselfer.

Some of the rebuilding alternatives include:

Individual parts - If the inspection procedures reveal that the engine block and most engine components are in reusable condition, purchasing individual parts may be the most economical alternative. The block, crankshaft and piston/connecting rod assemblies should all be inspected carefully. Even if the block shows little wear, the cylinder bores should be surface honed.

Short block - A short block consists of an engine block with a crankshaft and piston/connecting rod assemblies already installed. All new bearings are incorporated and all clearances will be correct. The existing camshaft, valve train components, cylinder head(s) and external parts can be bolted to the short block with little or no machine shop work necessary.

Long block - A long block consists of a short block plus an oil pump, oil pan, cylinder head(s), rocker arm cover(s), camshaft and valve train components, timing sprockets and chain or gears and timing cover. All components are installed with new bearings, seals and gaskets incorporated throughout. The installation of manifolds and external parts is all that's necessary.

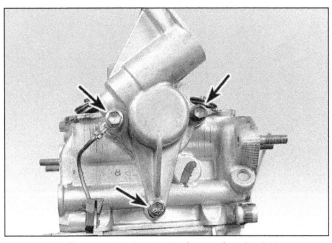

8.2 Remove the three bolts (arrows) and pull the distributor case off

8.3 Carefully slip the camshaft out of the rear (distributor end) of the cylinder head - guide the camshaft with your other hand and be careful not to damage the bearing surfaces in the head (8-valve engines)

Used Engine - While overhaul provides the best assurance of a like-new engine, used engines available from wrecking yards and importers are often a very simple and economical solution. Many used engines come with warranties., but always give any used engine a thorough diagnostic check-out before purchase. Check compression and also for signs of leakage. If possible, have the seller run the engine either in the vehicle or on a test stand so you can be sure it runs smoothly with no knocking or other noises.

Give careful thought to which alternative is best for you and discuss the situation with local automotive machine shops, auto parts dealers and experienced rebuilders before ordering or purchasing replacement parts.

7 Engine overhaul - disassembly sequence

1   It's much easier to disassemble and work on the engine if it's mounted on a portable engine stand. A stand can often be rented quite cheaply from an equipment rental yard. Before the engine is mounted on a stand, the flywheel/driveplate and rear oil seal housing should be removed from the engine.

2   If a stand isn't available, it's possible to disassemble the engine with it blocked up on the floor. Be extra careful not to tip or drop the engine when working without a stand.

3   If you're going to obtain a rebuilt engine, all external components must come off first, to be transferred to the replacement engine, just as they will if you're doing a complete engine overhaul yourself. These include:

Alternator and brackets
Emissions control components
Distributor (if equipped), ignition coil/igniter units (distributorless models), spark plug wires and spark plugs
Thermostat and housing cover
Water pump
Carburetor, TBI or SEFI components

Intake and exhaust manifolds
Oil filter
Engine mounts
Clutch and flywheel (manual transmission) or driveplate (automatics)
Engine rear plate

Note: *When removing the external components from the engine, pay close attention to details that may be helpful or important during installation. Note the installed position of gaskets, seals, spacers, pins, brackets, washers, bolts and other small items.*

4   If you're obtaining a short block, which consists of the engine block, crankshaft, pistons and connecting rods all assembled, then the cylinder head, oil pan and oil pump will have to be removed as well. See Engine rebuilding alternatives for additional information regarding the different possibilities to be considered.

5   If you're planning a complete overhaul, the engine must be disassembled and the internal components removed in the following order:

Camshaft cover
Intake and exhaust manifolds
Timing belt and sprockets
Rocker arms and shafts
Cylinder head
Oil pan
Oil pump
Piston/connecting rod assemblies
Crankshaft and main bearings

6   Before beginning the disassembly and overhaul procedures, make sure the following items are available. Also, refer to *Engine overhaul - reassembly sequence* for a list of tools and materials needed for engine reassembly.

Common hand tools
Small cardboard boxes or plastic bags for storing parts
Gasket scraper
Ridge reamer
Micrometers
Telescoping gauges
Dial indicator set

Valve spring compressor
Cylinder surfacing hone
Piston ring groove cleaning tool
Electric drill motor
Tap and die set
Wire brushes
Oil gallery brushes
Cleaning solvent

8 Cylinder head - disassembly

Note: *New and rebuilt cylinder heads are usually available for most engines at dealerships and auto parts stores. Because some specialized tools are needed to disassemble and inspect the cylinder head, and because some parts may not be available, it might be more practical and economical to buy a replacement head instead of disassembling, inspecting and reconditioning the old head.*

SOHC engines

Refer to illustrations 8.2, 8.3, 8.4, 8.5, 8.6a and 8.6b

1   Remove the rocker arms and shafts from the cylinder head (see Section 10 in Chapter 2A). Label the parts or store them

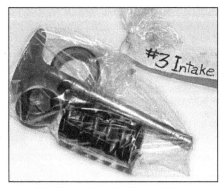

8.4 A small plastic bag, with an appropriate label, can be used to store the valve train components so they can be kept together and reinstalled in the original position

8.5 Use a valve spring compressor to compress the spring, then remove the keepers from the valve stem

8.6a The valve seals may be pulled off with pliers

8.6b If the valve won't pull through the guide, deburr the edge of the stem end and the area around the top of the keeper groove with a file

separately so they can be reinstalled in their original locations.

2   Remove the distributor case from the cylinder head **(see illustration)**.

3   On 8-valve engines, gently guide the camshaft out of the cylinder head **(see illustration)**. **Note:** *Refer to the* Rocker arms and shafts - removal and installation *procedure in part A for camshaft removal and installation on 16-valve engines.*

4   Before the valves are removed, arrange to label and store them, along with their related components, so they can be kept separate and reinstalled in the same valve guides they are removed from **(see illustration)**.

5   Compress the springs on the first valve with a spring compressor and remove the keepers **(see illustration)**. Carefully release the valve spring compressor and remove the retainer, the spring and the spring seat (if used).

6   Pull the valve out of the head, then remove the oil seal from the guide **(see illustration)**. If the valve binds in the guide (won't pull through), push it back into the head and deburr the area around the keeper groove

with a fine file or whetstone **(see illustration)**.

7   Repeat the procedure for the remaining valves. Remember to keep all the parts for each valve together so they can be reinstalled in the same locations.

8   Once the valves and related components have been removed and stored in an organized manner, the head should be thoroughly cleaned and inspected. If a complete engine overhaul is being done, finish the engine disassembly procedures before beginning the cylinder head cleaning and inspection process.

DOHC engines

Refer to illustration 8.10

9   Remove the camshafts and lifters (see Section 9 in Chapter 2B).

10   The remainder of disassembly is similar to the disassembly procedure for SOHC engines (see Steps 4 through 8) except that you'll need a different kind of valve spring compressor **(see illustration)**. **Caution:** *Be very careful not to nick or otherwise damage the lifter bores when compressing the valve springs.* **Note:** *If the upper (valve spring retainer) end of your spring compressor doesn't have cutouts on the side like the one shown in the accompanying illustration, you'll need to obtain an adapter (available at automotive parts stores).*

11   There are *two* valve springs per valve on DOHC engines. Be sure to keep each valve spring pair together. Also, note that each spring has a "top" and "bottom." The large-pitch end is the top of the spring; the small-pitch end is the bottom.

***

9   Cylinder head and camshaft - cleaning and inspection

***

1   Thorough cleaning of the cylinder head(s) and related valve train components, followed by a detailed inspection, will enable you to decide how much valve service work must be done during the engine overhaul. **Note:** *If the engine was severely overheated, the cylinder head is probably warped (see Step 12).*

Cleaning

2   Scrape all traces of old gasket material and sealing compound off the head gasket, intake manifold and exhaust manifold sealing surfaces. Be very careful not to gouge the cylinder head. Special gasket removal solvents can soften gaskets and make removal much easier; they're available at auto parts stores.

3   Remove all built up scale from the coolant passages.

4   Run a stiff wire brush through the various holes to remove deposits that may have formed in them.

5   Run an appropriate size tap into each of the threaded holes to remove corrosion and thread sealant that may be present. If compressed air is available, use it to clear the holes of debris produced by this operation. **Warning:** *Wear eye protection when using compressed air!*

6   Clean the rocker arm shaft oil holes with a wire and compressed air (if available). **Warning:** *Wear eye protection.*

7   Clean the cylinder head and with solvent and dry it thoroughly. Compressed air will speed the drying process and ensure that all holes and recessed areas are clean. **Note:** *Decarbonizing chemicals are available and may prove very useful when cleaning cylinder heads and valve train components. They are very caustic and should be used with caution. Be sure to follow the instructions on the container.*

8   Clean the rocker arms, rocker shafts and camshaft with solvent and dry them thoroughly (don't mix them up during the cleaning process). Compressed air will speed the drying process and can be used to clean out the oil passages.

9   Clean all the valve springs, spring seats, keepers and retainers with solvent and dry them thoroughly. Do the components from one valve at a time to avoid mixing up the parts.

10   Scrape off any heavy deposits that may have formed on the valves, then use a motorized wire brush to remove deposits from the

8.10 Using a spring compressor suitable for DOHC engines, compress the spring until the keepers can be removed with a small magnetic screwdriver or with needle-nose pliers

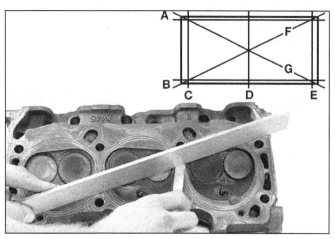

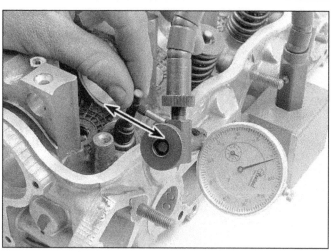

9.12 Check the cylinder head gasket surface for warpage by trying to slip a feeler gauge under the straightedge (see this Chapter's Specifications for the maximum warpage allowed and use a feeler gauge of that thickness)

9.14 A dial indicator can be used to determine the valve stem-to-guide clearance (move the valve stem as indicated by the arrows)

valve heads and stems. Again, make sure the valves don't get mixed up.

Inspection

Note: *Be sure to perform all of the following inspection procedures before concluding that machine shop work is required. Make a list of the items that need attention.*

Cylinder head

Refer to illustrations 9.12 and 9.14

11   Inspect the head very carefully for cracks, evidence of coolant leakage and other damage. If cracks are found, check with an automotive machine shop concerning repair. If repair isn't possible, a new cylinder head should be obtained.

12   Using a straightedge and feeler gauge, check the head gasket mating surface for warpage **(see illustration)**. If the warpage exceeds the limit listed in this Chapter's

Specifications, it can be resurfaced at an automotive machine shop.

13   Examine the valve seats in each of the combustion chambers. If they're pitted, cracked or burned, the head will require valve service that's beyond the scope of the home mechanic.

14   Check the valve stem-to-guide clearance by measuring the lateral movement of the valve stem with a dial indicator attached securely to the head **(see illustration)**. The valve must be in the guide and approximately 1/16-inch off the seat. The total valve stem movement indicated by the gauge needle must be divided by two to obtain the actual clearance. After this is done, if there's still some doubt regarding the condition of the valve guides they should be checked by an automotive machine shop (the cost should be minimal).

Valves

Refer to illustrations 9.15 and 9.16

15   Carefully inspect each valve face for uneven wear, deformation, cracks, pits and burned areas **(see illustration)**. Check the

valve stem for scuffing and galling and the neck for cracks. Rotate the valve and check for any obvious indication that it's bent. Look for pits and excessive wear on the end of the stem. The presence of any of these conditions indicates the need for valve service by an automotive machine shop.

16   Measure the margin width on each valve **(see illustration)**. Any valve with a margin narrower than specified will have to be replaced with a new one.

Valve components

Refer to illustrations 9.17 and 9.18

17   Check each valve spring for wear (on the ends) and pits. Measure the free length and compare it to the Specifications in this Chapter **(see illustration)**. Any springs that are shorter than specified have sagged and should not be reused. The tension of all springs should be checked with a special fixture before deciding that they're suitable for use in a rebuilt engine (take the springs to an automotive machine shop for this check).

18   Stand each spring on a flat surface and check it for squareness **(see illustration)**. If

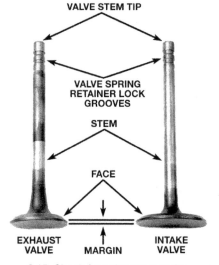

9.15 Check for valve wear at the points shown here

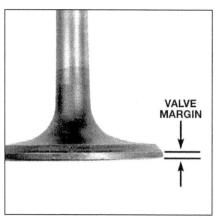

9.16 The margin width on each valve must be as specified (if no margin exists, the valve cannot be reused)

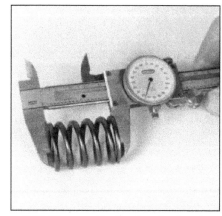

9.17 Measure the free length of each valve spring with a dial or vernier caliper

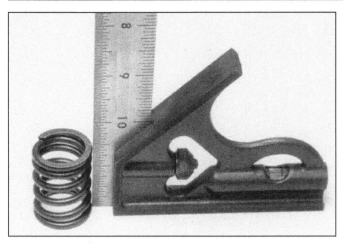

9.18 Check each valve spring for squareness

9.24 Check the cam lobes for pitting, wear and score marks - if scoring is excessive, as is the case here, replace the camshaft

any of the springs are distorted or sagged, replace all of them with new parts.
19 Check the spring retainers and keepers for obvious wear and cracks. Any questionable parts should be replaced with new ones, as extensive damage will occur if they fail during engine operation.

9.25 Measure the height of each lobe - if any lobe height is less than the specified minimum, replace the camshaft

Rocker arm components (SOHC engines only)

20 Check the rocker arm faces (the areas that contact the camshaft and valve stems) for pits, wear, galling, score marks and rough spots. Check the rocker arm pivot contact areas and shafts as well. Look for cracks in each rocker arm.
21 Check the rocker arm adjusting screws for damaged threads and nuts.
22 Any damaged or excessively worn parts must be replaced with new ones.
23 If the inspection process indicates that the valve components are in generally poor condition and worn beyond the limits specified, which is usually the case in an engine that's being overhauled, reassemble the valves in the cylinder head and refer to Section 10 for valve servicing recommendations.

Camshaft

Refer to illustrations 9.24, 9.25 and 9.26
24 Visually examine the camshaft journals and rocker arms. Check for score marks, pitting and evidence of overheating (blue, discolored areas) **(see illustration)**. If wear is excessive or damage is evident, the component will have to be replaced.

25 Using a micrometer or an accurate caliper, measure the cam lobe height and compare it to this Chapter's Specifications. If the lobe height is less than the minimum allowable, the camshaft is worn and must be replaced **(see illustration)**.
26 Using a micrometer or accurate caliper, measure the diameter of each journal and compare it to this Chapter's Specifications **(see illustration)**. If the journals are worn or damaged, replace the camshaft.
27 Using an inside micrometer or a telescoping gauge, measure each housing bore. Subtract the journal diameter measurements from the housing bore measurements to determine the bearing oil clearance. Compare it to this Chapter's Specifications. If the clearance is greater than the maximum, replace the camshaft and, if necessary, the cylinder head.

Lifters (DOHC engines only)

Refer to illustrations 9.28a, 9.28b, 9.30a, 9.30b and 9.30c
28 Inspect the contact (top) and sliding (side) surfaces of each lifter for excessive wear, pitting, scratches or other damage **(see illustrations)**. If a lifter pad is excessively worn or damaged, inspect the corresponding camshaft lobe.

9.26 Check the diameter of each camshaft bearing journal to pinpoint excessive wear and out-of-round conditions

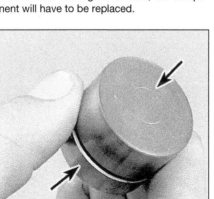

9.28a Inspect the indicated areas (arrows) of the valve lifters (DOHC engines)

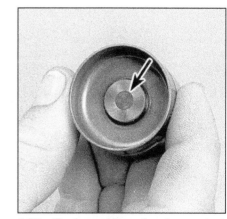

9.28b Also inspect the valve stem contact area (arrow) of the lifter

9.30a Measure the outside diameter of each lifter with a micrometer . . .

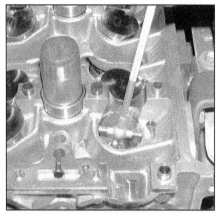

9.30b . . . measure the inside diameter of its corresponding bore in the cylinder head with a telescoping gauge . . .

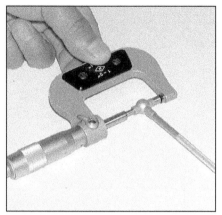

9.30c . . . measure the telescoping gauge with a micrometer, and then subtract the difference in diameter between the lifter bore and the lifter and compare your measurement to the lifter bore clearance listed in this Chapter's Specifications

29   Verify that each lifter freely moves up and down in its lifter bore. If it doesn't, the valve might stick open and damage the top of the piston and/or itself.

30   Measure the outside diameter of each lifter with a micrometer **(see illustration)**, measure the inside diameter of its corresponding lifter bore **(see illustrations)**, and then determine the lifter-to-bore clearance and compare your measurement to the lifter-to-bore clearance listed in this Chapter's Specifications. If the lifter-to-bore clearance is excessive, replace either the lifter or the cylinder head. If a new lifter (which will be within the specified standard range) will bring the clearance within the specified clearance, replace the lifter; if it won't, replace the cylinder head.

10   Valves - servicing

1   Because of the complex nature of the job and the special tools and equipment needed, servicing of the valves, the valve seats and the valve guides, commonly known as a valve job, should be done by a professional.

2   The home mechanic can remove and disassemble the head, do the initial cleaning and inspection, then reassemble and deliver it to a dealer service department or an automotive machine shop for the actual service work. Doing the inspection will enable you to see what condition the head and valvetrain components are in and will ensure that you

know what work and new parts are required when dealing with an automotive machine shop.

3   A dealer service department or automotive machine shop will remove the valves and springs, recondition or replace the valves and valve seats, recondition the valve guides, check and replace the valve springs, spring retainers and keepers as necessary, replace the valve seals with new ones, reassemble the valve components and make sure the installed spring height is correct. The cylinder head gasket surface will also be resurfaced if it's warped.

4   After the valve job has been performed by a professional, the head will be in like new condition. When the head is returned, be sure to clean it again before installation on the engine to remove any metal particles and abrasive grit that may still be present from the valve service or head resurfacing operations. Use compressed air, if available, to blow out all the oil holes and passages.

11   Cylinder head - reassembly

Refer to illustrations 11.3a, 11.3b, 11.5, 11.6 and 11.8

1   Regardless of whether or not the head was sent to an automotive repair shop for valve servicing, make sure it's clean before beginning reassembly.

2   If the head was sent out for valve servic-

ing, the valves and related components will already be in place. Begin the reassembly procedure with Step 8.

3   Install new seals on each of the valve guides. Using a hammer and a deep socket or seal installation tool, gently tap each seal into place until it's completely seated on the guide **(see illustrations)**. Don't twist or cock the seals during installation or they won't seal properly on the valve stems.

4   Beginning at one end of the head, lubricate and install the first valve. Apply molybase grease or clean engine oil to the valve stem.

5   Drop the spring seat or shim(s) (if used) over the valve guide and set the valve spring (s) and retainer in place **(see illustration)**. On DOHC engines, make sure that the inner and outer springs are installed with the large-pitch end at the top of the spring and the small-pitch end at the bottom.

6   Compress the springs with a valve spring compressor and carefully install the keepers in the groove, then slowly release the compressor and make sure the keepers seat

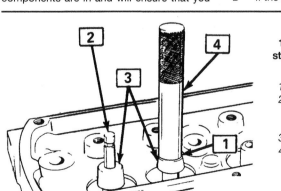

11.3a Make sure the new valve stem seals are seated against the tops of the valve guides

1   *Seal seated in tool*
2   *End of valve stem - be sure to deburr this area before installing the seal*
3   *Seal*
4   *Valve seal installation tool (if you don't have this tool, a deep socket also will work)*

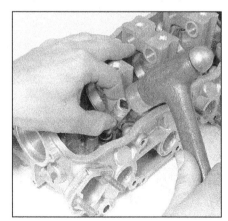

11.3b . . . but a deep socket and a hammer can be used if the tool isn't available - don't hammer on the seals once they're installed

11.5 Install the valve springs with the closely wound coils toward the cylinder head

11.6 Apply a small dab of grease to each keeper as shown here before installation - it'll hold the keepers in place on the valve stem as the spring is released

11.8 On 8-valve SOHC engines, be sure to check the valve spring installed height (the distance from the top of the seat/shims to the top of the spring)

properly. Apply a small dab of grease to each keeper to hold it in place if necessary **(see illustration)**.

7   Repeat the procedure for the remaining valves. Be sure to return the components to their original locations - don't mix them up!

8   Check the installed valve spring height with a ruler graduated in 1/32-inch increments or a dial caliper. If the head was sent out for service work, the installed height should be correct (but don't automatically assume that it is). The measurement is taken from the top of each spring seat or shim(s) to the bottom of the retainer **(see illustration)**. If the height is greater than specified, shims can be added under the springs to correct it. **Caution:** *Don't, under any circumstances, shim the springs to the point where the installed height is less than specified.*

9   Apply moly-base grease to the rocker arm faces and the camshaft, then install the camshaft, rocker arms, springs and shafts. On 16-valve engines, tighten the bearing caps, in the recommended sequence (see Chapter 2A, Section 10) to the torque listed in the Chapter 2A Specifications.

10   Reinstall the distributor case, using a new O-ring.

12   Pistons/connecting rods - removal

Refer to illustrations 12.1, 12.3 and 12.6
Note: *Prior to removing the piston/connecting rod assemblies, remove the cylinder head, the oil pan and the oil pickup by referring to the appropriate Sections in Chapter 2A or 2B.*

1   Use your fingernail to feel if a ridge has formed at the upper limit of ring travel (about 1/4-inch down from the top of each cylinder). If carbon deposits or cylinder wear have produced ridges, they must be completely removed with a special ridge reamer tool **(see illustration)**. Follow the manufacturer's instructions provided with the tool. Failure to remove the ridges before attempting to remove the piston/connecting rod assemblies may result in piston breakage.

2   After the cylinder ridges have been removed, turn the engine upside-down so the crankshaft is facing up.

3   Before the connecting rods are removed, check the endplay with feeler gauges. Slide them between the first connecting rod and the crankshaft throw until the

play is removed **(see illustration)**. The endplay is equal to the thickness of the feeler gauge(s). If the endplay exceeds the service limit, new connecting rods will be required. If new rods (or a new crankshaft) are installed, the endplay may fall under the specified minimum (if it does, the rods will have to be machined to restore it - consult an automotive machine shop for advice if necessary). Repeat the procedure for the remaining connecting rods.

4   Check the connecting rods and caps for identification marks. If they aren't plainly marked, use a small center punch to make the appropriate number of indentations on each rod and cap (1, 2, 3 or 4, depending on the cylinder with which they're associated).

5   Loosen each of the connecting rod cap nuts 1/2-turn at a time until they can be removed by hand. Remove the number one connecting rod cap and bearing insert. Don't drop the bearing insert out of the cap.

6   Slip a short length of plastic or rubber hose over each connecting rod cap bolt to protect the crankshaft journal and cylinder wall as the piston is removed **(see illustration)**.

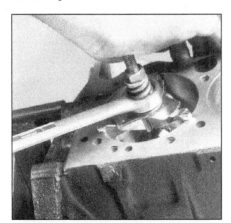

12.1 A ridge reamer is required to remove the ridge from the top of each cylinder - do this before removing the pistons!

12.3 Checking connecting rod endplay with a feeler gauge

12.6 To prevent damage to the crankshaft journals and cylinder walls, slip sections of rubber or plastic hose over the rod bolts before removing the pistons

13.1 Checking crankshaft endplay with a dial indicator

13.3 Checking crankshaft endplay with a feeler gauge

13.4a On SOHC engines, use a center-punch or number stamping dies to mark the main bearing caps to ensure installation in their original locations on the block (make the punch marks near one of the bolt heads)

7   Remove the bearing insert and push the connecting rod/piston assembly out through the top of the engine. Use a wooden hammer handle to push on the upper bearing surface in the connecting rod. If resistance is felt, double-check to make sure that the entire ridge was removed from the cylinder.
8   Repeat the procedure for the remaining cylinders.
9   After removal, reassemble the connecting rod caps and bearing inserts in their respective connecting rods and install the cap nuts finger tight. Leaving the old bearing inserts in place until reassembly will help prevent the connecting rod bearing surfaces from being accidentally nicked or gouged.
10   Don't separate the pistons from the connecting rods (see Section 17 for additional information).

13 Crankshaft - removal

Refer to illustrations 13.1, 13.3, 13.4a, 13.4b and 13.5
Note: *The crankshaft can be removed only after the engine has been removed from the vehicle. It's assumed that the flywheel or*

driveplate, crankshaft pulley, timing belt (SOHC engines) or timing chains (DOHC engine), oil pan, oil pump pickup, oil pump, piston/connecting rod assemblies and rear oil seal housing have already been removed.
1   Before the crankshaft is removed, check the endplay. Mount a dial indicator with the stem in line with the crankshaft and just touching one of the crank throws **(see illustration)**.
2   Push the crankshaft all the way to the rear and zero the dial indicator. Next, pry the crankshaft to the front as far as possible and check the reading on the dial indicator. The distance that it moves is the endplay. If it's greater than specified, check the crankshaft thrust surfaces for wear. If no wear is evident, new thrust bearings should correct the endplay.
3   If a dial indicator isn't available, feeler gauges can be used. Gently pry or push the crankshaft all the way to the front of the engine. Slip feeler gauges between the crankshaft and the front face of the thrust bearing to determine the clearance **(see illustration)**. The thrust bearings are located on both sides of the third main bearing saddle (not on the bearing cap).

4   On SOHC engines, check the main bearing caps to see if they're marked to indicate their locations. They should be numbered consecutively from the front of the engine to the rear. If they aren't, mark them with number stamping dies or a center punch **(see illustration)**. Main bearing caps generally have a cast-in arrow, which points to the front of the engine **(see illustration)**. Loosen the main bearing cap bolts 1/4-turn at a time each, until they can be removed by hand. Note if any stud bolts are used and make sure they're returned to their original locations when the crankshaft is reinstalled. Gently tap the caps with a soft-face hammer, then separate them from the engine block. If necessary, use the bolts as levers to remove the caps. Try not to drop the bearing inserts if they come out with the caps.
5   On the DOHC engine, there are no crankshaft main bearing caps. This engine uses a "lower crankcase," a single part that secures the crankshaft to the block, serves as the bearing caps, houses the lower half of

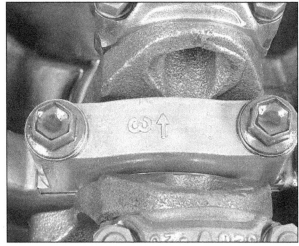

13.4b On SOHC engines, the arrow on the main bearing cap indicates the front of the engine

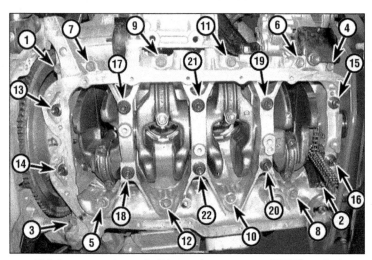

13.5 Lower crankcase bolt loosening sequence (DOHC engine)

each crank main bearing and reinforces the engine block. To detach the lower crankcase from the block, remove the bolts in the indicated sequence **(see illustrations)**.

6   Carefully lift the crankshaft out of the engine. It may be a good idea to have an assistant available, since the crankshaft is quite heavy. With the bearing inserts in place in the engine block and main bearing caps, return the caps to their respective locations on the engine block and tighten the bolts finger tight.

14   Engine block - cleaning

Refer to illustration 14.8
Caution: *The core plugs (also known as freeze or soft plugs) may be difficult or impossible to retrieve if they're driven into the block coolant passages.*

1   Remove the engine mount brackets and any other components still attached to the engine block.
2   Using a gasket scraper, remove all traces of gasket material from the engine block. Be very careful not to nick or gouge the gasket sealing surfaces.
3   Remove the main bearing caps and separate the bearing inserts from the caps and the engine block. Tag the bearings, indicating which cylinder they were removed from and whether they were in the cap or the block, then set them aside.
4   Remove all of the threaded oil gallery plugs from the block. The plugs are usually very tight - they may have to be drilled out and the holes retapped. Use new plugs when the engine is reassembled.
5   If the engine is extremely dirty it should be taken to an automotive machine shop to be steam cleaned or hot tanked.
6   After the block is returned, clean all oil holes and oil galleries one more time. Brushes specifically designed for this purpose are available at most auto parts stores. Flush the passages with warm water until the water runs clear, dry the block thoroughly and wipe all machined surfaces with a light, rust preventive oil. If you have access to compressed air, use it to speed the drying process and to blow out all the oil holes and galleries. **Warning:** *Wear eye protection when using compressed air!*
7   If the block isn't extremely dirty or sludged up, you can do an adequate cleaning job with hot soapy water and a stiff brush. Take plenty of time and do a thorough job. Regardless of the cleaning method used, be sure to clean all oil holes and galleries very thoroughly, dry the block completely and coat all machined surfaces with light oil.
8   The threaded holes in the block must be clean to ensure accurate torque readings during reassembly. Run the proper size tap into each of the holes to remove rust, corrosion, thread sealant or sludge and restore damaged threads **(see illustration)**. If possible, use compressed air to clear the holes of

14.8 All bolt holes in the block - particularly the main bearing cap and head bolt holes - should be cleaned and restored with a tap (be sure to remove debris from the holes after this is done)

debris produced by this operation. Now is a good time to clean the threads on the head bolts and the main bearing cap bolts as well.
9   Reinstall the main bearing caps and tighten the bolts finger tight.
10   Apply non-hardening sealant (such as Permatex no. 2 or Teflon pipe sealant) to the new oil gallery plugs and thread them into the holes in the block. Make sure they're tightened securely.
11   If the engine isn't going to be reassembled right away, cover it with a large plastic trash bag to keep it clean.

15   Engine block - inspection

Refer to illustrations 15.4a, 15.4b, 15.4c, 15.13a and 15.13b
1   Before the block is inspected, it should be cleaned as described in Section 14.
2   Visually check the block for cracks, rust and corrosion. Look for stripped threads in

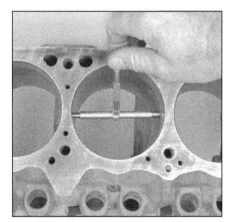

15.4b Use a telescoping gauge to measure the bore - the ability to "feel" when it is at the correct point will be developed over time, so work slowly and repeat the check until you're satisfied the bore measurement is accurate

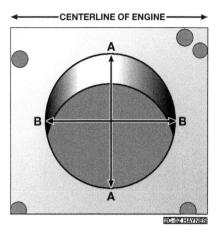

15.4a Measure the diameter of each cylinder at a right angle to the engine centerline (A), and parallel to the engine centerline (B). Out-of-round is the difference between A and B; taper is the difference between A and B at the top of the cylinder and A and B at the bottom of the cylinder

the threaded holes. It's also a good idea to have the block checked for hidden cracks by an automotive machine shop that has the special equipment to do this type of work. If defects are found, have the block repaired, if possible, or replaced.
3   Check the cylinder bores for scuffing and scoring.
4   Measure the diameter of each cylinder at the top (just under the ridge area), center and bottom of the cylinder bore, parallel to the crankshaft axis **(see illustrations)**.
5   Next, measure each cylinder's diameter at the same three locations across the crankshaft axis. Compare the results to the Specifications.
6   If the required precision measuring tools aren't available, the piston-to-cylinder clearances can be obtained, though not quite as accurately, using feeler gauge stock. Feeler gauge stock comes in 12-inch lengths and various thicknesses and is generally available at auto parts stores.

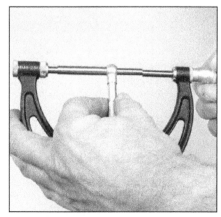

15.4c The gauge is then measured with a micrometer to determine the bore size

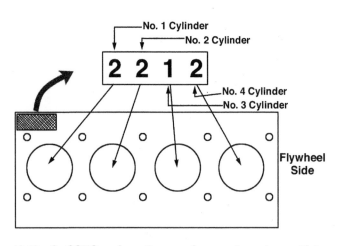

15.13a On SOHC engines, there are four numbers stamped into the top of the engine block near the front right corner; each number indicates the inner diameter of its corresponding cylinder (see Specifications)

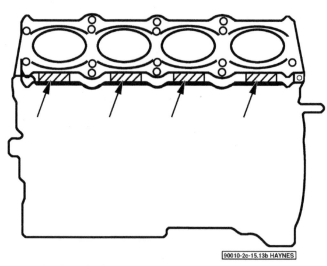

90010-2c-15.13b HAYNES

15.13b On DOHC engines, there is a red or blue paint mark adjacent to each cylinder; each paint mark indicates the inner diameter of its corresponding cylinder (see Specifications)

7   To check the clearance, select a feeler gauge and slip it into the cylinder along with the matching piston. The piston must be positioned exactly as it normally would be. The feeler gauge must be between the piston and cylinder on one of the thrust faces (at a 90-degree angle to the piston pin bore).

8   The piston should slip through the cylinder (with the feeler gauge in place) with moderate pressure.

9   If it falls through or slides through easily, the clearance is excessive and a new piston will be required. If the piston binds at the lower end of the cylinder and is loose toward the top, the cylinder is tapered. If tight spots are encountered as the piston/feeler gauge is rotated in the cylinder, the cylinder is out-of-round.

10   Repeat the procedure for the remaining pistons and cylinders.

11   If the cylinder walls are badly scuffed or scored, or if they're out-of-round or tapered beyond the limits given in the Specifications, have the engine block rebored and honed at an automotive machine shop. If a rebore is done, oversize pistons and rings will be required.

12   If the cylinders are in reasonably good condition and not worn to the outside of the limits, and if the piston-to-cylinder clearances are still within the correct range (see Section 17), then they don't have to be rebored. Honing is all that's necessary (see Section 16).

13   On SOHC engines, the front upper right corner of the engine block is stamped with four numbers, each of them a "1" or "2," to indicate the inner diameter of each cylinder (**see illustration**). On DOHC engines, there is a red or blue paint mark adjacent to each cylinder block indicating the inner diameter of each cylinder (**see illustration**). These numbers or paint marks are used at the factory to match the pistons to the cylinders. The origi-

nal pistons on all engines are also marked, with either a "1" or a "2," which indicates their outside diameter (see Section 17). On SOHC engines, a piston with a "1" marking must be matched to a cylinder with a "1" marking and a piston with a "2" marking must be matched to a cylinder with a "2" marking. On DOHC engines, a piston with a "1" marking must be matched to a cylinder with a red paint mark and a piston with a "2" marking must be matched to a cylinder with a blue paint mark. Refer to the Specifications for the dimensions corresponding to these number or color codes. When you reassemble the engine, you'll need this information to determine whether the piston-to-cylinder clearance is still within the correct range.

16 Cylinder honing

Refer to illustrations 16.3a and 16.3b

1   Prior to engine reassembly, the cylinder bores must be honed so the new piston rings will seat correctly and provide the best possi-

16.3a A "bottle brush" hone is the easiest type of hone to use

ble combustion chamber seal. **Note:** *If you don't have the tools or don't want to tackle the honing operation, most automotive machine shops will do it for a reasonable fee.*

2   Before honing the cylinders, install the main bearing caps and tighten the bolts to the torque listed in this Chapter's Specifications.

3   Two types of cylinder hones are commonly available - the flex hone or "bottle brush" type and the more traditional surfacing hone with spring-loaded stones. Both will do the job, but for the less experienced mechanic the "bottle brush" hone will probably be easier to use. You'll also need some kerosene or honing oil, rags and an electric drill motor. Proceed as follows:

a) *Mount the hone in the drill motor, compress the stones and slip it into the first cylinder* (**see illustration**). *Be sure to wear safety goggles or a face shield!*

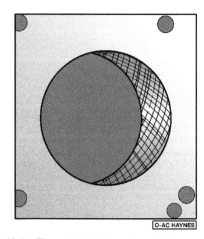

O-AC HAYNES

16.3b The cylinder hone should leave a smooth, crosshatch pattern with the lines intersecting at approximately a 60-degree angle

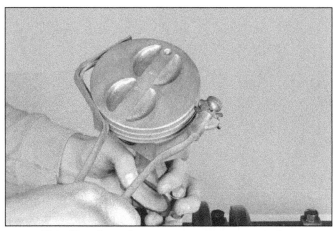

17.4a The piston ring grooves can be cleaned with a special tool, as shown here . . .

17.4b . . . or a section of a broken ring

b) *Lubricate the cylinder with plenty of honing oil, turn on the drill and move the hone up-and-down in the cylinder at a pace that will produce a fine crosshatch pattern on the cylinder walls. Ideally, the crosshatch lines should intersect at approximately a 60-degree angle **(see illustration)**. Be sure to use plenty of lubricant and don't take off any more material than is absolutely necessary to produce the desired finish. **Note:** Piston ring manufacturers may specify a smaller crosshatch angle than the traditional 60-degree - read and follow any instructions included with the new rings.*

c) *Don't withdraw the hone from the cylinder while it's running. Instead, shut off the drill and continue moving the hone up-and down in the cylinder until it comes to a complete stop, then compress the stones and withdraw the hone. If you're using a "bottle brush" type hone, stop the drill motor, then turn the chuck in the normal direction of rotation while withdrawing the hone from the cylinder.*

d) *Wipe the oil out of the cylinder and repeat the procedure for the remaining cylinders.*

4   After the honing job is complete, chamfer the top edges of the cylinder bores with a small file so the rings won't catch when the pistons are installed. Be very careful not to nick the cylinder walls with the end of the file.

5   The entire engine block must be washed again very thoroughly with warm, soapy water to remove all traces of the abrasive grit produced during the honing operation. **Note:** *The bores can be considered clean when a lint-free white cloth - dampened with clean engine oil- used to wipe them out doesn't pick up any more honing residue, which will show up as gray areas on the cloth.* Be sure to run a brush through all oil holes and galleries and flush them with running water.

6   After rinsing, dry the block and apply a coat of light rust preventive oil to all machined surfaces. Wrap the block in a plastic trash bag to keep it clean and set it aside until reassembly.

17   Pistons/connecting rods - inspection

Refer to illustrations 17.4a, 17.4b, 17.10, 17.11a and 17.11b

1   Before the inspection process can be carried out, the piston/connecting rod assemblies must be cleaned and the original piston rings removed from the pistons. **Note:** *Always use new piston rings when the engine is reassembled.*

2   Using a piston ring installation tool, carefully remove the rings from the pistons. Be careful not to nick or gouge the pistons in the process.

3   Scrape all traces of carbon from the top of the piston. A hand held wire brush or a piece of fine emery cloth can be used once the majority of the deposits have been scraped away. Do not, under any circumstances, use a wire brush mounted in a drill motor to remove deposits from the pistons. The piston material is soft and may be eroded away by the wire brush.

4   Use a piston ring groove-cleaning tool to remove carbon deposits from the ring grooves. If a tool isn't available, a piece broken off the old ring will do the job. Be very careful to remove only the carbon deposits - don't remove any metal and do not nick or scratch the sides of the ring grooves **(see illustrations)**.

5   Once the deposits have been removed, clean the piston/rod assemblies with solvent and dry them with compressed air (if available). Make sure the oil return holes in the back sides of the ring grooves are clear.

6   If the pistons and cylinder walls aren't damaged or worn excessively, and if the engine block is not rebored, new pistons won't be necessary. Normal piston wear appears as even vertical wear on the piston thrust surfaces and slight looseness of the top ring in its groove. New piston rings, however, should always be used when an engine is rebuilt.

7   Carefully inspect each piston for cracks around the skirt, at the pin bosses and at the ring lands.

8   Look for scoring and scuffing on the thrust faces of the skirt, holes in the piston crown and burned areas at the edge of the crown. If the skirt is scored or scuffed, the engine may have been suffering from overheating and/or abnormal combustion, which caused excessively high operating temperatures. The cooling and lubrication systems should be checked thoroughly. A hole in the piston crown is an indication that abnormal combustion (preignition) was occurring. Burned areas at the edge of the piston crown are usually evidence of spark knock (detonation). If any of the above problems exist, the causes must be corrected or the damage will occur again. The causes may include intake air leaks, incorrect fuel/air mixture, incorrect ignition timing and EGR system malfunctions.

9   Corrosion of the piston, in the form of small pits, indicates that coolant is leaking into the combustion chamber and/or the crankcase. Again, the cause must be corrected or the problem may persist in the rebuilt engine.

10   Measure the piston ring side clearance by laying a new piston ring in each ring groove and slipping a feeler gauge in beside it **(see illustration)**. Check the clearance at three or four locations around each groove.

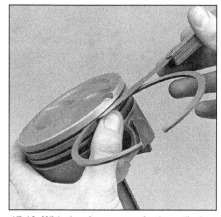

17.10 With the ring square in the cylinder, measure the end gap with a feeler gauge

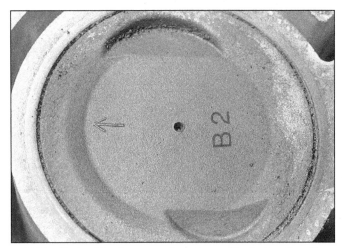

17.11a　Each piston has a number stamped in the top (1 or 2) - the arrow points to front of the engine and the letter (B) is a production code; note the arrow, which faces toward the front of the engine

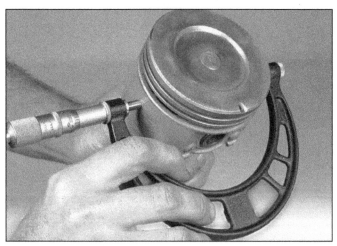

17.11b　Measure the piston diameter at a 90-degree angle to the piston pin and in line with it

Be sure to use the correct ring for each groove - they are different. If the side clearance is greater than specified, new pistons will have to be used.

11　Check the piston-to-bore clearance by measuring the bore (see Section 15) and the piston diameter. Either of two standard size pistons may be fitted at the factory. Each piston has a number stamped on top **(see illustration)**. There are also numbers stamped on the block on SOHC engines **(see illustration 15.13a)** or paint marks on the block next to the cylinders on DOHC engines **(see illustration 15.13b)**. The numbers (either 1 or 2) on the pistons and block designate the bore size. Make sure the pistons and bores are correctly matched. Measure the piston across the skirt, at a 90-degree angle to and in line with the piston pin **(see illustration)**. Subtract the piston diameter from the bore diameter to obtain the clearance. If it's greater than specified, the block will have to be rebored and new pistons and rings installed.

12　Check the piston-to-rod clearance by twisting the piston and rod in opposite directions. Any noticeable play indicates excessive wear, which must be corrected. The piston/connecting rod assemblies should be taken to an automotive machine shop to have the pistons and rods resized and new pins installed.

13　If the pistons must be removed from the connecting rods for any reason, they should be taken to an automotive machine shop. While they are there have the connecting rods checked for bend and twist, since automotive machine shops have special equipment for this purpose. **Note:** *Unless new pistons and/or connecting rods must be installed, do not disassemble the pistons and connecting rods.*

14　Check the connecting rods for cracks and other damage. Temporarily remove the rod caps, lift out the old bearing inserts, wipe the rod and cap bearing surfaces clean and inspect them for nicks, gouges and scratches. After checking the rods, replace

the old bearings, slip the caps into place and tighten the nuts finger tight. **Note:** *If the engine is being rebuilt because of a connecting rod knock, be sure to install new rods.*

18　Crankshaft - inspection

Refer to illustrations 18.1, 18.3, 18.4, 18.6 and 18.8

1　Remove all burrs from the crankshaft oil holes with a stone, file or scraper **(see illustration)**.

2　Check the main and connecting rod bearing journals for uneven wear, scoring, pits and cracks.

3　Rub a penny across each journal several times **(see illustration)**. If a journal picks up copper from the penny, it's too rough and must be reground.

4　Clean the crankshaft with solvent and dry it with compressed air (if available). Be sure to clean the oil holes with a stiff brush **(see illustration)** and flush them with solvent.

18.1　The oil holes should be chamfered so sharp edges don't gouge or scratch the new bearings

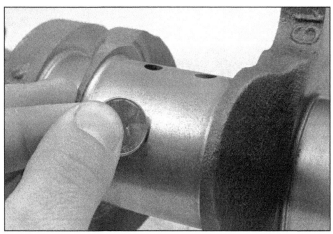

18.3　Rubbing a penny lengthwise on each journal will reveal its condition - if copper rubs off and is embedded in the crankshaft, the journals should be reground

18.4 Use a wire or stiff plastic bristle brush to clean the oil passages in the crankshaft

18.6 Measure the diameter of each crankshaft journal at several points to detect taper and out-of-round conditions

5   Check the rest of the crankshaft for cracks and other damage. It should be magnafluxed to reveal hidden cracks - an automotive machine shop will handle the procedure.

6   Using a micrometer, measure the diameter of the main and connecting rod journals and compare the results to the Specifications **(see illustration)**. By measuring the diameter at a number of points around each journal's circumference, you'll be able to determine whether or not the journal is out-of-round. Take the measurement at each end of the journal, near the crank throws, to determine if the journal is tapered.

7   If the crankshaft journals are damaged, tapered, out-of-round or worn beyond the limits given in the Specifications, have the crankshaft reground by an automotive machine shop. Be sure to use the correct size bearing inserts if the crankshaft is reconditioned.

8   Check the oil seal journals at each end of the crankshaft for wear and damage. If the seal has worn a groove in the journal, or if it's

nicked or scratched **(see illustration)**, the new seal may leak when the engine is reassembled. In some cases, an automotive machine shop may be able to repair the journal by pressing on a thin sleeve. If repair isn't feasible, a new or different crankshaft should be installed.

9   Refer to Section 19 and examine the main and rod bearing inserts.

19   Main and connecting rod bearings - inspection and selection

Inspection

Refer to illustration 19.1

1   Even though the main and connecting

18.8 If the seals have worn grooves in the crankshaft journals, or if the seal contact surfaces are nicked or scratched, the new seals will leak

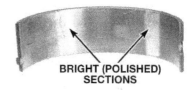

FATIGUE FAILURE

IMPROPER SEATING

SCRATCHED BY DIRT

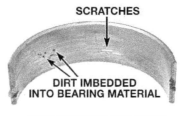

LACK OF OIL

EXCESSIVE WEAR

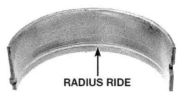

TAPERED JOURNAL

19.1 Typical bearing failures

rod bearings should be replaced with new ones during the engine overhaul, the old bearings should be retained for close examination, as they may reveal valuable information about the condition of the engine **(see illustration)**.

2   Bearing failure occurs because of lack of lubrication, the presence of dirt or other foreign particles, overloading the engine and corrosion. Regardless of the cause of bearing failure, it must be corrected before the engine is reassembled to prevent it from happening again.

3   When examining the bearings, remove them from the engine block, the main bearing caps, the connecting rods and the rod caps and lay them out on a clean surface in the same general position as their location in the engine. This will enable you to match any bearing problems with the corresponding crankshaft journal.

4   Dirt and other foreign particles get into the engine in a variety of ways. It may be left in the engine during assembly, or it may pass through filters or the PCV system. It may get into the oil, and from there into the bearings. Metal chips from machining operations and normal engine wear are often present. Abrasives are sometimes left in engine components after reconditioning, especially when parts are not thoroughly cleaned using the proper cleaning methods. Whatever the source, these foreign objects often end up embedded in the soft bearing material and are easily recognized. Large particles will not embed in the bearing and will score or gouge the bearing and journal. The best prevention for this cause of bearing failure is to clean all parts thoroughly and keep everything spotlessly clean during engine assembly. Frequent and regular engine oil and filter changes are also recommended.

5   Lack of lubrication (or lubrication breakdown) has a number of interrelated causes. Excessive heat (which thins the oil), overloading (which squeezes the oil from the bearing face) and oil leakage or throw off (from excessive bearing clearances, worn oil pump or high engine speeds) all contribute to lubri-

cation breakdown. Blocked oil passages, which usually are the result of misaligned oil holes in a bearing shell, will also oil starve a bearing and destroy it. When lack of lubrication is the cause of bearing failure, the bearing material is wiped or extruded from the steel backing of the bearing. Temperatures may increase to the point where the steel backing turns blue from overheating.

6   Driving habits can have a definite effect on bearing life. Full throttle, low speed operation (lugging the engine) puts very high loads on bearings, which tends to squeeze out the oil film. These loads cause the bearings to flex, which produces fine cracks in the bearing face (fatigue failure). Eventually the bearing material will loosen in pieces and tear away from the steel backing. Short-trip driving leads to corrosion of bearings because insufficient engine heat is produced to drive off the condensed water and corrosive gases. These products collect in the engine oil, forming acid and sludge. As the oil is carried to the engine bearings, the acid attacks and corrodes the bearing material.

7   Incorrect bearing installation during engine assembly will lead to bearing failure as well. Tight fitting bearings leave insufficient bearing oil clearance and will result in oil starvation. Dirt or foreign particles trapped behind a bearing insert result in high spots on the bearing which lead to failure.

Selection

8   If the original bearings are worn or damaged, or if the oil clearances are incorrect (see Sections 22 or 24), new bearings will have to be purchased. It is rare during a thorough rebuild of an engine with many miles on it that new replacement bearings would not be employed. However, if the crankshaft has been reground, new undersize bearings must be installed.

9   The automotive machine shop that reconditions the crankshaft will provide or help you select the correct size bearings. Depending on how much material has to be ground from the crankshaft to restore it, different undersize bearings are required.

Crankshafts are normally ground in increments of 0.010-inch. Sometimes the amount of material machined on a crankshaft will differ between the mains and rod journals, especially if a rod journal was damaged. Markings on most reground crankshafts indicate how much was machined, such as "10-10", meaning that 0.010-inch was removed from both the rod and main journals. Such a crankshaft would require 0.010-inch undersize bearings, a common replacement bearing size.

10   Regardless of how the bearing sizes are determined, use the oil clearance, measured with Plastigage, as the final guide to ensure the bearings are the right size. If you have any questions or are unsure which bearings to use, get help from your machine shop or a dealer parts or service department.

20   Engine overhaul - reassembly sequence

Refer to illustrations 20.2a, 20.2b and 20.2c

1   Before beginning engine reassembly, make sure you have all the necessary new parts, gaskets and seals as well as the following items on hand:

> *Common hand tools*
> *A 1/2-inch drive torque wrench*
> *Piston ring installation tool*
> *Piston ring compressor*
> *Short lengths of rubber or plastic hose to fit over connecting rod bolts*
> *Plastigage*
> *Feeler gauges*
> *A fine-tooth file*
> *New engine oil*
> *Engine assembly lube or moly-base grease*
> *Gasket sealant*
> *Thread locking compound*

2   In order to save time and avoid problems, engine reassembly must be done in the following general order **(see illustrations)**:

> *Piston rings*
> *Crankshaft and main bearings*
> *Piston/connecting rod assemblies*

20.2a Carburetor equipped engine - left side view

20.2b Carburetor equipped engine - front view

20.2c Carburetor equipped engine - right side view

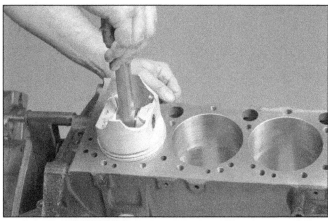

21.3 When checking piston ring end gap, the ring must be square in the cylinder bore (this is done by pushing the ring down with the top of a piston as shown)

Oil pump
Camshaft, rocker arms and shafts
  (SOHC engines)
Oil pan
Cylinder head
Lifters and camshafts (DOHC engine)
Timing belt (SOHC engines) or timing
  chains (DOHC engine) and sprockets
Timing belt/chain cover
Intake and exhaust manifolds
Camshaft cover
Engine rear plate
Flywheel/driveplate

21 Piston rings - installation

Refer to illustrations 21.3, 21.4, 21.5, 21.9a,
21.9b, 21.12a and 21.12b

1   Before installing the new piston rings, the ring end gaps must be checked. It's assumed that the piston ring side clearance has been checked and verified correct (see Section 17).

2   Lay out the piston/connecting rod assemblies and the new ring sets so the ring sets will be matched with the same piston and cylinder during the end gap measurement and engine assembly.

3   Insert the top (number one) ring into the first cylinder and square it up with the cylinder walls by pushing it in with the top of the piston (see illustration). The ring should be near the bottom of the cylinder, at the lower limit of ring travel.

4   To measure the end gap, slip feeler gauges between the ends of the ring until a gauge equal to the gap width is found (see illustration). The feeler gauge should slide between the ring ends with a slight amount of drag. Compare the measurement to the Specifications. If the gap is larger or smaller than specified, double-check to make sure you have the correct rings before proceeding.

5   If the gap is too small, it must be enlarged or the ring ends may come in contact with each other during engine operation, which can cause serious damage to the engine. The end gap can be increased by filing the ring ends very carefully with a fine file. Mount the file in a vise equipped with soft jaws, slip the ring over the file with the ends contacting the file face and slowly move the ring to remove material from the ends. When performing this operation, file only from the outside in (see illustration).

6   Excess end gap isn't critical unless it's greater than 0.040-inch. Again, double-check to make sure you have the correct rings for your engine.

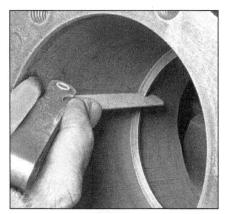

21.4 With the ring square in the cylinder, measure the end gap with a feeler gauge

7   Repeat the procedure for each ring that will be installed in the first cylinder and for each ring in the remaining cylinders. Remember to keep rings, pistons and cylinders matched up.

8   Once the ring end gaps have been checked/corrected, the rings can be installed on the pistons.

9   The oil control ring (lowest one on the piston) is usually installed first. It's composed of three separate components. Slip the

21.5 If the end gap is too small, clamp a file in a vise and file the ring ends (from the outside in only) to enlarge the gap slightly

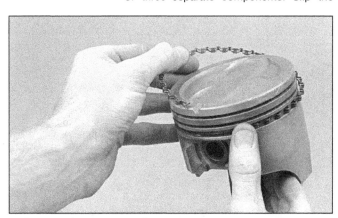

21.9a Installing the spacer/expander in the oil control ring groove

21.9b DO NOT use a piston ring installation tool when installing the oil ring side rails

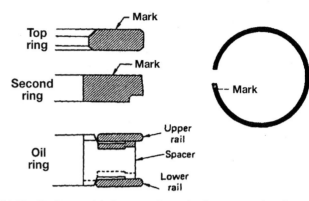

21.12a Earlier models have marks on both compression rings to indicate the top; later models may not have marks on the top rings - always follow the piston ring manufacturer's instructions when installing the rings

spacer/expander into the groove (see illustration). If an anti-rotation tang is used, make sure it's inserted into the drilled hole in the ring groove. Next, install the lower side rail. Don't use a piston ring installation tool on the oil ring side rails, as they may be damaged. Instead, place one end of the side rail into the groove between the spacer/expander and the ring land, hold it firmly in place and slide a finger around the piston while pushing the rail into the groove (see illustration). Next, install the upper side rail in the same manner.

10   After the three oil ring components have been installed, check to make sure that both the upper and lower side rails can be turned smoothly in the ring groove.

11   The number two (middle) ring is installed next. It's usually stamped with a mark, which must face up, toward the top of the piston. **Note:** *Always follow the instructions printed on the ring package or box - different manufacturers may require different approaches. Do not mix up the top and middle rings, as they have different cross sections.*

12   Use a piston ring installation tool and make sure the identification mark is facing the top of the piston, then slip the ring into

the middle groove on the piston (see illustrations). Don't expand the ring any more than necessary to slide it over the piston.

13   Install the number one (top) ring in the same manner. Make sure the mark is facing up. Be careful not to confuse the number one and number two rings. **Note:** *Later model engines may not have a mark on the top ring. These unmarked rings can be installed with either side facing up.*

14   Repeat the procedure for the remaining pistons and rings.

22   Crankshaft - installation and main bearing oil clearance check

1   Crankshaft installation is the first step in engine reassembly. It's assumed at this point that the engine block and crankshaft have been cleaned, inspected and repaired or reconditioned.

2   Position the engine with the bottom facing up.

3   Remove the main bearing cap bolts and lift out the caps. Lay them out in the proper

order to ensure correct installation.

4   If they're still in place, remove the original bearing inserts from the block and the main bearing caps. Wipe the bearing surfaces of the block and caps with a clean, lint-free cloth. They must be kept spotlessly clean.

Main bearing oil clearance check

Refer to illustrations 22.5, 22.6, 22.11 and 22.15

5   Clean the backsides of the new main bearing inserts and lay one in each main bearing saddle in the block. If one of the bearing inserts from each set has a large groove in it, make sure the grooved insert is installed in the block. Lay the other bearing from each set in the corresponding main bearing cap. Make sure the tab on the bearing insert fits into the recess in the block or cap. **Caution:** *The oil holes in the block must line up with the oil holes in the bearing insert* (see illustration). *Do not hammer the bearing into place and don't nick or gouge the bearing faces. No lubrication should be used at this time.*

21.12b Installing the compression rings with a ring expander - the mark (arrow) must face up

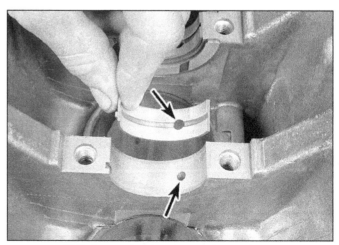

22.5 Ensure that the oil holes (arrows) align

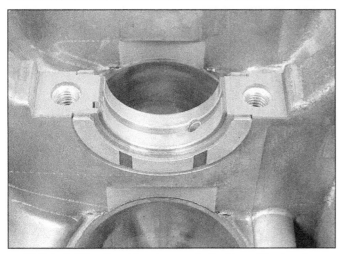

22.6 Install the thrust bearings on both sides of the bearing saddle in the block; be sure the oil grooves face out

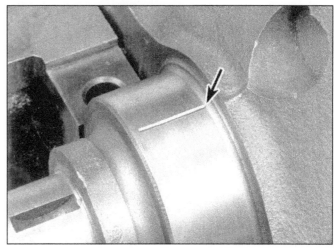

22.11 Lay the Plastigage strips (arrow) on the main bearing journals, parallel to the crankshaft centerline

6   The flanged thrust bearings must be installed in the number three saddle **(see illustration)**.

7   Clean the faces of the bearings in the block and the crankshaft main bearing journals with a clean, lint-free cloth.

8   Check or clean the oil holes in the crankshaft, as any dirt here can go only one way - straight through the new bearings.

9   Once you're certain the crankshaft is clean, carefully lay it in position in the main bearings.

10   Before the crankshaft can be permanently installed, the main bearing oil clearance must be checked.

11   Cut several pieces of the appropriate size Plastigage (they must be slightly shorter than the width of the main bearings) and place one piece on each crankshaft main bearing journal, parallel with the journal axis **(see illustration)**.

12   Clean the faces of the bearings in the caps or lower crankcase. On SOHC engines, install the caps in their respective positions (don't mix them up) with the arrows pointing toward the front of the engine. On DOHC engines, install the lower crankcase. Don't disturb the Plastigage.

13   On SOHC engines, start with the center main and work out toward the ends, tightening the main bearing cap bolts, in three steps, to the torque listed in this Chapter's Specifications. On DOHC engines, tighten the lower crankcase bolts gradually and evenly, in the opposite order of the sequence indicated in **illustration 13.5**, to the torque listed in this Chapter's Specifications. Don't rotate the crankshaft at any time during this operation.

14   Remove the bolts and carefully lift off the main bearing caps or lower crankcase. Keep the caps in order. Don't disturb the Plastigage or rotate the crankshaft. If any of the main bearing caps are difficult to remove, tap them gently from side-to-side with a soft-face hammer to loosen them.

15   Compare the width of the crushed Plastigage on each journal to the scale printed on the Plastigage envelope to obtain the main bearing oil clearance **(see illustration)**. Check the Specifications to make sure it's correct.

16   If the clearance is not as specified, the bearing inserts may be the wrong size (which means different ones will be required). Before deciding that different inserts are needed, make sure that no dirt or oil was between the bearing inserts and the caps or block when the clearance was measured. If the Plastigage was wider at one end than the other, the journal may be tapered (refer to Section 18).

17   Carefully scrape all traces of the Plastigage material off the main bearing journals and/or the bearing faces. Use your fingernail or the edge of a credit card - don't nick or scratch the bearing faces.

Final crankshaft installation

18   Carefully lift the crankshaft out of the engine.

19   Clean the bearing faces in the block, then apply a thin, uniform layer of moly-base grease or engine assembly lube to each of the bearing surfaces and the thrust bearings.

20   Make sure the crankshaft journals are clean, then lay the crankshaft back in place in the block.

21   Clean the faces of the bearings in the caps, then apply lubricant to them.

22   On SOHC engines, install the caps in their respective positions with the arrows pointing toward the front of the engine. On DOHC engines, install the lower crankcase.

23   Install the bearing cap or lower crankcase bolts and hand-tighten them.

24   On SOHC engines, tighten all except the thrust bearing cap bolts to the torque listed in this Chapter's Specifications (work from the center out and approach the final torque in three steps). Tighten the thrust bearing cap bolts to 10-to-12 ft-lbs.

25   On DOHC engines, tighten the lower crankcase bolts gradually and evenly, in the indicated sequence **(see illustration 22.13)**, to the torque listed in this Chapter's Specifications.

22.15 Compare the width of the crushed Plastigage to the scale on the envelope to determine the main bearing oil clearance (always take the measurement at the widest point of the Plastigage); be sure to use the correct scale - standard and metric ones are included

26   Tap the ends of the crankshaft forward and backward with a lead or brass hammer to line up the main bearing and crankshaft thrust surfaces.

27   Retighten all main bearing cap bolts to the torque listed in this Chapter's Specifications, starting with the center main and working out toward the ends on SOHC engines, or reversing the indicated bolt loosening sequence on DOHC engines **(see illustration 13.5)**.

28   On manual transmission equipped models, install a new pilot bearing in the end of the crankshaft (see Chapter 8).

29   Rotate the crankshaft a number of times by hand to check for any obvious binding.

30   The final step is to check the crankshaft endplay with a feeler gauge or a dial indicator as described in Section 13. The endplay should be correct if the crankshaft thrust faces aren't worn or damaged and new bearings have been installed.

23.3 Support the seal retainer on two wood blocks and carefully drive out the seal

23.4 Use a wood block (to spread the load) and hammer to drive the seal into the cover

23   Rear main oil seal installation

Refer to illustrations 23.3 and 23.4

1   All models are equipped with a one-piece seal that fits into a housing attached to the block. The crankshaft must be installed first and the main bearing caps bolted in place, then the new seal should be installed in the housing and the housing bolted to the block.

2   Before installing the seal and housing, check the seal contact surface very carefully for scratches and nicks that could damage the new seal lip and cause oil leaks. If the crankshaft is damaged, the only alternative is a new or different crankshaft.

3   The old seal can be removed from the housing with a hammer and punch by driving it out from the backside **(see illustration)**. Be sure to note how far it's recessed into the housing bore before removing it; the new seal will have to be recessed an equal amount. Be very careful not to scratch or otherwise damage the bore in the housing or oil leaks could develop.

4   Make sure the housing is clean, then apply a thin coat of engine oil to the outer edge of the new seal. The seal must be pressed squarely into the housing bore, so hammering directly on it is not recommended. If you don't have access to a press, tap the seal into place with a hammer and a block of wood **(see illustration)**. The block of wood must be thick enough to distribute the force evenly around the entire circumference of the seal. Work slowly and make sure the seal enters the bore squarely.

5   The seal lips must be lubricated with moly-base grease or engine assembly lube before the seal/housing is slipped over the crankshaft and bolted to the block. Use a new gasket - no sealant is required - and make sure the dowel pins are in place before installing the housing.

6   Tighten the bolts a little at a time to the torque specified in this Chapter.

24   Pistons/connecting rods - installation and rod bearing oil clearance check

1   Before installing the piston/connecting rod assemblies, the cylinder walls must be perfectly clean, the top edge of each cylinder must be chamfered, and the crankshaft must be in place.

2   Remove the cap from the end of the number one connecting rod (refer to the marks made during removal). Remove the original bearing inserts and wipe the bearing surfaces of the connecting rod and cap with a clean, lint-free cloth. They must be kept spotlessly clean.

Connecting rod bearing oil clearance check

Refer to illustrations 24.5, 24.11, 24.13 and 24.17

3   Clean the backside of the new upper bearing insert, then lay it in place in the connecting rod. Make sure the tab on the bearing fits into the recess in the rod. Don't hammer the bearing insert into place and be very careful not to nick or gouge the bearing face. Don't lubricate the bearing at this time.

4   Clean the backside of the other bearing insert and install it in the rod cap. Again, make sure the tab on the bearing fits into the recess in the cap, and don't apply any lubricant. It's critically important that the mating surfaces of the bearing and connecting rod are perfectly clean and oil free when they're assembled.

5   Position the piston ring gaps around the piston as shown **(see illustration)**.

6   Slip a section of plastic or rubber hose over each connecting rod cap bolt.

7   Lubricate the piston and rings with clean engine oil and attach a piston ring compressor to the piston. Leave the skirt protruding about 1/4-inch to guide the piston into the cylinder. The rings must be compressed until they're flush with the piston.

8   Rotate the crankshaft until the number one connecting rod journal is at BDC (bottom dead center) and apply a coat of engine oil to the cylinder walls.

9   If you're installing the standard pistons, make sure that the pistons are matched to the cylinders **(SOHC engines, see illustration 15.13a; DOHC engines, see illustration**

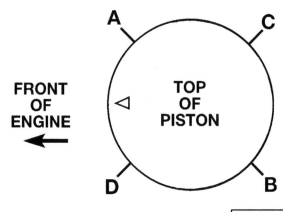

24.5 Ring end gap positions

A   *Oil ring rail gap - lower*
B   *Oil ring rail gap - upper*
C   *Top compression ring gap*
D   *Second compression ring gap and oil ring spacer gap*

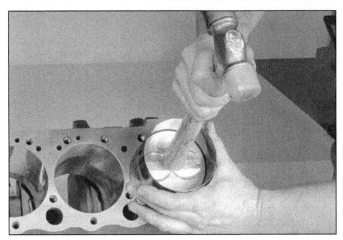

24.11 The piston can be driven (gently) into the cylinder bore with the end of a wooden or plastic hammer handle

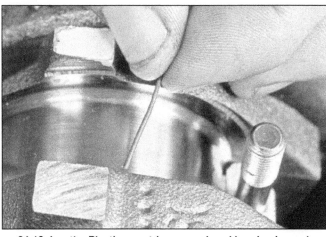

24.13 Lay the Plastigage strips on each rod bearing journal, parallel to the crankshaft centerline

15.13b). Remember: On SOHC engines, a piston with a "1" marking must be matched to a cylinder with a "1" marking and a piston with a "2" marking must be matched to a cylinder with a "2" marking. On DOHC engines, a piston with a "1" marking must be matched to a cylinder with a red paint mark and a piston with a "2" marking must be matched to a cylinder with a blue paint mark. With the arrow on top of the piston **(see illustration 17.11a)** facing the front of the engine, gently insert the piston/connecting rod assembly into the number one cylinder bore and rest the bottom edge of the ring compressor on the engine block.

10   Tap the top edge of the ring compressor to make sure it's contacting the block around its entire circumference.

11   Gently tap on the top of the piston with the end of a wooden or plastic hammer handle **(see illustration)** while guiding the end of the connecting rod into place on the crankshaft journal. The piston rings may try to pop out of the ring compressor just before entering the cylinder bore, so keep some downward pressure on the ring compressor. Work slowly, and if any resistance is felt as the piston enters the cylinder, stop immediately. Find out what's hanging up and fix it before proceeding. Do not, for any reason, force the piston into the cylinder - you might break a ring and/or the piston.

12   Once the piston/connecting rod assembly is installed, the connecting rod bearing oil clearance must be checked before the rod cap is permanently bolted in place.

13   Cut a piece of the appropriate size Plastigage slightly shorter than the width of the connecting rod bearing and lay it in place on the number one connecting rod journal, parallel with the journal axis **(see illustration)**.

14   Clean the connecting rod cap bearing face, remove the protective hoses from the connecting rod bolts and install the rod cap. Make sure the mating mark on the cap is on the same side as the mark on the connecting rod.

15   Install the nuts and tighten them to the torque listed in this Chapter's Specifications, working up to it in three steps. **Note:** *Use a thin-wall socket to avoid erroneous torque readings that can result if the socket is wedged between the rod cap and nut. If the socket tends to wedge itself between the nut and the cap, lift up on it slightly until it no longer contacts the cap. Do not rotate the crankshaft at any time during this operation.*

16   Remove the nuts and detach the rod cap, being very careful not to disturb the Plastigage.

17   Compare the width of the crushed Plastigage to the scale printed on the Plastigage envelope to obtain the oil clearance **(see illustration)**. Compare it to the Specifications to make sure the clearance is correct.

18   If the clearance is not as specified, the bearing inserts may be the wrong size (which means different ones will be required). Before deciding that different inserts are needed, make sure that no dirt or oil was between the bearing inserts and the connecting rod or cap when the clearance was measured. Also, recheck the journal diameter. If the Plastigage was wider at one end than the other, the journal may be tapered (refer to Section 18).

Final connecting rod installation

19   Carefully scrape all traces of the Plastigage material off the rod journal and/or bearing face. Be very careful not to scratch the bearing - use your fingernail or the edge of a credit card.

20   Make sure the bearing faces are perfectly clean, then apply a uniform layer of clean moly-base grease or engine assembly lube to both of them. You'll have to push the piston into the cylinder to expose the face of the bearing insert in the connecting rod - be sure to slip the protective hoses over the rod bolts first.

21   Slide the connecting rod back into place on the journal, remove the protective hoses from the rod cap bolts, install the rod cap and tighten the nuts to the torque listed in this

24.17 Measuring the width of the crushed Plastigage to determine the rod bearing oil clearance (be sure to use the correct scale - standard and metric ones are included)

Chapter's Specifications. Again, work up to the torque in three steps.

22   Repeat the entire procedure for the remaining pistons/connecting rods.

23   The important points to remember are:

a) *Keep the backsides of the bearing inserts and the insides of the connecting rods and caps perfectly clean when assembling them.*

b) *Make sure you have the correct piston/rod assembly for each cylinder.*

c) *The arrow on the piston must face the front of the engine.*

d) *Lubricate the cylinder walls with clean oil.*

e) *Lubricate the bearing faces when installing the rod caps after the oil clearance has been checked.*

24   After all the piston/connecting rod assemblies have been properly installed, rotate the crankshaft a number of times by hand to check for any obvious binding.

25   As a final step, the connecting rod end-play must be checked. Refer to Section 12

for this procedure.

26   Compare the measured endplay to the Specifications to make sure it's correct. If it was correct before disassembly and the original crankshaft and rods were reinstalled, it should still be right. If new rods or a new crankshaft were installed, the endplay may be inadequate. If so, the rods will have to be removed and taken to an automotive machine shop for resizing.

27   Install the remaining components in the reverse order of removal. Refer to Section 20 and Part A of this Chapter for additional information.

25   Initial start-up and break-in after overhaul

Warning: *Have a fire extinguisher handy when starting the engine for the first time.*

1   Once the engine has been installed in the vehicle, double-check the engine oil and coolant levels.

2   With the spark plugs out of the engine and the ignition and fuel (fuel-injected models) systems disabled (see Section 3), crank the engine until oil pressure registers on the gauge.

3   Install the spark plugs, hook up the plug wires and restore the ignition system functions (see Section 3).

4   Start the engine. It may take a few moments for the fuel system to build up pressure, but the engine should start without a great deal of effort. **Note:** *If backfiring occurs through the carburetor or throttle body, recheck the valve timing and ignition timing.*

5   After the engine starts, it should be allowed to warm up to normal operating temperature. While the engine is warming up, make a thorough check for fuel, oil and coolant leaks.

6   Shut the engine off and recheck the engine oil and coolant levels.

7   Drive the vehicle to an area with minimum traffic, accelerate at full throttle from 30 to 50 mph, then allow the vehicle to slow to 30 mph with the throttle closed. Repeat the procedure 10 or 12 times. This will load the piston rings and cause them to seat properly against the cylinder walls. Check again for oil and coolant leaks.

8   Drive the vehicle gently for the first 500 miles (no sustained high speeds) and keep a constant check on the oil level. It is not unusual for an engine to use oil during the break-in period.

9   At approximately 500 to 600 miles, change the oil and filter.

10   For the next few hundred miles, drive the vehicle normally. Do not pamper it or abuse it.

11   After 2000 miles, change the oil and filter again and consider the engine broken in.

Chapter 3
Cooling, heating and air conditioning systems

Contents

Specification

General

Coolant capacity	See Chapter 1
Radiator pressure cap rating	12.8 psi
Thermostat rating	
Samurai	
Thermostat "A"*	
Starts to open	179 degrees F
Fully open	203 degrees F
Thermostat "B"*	
Starts to open	190 degrees F
Fully open	212 degrees F
Sidekick/Tracker/Vitara	
DOHC engine	
Starts to open	179 degrees F
Fully open	203 degrees F
SOHC engine	
Starts to open	190 degrees F
Fully open	212 degrees F

*Either thermostat could be installed, depending upon vehicle's intended usage.

Torque specifications

Ft-lb (unless otherwise indicated)

Thermostat housing bolts
    Samurai .... 84 to 144 in-lbs
    Sidekick/Tracker/Vitara .... 90 to 138 in-lbs
Water pump-to-block bolts
    Samurai .... 90 to 108 in-lbs
    Sidekick/Tracker/Vitara
        1987 through 1995... 90 to 108 in-lbs
        1996 through 1998... 102 in-lbs
        1999 on
            SOHC engine... 96 in-lbs
            DOHC engine 19.5
Fan clutch-to-water pump nuts.. 96 in-lbs

1   General information

Refer to illustration 1.2

Engine cooling system

All vehicles covered by this manual employ a pressurized engine cooling system with thermostatically controlled coolant circulation. An impeller type water pump mounted on the front of the block pumps coolant through the engine. The coolant flows around each cylinder and toward the rear of the engine. Cast-in coolant passages direct coolant around the intake and exhaust ports, near the spark plug areas and in close proximity to the exhaust valve guides.

A wax pellet type thermostat is located in a housing near the front of the engine. During warm up, the closed thermostat prevents coolant from circulating through the radiator. As the engine nears normal operating tem-

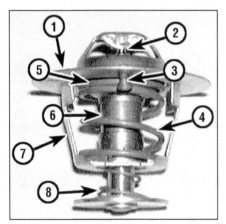

1.2 Typical thermostat

1   Flange
2   Piston
3   Jiggle valve
4   Main coil spring
5   Valve seat
6   Valve
7   Frame
8   Secondary coil spring

perature, the thermostat opens and allows hot coolant to travel through the radiator, where it's cooled before returning to the engine **(see illustration)**.

The cooling system is sealed by a pressure type radiator cap, which raises the boiling point of the coolant and increases the cooling efficiency of the radiator. If the system pressure exceeds the cap pressure relief value, the excess pressure in the system forces the spring-loaded valve inside the cap off its seat and allows the coolant to escape through the overflow tube into a coolant reservoir. When the system cools the excess coolant is automatically drawn from the reservoir back into the radiator.

The coolant reservoir serves as both the point at which fresh coolant is added to the cooling system to maintain the proper fluid level and as a holding tank for overheated coolant.

This type of cooling system is known as a closed design because coolant that escapes past the pressure cap is saved and reused.

Heating system

The heating system consists of a blower fan and heater core located in the heater box, the hoses connecting the heater core to the engine cooling system and the heater/air conditioning control head on the dashboard. Hot engine coolant is circulated through the heater core. When the heater mode is activated, a flap door opens to expose the heater box to the passenger compartment. A fan switch on the control head activates the blower motor, which forces air through the core, heating the air.

Air conditioning system

The air conditioning system consists of a condenser mounted in front of the radiator, an evaporator mounted adjacent to the heater core, a compressor mounted on the engine, a filter-drier which contains a high pressure relief valve and the plumbing connecting all of the above components.

A blower fan forces the warmer air of the

passenger compartment through the evaporator core (sort of a radiator-in-reverse), transferring the heat from the air to the refrigerant. The liquid refrigerant boils off into low pressure vapor, taking the heat with it when it leaves the evaporator.

2   Antifreeze - general information

Warning: *Do not allow antifreeze to come in contact with your skin or painted surfaces of the vehicle. Rinse off spills immediately with plenty of water. NEVER leave antifreeze lying around in an open container or in a puddle in the driveway or on the garage floor. Children and pets are attracted by it's sweet smell. Antifreeze is fatal if ingested.*

The cooling system should be filled with a water/ethylene glycol based antifreeze solution, which will prevent freezing down to at least -20-degrees F, or lower if local climate requires it. It also provides protection against corrosion and increases the coolant boiling point.

The cooling system should be drained, flushed and refilled at the specified intervals (see Chapter 1). Old or contaminated antifreeze solutions are likely to cause damage and encourage the formation of rust and scale in the system. Use distilled water with the antifreeze.

Before adding antifreeze, check all hose connections, because antifreeze tends to leak through very minute openings. Engines don't normally consume coolant, so if the level goes down, find the cause and correct it.

The exact mixture of antifreeze-to-water which you should use depends on the relative weather conditions. The mixture should contain at least 50 percent antifreeze, but should never contain more than 70 percent antifreeze. Consult the mixture ratio chart on the antifreeze container before adding coolant. Hydrometers are available at most auto parts stores to test the coolant. Use antifreeze which meets the vehicle manufacturer's specifications.

3.8 Loosen the hose clamp at the thermostat housing and disconnect the hose from the thermostat (DOHC engine shown)

3.10a On SOHC engines, unsnap the mixture control valve (if equipped) from its bracket, then remove the thermostat housing cover bolts (and, on some models, the wiring bracket)

A Mixture control valve (some older SOHC engines)
B Thermostat housing cover bolts

3.10b On DOHC engines, remove the thermostat cover bolts (arrows) and then remove the cover

3 Thermostat - check and replacement

Warning: *Do not remove the radiator cap, drain the coolant or replace the thermostat until the engine has cooled completely.*

Check

1   Before assuming the thermostat is to blame for a cooling system problem, check the coolant level, drivebelt tension (see Chapter 1) and temperature gauge operation.

2   If the engine seems to be taking a long time to warm up (based on heater output or temperature gauge operation), the thermostat is probably stuck open. Replace the thermostat with a new one.

3   If the engine runs hot, use your hand to check the temperature of the upper radiator hose. If the hose isn't hot, but the engine is, the thermostat is probably stuck closed, preventing the coolant inside the engine from escaping to the radiator. Replace the thermostat. **Caution:** *Don't drive the vehicle without a thermostat. The computer may stay in open loop and emissions and fuel economy will suffer.*

4   If the upper radiator hose is hot, it means that the coolant is flowing and the thermostat is open. Consult the Troubleshooting Section at the front of this manual for cooling system diagnosis.

Replacement

Refer to illustrations 3.8, 3.10a, 3.10b, 3.13a and 3.13b

5   Disconnect the negative battery cable from the battery.

6   Drain the cooling system (see Chapter 1). If the coolant is relatively new or in good condition (see Chapter 1), save it and reuse it.

7   To locate the thermostat housing, follow the upper radiator hose to the engine.

8   Loosen the hose clamp **(see illustration)**, then detach the hose from the fitting. If it's stuck, grasp it near the end with a pair of

adjustable pliers and twist it to break the seal, then pull it off. If the hose is old or deteriorated, cut it off and install a new one.

9   If the outer surface of the large fitting that mates with the hose is deteriorated (corroded, pitted, etc.) it may be damaged further by hose removal. If it is, the thermostat housing cover will have to be replaced.

10   Remove the bolts **(see illustrations)** and detach the housing cover. If the cover is stuck, tap it with a soft-face hammer to jar it loose. Be prepared for some coolant to spill as the gasket seal is broken.

11   Note how it's installed (which end is facing up), then remove the thermostat.

12   Stuff a rag into the engine opening, then remove all traces of old gasket material and sealant from the housing and cover with a gasket scraper. Remove the rag from the opening and clean the gasket mating surfaces with lacquer thinner or acetone.

13   Install the new thermostat in the hous-

3.13a Install the thermostat, with the spring end down, into the stepped seat in the housing (SOHC engine shown, DOHC engines similar)

ing. Make sure the correct end faces up - the spring end is directed into the engine **(see illustration)**. Also, on late-model thermostat housings in which the thermostat is installed horizontally, make sure that the air bleed hole is at the top. On DOHC engines, make sure that the air bleed hole is aligned with the mark on the thermostat housing **(see illustration)**.

14   Apply a thin, uniform layer of RTV sealant to both sides of the new gasket and position it on the housing.

15   Install the cover and bolts. Tighten the bolts to the torque listed in this Chapter's Specifications.

16   Reattach the hose to the fitting and tighten the hose clamp securely.

17   Refill the cooling system (see Chapter 1).

18   Start the engine and allow it to reach normal operating temperature, then check for leaks and proper thermostat operation (as described in Steps 2 through 4).

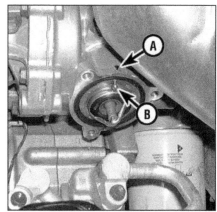

3.13b If the thermostat is installed on its side (rather than vertically), make sure that the air valve hole is at the top and, on DOHC engines (shown) make sure that the hole is aligned with the mark on the thermostat housing

A Index mark
B Air bleed hole

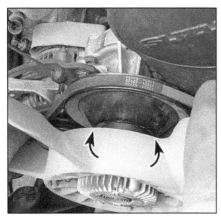

4.3 Rock the fan back and forth to check for play in the fan clutch

4 Engine cooling fan and clutch - check and replacement

Warning: *To avoid possible injury or damage, DO NOT operate the engine with a damaged fan. Do not attempt to repair fan blades - replace a damaged fan with a new one.*

Check

Auxiliary electric fan (air conditioned models only)

1   To test the motor, unplug the electrical connector at the motor and use fused jumper wires to connect the fan directly to the battery. If the fan still doesn't work, replace the motor.
2   If the motor tested OK, the fault lies in the condenser fan motor relay, the dual switch, the air conditioner amplifier or the wiring which connects the components (see the wiring diagrams in Chapter 12). Carefully check all wiring and connections. If no obvious problems are found, further diagnosis should be done by a dealer service department or repair shop.

Belt-driven fan with viscous clutch

Refer to illustration 4.3

3   Disconnect the negative battery cable

4.15 The fan clutch is mounted with four nuts - three are visible in this photo (arrows)

4.9 Unplug the electrical connector for the condenser fan motor (air-conditioned models)

and rock the fan back and forth by hand to check for excessive play in the fan clutch **(see illustration)**.
4   With the engine cold, turn the fan blades by hand. The fan should turn freely.
5   Visually inspect for substantial fluid leakage from the clutch assembly. If problems are noted, replace the clutch assembly.
6   With the engine completely warmed up, turn off the ignition switch and disconnect the negative battery cable from the battery. Turn the fan by hand. Some drag should be evident. If the fan turns easily, replace the fan clutch.

Removal and installation

Auxiliary electric fan (air conditioned models only)

Refer to illustrations 4.9 and 4.10

7   Disconnect the negative battery cable from the battery.
8   Remove the front bumper and grille (see Chapter 11).
9   Insert a small screwdriver into the connector to lift the lock tab and unplug the fan wire harness **(see illustration)**.
10   Unbolt the fan bracket and shroud assembly **(see illustration)** and then carefully lift it out of the vehicle.

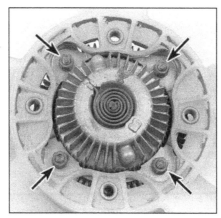

4.18 If you intend to replace the fan or clutch, remove the mounting nuts (arrows)

4.10 To detach the condenser fan assembly from the condenser, remove these four bolts (arrows) (two lower bolts not visible)

11   To detach the fan from the motor, remove the motor shaft fastener.
12   To remove the bracket from the fan motor, remove the mounting nuts.
13   Installation is the reverse of removal.

Belt-driven fan with viscous clutch

Refer to illustrations 4.15 and 4.18

14   Disconnect the negative battery cable. Remove the fan shroud mounting screws and detach the shroud.
15   Remove the bolts/nuts attaching the fan/clutch assembly to the water pump hub **(see illustration)**.
16   Lift the fan/clutch assembly (and shroud, if necessary) out of the engine compartment.
17   Carefully inspect the fan blades for damage and defects. Replace it if necessary.
18   At this point, the fan may be unbolted from the clutch, if necessary **(see illustration)**. If the fan clutch is stored, position it with the radiator side facing down.
19   Installation is the reverse of removal Be sure to tighten the fan and clutch mounting nuts/bolts evenly and securely.

5 Radiator - removal and installation

Refer to illustrations 5.3, 5.5a, 5.5b, 5.5c, 5.6, 5.8 and 5.13
Warning: *Wait until the engine is completely cool before beginning this procedure.*
1   Disconnect the negative battery cable from the battery.
2   Drain the cooling system (see Chapter 1). If the coolant is relatively new or in good condition, save it and reuse it.
3   Loosen the hose clamps **(see illustration)**, then detach the radiator hoses from the fittings. If they're stuck, grasp each hose near the end with a pair of water pump pliers and twist it to break the seal, then pull it off - be careful not to distort the radiator fittings! If the hoses are old or deteriorated, cut them off and install new ones.

5.3 Before removing the fan shroud or the radiator, loosen the hose clamps (arrows) that secure the inlet and outlet hoses to the radiator and disconnect both hoses

5.5a On earlier models, unbolt the fan shroud - the bolts are located at each corner . . .

1   Shroud bolts   2   Radiator mounting bolts

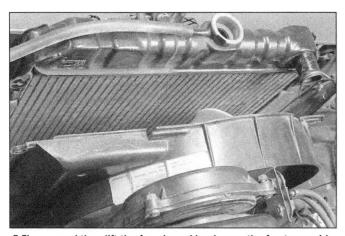

5.5b . . . and then lift the fan shroud back over the fan to provide clearance for radiator removal

5.5c On later models, the fan shroud is secured to the radiator by four push fasteners (arrows) (two lower fasteners, one at each corner, not visible in this photo)

4   Disconnect the coolant reservoir hose from the radiator filler neck.

5   On older models, remove the screws that attach the shroud to the radiator and slide the shroud toward the engine (see illus-trations). On later models, remove the four push fasteners that attach the shroud to the radiator (see illustration).

6   If the vehicle is equipped with an auto-matic transmission, disconnect the cooler lines from the radiator (see illustration). Use a drip pan to catch spilled fluid.

7   Plug the lines and fittings.

8   Remove the radiator mounting bolts (see illustration).

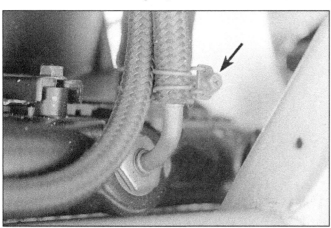

5.6 On automatic transmission equipped models, disconnect the cooling line (arrow) from each side of the bottom of the radiator

5.8 Air conditioned models have the radiator mounting bolts on the ends (arrow)

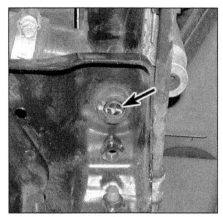

5.13 When installing the radiator, make sure that each of the two locating pins on the bottom of the radiator is fully seated in its corresponding rubber insulator (arrow)

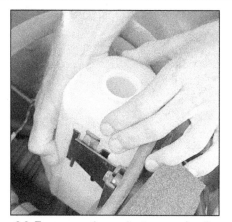

6.2 To remove the coolant reservoir, slip it off the bracket by pulling straight up (the location of the reservoir varies from model to model, but you'll always find it in the engine compartment)

7.4a The water pump weep hole (arrow) will drip coolant when the seal on the pump shaft fails (pulley removed for clarity) (SOHC engines)

9   Carefully lift out the radiator. Don't spill coolant on the vehicle or scratch the paint.

10   With the radiator removed, it can be inspected for leaks and damage. If it needs repair, have a radiator shop or dealer service department perform the work as special techniques are required.

11   Bugs and dirt can be removed from the radiator with compressed air and a soft brush. Don't bend the cooling fins as this is done.

12   Inspect the rubber radiator insulators for deterioration and make sure there's nothing in them when the radiator is installed.

13   Installation is the reverse of the removal procedure. Make sure that the radiator mounting pins are seated in the rubber insulators **(see illustration)**.

14   After installation, fill the cooling system with the proper mixture of antifreeze and water. Refer to Chapter 1 if necessary.

15   Start the engine and check for leaks. Allow the engine to reach normal operating temperature, indicated by the upper radiator hose becoming hot. Recheck the coolant level and add more if required.

16   If you're working on an automatic transmission equipped vehicle, check and add fluid as needed.

6   Coolant reservoir - removal and installation

Refer to illustration 6.2

1   Detach the reservoir cap.

2   Remove the mounting bolt located on the bottom of the reservoir (if equipped) and lift the reservoir straight up from its mounting bracket **(see illustration)**.

3   Pour the contents of the reservoir into a clean container.

4   Installation is the reverse of removal.

5   Refill the container with the proper mixture of antifreeze and water. Refer to Chapter 1 if necessary.

7   Water pump - check

Refer to illustrations 7.4a, 7.4b and 7.5

1   A failure in the water pump can cause serious engine damage due to overheating.

2   There are three ways to check the operation of the water pump while it's installed on the engine. If the pump is defective, it should be replaced with a new or rebuilt unit.

3   With the engine running at normal operating temperature, squeeze the upper radiator hose. If the water pump is working properly, a pressure surge should be felt as the hose is released. **Warning:** *Keep your hands away from the fan blades!*

4   Water pumps are equipped with weep or vent holes. If a failure occurs in the pump seal, coolant will leak from the hole. In most cases you'll need a flashlight to find the hole on the water pump from underneath to check for leaks **(see illustrations)**.

5   If the fan or water pump shaft bearings fail there may be a howling sound at the front of the engine while it's running. Shaft wear can be felt if the fan or water pump pulley is rocked up and down **(see illustration)**. Don't mistake drivebelt slippage, which causes a squealing sound, for water pump bearing failure.

7.4b Water pump weep hole location (arrow) (DOHC engines)

8   Water pump - replacement

Refer to illustrations 8.8, 8.9a and 8.9b

Warning: *Wait until the engine is completely cool before beginning this procedure. On air conditioned vehicles, do not disconnect any refrigerant lines/fittings unless the system has been depressurized by an air conditioning technician.*

1   On DOHC engines, we recommend draining the oil and removing the oil filter (see Chapter 1). Some of the water pump bolts are extremely difficult to reach with the oil filter installed.

2   Drain the cooling system (see Chapter 1). If the coolant is relatively new or in good condition, save it and reuse it.

3   Loosen the clamps and detach the hoses from the water pump. If they're stuck, grasp each hose near the end with a pair of adjustable pliers and twist it to break the seal, then pull it off. If the hoses are deteriorated, cut them off and install new ones.

4   Remove the cooling fan and shroud (see Sections 4 and 5).

5   Remove the drivebelt(s) (see Chapter 1) and the pulley at the end of the water pump shaft.

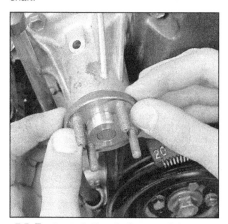

7.5 Try to rock the shaft up and down to check for play

8.8 On DOHC engines, remove the heater outlet hose from the heater outlet pipe (A), remove the heater outlet pipe retaining bolt (B) and then remove the heater outlet pipe; arrow C indicates the fourth water pump bolt, which is not visible in illustration 8.9b

8.9a Water pump mounting bolts (arrows) on SOHC engines; when installing the pump, insert new rubber seals in the ridge gaps at the top and bottom of the pump

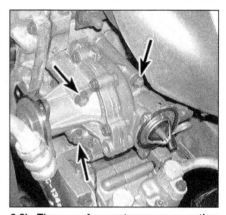

8.9b There are four water pump mounting bolts on DOHC engines; three bolts (arrows) are shown in this photo and the fourth, not visible in this photo, is underneath (see illustration 8.8)

6 On SOHC engines, remove all accessory brackets from the water pump. When removing the power steering pump and air conditioning compressor, (if equipped) don't disconnect the hoses. Tie the units aside with the hoses attached.

7 On SOHC engines, remove the timing belt and idler assembly (see Chapter 2A).

8 On DOHC engines, disconnect the heater outlet hose, remove the heater outlet pipe bolt **(see illustration)** and remove the heater outlet pipe.

9 Remove the bolts **(see illustrations)** and detach the water pump from the engine. Note the locations of the various lengths and different types of bolts as they're removed to ensure correct installation.

10 Clean the bolt threads and the threaded holes in the engine to remove corrosion and sealant.

11 Compare the new pump to the old one to make sure they're identical.

12 Remove all traces of old gasket material from the engine with a gasket scraper.

13 Clean the engine and new water pump mating surfaces with lacquer thinner or acetone.

14 Apply a thin coat of RTV sealant to the engine side of the new gasket.

15 Apply a thin layer of RTV sealant to the gasket mating surface of the new pump, then carefully mate the gasket and the pump. Slip a couple of bolts through the pump mounting holes to hold the gasket in place.

16 Carefully attach the pump and gasket to the engine and thread the bolts into the holes finger tight.

17 Install the remaining bolts (if they also hold an accessory bracket in place, be sure to reposition the bracket at this time). Tighten them to the torque listed in this Chapter's Specifications in 1/4-turn increments. Don't overtighten them or the pump may be distorted.

18 Reinstall all parts removed for access to the pump.

19 Refill the cooling system and check the drivebelt tension (see Chapter 1). Run the engine and check for leaks.

9 Coolant temperature sending unit - check and replacement

Refer to illustrations 9.1a and 9.1b
Warning: *Wait until the engine is completely cool before beginning this procedure.*

1 The coolant temperature indicator system consists of a temperature gauge mounted in the instrument panel and a coolant temperature sending unit mounted on the cylinder head **(see illustration)** on earlier models, or on the intake manifold on later models. Some vehicles have more than one coolant temperature sending unit, but only one is used for the coolant temperature gauge on the dash. Other temperature sending units are used as information sensors for the air-conditioning system or for the Transmission Control Module (TCM). On fuel-injected vehicles, there is an Engine Coolant Temperature (ECT) *sensor* that monitors the temperature of the engine coolant for the Powertrain Control Module (PCM). On fuel-injected SOHC engines, the ECT sensor is located in the side of the intake manifold; on DOHC engines, the ECT sensor is screwed into the coolant outlet pipe elbow at the back of the cylinder head **(see illustration)**. How can you tell the difference between a temperature sending unit and an ECT sensor? Count the number of wires to the sending unit/sensor connector: Coolant temperature sending units have one wire. ECT sensors have two or three wires. On 1996 and later (OBD-II) models, there is no separate coolant temperature sending unit; on these models, the ECT sensor functions as the sending unit for the temperature gauge and as the information sensor for the PCM. For information regarding checking and replacing the ECT sensor, refer to Chapter 6. **Warning:** *If the vehicle is equipped with an auxiliary electric cooling fan, stay clear of the fan blades, which can come on at any time.*

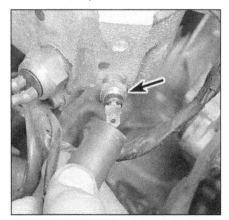

9.1a On carbureted models, the coolant temperature sending unit is located just below the thermostat housing (arrow) - upper radiator hose removed for clarity; on pre-OBD-II models, the temperature sending unit is located on the intake manifold

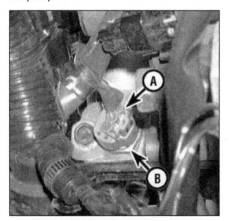

9.1b On OBD-II models, the Engine Coolant Temperature (ECT) sensor is located on the side of the intake manifold (SOHC engines, not shown) or on the coolant outlet pipe elbow on the back of the cylinder head (DOHC engines)

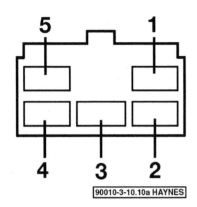

10.10a Blower motor switch connector
(Cami connector) terminal guide (1989
through 1997 Sidekick/Tracker models)

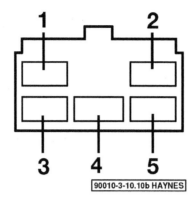

10.10b Blower motor switch connector
(Iwata connector) terminal guide (1989
through 1997 Sidekick/Tracker models)

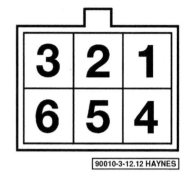

10.12 Blower motor switch
connector terminal guide
(1998 Sidekick/Tracker models)

2   If an overheating indication occurs, check the coolant level in the system and then make sure the wiring between the gauge and the switch is secure and all fuses are intact.

3   Test the circuit by grounding the wire to the switch while the ignition is on (engine not running for safety). If the gauge deflects full scale, replace the sending unit.

4   If the sending unit must be replaced, simply unscrew it from the engine and install the replacement. Use sealant on the threads. Make sure the engine is cool before removing the defective sending unit. There will be some coolant loss as the unit is removed, so be prepared to catch it. Check the level after the replacement has been installed.

10   Blower motor unit and circuit - check and component replacement

General description

1   On Samurai models, the blower unit is above and to the right of the throttle pedal. On Sidekick and Tracker models, the blower unit is mounted in the dash, in front of the glove box. The blower motor unit circuit consists of the battery, a switch on the dash-mounted air conditioning and/or heating control assembly, a resistor, and a relay (1999 and later Vitara/Tracker models only).

Check

Blower motor circuit and blower motor unit

2   Check the fuses and all connections in the circuit for looseness and corrosion. Make sure that the battery is fully charged.

3   Using T-pins, backprobe the blower unit electrical connector terminals and then attach the leads of a voltmeter to the T-pins with alligator clips. Turn the ignition key to ON; it is not necessary to start the engine. Move the blower motor switch through each of its positions and note the voltage readings. As the switch is turned to a numerically higher position, the resistance of the circuit should decrease, the voltage should

increase, and the blower fan speed should increase. If the blower unit doesn't come on at all, check the blower speed switch. If the blower unit comes on but the speed of the fan doesn't increase at each position as the switch is turned up, check the switch and the blower resistor.

4   To check the blower unit, unplug the electrical connector to the blower motor (see below). Using an ohmmeter, verify that there is continuity between the two terminals of the blower motor connector.

5   If there is no continuity, there is an open circuit in the blower winding. Replace the blower unit.

6   If there is continuity, the blower winding is okay. Using a pair of jumper wires, connect the battery, or a battery voltage source, to the blower unit connector terminals. Connect the positive terminal of the battery to the power terminal of the blower unit with one jumper and ground the other terminal with the other jumper. Power up the blower unit with battery voltage and note whether the blower unit comes on.

a) *If the blower unit does not come on, replace the blower unit.*

b) *If the blower unit comes on when jumped, but did not come on when switched on at the dash, check the blower motor switch.*

Blower motor switch

7   Remove the blower switch (see below).

Samurai models

8   Using an ohmmeter or a self-powered test light, check the continuity between the indicated terminals at each switch position:

a) *At the OFF position, there should be no continuity between any of the terminals.*

b) *At the first position, there should be continuity between the terminals for the black and the blue/red wires.*

c) *At the second position, there should be continuity between the terminals for the black and the blue/yellow wires.*

d) *At the third position, there should be continuity between the terminals for the black and the blue wires.*

9   If the continuity is incorrect for any of these terminal pairs, replace the blower switch.

1987 through 1997 Sidekick/Tracker models

Refer to illustrations 10.10a and 10.10b

10   Using an ohmmeter or a self-powered test light, check the continuity between the indicated terminals at each switch position **(see illustrations)**:

a) *At the OFF position, there should be no continuity between any of the terminals.*

b) *At the LOW position, there should be continuity between terminals 1 (light green wire) and 2 (pink/black wire).*

c) *At the M1 position, there should be continuity between terminals 1 (light green wire), 2 (pink/black wire) and 3 (pink/blue wire).*

d) *At the M2 position, there should be continuity between terminals 1 (light green wire), 2 (pink/black wire) and 4 (pink/green wire)*

e) *At the HIGH position, there should be continuity between terminals 1 (light green wire), 2 (pink/black wire) and 5 (pink wire).*

11   If the continuity is incorrect for any of these terminal pairs, replace the blower switch.

1998 Sidekick/Tracker models

Refer to illustration 10.12

12   Using an ohmmeter or a self-powered test light, check the continuity between the indicated terminals at each switch position **(see illustration)**:

a) *At the OFF position, there should be no continuity between any of the terminals.*

b) *At the LOW position, there should be continuity between terminals 1 and 4.*

c) *At the M1 position, there should be continuity between terminals 1, 4 and 5.*

d) *At the M2 position, there should be continuity between terminals 1, 4 and 6.*

e) *At the HIGH position, there should be continuity between terminals 1, 4 and 3.*

13   If the continuity is incorrect for any of these terminal pairs, replace the blower switch.

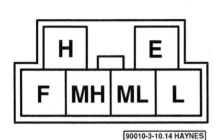

10.14 Blower motor switch connector terminal guide (1999 and later Vitara/Tracker models)

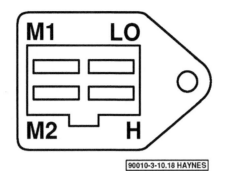

10.18 Blower resistor connector terminal guide (1987 through 1998 Sidekick/Tracker models)

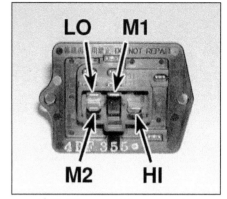

10.20 Blower resistor connector terminal guide (1999 and later Vitara/Tracker models)

1999 and later Vitara/Tracker models

Refer to illustration 10.14

14   Using an ohmmeter or a self-powered test light, check the continuity between the indicated terminals at each switch position **(see illustration)**:

a) *At the OFF position, there should be no continuity between any of the terminals.*
b) *At the LOW position, there should be continuity between terminals E, F and L.*
c) *At the M1 position, there should be continuity between terminals E, F and ML.*
d) *At the M2 position, there should be continuity between terminals E, F and MH.*
e) *At the HIGH position, there should be continuity between terminals E, F and H.*

15   If the continuity is incorrect for any of these terminal pairs, replace the blower switch.

Heater blower resistor

16   Remove the blower resistor (see below). **Note:** *You can also test the blower resistor while it's still installed, but it's easy to remove, and it's easier to see the connector terminals with the resistor removed from the heater box.*

Samurai models

17   Using an ohmmeter, measure the resistance between the resistor connector terminals for the blue/white and the blue/black wires. There should be several ohms resistance. If there is no resistance, or high resistance, replace the resistor.

1987 through 1998 Sidekick/Tracker models

Refer to illustration 10.18

18   Using an ohmmeter, measure the resistance between the resistor connector terminals for the indicated terminals **(see illustration)**:

a) *Between terminals H and LO, there should be about 1.8 ohms of resistance..*
b) *Between terminals H and M1, there should be about 1.0 ohm of resistance.*
c) *Between terminals H and M2, there should be about 0.5 ohm of resistance.*

19   If the resistance is incorrect for any of these terminal pairs, replace the heater blower resistor.

1999 and later Vitara/Tracker models

Refer to illustration 10.20

20   Using an ohmmeter, measure the resistance between the resistor connector terminals for the indicated terminals **(see illustration)**:

a) *Between terminals H and LO, there should be about 2.0 ohms of resistance..*
b) *Between terminals H and M1, there should be about 1.0 ohm of resistance.*
c) *Between terminals H and M2, there should be about 0.4 ohm of resistance.*

21   If the resistance is incorrect for any of these terminal pairs, replace the heater blower resistor.

Heater blower relay (1999 and later models only)

Refer to illustration 10.22

22   Remove the heater blower relay (see below).

23   Connect the leads of an ohmmeter to terminals 3 and 4 of the relay and then, using a pair of jumper cables, hook up the positive battery terminal to terminal 1 of the relay connector and the negative terminal of the battery to terminal 1 **(see illustration)**. With the

relay powered up by battery voltage, there should be continuity between terminals 3 and 4. If there isn't, replace the relay.

Replacement

Samurai and pre-1999 Sidekick/Tracker models

Blower motor switch

24   Remove the air conditioning and heater control assembly (see Section 12) and remove the blower motor switch from the back of the assembly.

25   Installation is the reverse of removal.

Blower resistor

26   The blower resistor is located on the blower case. Unplug the resistor electrical connector and then remove the two blower resistor mounting screws and remove the resistor.

27   Installation is the reverse of removal.

Blower unit

Refer to illustrations 10.28 and 10.30

28   Disconnect the blower motor electrical connector and hose, if equipped **(see illustration)**.

29   Remove the three screws holding the blower motor to the heater unit housing.

10.22 Heater blower relay connector terminal guide (1999 and later Vitara/Tracker models)

10.28 Unplug the electrical connector from the motor, remove the three screws, then maneuver the blower unit out from under the dash (Samurai models)

10.30 The fan is attached to the motor shaft with a nut (arrow) (Samurai shown, other models similar)

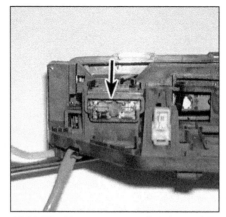

10.32 The blower motor switch (arrow), is secured to the air conditioning and heater control assembly by a pair of locking tangs; to remove the switch, depress the tangs and pull the switch straight out of the assembly

10.34 The blower resistor (arrow) is located on the blower case

10.35 To remove the blower resistor, unplug the electrical connector and then remove the two blower resistor mounting screws (arrows)

30   If you are replacing the motor, detach the fan and transfer it to the new motor **(see illustration)**.

31   Installation procedures are the reverse of those for removal. Run the fan and check for proper operation.

Vitara and Tracker models

Blower motor switch

Refer to illustration 10.32

32   Remove the air conditioning and heater control assembly (see Section 12) and remove the blower motor switch **(see illustration)** from the back of the assembly.

33   Installation is the reverse of removal.

Blower resistor

Refer to illustrations 10.34 and 10.35

34   The blower resistor **(see illustration)** is located on the blower case. To access the blower resistor, remove the glove box (see Section 19 in Chapter 11).

35   Unplug the electrical connector and then remove the two blower resistor mounting screws **(see illustration)** and remove the

resistor.

36   Installation is the reverse of removal.

Blower unit

Refer to illustrations 10.41 and 10.42

37   Disconnect the negative battery cable.

38   Disable the airbag system, if equipped (see Chapter 12).

39   Open the glove box, remove the retaining screw and remove the glove box (see Section 19 in Chapter 11).

40   Remove the Powertain Control Module (PCM) (see Chapter 6).

41   Remove the PCM mounting bracket **(see illustration)**.

42   Unplug the blower motor electrical connector **(see illustration)**, remove the blower motor retaining screws and then remove the blower motor unit.

43   Installation is the reverse of removal.

Relay

Refer to illustration 10.44

44   The relay **(see illustration)** is located to

the right of the heater case. To access the relay, remove the glove box (see Section 19 in Chapter 11).

45   To remove the relay, unplug the electrical connector, remove the retaining screw and lift out the relay.

46   Installation is the reverse of removal.

10.41 To detach the Powertain Control Module (PCM) mounting bracket (lower arrow) from the dash on Sidekick/Tracker models, remove these four screws (arrows)

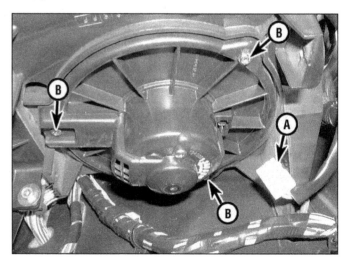

10.42 To remove the blower motor unit on Sidekick and Tracker models, unplug the electrical connector (A) and then remove the three blower unit retaining screws (B)

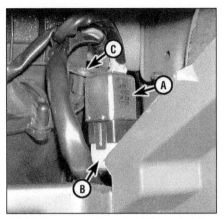

10.44 To remove the relay (A), unplug the electrical connector (B) and then remove the retaining screw (C)

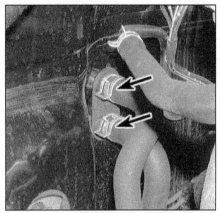

11.4 To disconnect the heater hoses from the heater core pipes, loosen these hose clamps (arrows), slide back the clamps and pull off the hoses

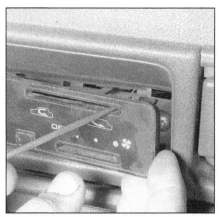

12.3a On Samurai models, slip an L-hook through the slotted openings and gently pull the faceplate off . . .

11 Heater core - removal and installation

Refer to illustration 11.4

Warning: *1996 and later models are equipped with an airbag system. Always disable the airbag(s) prior to working in the vicinity of the steering column, instrument panel or airbag components to avoid the possibility of accidental deployment of the airbags, which could cause personal injury (see Chapter 12).*

1 On vehicles with air conditioning, have the air conditioning system discharged by a dealer service department or by an automotive air conditioning shop.

2 On vehicles with airbags, disable the airbag system (see Chapter 12).

3 Drain the cooling system (see Chapter 1).

4 Working in the engine compartment, disconnect the heater hoses where they enter the firewall **(see illustration)**.

5 Remove the instrument panel and, if equipped, the center console (see Chapter 11).

6 If the vehicle is equipped with air conditioning, remove the evaporator unit (see Section 17).

7 Label and detach the air ducts, wiring and controls still attached to the heater housing. On 1999 and later models, remove the air conditioning controller, if equipped, and, on vehicles with airbags, remove the Sensing and Diagnostic Module (SDM).

8 Unbolt the heater housing unit and lift it from the vehicle.

9 Remove any remaining heater core pipe clamps and then remove the rubber grommet where the two heater core pipes protrude through the firewall.

10 Remove the screws and/or clips that secure the two halves of the heater housing and then separate the two halves of the housing. Take out the old heater core and install the new unit.

11 Reassemble the heater housing unit and check the operation of the air control flaps. If any parts bind, correct the problem before installation.

12 Reinstall the remaining parts in the

reverse order of removal.

13 Refill the cooling system, reconnect the battery and run the engine. Check for leaks and proper system operation.

12 Air conditioning and heater control assembly - removal and installation

Samurai models and 1987 through 1998 Sidekick/Tracker models

Refer to illustrations 12.3a, 12.3b, 12.3c, 12.4a, 12.4b, 12.4c and 12.6

1 Disconnect the negative cable from the battery.

2 Remove the radio, if equipped (see Chapter 12).

3 Pull off the control knobs and faceplate **(see illustrations)**.

4 Remove the mounting screws located on the front of the control assembly **(see**

12.3b . . . then reach behind the faceplate and remove the light socket

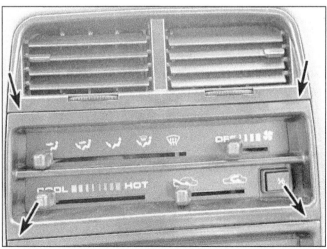

12.3c On Tracker and Sidekick models, gently grip the trim panel (arrows) and pull it off

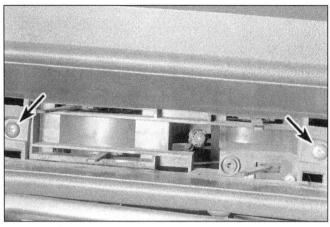

12.4a Samurai control mounting screw locations (arrows)

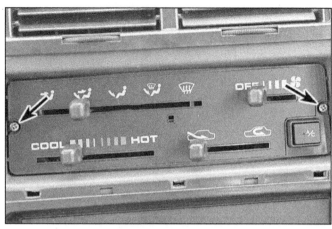

12.4b Sidekick and Tracker control mounting
screw locations (arrows)

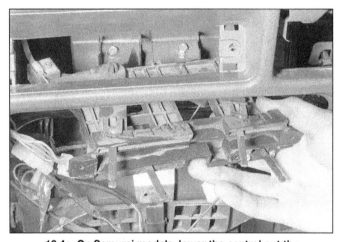

12.4c On Samurai models, lower the control out the
bottom of the dash

12.6 Separate the control cables (1) from the operating levers (2)

illustrations). On Samurai models, push the control assembly back into the dash and lower it out the bottom **(see illustration)**.
5   Pull the control out slightly. On Tracker and Sidekick models it may be necessary to disconnect the cables at the operating ends before this is possible.

6   Clearly mark the positions of the cable housings relative to the adjustment clips, detach the cables and wiring from the control assembly **(see illustration)** and lift the assembly from the dash.
7   To install the unit, reverse the above procedure.

8   To adjust the cables, remove the adjustment clips at the lever ends and move the cable housings to the positions marked previously. Fasten the clips and check for stiffness or binding through the full range of operation.
9   Run the engine and check for proper functioning of the heater (and air conditioning, if equipped).

1999 and later Vitara/Tracker models

Refer to illustrations 12.12, 12.14, 12.15, 12.16a, 12.16b, 12.16c and 12.16d
10   Disconnect the negative battery cable.
11   Disable the airbag system (see Chapter 12).
12   Pull off the heater control lever knobs **(see illustration)**.
13   Remove the instrument cluster trim panel, the ashtray, the center trim bezel and the glove box (see Section 19 in Chapter 11).
14   Remove the air conditioning and heater control assembly mounting screws **(see illustration)**.
15   Pull out the air conditioning and heater control assembly and unplug the electrical connectors for the blower motor switch, the

12.12 To remove the heater control lever knobs, pry up the locking tang on each knob and then pull the knob straight off the lever

12.14 To detach the air conditioning and heater control assembly from the dash, remove these three mounting screws (arrows)

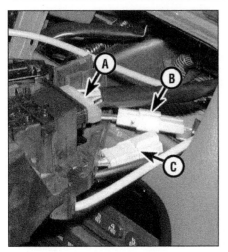

12.15 Pull out the air conditioning and heater control assembly far enough to unplug the electrical connectors for the blower motor switch (A), the illumination bulb (B) and the air conditioning system (C), but don't try to pull out the assembly until you have disconnected the three cables

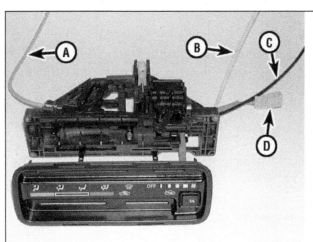

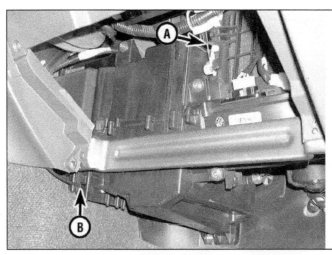

12.16a Trace the three cables from their respective slider levers on the backside of the air conditioning and heater control assembly to the levers they actuate on the heater case . . .

A   Vent control cable
B   Fresh air control cable
C   Temperature control cable
D   Heater blower motor switch connector

12.16b . . . locate the fresh air control cable (A) and the temperature control cable (B) on the right side of the heater case (the vent cable, not shown in this photo, is attached to a lever on the left side of the heater case) . . .

illumination bulb and the air conditioning system **(see illustration)**. Do not try to pull the air conditioning and heater control assembly out of the dash yet.

16   Trace the three cables **(see illustration)** from their respective slider levers on the backside of the air conditioning and heater control assembly to the levers they actuate on the heater case and then unclip and disengage each cable from its lever. The vent cable is attached to a lever on the left side of the heater case. To access the lever end of the vent cable, remove the lower steering column cover (see Section 19 in Chapter 11). The fresh air control cable is on the right side of the heater case, and the temperature control cable is attached to a lever on the front right lower corner of the heater case **(see illustration)**. Make sure that you clearly label each cable so that it's not accidentally reconnected to the wrong lever during reassembly, and then unclamp each cable **(see illustrations)**.

17   Installation is the reverse of removal. Adjust the cables before installing any dash trim panels.

Cable adjustment

18   Remove the lower steering column cover and the glove box (see Section 19 in Chapter 11).

19   To adjust the fresh air control cable, slide the control lever slider (on the air conditioning and heater control assembly) all the way to the right, disengage the fresh air cable from the clamp **(see illustrations 12.16c and 12.16d)**, push the fresh air control lever (on the heater case) all the way up, and then reclamp the cable.

20   The other two cables are adjusted in essentially the same manner as the fresh air cable.

12.16c . . . locate the cable clip (arrow) . . .

13   Air conditioning system - check and maintenance

Refer to illustration 13.7
Warning: *The air conditioning system is under high pressure. Do not loosen any hose fittings or remove any components until after the system has been discharged by a dealer service department or service station. Always wear eye protection when disconnecting or charging air conditioning system fittings.*

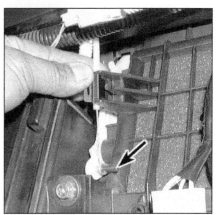

12.16d . . . then unbend the clip and disengage each cable from the pin (arrow) on its corresponding lever (fresh air control cable shown, other cables are secured by a similar clip and are looped over a pin on a lever in the same manner)

Note: *The air conditioning system on 1995 and later models uses the non-ozone depleting refrigerant, referred to as R-134a. The R-134a refrigerant and its lubricating oil are not compatible with the R-12 system, and under no circumstances should the two different types of refrigerant and lubricating oil be inter-*

13.7 The sight glass is located on the top of the receiver/drier (arrow)

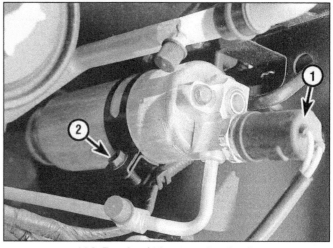

14.3 Receiver/drier mounting details

1   Pressure switch connector
2   Mounting bracket pinch bolt

mixed. If mixed, it could result in costly compressor failure due to improper lubrication.

1   The following maintenance checks should be performed on a regular basis to ensure the air conditioner continues to operate at peak efficiency.

   a) *Check the compressor drivebelt. If it's worn or deteriorated, replace it (see Chapter 1).*
   b) *Check the drivebelt tension and, if necessary, adjust it (see Chapter 1).*
   c) *Check the system hoses. Look for cracks, bubbles, hard spots and deterioration. Inspect the hoses and all fittings for oil bubbles and seepage. If there's any evidence of wear, damage or leaks, replace the hose(s).*
   d) *Inspect the condenser fins for leaves, bugs and other debris. Use a "fin comb" or compressed air to clean the condenser.*
   e) *Make sure the system has the correct refrigerant charge.*

2   It's a good idea to operate the system for about 10 minutes at least once a month, particularly during the winter. Long term non-use can cause hardening, and subsequent failure, of the seals.

3   Because of the complexity of the air conditioning system and the special equipment necessary to service it, in-depth troubleshooting and repairs are not included in this manual. However, simple checks and component replacement procedures are provided in this Chapter.

4   The most common cause of poor cooling is simply a low system refrigerant charge. If a noticeable drop in cool air output occurs, one of the following quick checks will help you determine if the refrigerant level is low.

5   Warm the engine up to normal operating temperature.

6   Place the air conditioning temperature selector at the coldest setting and put the blower at the highest setting. Open the doors (to make sure the air conditioning system

doesn't cycle off as soon as it cools the passenger compartment).

7   With the compressor engaged - the clutch will make an audible click and the center of the clutch will rotate - inspect the sight glass **(see illustration)**. If the refrigerant looks foamy, it's low. Have a dealer service department or automotive air conditioning shop charge the system.

8   If there's no sight glass, feel the inlet and outlet pipes at the compressor. One side should be cold and one hot. If there's no perceptible difference between the two pipes, there's something wrong with the compressor or the system. It might be a low charge. It might be something else. Take the vehicle to a dealer service department or an automotive air conditioning shop.

---

14   Air conditioning system receiver/drier - removal and installation

---

Pre-1999 models

Refer to illustration 14.3

Warning: *The air conditioning system is under high pressure. DO NOT disassemble any part of the system (hose, compressor, line fittings, etc.) until after the system has been depressurized by a dealer service department or service station.*

1   Have the air conditioning system discharged (see **Warning** above).

2   Disconnect the negative battery cable from the battery.

3   Unplug the electrical connector from the pressure switch near the top of the receiver/drier **(see illustration)**.

4   Remove the bolts and disconnect the refrigerant lines from the top of the receiver/drier.

5   Plug the open fittings to prevent entry of dirt and moisture.

6   Loosen the mounting bracket pinch bolt

and lift the receiver/drier out.

7   If a new receiver/drier is being installed, remove the Schrader valve and pour the oil out into a measuring cup, noting the amount. Add fresh refrigerant oil to the new receiver/drier equal to the amount removed from the old unit, plus one ounce.

8   Installation is the reverse of removal.

9   Take the vehicle back to the shop that discharged it. Have the air conditioning system evacuated, charged and leak tested.

1999 and later models

Refer to illustration 14.10

10   On these models, the receiver/drier **(see illustration)** is welded to the condenser and cannot be replaced separately. However, the filter and dryer elements inside the receiver/drier can be replaced.

11   Have the system discharged by a dealer service department or by an automotive air conditioning shop.

12   Remove the condenser (see Section 16).

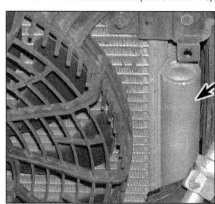

14.10 On 1999 and later Vitara/Tracker models, the receiver/drier (arrow) is welded to the condenser and cannot be replaced separately; however, the filter and drier inside can be replaced by unscrewing the top and pulling them out (but NOT while the system is charged!)

15.5 Never remove the refrigerant lines-to-compressor bolts (arrows) unless the system has been discharged by a service technician

15.6a To detach the compressor from its mounting bracket, remove these bolts (arrows) from the front . . .

13   Unscrew the plug from the top of the receiver/drier unit with a wrench and then pull out the filter and drier with a pair of long needle nose pliers.

14   Before installing a new drier, be sure to add 40 cc of refrigerant oil. Do not remove the drier from its plastic bag until right before inserting it into the receiver/drier. When inserting the new drier into the receiver/drier unit, make sure that the double-layer portion is facing down, toward the bottom of the receiver/drier unit.

15   Installation is otherwise the reverse of removal.

16   When you're done, have the system evacuated and recharged by a dealer service department or by an automotive air conditioning shop.

15 Air conditioning system compressor - removal and installation

Refer to illustrations 15.5, 15.6a and 15.6b
Warning: *The air conditioning system is under high pressure. DO NOT disassemble any part of the system (hoses, compressor, line fittings, etc.) until after the system has been depressurized by a dealer service department or service station.*
Note: *The receiver/drier (see Section 14) should be replaced whenever the compressor is replaced.*

1   Have the A/C system discharged (see **Warning** above).
2   Disconnect the negative battery cable from the battery.
3   Disconnect the compressor clutch wiring harness.
4   Remove the drivebelt (see Chapter 1).
5   Disconnect the refrigerant lines from the top of the compressor **(see illustration)**. Plug the open fittings to prevent entry of dirt and moisture.
6   Unbolt the compressor from the mounting brackets **(see illustrations)** and lift it out of the vehicle.
7   If a new compressor is being installed, follow the directions with the compressor regarding the draining of excess oil prior to installation.
8   The clutch may have to be transferred from the original to the new compressor.
9   Installation is the reverse of removal. Replace all O-rings with new ones specifi-

cally made for A/C system use and lubricate them with refrigerant oil.
10   Have the system evacuated, recharged and leak tested by the shop that discharged it.

16 Air conditioning system condenser - removal and installation

Refer to illustrations 16.4, 16.5a and 16.5b
Warning: *The air conditioning system is under high pressure. DO NOT disassemble any part of the system (hoses, compressor, line fittings, etc.) until after the system has been depressurized by a dealer service department or service station.*
Note: *The receiver/drier (see Section 14) should be replaced whenever the condenser is replaced.*

1   Have the air conditioning system discharged (see **Warning** above).
2   Disconnect the negative cable from the battery.
3   Remove the grille and the front bumper (see Chapter 11).
4   Unplug the electrical connector **(see illustration)** for the dual-pressure switch.

15.6b . . . and this bolt (arrow) from underneath, at the rear of the compressor (1999 and later model shown, earlier models similar)

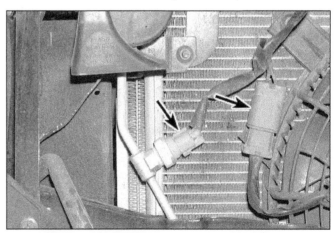

16.4 Unplug the electrical connectors for the dual-pressure switch (left arrow) and the condenser fan (right arrow)

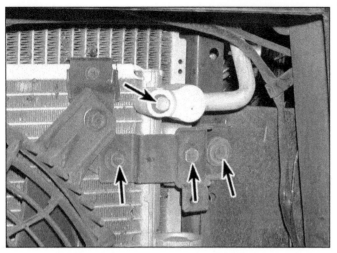

16.5a To disconnect the delivery hose from the condenser, remove this bolt (upper arrow). The condenser is attached to the vehicle by four brackets, one at each corner; the three lower arrows indicate the condenser bracket bolts at the upper right corner

16.5b To disconnect the outlet pipe from the condenser, remove this bolt (arrow)

5   Disconnect the refrigerant lines from the condenser **(see illustrations)**.
6   Remove the auxiliary electric fan (see Section 4).
7   Remove the mounting bolts from the condenser brackets **(see illustration 16.5a)**.
8   Lift the condenser out of the vehicle and plug the lines to keep dirt and moisture out.

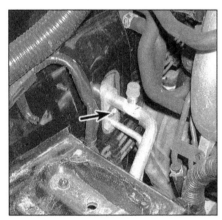

17.4 To disconnect the evaporator lines from the evaporator, remove this bolt (arrow)

9   If the original condenser will be reinstalled, store it with the line fittings on top to prevent oil from draining out.
10   If a new condenser is being installed, pour one ounce of refrigerant oil into it prior to installation.
11   Reinstall the components in the reverse order of removal. Be sure the rubber insulator pads are in place.
12   Have the system evacuated, recharged and leak tested by the shop that discharged it.

---

17   Air conditioning system evaporator - removal and installation

---

Refer to illustration 17.4
Warning 1: *1996 and later models are equipped with an airbag system. Always disable the airbags prior to working in the vicinity of the steering column, instrument panel or airbag components to avoid the possibility of accidental deployment of the airbags, which could cause personal injury (see Chapter 12).*
Warning 2: *The air conditioning system is under high pressure. DO NOT disassemble any part of the system until the system has*

been depressurized by a dealer service department of service station.
1   Have the air conditioning system discharged (see **Warning** above).
2   Disconnect the negative battery cable from the battery.
3   Disable the airbag module, if equipped (see Chapter 12).
4   Disconnect the evaporator inlet and outlet lines at the firewall **(see illustration)**.
5   Remove the evaporator case mounting nut from the firewall.
6   Remove the instrument panel center brace.
7   Remove the blower case.
8   Loosen the evaporator case-to-heater case retaining band and slide the band onto the heater case.
9   Disconnect the electrical connectors from the evaporator case and remove the drain hose.
10   Remove the evaporator case mounting bolts and remove the case from the vehicle.
11   Separate the two case halves, remove the evaporator pipe clamp and remove the evaporator core from the case.
12   Remove the expansion valve and thermistor from the evaporator core.
13   Installation is the reverse of removal.

Chapter 4
Fuel and exhaust systems

Contents

Specifications

Carburetor

Choke plate-to-bore clearance
  At 77-degrees F (25-degrees C) ambient temperature ... 0.004 to 0.023 in (0.1 to 0.6 mm)
  At 104-degrees F (40-degrees C) ambient temperature ... 0.05 to 0.11 in (1.3 to 2.8 mm)

Electronic Fuel Injection (EFI)

Fuel pressure at idle ... 24 to 30 psi
Fuel injector coil resistance ... 1.0 to 2.0 ohms (at 68-degrees F, 20-degrees C)
Fuel injector leakage rate ... One drop per minute or less

Sequential multiport fuel injection

Fuel pressure (engine idling, vacuum hose connected to pressure regulator) ... 29.8 to 37 psi
Fuel injector resistance ... 13.0 to 16.0 ohms
Fuel injector leakage rate ... One drop per minute or less

Throttle cable freeplay

Carbureted models
  Cold engine ... 3/8 to 5/8 in (10 to 15 mm)
  Warm engine ... 1/8 to 3/16 in (3 to 5 mm)
Electronic Fuel Injection (EFI) models ... 3/8 to 5/8 in (10 to 15 mm)
Sequential multiport fuel injection models
  Accelerator pedal freeplay ... 5/64 to 9/32 in (2 to 7 mm)
  Throttle lever-to-lever stopper clearance ... 1/32 to 5/64 in (0.5 to 2.0 mm)

Torque specifications

Ft-lbs (unless otherwise indicated)

Fuel pump-to-cylinder head nuts (carbureted models) ... 84 to 138 in-lbs
Carburetor/ throttle body injection unit mounting nuts/bolts ... 13.5 to 20
Throttle body mounting bolts (sequential multiport fuel injection) ... 18
Fuel pressure regulator bolts (sequential multiport fuel injection) ... 90 in-lbs
Fuel rail mounting bolts (sequential multiport fuel injection) ... 17
Exhaust pipe-to-exhaust manifold bolts ... 37

1   General information

The fuel system consists of a rear mounted tank, combination metal and rubber fuel hoses, an engine-mounted mechanical pump or an in-tank electric pump, and either a two-stage, two-venturi carburetor or an electronic fuel injection system. Electronic Fuel Injection (EFI) (sometimes referred to as single-point or throttle body injection) is used on most earlier fuel-injected models. Some 1992 through 1995 and all 1996 and later models are equipped with sequential multi-port fuel injection.

The exhaust system consists of the exhaust manifold, the catalytic converter, the muffler/tailpipe assembly, and the pipes connecting these components.

The emission control systems modify the functions of both the exhaust and fuel systems. There may be some cross-references throughout this Chapter to Sections in Chapter 6 because the emissions control systems are an integral part of the fuel and exhaust systems. **Warning:** *Exercise extreme caution when dealing with either the fuel or the exhaust system. Fuel is a primary element for combustion. Be very careful! The exhaust system is also an area for exercising caution as it operates at very high temperatures. Serious burns can result from even momentary contact with any part of the exhaust system and the fire potential is ever present.*

2   Fuel pressure relief procedure

Warning: *Gasoline is extremely flammable, so extra precautions must be taken when working on any part of the fuel system. Do not smoke or allow open flames or bare light bulbs in the work area. Also, do not work in a garage if a natural gas-type appliance is present. Always keep a dry chemical (Class B) fire extinguisher near the work area.*

1   Before servicing any component on the fuel system, relieve the residual fuel pressure to minimize the risk of fire and personal injury.

2.5 To relieve the fuel system pressure on 1990 and earlier fuel-injected vehicles, you'll need one wrench to loosen the service bolt on top of the fuel filter and another wrench to hold the special banjo bolt into which the service bolt is installed

2   Disconnect the cable from the negative terminal of the battery.

Carbureted vehicles

3   Unscrew the fuel filler cap to release the pressure caused by fuel vapor.

1990 and earlier fuel-injected models

Refer to illustration 2.5

4   Perform Step 2, then raise the vehicle and support it securely on jackstands.
5   Position a container under the fuel filter. Cover the fuel filter with a rag and, using a backup wrench to prevent the fuel filter union bolt from turning, slowly loosen the plug bolt and allow the pressurized fuel to escape **(see illustration)**.
6   When the pressure has been relieved completely, tighten the plug bolt securely.
7   Reconnect the battery cable only after all work to the fuel system has been completed, then start the engine and check the fittings for leaks.

1991 and later fuel-injected models

Refer to illustrations 2.9a, 2.9b and 2.9c

8   This procedure must be performed with the engine cold (to prevent damage to the catalytic converter) and the transmission in Park (automatic) or Neutral (manual).
9   Unplug the electrical connector from the fuel pump relay located next to the PCM under the left (Sidekick/Tracker) or right (Samurai) end of the instrument panel **(see illustrations)**. The fuel pump relay on 1996 through 1998 models is located behind the center of the instrument panel near the radio and heater controls **(see illustration)**. The fuel pump relay on most models is usually located next to the main relay.
10   Remove the fuel tank cap to release the pressure in the tank, then reinstall it.
11   Start the engine and let it run until it stalls from lack of fuel. Crank the starter two or three times in short (three-second) bursts to release any residual fuel pressure.

3   Fuel lines and fittings - inspection and replacement

Warning: *Gasoline is extremely flammable, so extra precautions must be taken when working on any part of the fuel system. Do not smoke or allow open flames or bare light bulbs in the work area. Also, do not work in a garage if a natural gas-type appliance is present. Always keep a dry chemical (Class B) fire extinguisher near the work area.*

Inspection

1   Once in a while, you will have to raise the vehicle to service or replace some component (an exhaust pipe hanger, for example). Whenever you work under the vehicle, always inspect the fuel lines and fittings for possible damage or deterioration.
2   Check all hoses and pipes for cracks, kinks, deformation or obstructions.
3   Make sure all hose and pipe clips attach

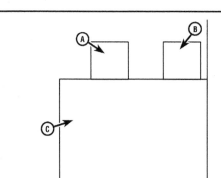

2.9a On 1992 through 1995 Sidekick/Tracker models, the fuel pump relay (A) is located behind the left side fascia next to the main relay (B) and the PCM (C)

2.9b On Samurai models, the fuel pump relay (A) is located above the PCM (C) next to the main relay (B) behind the right side instrument panel

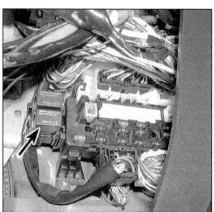

2.9c On 1996 through 1998 models, the fuel pump relay (arrow) is located on the right side of the heater/evaporator housing (behind the center of the instrument panel)

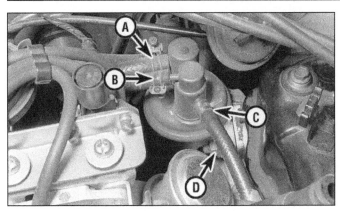

4.3 The fuel pump on carburetor-equipped models is mounted on the right rear corner of the cylinder head

A	Fuel inlet hose	C Fuel outlet hose
B	Fuel return hose	D Mounting nut (1 of 2)

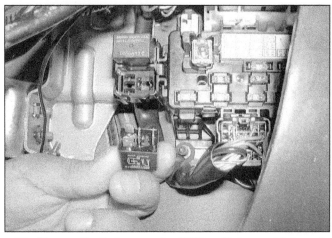

4.9a Remove the fuel pump relay (1999 and later model shown)

their associated hoses or pipes securely to the underside of the vehicle.

4   Verify all hose clamps attaching rubber hoses to metal fuel lines or pipes are snug enough to assure a tight fit between the hoses and pipes.

Replacement

5   If you must replace any damaged sections, use original equipment replacement hoses or pipes constructed from exactly the same material as the section you are replacing. Do not install substitutes constructed from inferior or inappropriate material or you could cause a fuel leak or a fire.

6   Always, before detaching or disassembling any part of the fuel line system, note the routing of all hoses and pipes and the orientation of all clamps and clips to assure that replacement sections are installed in exactly the same manner.

7   Before detaching any part of the fuel system, be sure to relieve the fuel tank pressure (see Section 2).

8   While you're under the vehicle, it's a good idea to check the condition of the fuel filter - make sure that it's not clogged or damaged (see Chapter 1).

---

4   Fuel pump - check

Warning: *Gasoline is extremely flammable, so extra precautions must be taken when working on any part of the fuel system. Do not smoke or allow open flames or bare light bulbs in the work area. Also, do not work in a garage if a natural gas-type appliance is present. Always keep a dry chemical (Class B) fire extinguisher near the work area.*
Note: *The following checks assume the fuel filter is in good condition. If you doubt it's condition, install a new one (see Chapter 1).*

1   Check that there is adequate fuel in the fuel tank. If you doubt the reading on the gauge, insert a long wooden dowel at the filler opening; it will serve as a dipstick.

Mechanical pump (carbureted vehicles)

Refer to illustration 4.3

2   Raise the vehicle and support it securely on jackstands. With the engine running, examine all fuel lines between the fuel tank and fuel pump for leaks, loose connections, kinks or flattening in the rubber hoses. Do this quickly, before the engine gets hot. Air leaks upstream of the fuel pump can seriously affect the pump's output. Shut off the engine.

3   Check the body of the pump for leaks **(see illustration)**.

4   Remove the fuel filler cap to relieve the fuel tank pressure. Disconnect the fuel line at the carburetor. Disconnect the primary wiring connector to the ignition coil so the engine can be cranked without it firing. Place an approved gasoline container at the end of the detached fuel line and have an assistant crank the engine for several seconds. There should be a strong spurt of gasoline from the line on every second revolution.

5   If little or no gasoline emerges from the line during engine cranking, either the fuel line is clogged or the fuel pump is not working properly. Disconnect the fuel feed line from the pump and blow air through it to be sure that the line is clear. If the line is not clogged, then the pump is suspect and needs to be replaced with a new one.

Electric pump (fuel injected vehicles)

General check

Refer to illustrations 4.9a, 4.9b, 4.9c and 4.9d

6   Although the best way to check the operation of the fuel pump is with a fuel pressure gauge (see step 10), to do so on these fuel injected vehicles requires the use of special adapters not normally available to the home mechanic. It is possible, however, to determine if the fuel pump is receiving power and rotating, which is usually a pretty good indication that it is pumping fuel.

7   Turn the ignition key to the ON position, but don't crank the engine. Remove the fuel filler cap, put your ear next to the opening and listen for the fuel pump working, which is characterized by a whirring sound. After listening, turn the ignition switch to OFF.

8   If the fuel pump is not working, check the fuse. If the fuse is okay, find the fuel pump relay, which is located under the left side of the dash on Sidekick/Tracker models or under the right side of the dash on Samurai models **(see illustrations 2.9a through 2.9c)**.

9   Remove the fuel pump relay **(see illustration)** and test it as follows:

a) *On 1990 and earlier models, check the resistance across terminals G and F (see illustration); the ohmmeter should indicate infinite resistance. Install a jumper wire from the battery (+) to terminal H and another jumper wire (-) to terminal J. Resistance across terminals G and F should be zero. Also, check the resistance, without any battery voltage applied, between H and J and then F and I. Continuity should exist. If the test results are incorrect, replace the relay.*

b) *On 1991 through 1998 models, check the resistance across terminals A and B* **(see illustration)**; *the ohmmeter should indicate infinite resistance. Next, install a jumper wire from the battery (+) to terminal C and another jumper wire (-) to terminal D. Resistance across terminals A and B should be zero. If the test results are incorrect, replace the relay.*

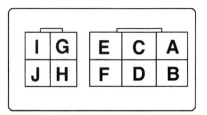

4.9b Fuel pump relay terminal guide (1990 and earlier models)

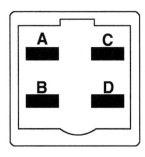

4.9c Fuel pump relay terminal guide (1991 through 1998 models)

c) *On 1999 and later models, check resistance across terminals A and B (see illustration); the ohmmeter should read infinite resistance. Then check the resistance across terminals C and D; there should be 63 to 77 ohms. Next, install a jumper wire from the battery (-) to terminal C and another jumper wire (+) to terminal D. Resistance across terminals A and B should be zero. If the test results are incorrect, replace the relay.*

If the fuel pump is still not working, check the fuel pump relay circuit (see the wiring diagrams at the end of Chapter 12).

Fuel pressure check

10   Relieve the fuel pressure (see Section 2).
11   Install a fuel pressure gauge. On models with sequential multiport fuel injection, disconnect the fuel feed line from the fuel rail and install the gauge using a special hose and adapter. On models with EFI, disconnect the fuel inlet fitting from the fuel filter and install the gauge using a banjo-type connector.
12   Turn the ignition switch to ON. The fuel pump should run for about two seconds. Turn the key off and on about four times (pausing about two seconds each time) until the fuel pressure reading on the gauge stabilizes. It should be about 3 to 10 psi above the range listed in this Chapter's Specifications.
13   Start the engine and let it idle at normal operating temperature. The pressure should be within the range listed in this Chapter's Specifications. If all the pressure readings are within the limits listed in this Chapter's Specifications, the system is operating normally.
14   On models with sequential multiport fuel injection, if the pressure did not drop by 3 to 10 psi after starting the engine, apply 12 to 14 inches of vacuum to the pressure regulator. If the pressure now drops, repair the vacuum source to the pressure regulator. If the pressure does not drop, replace the regulator.
15   If the pressure is higher than specified, check for a faulty fuel pressure regulator or a pinched or clogged fuel return hose or pipe.
16   If the pressure is lower than specified, check the following:

a) *Inspect the fuel filter - make sure it's not clogged.*

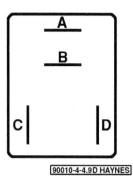

90010-4-4.9D HAYNES

4.9d Fuel pump relay terminal guide (1999 and later models)

b) *Look for a pinched or clogged fuel hose between the fuel tank and fuel rail.*
c) *Pinch the fuel return hose; if the pressure increases, replace the fuel pressure regulator.*
d) *Look for leaks in the fuel feed line/hose.*
e) *Check for leaking injectors.*
f) *Check the in-tank fuel pump check valve.*

17   If the fuel pressure is low, there are no problems with any of the above-listed components and the fuel pump is operating (see Step 6), the problem is one of the following:

a) *The fuel pump is not receiving sufficient voltage (check the battery and charging system - see Chapter 5).*
b) *There is a poor electrical connection in the fuel pump circuit (a bad ground connection is very likely).*
c) *The fuel pump itself is faulty.*

18   Turn off the engine and note the reading on the gauge. After one minute, the pressure should not drop to less than 21 psi on EFI models or 25 psi on models with sequential multiport fuel injection. If the fuel pressure drops rapidly after turning off the engine, there's a leak in the fuel system. If there are no external leaks, the leak is probably at the fuel pump (likely a leaking outlet check valve) or at the fuel injector(s).
19   After the testing is done, relieve the fuel pressure (see Section 2) and remove the fuel gauge.

5   Fuel pump - removal and installation

Warning: *Gasoline is extremely flammable, so extra precautions must be taken when working on any part of the fuel system. Do not smoke or allow open flames or bare light bulbs in the work area. Also, do not work in a garage if a natural gas-type appliance is present. Always keep a dry chemical (Class B) fire extinguisher near the work area.*

1   Disconnect the cable from the negative terminal of the battery.
2   Relieve the fuel system pressure (see Section 2).

Mechanical pump (carbureted vehicles)

3   Relieve the fuel tank pressure by removing the fuel filler cap.
4   Locate the fuel pump mounted on the right rear corner of the cylinder head **(see illustration 4.3)**. Place rags underneath the pump to catch any spilled fuel.
5   Loosen the hose clamps and slide them down the hoses, past the fittings. Disconnect the hoses from the pump, using a twisting motion as you pull them from the fittings. Immediately plug the hoses to prevent leakage of fuel and the entry of dirt.
6   Unscrew the fasteners that retain the pump to the cylinder head, then detach the pump from the head. Remove the pump pushrod and inspect it for wear, replacing it if necessary. Coat it with clean engine oil before installing it.
7   Using a gasket scraper or putty knife, remove all traces of old gasket material from the mating surfaces on the cylinder head (and the fuel pump, if the same one will be reinstalled). While scraping, be careful not to gouge the soft aluminum surfaces.
8   Installation is the reverse of the removal procedure, but be sure to use a new gasket and tighten the mounting fasteners to the specified torque.

Electric pump (fuel injected vehicles)

Refer to illustrations 5.10a and 5.10b
9   Remove the fuel tank (see Section 6).
10   Unscrew the six bolts that secure the fuel pump flange **(see illustration)**, then lift the assembly out of the fuel tank **(see illustration)**.
11   Remove the pump from its bracket and disconnect the electrical connectors, noting their positions.
12   Push the nozzle of the new pump into the coupling hose, then insert the bottom of the pump into its bracket. Connect the wires to the proper terminals. If the filter at the bottom of the pump appears dirty, replace it.
13   The remainder of Installation is the reverse of removal. Be sure to replace the pump unit sealing O-ring if it shows any signs of deterioration.

5.10a To detach the fuel pump/fuel level gauge sending unit from the fuel tank, remove these six bolts (arrows) . . .

5.10b . . . and then pull the assembly straight up (fuel tank removed for clarity)

6 Fuel tank - removal and installation

Refer to illustrations 6.6, 6.7a, 6.7b, 6.9 and 6.10

Note: *The following procedure is much easier to perform if the fuel tank is empty. Some tanks have a drain plug for this purpose. If the tank does not have a drain plug, use a siphoning kit (available at most auto parts stores) and drain the fuel into an approved gasoline container.*

Warning: *Gasoline is extremely flammable, so extra precautions must be taken when working on any part of the fuel system. Do not smoke or allow open flames or bare light bulbs near the work area. Also, do not work in a garage if a natural gas-type appliance is present. When performing any work on the fuel tank, wear safety glasses and have a dry chemical (Class B) fire extinguisher on hand. If you spill any fuel on your skin, rinse it off immediately with soap and water.*

1   Remove the fuel tank filler cap to relieve fuel tank pressure.
2   Relieve the fuel system pressure (see Section 2).
3   Detach the cable from the negative terminal of the battery.
4   Raise the vehicle and place it securely

6.6 Loosen the hose clamp on the fuel filler neck and slide the hose off the tank (early model shown, later models similar)

6.7b . . . and these two (arrows) on the right

on jackstands.
5   If the tank has a drain plug, remove it and allow the fuel to collect in an approved gasoline container. If not, use a siphoning kit (available at most auto parts stores) and drain the fuel into an approved gasoline container.
6   Loosen the fuel filler neck hose clamp **(see illustration)** and then disconnect the filler neck hose from the fuel tank.
7   Remove the four fuel tank skid plate bolts **(see illustrations)**.

6.7a The fuel tank skid plate is secured to the underside of the vehicle with four bolts: these two (arrows) on the left . . .

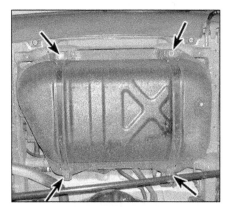

6.9 To release the fuel tank from the underside of the vehicle, remove these four retaining strap bolts (arrows)

8   Support the fuel tank with a floor jack. Position a piece of wood between the jack head and the fuel tank to protect the tank.
9   Remove the four fuel tank retaining strap bolts **(see illustration)**.
10   Carefully lower the tank just enough to disconnect the electrical connectors and hoses from the tank **(see illustration)**. **Note:** *The fuel and vapor hoses are different diameters to prevent confusion during reassembly. But if you have any doubts, clearly label the hoses and their corresponding pipes. Be sure to plug the hoses to prevent leakage and contamination of the fuel system.*
11   Remove the tank from the vehicle.
12   Installation is the reverse of removal.

7 Fuel tank cleaning and repair - general information

1   All repairs to the fuel tank or filler neck should be carried out by a professional who has experience in this critical and potentially dangerous work. Even after cleaning and flushing of the fuel system, explosive fumes can remain and ignite during repair of the tank.

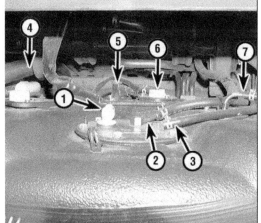

6.10 Typical 1999 and later hoses and electrical connectors (2001 model shown, earlier models similar, except that pre-1999 models are not equipped with OBD-II evaporative control system)

1   *Electrical connection for in-tank fuel pump and fuel level gauge sending unit*
2   *Fuel supply hose*
3   *Fuel return hose*
4   *Vapor control valve*
5   *Fuel tank pressure sensor*
6   *Fuel cut valve*
7   *Recirculation hose*

8.1 Typical air cleaner installation details (carbureted models)

A   Clips
B   Air cleaner mounting fasteners

2   If the fuel tank is removed from the vehicle, it should not be placed in an area where sparks or open flames could ignite the fumes coming out of the tank. Be especially careful inside garages where a natural gas-type appliance is located, because the burner could cause an explosion.

8   Air cleaner assembly - removal and installation

Carbureted vehicles

Refer to illustration 8.1

1   Unlatch the clips that secure the top of the air cleaner to the main housing **(see illustration)**.
2   Remove the nut that secures the air intake case to the carburetor, then lift off the case and disconnect the vacuum line from the underside. Unplug the breather hose from the rocker arm cover and remove the air intake case.

8.7 To detach the air cleaner housing from the inner fender on multiport fuel injected models, remove these bolts (arrows)

3   Disconnect the hot and fresh air intake hoses from the air cleaner. Remove the three mounting nuts and washers that secure the air cleaner to the inner fender panel **(see illustration 8.1)** and remove the air cleaner from the vehicle.
4   Installation is the reverse of removal.

Fuel injected vehicles

Refer to illustrations 8.5 and 8.7

5   Remove the air intake duct **(see illustration)**.
6   Unscrew or unclip the air cleaner housing cover and then remove the air filter element (see Chapter 1).
7   Unbolt the air cleaner case from the inner fender panel **(see illustration)**.
8   Installation is the reverse of removal.

9   Throttle cable - removal, installation and adjustment

1   Disconnect the cable from the negative terminal of the battery.

9.4 Pass the cable through the slot (arrow) and slide the cable end out of the accelerator pedal linkage (carbureted and throttle body injection vehicles)

8.5 Typical air intake duct installation details (multiport fuel-injected model shown; throttle body injection models similar, except that air cleaner housing and duct are located on driver's side of engine compartment)

1   MAF sensor electrical connector
2   Wiring harness clip (pull straight out to detach harness from clip)
3   Air intake duct hose clamp
4   Air cleaner housing cover clip (one at each corner of housing); earlier models use screws instead of clips

Carbureted vehicles and vehicles with throttle body injection

Removal

Refer to illustration 9.4

2   Unscrew the locknut on the threaded portion of the throttle cable at the carburetor or throttle body.
3   Pass the cable through the slot in the throttle lever and slide the cable end from the lever.
4   Working inside the vehicle, disconnect the cable from the accelerator pedal by passing the cable through the slot at the top of the pedal linkage **(see illustration)**.
5   Pull the cable through the firewall into the engine compartment.

Installation and adjustment

Refer to illustration 9.7

6   Installation is the reverse of the removal procedure, but it is important to adjust the cable to obtain the correct amount of free play.

9.7 Push on the cable and measure the freeplay - it should be as listed in this Chapter's Specifications (carbureted and throttle body injection vehicles)

9.9 To disengage the throttle cable from the throttle linkage cam, rotate the cam clockwise and slide out the cable end plug (multiport fuel injection vehicles)

9.10 To disengage the throttle cable from the bracket on the air intake plenum, loosen the locknut (upper arrow) and then loosen the adjuster nut (lower arrow) (multiport fuel injection vehicles)

7   Lightly push on the cable between the throttle lever and the cable bracket (see illustration). Measure the amount the cable deflects and compare it with the figure listed in the Specifications section.

8   To adjust cable free play, screw the adjuster nut up or down on the threaded portion of the cable casing, as necessary, to obtain the correct amount of freeplay. If you adjust the cable when the engine is cold, be sure to recheck the adjustment after it has reached normal operating temperature.

Vehicles with sequential multiport fuel injection

Removal
Refer to illustrations 9.9, 9.10, 9.11 and 9.12

9   Rotate the throttle linkage cam clockwise and disengage the end of the throttle cable from the cam (see illustration).

10   Loosen the locknut and the adjuster nut at the cable bracket on the air intake plenum (see illustration).

11   Disengage the cable from the bracket on the air intake plenum and from the clip at the rear of the camshaft cover (see illustration).

12   Working inside the vehicle, disengage the cable end from the accelerator pedal (see illustration), squeeze the tangs together on the cable retaining clip at the firewall, and then disengage the retainer from the firewall.

13   Pull out the cable through the firewall from the engine compartment side.

Installation and adjustment
Refer to illustrations 9.14 and 9.15

14   With the throttle valve closed, measure accelerator pedal freeplay (see illustration). If the pedal freeplay is incorrect, adjust it by loosening the locknut and turning the adjuster nut at the cable bracket on the air intake plenum (see illustration 9.11).

15   Have an assistant push the accelerator pedal all way to the floor until it bottoms against the stopper bolt (see illustration 9.14). In the engine compartment, measure

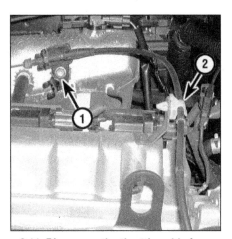

9.11 Disengage the throttle cable from the bracket on the air intake plenum (1) and then pull the cable out of the clip (2) at the rear end of the camshaft cover (multiport fuel injection vehicles)

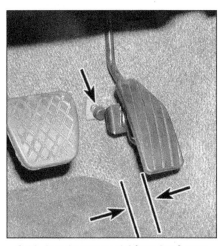

9.14 Accelerator pedal freeplay (lower arrows) is the distance you can depress the accelerator pedal before you feel resistance; to adjust the throttle lever-to-lever stopper clearance, turn the pedal stopper bolt (upper arrow) in or out

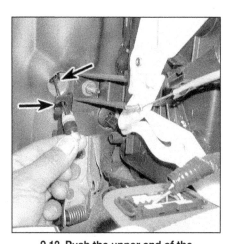

9.12 Push the upper end of the accelerator pedal forward and disengage the throttle cable by sliding it out of the slot (lower arrow) at the top of the pedal assembly and then squeeze the tangs of the retainer (upper arrow) together and pull the cable through the firewall from the engine compartment side (multiport fuel injection vehicles)

9.15 To measure the throttle lever-to-lever stopper clearance, insert a feeler gauge into the gap (arrows) between the throttle lever and the lever stopper

10.2 Using a feeler gauge (arrow), check the clearance of the choke plate to the carburetor bore

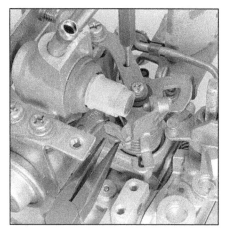

10.5 With the idle-up actuator removed, rotate the fast idle cam to align the hole in the cam with the hole in the bracket - insert a punch through the two holes to hold the cam in this position - the choke lever can now be adjusted with a pair of pliers

10.10 With the carburetor cool, make sure the mark on the fast idle cam is in alignment with the center of the cam follower

the clearance between the throttle lever and the lever stopper **(see illustration)**. Compare your measurement against the throttle lever-to-lever stopper clearance listed in this Chapter's Specifications. If the throttle stop clearance is incorrect, adjust it by turning the pedal stopper bolt inside the vehicle **(see illustration 9.14)**.

10 Carburetor - check and adjustment

Warning: *Gasoline is extremely flammable, so extra precautions must be taken when working on any part of the fuel system. Do not smoke or allow open flames or bare light bulbs in the work area. Also, do not work in a garage if a natural gas-type appliance is present. Always keep a dry chemical (Class B) fire extinguisher near the work area.*

Note: *If your vehicle's engine is hard to start or does not start at all, has an unstable idle or poor driveability and you suspect the carburetor is malfunctioning, it's best to first take the vehicle to a dealer service department that has the equipment necessary to diagnose this highly complicated system. The following procedures are intended to help the home mechanic verify proper operation of components, make minor adjustments, and replace some components. They are not intended as troubleshooting procedures.*

Choke check and adjustment

Refer to illustrations 10.2 and 10.5

1   Remove the air intake case (see Section 8 if necessary). With the engine stopped and cold, hold the throttle open and depress the choke plate with your finger - it should move smoothly. The choke should be almost fully closed if the ambient air temperature is below 77-degrees F (25-degrees C) and the engine is cold.

2   Check the clearance between the choke plate and the carburetor bore **(see illustration)** and compare your reading with those

listed in the Specifications section. If the clearances are too small or too large, lubricate the choke linkage and take another measurement.

3   Start the engine and allow it to warm to normal operating temperature. Depress and release the accelerator once. The choke should now be fully open. If it's not, and the plate-to-bore clearance is correct, there's a problem with the choke linkage or operating mechanism.

4   If the plate-to-bore clearance is not as specified, remove the carburetor (see Section 11) and adjust the choke lever as follows.

5   Remove the idle-up actuator from the carburetor, turn the fast idle cam counterclockwise and insert a pin into the cam and bracket to lock them into place **(see illustration)**.

6   Using a pair of pliers, bend the choke lever up or down until the choke-to-bore clearance is set to the specified amount. Bending the tab up causes the choke valve to close, while bending it down allows it to open a little more **(see illustration 10.5)**.

7   Reinstall the carburetor and check the choke again.

Accelerator pump

8   Remove the air intake case (see Section 8). With the engine Off, operate the throttle linkage through its full range of travel while looking down the throat of the carburetor (you may have to hold the choke plate open). A healthy stream of fuel should squirt out of the pump discharge nozzle. If fuel just dribbles out or there is no squirt at all, the accelerator pump is defective or the discharge passage is clogged. In either case, an overhaul of the carburetor is required.

Fast idle adjustment

Refer to illustrations 10.10 and 10.11

9   Allow the engine to cool for at least four

hours, then remove the carburetor and let it cool off for another hour. The ambient temperature should be between 71 and 82-degrees F.

10   After the carburetor has cooled, check to see that the mark on the fast idle cam and the center of the cam follower are in alignment **(see illustration)**.

11   Disconnect the hose from the idle-up actuator and connect a hand-held vacuum pump to the port. Apply a vacuum of approximately 16-in Hg. and measure the clearance between the actuator rod and the idle-up adjusting screw **(see illustration)**. It should be 0.10 to 0.12 in (2.5 to 3.0 mm). If it isn't, turn the idle-up adjusting screw accordingly. Reinstall the carburetor.

Idle-up adjustment

12   Run the engine until it reaches normal operating temperature. Connect a tachometer and verify that the idle speed is as specified (see Chapter 1).

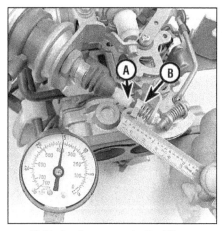

10.11 Apply vacuum to the idle-up actuator and measure the clearance between the actuator rod and the idle up adjusting screw

A   Actuator rod
B   idle-up adjusting screw

13   Turn the parking lamps On and check to see that the idle-up actuator rod moves down. Now turn the headlights On and check the engine rpm (the heater fan, rear defogger and air conditioner must be turned Off). It should increase to 900-1000 rpm. If it doesn't increase, turn the adjusting screw **(see illustration 10.11)**.

11 Carburetor - removal and installation

Warning: *Gasoline is extremely flammable so extra precautions must be taken when working on any part of the fuel system. DO NOT smoke or allow open flames or bare light bulbs in or near the work area. Also, don't work in a garage if a natural gas appliance is present. Always keep a dry chemical (Class B) fire extinguisher near the work area.*

Removal

1   Remove the fuel filler cap to relieve fuel tank pressure.
2   Remove the air cleaner from the carburetor. Be sure to label all vacuum hoses attached to the air cleaner housing.
3   Disconnect the throttle cable from the throttle lever (see Section 9).
4   If the vehicle is equipped with an automatic transmission, disconnect the TV cable from the throttle lever.
5   Clearly label all vacuum hoses and fittings, then disconnect the hoses.
6   Disconnect the fuel line from the carburetor.
7   Label the wires and terminals, then unplug all wire harness connectors.
8   Remove the mounting fasteners and detach the carburetor from the intake manifold. Remove the carburetor mounting gasket. Stuff a shop rag into the intake manifold openings.

Installation

9   Use a gasket scraper to remove all traces of gasket material and sealant from the intake manifold (and the carburetor, if it's being reinstalled), then remove the shop rag from the manifold openings. Clean the mating surfaces with lacquer thinner or acetone.
10   Place a new gasket on the intake manifold.
11   Position the carburetor on the gasket and install the mounting fasteners.
12   To prevent carburetor distortion or damage, tighten the fasteners to the specified torque in a criss-cross pattern, 1/4-turn at a time.
13   The remaining installation steps are the reverse of removal.
14   Check and, if necessary, adjust the idle speed (Chapter 1).
15   If the vehicle is equipped with an automatic transmission, refer to Chapter 7B for the kickdown cable adjustment procedure.
16   Start the engine and check carefully for fuel leaks.

12 Carburetor - diagnosis and overhaul

Refer to illustration 12.6
Warning: *Gasoline is extremely flammable, so extra precautions must be taken when working on any part of the fuel system. DO NOT smoke or allow open flames or bare light bulbs in or near the work area. Also, don't work in a garage if a natural gas appliance is present. Always keep a dry chemical (Class B) fire extinguisher near the work area.*

Diagnosis

1   A thorough road test and check of carburetor adjustments should be done before any major carburetor service work. Follow the procedures in Section 10. Specifications for some adjustments are listed on the Vehicle Emissions Control Information (VECI) label found in the engine compartment.
2   Carburetor problems usually show up as flooding, hard starting, stalling, severe backfiring and poor acceleration. A carburetor that's leaking fuel and/or covered with wet looking deposits definitely needs attention.
3   Some performance complaints directed at the carburetor are actually a result of loose, out-of-adjustment or malfunctioning engine or electrical components. Others develop when vacuum hoses leak, are disconnected or are incorrectly routed. The proper approach to analyzing carburetor problems should include the following items:

a)   *Inspect all vacuum hoses and actuators for leaks and correct installation (see Chapters 1 and 6).*
b)   *Tighten the intake manifold and carburetor mounting nuts/bolts to the torques specified in this Chapter and Chapter 2A.*
c)   *Perform a cylinder compression test (see Chapter 2B).*
d)   *Clean or replace the spark plugs as necessary (see Chapter 1).*
e)   *Check the spark plug wires (see Chapter 1).*
f)   *Inspect the ignition primary wires.*
g)   *Check the ignition timing (follow the instructions printed on the Emissions Control Information label).*
h)   *Check the fuel pump (see Chapter 4).*
i)   *Check the heat control valve in the air cleaner for proper operation (see Chapter 1).*
j)   *Check/replace the air filter element (see Chapter 1).*
k)   *Check the PCV system (see Chapter 6).*
l)   *Check/replace the fuel filter (see Chapter 1). Also, the strainer in the tank could be restricted.*
m)   *Check for a plugged exhaust system.*
n)   *Check EGR valve operation (see Chapter 1).*
o)   *Check the choke - it should be completely open at normal engine operating temperature (see Section 10).*
p)   *Check for fuel leaks and kinked or dented fuel lines (see Chapters 1 and 4).*
q)   *Check accelerator pump operation with the engine off (remove the air cleaner cover and operate the throttle as you look into the carburetor throat - you should see a stream of gasoline enter the carburetor).*
r)   *Check for incorrect fuel or bad gasoline.*
s)   *Check the valve clearances (see Chapter 1).*
t)   *Have a dealer service department or repair shop check the electronic engine and carburetor controls.*

4   Diagnosing carburetor problems may require that the engine be started and run with the air cleaner off. While running the engine without the air cleaner, backfires are possible. This situation is likely to occur if the carburetor is malfunctioning, but just the removal of the air cleaner can lean the fuel/air mixture enough to produce an engine backfire. **Warning:** *Do not position any part of your body, especially your face, directly over the carburetor during inspection and servicing procedures. Wear eye protection!*

Overhaul

5   Once it's determined that the carburetor needs an overhaul, several options are available. If you're going to attempt to overhaul the carburetor yourself, first obtain a good quality carburetor rebuild kit (which will include all necessary gaskets, internal parts, instructions and a parts list). You'll also need some special solvent and a means of blowing out the internal passages of the carburetor with air.
6   An alternative is to obtain a new or rebuilt carburetor. They are readily available from dealers and auto parts stores. Make absolutely sure the exchange carburetor is identical to the original. A tag is usually attached to the top of the carburetor or a number is stamped on the float bowl **(see illustration)**. It will help determine the exact type of carburetor you have. When obtaining a rebuilt carburetor or a rebuild kit, make sure the kit or carburetor matches your application

12.6 Before buying a carburetor rebuild kit, look for a number like this and write it down - it will ensure that you get the correct rebuild kit for the carburetor

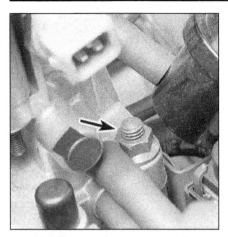

14.6 Remove the two nuts on the backside of the throttle body injection unit (arrow) - the other mounting nut is not visible in this photo

exactly. Seemingly insignificant differences can make a large difference in engine performance.

7   If you choose to overhaul your own carburetor, allow enough time to disassemble it carefully, soak the necessary parts in the cleaning solvent (usually for at least one-half day or according to the instructions listed on the carburetor cleaner) and reassemble it, which will usually take much longer than disassembly. When disassembling the carburetor, match each part with the illustration in the carburetor kit and lay the parts out in order on a clean work surface. Overhauls by inexperienced mechanics can result in an engine which runs poorly or not at all. To avoid this, use care and patience when disassembling the carburetor so you can reassemble it correctly.

8   Because carburetor designs are constantly modified by the manufacturer in order to meet increasingly more stringent emissions regulations, it isn't feasible to include a step-by-step overhaul of each type. You'll receive a detailed, well illustrated set of instructions with any carburetor overhaul kit; they will apply in a more specific manner to the carburetor on your vehicle.

13   Electronic Fuel Injection (EFI) - general information

The Electronic Fuel Injection (EFI) system provides optimum mixture ratios at all stages of combustion and offers immediate throttle response characteristics. It also enables the engine to run at the leanest possible air/fuel mixture ratio, reducing exhaust gas emissions.

A throttle body injection unit replaces a conventional carburetor atop the intake manifold. It is controlled by the Powertrain Control Module (PCM), which monitors engine performance and adjusts the air/fuel mixture accordingly (see Chapter 6 for a complete description of the fuel control system).

An electric fuel pump - located in the fuel tank with the fuel gauge sending unit - pumps fuel to the fuel injection system through the fuel feed line and an inline fuel filter. A pressure regulator keeps fuel available at a constant pressure. Fuel in excess of injector needs is returned to the fuel tank by a separate line.

The basic throttle body injection unit consists of the throttle body housing, a fuel injector, a fuel pressure regulator, the throttle opener (which controls the throttle valve opening so that it's a little bit wider when the engine is starting than when it's at an idle), the throttle position sensor, an air valve (which lets an additional amount of air past the throttle valve during cold engine operation) and the idle speed control solenoid valve (which controls the idle speed according to the PCM).

The fuel injector is a solenoid-operated device controlled by the PCM. The PCM turns on the solenoid, which lifts a normally closed needle valve off its seat. The fuel, which is under pressure, is injected in a conical spray pattern at the walls of the throttle body bore above the throttle valve. The fuel which is not used by the injector passes through the pressure regulator before being returned to the fuel tank.

14   Throttle body injection unit - removal and installation

Refer to illustration 14.6
Warning: *Gasoline is extremely flammable, so extra precautions must be taken when working on any part of the fuel system. Do not smoke or allow open flames or bare light bulbs in the work area. Also, do not work in a garage if a natural gas-type appliance is present. Always keep a dry chemical (Class B) fire extinguisher near the work area.*

Removal

1   Disconnect the cable from the negative terminal of the battery.
2   Relieve the fuel system pressure (see Section 2).
3   Following the procedure described in Section 9, disconnect the throttle cable from the throttle lever at the throttle body injection unit.
4   Disconnect the fuel feed and return lines from the throttle body injection unit.
5   Label and disconnect any electrical connectors and vacuum hoses.
6   Remove the attaching bolts and lift the throttle body injection unit from the intake manifold **(see illustration)**. On automatic transmission models, also remove the PTC heater from underneath the throttle body injection unit and check the condition of the heating element (grid). If it is burned or has holes in it, replace it with a new one.

Installation

7   Using a gasket scraper or a putty knife,

remove all traces of old gasket material and sealant from the intake manifold (and throttle body, if the same one will be installed). While scraping, be careful not to gouge the soft aluminum surfaces.
8   Installation is the reverse of the removal procedure, but be sure to use a new throttle body injection unit base gasket, and tighten the bolts to the specified torque.

15   Electronic Fuel Injection (EFI) system - component check and replacement

Warning: *Gasoline is extremely flammable, so extra precautions must be taken when working on any part of the fuel system. Do not smoke or allow open flames or bare light bulbs in the work area. Also, do not work in a garage if a natural gas-type appliance is present. Always keep a dry chemical (Class B) fire extinguisher near the work area.*

Fuel injector

Refer to illustrations 15.2 and 15.11
1   Unbolt the air intake case from the top of the throttle body and move it aside.
2   Start the engine and carefully peer down into the throttle body (wear safety goggles), checking the injector spray pattern, using a timing light to illuminate the pattern. It should be an even, conical pattern **(see illustration)** - if it isn't, the injector must be replaced with a new one.
3   Shut off the engine and make sure the injection of fuel stops as well. The injector should not leak more than one drop per minute-if it does, replace it.
4   Disconnect the electrical connector from the injector and, using an ohmmeter, measure the resistance across the injector coils. If the resistance measured is not as listed in the Specifications section, replace the injector.
5   To replace the fuel injector, begin by

15.2 The fuel injector should emit a strong, conical spray of fuel against the walls of the throttle body injection unit bore

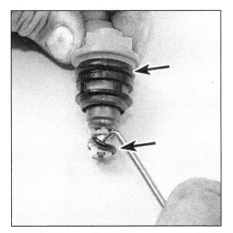

15.11 Remove the O-rings from the fuel injector (arrows)

relieving the fuel system pressure (see Section 2).

6   Disconnect the cable from the negative terminal of the battery.

7   Disconnect the fuel feed line from the throttle body. Remove the two injector cover screws and lift the cover from the injector.

8   Disconnect the injector electrical connector, release the harness clamp and dislodge the grommet from the throttle body housing.

9   To remove the injector from the throttle body, carefully direct compressed air into the fuel inlet port while pulling up on the injector. **Caution:** *Apply the compressed air gradually, using only enough to ease the injector out of the throttle body. Do not exceed 85 psi, or damage to the injector and other components may occur. Also, once the injector has been removed, handle it carefully and don't immerse it in solvent to clean it.*

10   Check the fuel filters on the injector for dirt particles. If there is any residue, clean the filters and check the fuel tank and lines for contamination.

11   Before installing the injector, lubricate the O-rings with light oil (if you are reinstalling the same injector, use new O-rings). Push the injector firmly into its bore, making sure the wiring harness is pointing toward its slot in the throttle body housing **(see illustration)**. Push the grommet on the wiring harness into the slot.

12   Install the injector cover and tighten the screws securely.

13   Hook up the cable to the negative battery terminal. Connect the injector electrical connector and pressurize the fuel system by turning the ignition key to the ON position. Check the fuel feed line and the injector for leakage.

14   Install the air intake case.

Fuel pressure regulator

Note: *Since checking the fuel pressure requires the use of some special adapters not normally available to the home mechanic, the fuel pressure should be checked by a dealer service department or other service facility*

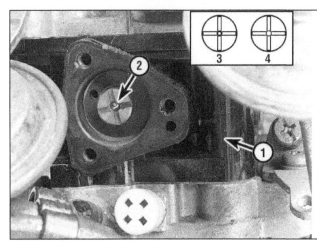

15.29 With the cap removed you can check the position of the air valve - it should be open when the engine is cool and closed when it's hot

1   *Throttle body*
2   *Air valve*
3   *Air valve open position (pintle backed-off)*
4   *Air valve closed position (pintle blocking hole)*

with the necessary hardware. If the regulator has been determined to be faulty, replace it using the following procedure.

15   Disconnect the cable from the negative terminal of the battery.

16   Relieve the fuel system pressure (see Section 2).

17   Loosen the hose clamp on the fuel return hose and disconnect the hose from the pressure regulator. Also disconnect the vacuum hose from its port on the regulator.

18   Remove the screws that secure the regulator to the throttle body, then pull the regulator straight out.

19   Coat the O-ring on the new fuel pressure regulator with light oil and push the regulator straight into the throttle body to install it. Tighten the screws securely.

20   Install the fuel return line, tightening the hose clamp securely. Connect the vacuum hose to its port on the regulator.

21   Connect the cable to the negative battery terminal, pressurize the fuel system by turning the ignition switch to ON and check around the regulator for fuel leaks.

Throttle opener

22   To check the throttle opener, disconnect the vacuum hose from it and connect a hand-held vacuum pump. Apply vacuum and verify that the plunger retracts.

23   If the throttle opener diaphragm does not hold vacuum or the plunger doesn't retract, replace it.

24   To replace the throttle opener, detach it along with its bracket by removing the two screws securing the bracket to the throttle body assembly. Position the new throttle opener and bracket on the throttle body and install the screws, tightening them securely.

25   To adjust the throttle opener, start the engine and allow it to warm up to normal operating temperature. Make sure all electrical accessories are turned off.

26   Connect a tachometer to the engine, following the manufacturer's instructions.

27   Disconnect the vacuum hose from the throttle opener diaphragm and plug it. Check the engine rpm - it should be approximately 1700 to 1800 rpm. If it isn't, adjust the rpm by turning the opener adjusting screw located

directly behind the throttle opener. Reconnect the vacuum hose.

Air valve

Refer to illustration 15.29

28   The air valve permits an additional amount of air to bypass the throttle plate when the engine is cold (when the coolant is less than 140-degrees F). This raises the engine rpm to a fast idle. If the engine doesn't run at a fast idle when it is cold, or if it doesn't idle down when warm, check the operation of the air valve as follows.

29   With the engine cold, remove the air valve cap from the throttle body injection unit. Look inside and confirm that the air valve is open **(see illustration)**.

30   Install the air valve cap and run the engine until it reaches normal operating temperature. Remove the cap once again - the valve should be closed.

31   If the air valve doesn't operate as described, it should be replaced with a new one. When installing the air valve cap, use a new gasket.

---

16   Sequential multiport fuel injection system - general information

Some 1992 through 1995 and all 1996 and later models are equipped with a sequential multiport fuel injection system. The sequential multiport fuel injection system consists of three basic subsystems: the fuel system, the air induction system and the electronic control system.

Fuel system

An electric fuel pump located inside the fuel tank supplies fuel under constant pressure to the fuel rail, which distributes pressurized fuel evenly to all injectors. From the fuel rail, fuel is injected into the intake ports, just above the intake valves, by four fuel injectors. The amount of fuel supplied by the injectors is precisely controlled by a Powertrain Control Module (PCM). The PCM opens each injector during the intake stroke of its corresponding cylinder for the precise duration of time necessary to produce the optimal air/fuel mixture for the current operating conditions. A fuel pressure regulator controls fuel pressure in relation to intake manifold vacuum. A fuel filter between the fuel pump and the fuel rail filters fuel to protect the components of the system.

Air induction system

The air induction system consists of the air filter housing, a Mass Airflow (MAF) sensor and a throttle body. During engine operation, the MAF sensor measures the amount of air flowing into the engine. This information helps the PCM determine the amount of fuel to be injected by the injectors (open-time duration of the injectors). The throttle valve inside the throttle body is controlled by the driver. As the throttle valve opens, the amount of air passing through the intake manifold increases, and the MAF sensor directs the PCM to increase the amount of fuel delivered to the intake ports.

Electronic control system

The electronic control system controls the fuel, ignition and many emissions systems by means of a Powertrain Control Module (PCM), a small but powerful microcomputer. The PCM receives signals from an array of information sensors which monitor such variables as intake air volume, throttle position, engine coolant temperature, intake air temperature, exhaust oxygen content, camshaft position, crankshaft position, engine rpm, vehicle speed, etc. The PCM uses the information it gathers from the information sensors to calculate such things as the duration of the "on-time" for the injectors, the ignition timing and the ignition advance curve. Then it uses an array of "output actuators" to effect the necessary changes. Some of the output actuators are: the fuel pump control system, the heaters for the oxygen sensors, the idle air control system, the ignition control system, the Exhaust Gas Recirculation (EGR) system and the Evaporative Emission (EVAP) control system. For more information regarding the PCM, the information sensors and the output actuators, see Chapter 6.

17   Sequential multiport fuel injection system - check

Refer to illustration 17.7

Warning: *Gasoline is extremely flammable, so extra precautions must be taken when working on any part of the fuel system. Do not smoke or allow open flames or bare light bulbs near the work area. Also, do not work in a garage if a natural gas-type appliance is present. While performing any work on the fuel system, wear safety glasses and have a dry chemical (Class B) fire extinguisher on hand. If you spill any fuel on your skin, rinse it off immediately with soap and water.*

1   Check the ground wire connections for tightness. Check all wiring and electrical con-

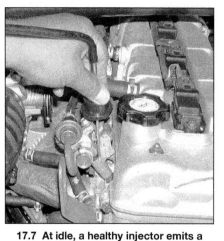

17.7   At idle, a healthy injector emits a crisp clicking sound. An automotive stethoscope is a handy tool for listening to injectors. If you don't have a stethoscope, use a large screwdriver: touch the tip of the screwdriver to the injector and, putting your ear up to the handle, "listen" to the injector (the screwdriver shaft transmits the clicking sound of the injector to the handle)

nectors that are related to the system. Loose electrical connectors and poor grounds can cause many problems that resemble more serious malfunctions.

2   Verify that the battery is fully charged - the PCM and information sensors need a stable and adequate voltage to accurately meter the fuel.

3   Inspect the air filter element - a dirty or partially blocked filter will severely impede performance and economy (see Chapter 1).

4   If a blown fuse is found, replace it and see if it blows again. If it does, look for a grounded wire in the harness related to the system.

5   Inspect the air intake duct, from the MAF sensor to the intake manifold, for leaks, which will result in an excessively lean mixture. Also inspect the condition of all vacuum hoses connected to the intake manifold.

6   Remove the air intake duct from the throttle body and look for dirt, carbon, sludge or other residue build-up. Deposits usually form in the throttle valve area. These deposits frequently cause driveability problems. Open the throttle plate by hand and clean any deposits in the throttle bore with a toothbrush and aerosol carburetor cleaner. Be sure the cleaner is safe for use with oxygen sensors and catalytic converters.

7   With the engine running, place a screwdriver or a stethoscope against each injector, one at a time, and listen through the handle for a clicking sound **(see illustration)**.

8   If an injector isn't operating (or sounds different than the others), turn off the engine and unplug the electrical connector from the injector. Check the resistance across the terminals of the injector and compare your reading with the resistance value listed in this Chapter's Specifications. If the resistance

isn't as specified, replace the injector with a new one.

9   Check the fuel pump and fuel pressure (see Section 4).

10   Clogged fuel injectors frequently cause driveability problems. Many repair shops have the equipment necessary to pressure-clean injectors and will perform this service at a reasonable cost. Several companies sell gasoline additives that help prevent injector clogs and, in some cases, can clean small clogs from injectors. Fuel additives are available inexpensively from auto parts stores.

11   The remainder of the system checks should be left to a dealer service department or other qualified repair shop, as there is a chance that the control unit may be damaged if not performed properly.

18   Throttle body - check, removal and installation

Check

1   Verify that the throttle linkage operates smoothly.

2   Inspect the throttle bore for deposits and clean, as necessary (see the previous Section). With the engine at idle, check for vacuum leaks at the throttle body base gasket, vacuum connections and at the throttle shaft by spraying aerosol carburetor cleaner in these areas. If the idle speed increases and smoothes out when spraying in one of these areas, you've located a vacuum leak (the carburetor cleaner temporarily seals the leak). Replace the base gasket or throttle body or repair the faulty vacuum connection, as necessary.

Removal and installation

Refer to illustration 18.8

3   Detach the cable from the negative terminal of the battery.

4   Drain the radiator (see Chapter 1).

5   Loosen the hose clamps and remove the air intake duct **(see illustration 8.5)**.

6   Detach the accelerator cable from the throttle linkage cam **(see illustration 9.9)**, and then detach the throttle cable bracket **(see illustration 9.11)** and set it aside (it's not necessary to detach the throttle cable from the bracket).

7   If your vehicle is equipped with an automatic transmission, detach the kickdown cable from the throttle linkage (see Chapter 7B), detach the cable brackets from the engine and set the cable and brackets aside.

8   Clearly label, then detach, the breather and coolant hoses from the throttle body **(see illustration)**.

9   Unplug the electrical connectors from the throttle position sensor and the idle air control valve **(see illustration 18.8)**.

10   Remove the throttle body mounting bolts **(see illustration 18.8)** and detach the throttle body and gasket from the air intake plenum.

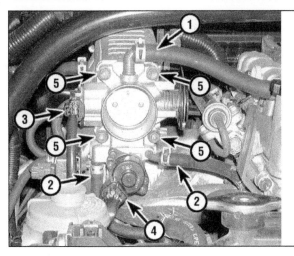

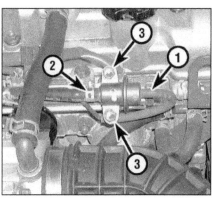

18.8 Throttle body installation details (sequential multiport fuel injection)

1   Crankcase breather hose
2   Coolant hoses
3   Throttle Position Sensor (TPS) electrical connector
4   Idle Air Control (IAC) valve electrical connector
5   Throttle body mounting bolts

19.3 Fuel pressure regulator installation details (sequential multiport fuel injection

1   Vacuum hose
2   Fuel return hose
3   Fuel pressure regulator mounting bolts

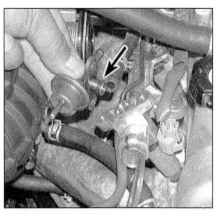

19.7 After removing the fuel pressure regulator, be sure to discard the old O-ring (arrow) and replace it with a new one

11   Using compressed air, thoroughly clean the throttle body casting, then blow out all passages with compressed air. **Caution:** *Do not clean the throttle position sensor or other components with anything. Just use compressed air or wipe them off carefully with a clean soft cloth. Do not use drill bits or wire to clean the passages in the casting.*

12   Installation of the throttle body is the reverse of removal. Be sure to tighten the throttle body mounting bolts to the torque listed in this Chapter's Specifications.

19   Fuel pressure regulator - check, removal and installation

Warning: *Gasoline is extremely flammable, so extra precautions must be taken when working on any part of the fuel system. Do not smoke or allow open flames or bare light bulbs near the work area. Also, do not work in a garage if a natural gas-type appliance is present. While performing any work on the fuel system, wear safety glasses and have a dry chemical (Class B) fire extinguisher on hand. If you spill any fuel on your skin, rinse it off immediately with soap and water.*

Check

1   Refer to the fuel pump/fuel pressure check procedure (see Section 4).

Removal and installation

Refer to illustrations 19.3 and 19.7

2   Relieve the fuel pressure (see Section 2) and then detach the cable from the negative terminal of the battery.

3   Detach the vacuum hose from the fuel pressure regulator **(see illustration)**.

4   Place a metal container or shop towel under the fuel return line. Also cover the line with a rag to catch any fuel that may spray out.

5   Loosen the hose clamp and detach the fuel return hose from the fuel pressure regulator pipe **(see illustration 19.3)**.

6   Remove the pressure regulator mounting bolts **(see illustration 19.3)** .

7   Detach the pressure regulator from the fuel rail **(see illustration)** and discard the old O-ring.

8   Installation is the reverse of removal. Be sure to use a new O-ring and tighten the pressure regulator mounting bolts to the torque listed in this Chapter's Specifications.

20   Fuel rail and fuel injectors - check, removal and installation

Warning: *Gasoline is extremely flammable, so extra precautions must be taken when working on any part of the fuel system. Do not smoke or allow open flames or bare light bulbs near the work area. Also, do not work in a garage if a natural gas-type appliance is present. While performing any work on the fuel system, wear safety glasses and have a dry chemical (Class B) fire extinguisher on hand. If you spill any fuel on your skin, rinse it off immediately with soap and water.*

Check

1   Refer to Section 17.

Removal and installation

Refer to illustrations 20.5, 20.7, 20.8, 20.9a, 20.9b and 20.10

2   Relieve the fuel pressure (see Section 2).

3   Detach the cable from the negative terminal of the battery.

4   On 1.6L engines, remove the front intake manifold stiffener bracket.

5   On 2.0L engines, remove the PCV hose and breather hose **(see illustration)**.

6   Disconnect the fuel supply hose banjo fitting from the rear end of the fuel rail and then disconnect the fuel return hose from the fuel pressure regulator **(see illustration 20.5)**.

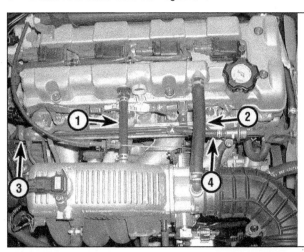

20.5 Fuel rail assembly installation details (sequential multiport fuel injection)

1   PCV hose
2   Crankcase breather hose
3   Fuel delivery hose banjo fitting (discard old sealing washers and install new ones when reattaching banjo fitting)
4   Fuel return hose (make sure that clamp is still tight enough to seal connection when reattaching)

20.7 Disconnect all four fuel injector connectors (sequential multiport fuel injection)

20.8 To detach the fuel rail from the cylinder head, remove these three bolts (arrows) (2.0L fuel rail shown, 1.6L fuel rail similar)

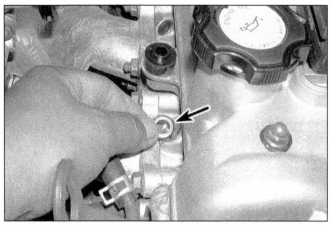

20.9a Be sure to remove - and discard - the old cushion (seal) from each injector bore; always use new cushions when installing the injectors

20.9b Remove the three fuel rail insulators (arrows) and inspect them for wear; if they're still in good condition, they can be reused

20.10 To disengage an injector from the fuel rail, pull it straight out; always discard the O-ring (left arrow) and seal (right arrow) and replace them before installing the injector

7   Unplug all four fuel injector electrical connectors **(see illustration)** and set the injector wire harness aside.

8   Remove the three fuel pressure regulator mounting bolts **(see illustration)** and then remove the fuel rail/injector assembly from the cylinder head by pulling on it while wiggling it back and forth.

9   Remove the old cushions (seals) from the injector bores **(see illustration)** and discard them. Be sure to use new cushions when reinstalling the injectors. Then remove the four fuel rail insulators **(see illustration)** from the cylinder head and set them aside (the insulators can be reused as long as they're in good condition).

10   Remove the fuel injectors from the fuel rail **(see illustration)**, set them aside in a clearly labeled storage container.

11   If you are replacing the injector(s), discard the old injector, the grommet and the O-ring **(see illustration 20.10)**. If you are simply replacing leaking injector O-rings and intend to re-use the same injectors, remove the old grommet and O-ring and discard them.

12   Installation is the reverse of removal. Be sure to tighten the fuel rail mounting bolts to the torque listed in this Chapter's Specifications.

21   Exhaust system servicing - general information

Refer to illustrations 21.1a, 21.1b and 21.1c
Warning: *Inspection and repair of exhaust system components should be done only after enough time has elapsed after driving the vehicle to allow the system components to cool completely. Also, when working under the vehicle, make sure it is securely supported on jackstands.*

1   The exhaust system consists of the exhaust manifold, the catalytic converter, the muffler, the tailpipe and all connecting pipes, brackets, hangers and clamps **(see illustrations)**. The exhaust system is attached to the body with mounting brackets and rubber hangers. If any of the parts are improperly installed, excessive noise and vibration will be transmitted to the body.

2   Conduct regular inspections of the exhaust system to keep it safe and quiet. Look for any damaged or bent parts, open seams, holes, loose connections, excessive corrosion or other defects which could allow exhaust fumes to enter the vehicle. Deterio-

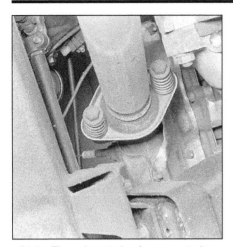

21.1a The exhaust pipe is connected to the exhaust manifold with two spring-loaded bolts - there is also a sealing ring or gasket (depending on the model) under the flange which should be replaced whenever the pipe is unbolted from the manifold

21.1b Here's a typical exhaust system hanger - these should be inspected for cracks and replaced if any are found

21.1c When unbolting exhaust system joints such as this catalytic converter-to-exhaust pipe flange, use a penetrating oil on the fasteners to ease removal

rated exhaust system components should not be repaired; they should be replaced with new parts.

3   If the exhaust system components are extremely corroded or rusted together, welding equipment will probably be required to remove them. The convenient way to accomplish this is to have a muffler repair shop remove the corroded sections with a cutting torch. If, however, you want to save money by doing it yourself (and you don't have a welding outfit with a cutting torch), simply cut off the old components with a hacksaw. If you have compressed air, special pneumatic cutting chisels can also be used. If you do decide to tackle the job at home, be sure to wear safety goggles to protect your eyes from metal chips and work gloves to protect your hands.

4   Here are some simple guidelines to follow when repairing the exhaust system:

a) *Work from the back to the front when removing exhaust system components.*

b) *Apply penetrating oil to the exhaust system component fasteners to make them easier to remove.*

c) *Use new gaskets, hangers and clamps when installing exhaust systems components.*

d) *Apply anti-seize compound to the threads of all exhaust system fasteners during reassembly.*

e) *Be sure to allow sufficient clearance between newly installed parts and all points on the underbody to avoid overheating the floor pan and possibly damaging the interior carpet and insulation. Pay particularly close attention to the catalytic converter and heat shield.*

Notes

Chapter 5
Engine electrical systems

Contents

Specifications

Ignition coil

1986 through 1989 (all models)	
Primary resistance	1.35 to 1.65 ohms
Secondary resistance	11.0 to 14.5 k-ohms
1990 and 1991	
Samurai	
Primary resistance	0.90 to 1.10 ohms
Secondary resistance	10.2 to 13.8 k-ohms
Sidekick/Tracker	
Primary resistance	1.35 to 1.65 ohms
Secondary resistance	11.0 to 14.5 k-ohms
1992	
Samurai	
Primary resistance	0.90 to 1.10 ohms
Secondary resistance	10.2 to 13.8 k-ohms
Sidekick	
Throttle body injection	
Primary resistance	1.35 to 1.65 ohms
Secondary resistance	11.0 to 14.5 k-ohms
Multiport fuel injection	
Primary resistance	0.72 to 0.88 ohms
Secondary resistance	10.2 to 14.0 k-ohms
1993	
Samurai	
Primary resistance	0.90 to 1.10 ohms
Secondary resistance	10.2 to 13.8 k-ohms
Sidekick	
Primary resistance	0.72 to 0.88 ohms
Secondary resistance	10.2 to 14.0 k-ohms

Ignition coil (continued)

1994 and 1995
    Samurai
        Primary resistance .... 0.90 to 1.10 ohms
        Secondary resistance .... 10.2 to 13.8 k-ohms
    Sidekick
        8-valve
            Primary resistance .... 0.72 to 0.88 ohms
            Secondary resistance .... 10.2 to 14.0 k-ohms
        16-valve
            Primary resistance .... 1.08 to 1.32 ohms
            Secondary resistance .... 22.1 to 30.0-ohms
1996 through 1998 (Sidekick/Tracker)
    Primary resistance.. 0.7 to 0.9 ohms
    Secondary resistance.. 13.0 to 18.0 k-ohms
1999 on (Vitara/Tracker)
    Primary resistance .... Specifications not available
    Secondary resistance... Specifications not available

Distributor

Generator (pick-up coil)* resistance
    Samurai
        1986 through 1989... 130 to 190 ohms
        1990 on .... Specifications not available
    Sidekick/Tracker .... Specifications not available
Generator (pick-up coil)* air gap
    Samurai
        1986 through 1989... 0.008 to 0.016 in (0.2 to 0.4 mm)
        1990 through 1992... Specifications not available
        1993 on .... Not adjustable
    Sidekick/Tracker
        1989 through 1991... 0.008 to 0.016 in (0.2 to 0.4 mm)
        1992 on .... Not adjustable
Igniter (ignition module)
    Samurai
        1986 through 1989... See Section 11
        1990 and later .... Specifications not available
    Sidekick/Tracker
        1991 through 1995
            No voltage applied to G and IB terminals .... No continuity
            Voltage applied to G and IB terminals.................................... Continuity
        1996 through 1998
            No voltage applied to No. 5 terminal..................................... No continuity
            Voltage applied to No. 5 terminal.. Continuity

*Referred to as the Camshaft Position (CMP) sensor on 1996 and later models (see Chapter 6 for more information).

Alternator

Standard charging voltage .... 14.2 to 14.8 volts
    Brush length
        Samurai
            Standard .... 0.43 in (11 mm)
            Minimum .... 0.20 in (5 mm)
        Sidekick/Tracker
            Standard .... 0.63 in (16 mm)
            Minimum .... 0.08 in (2 mm)

1   General information

The engine electrical systems include all ignition, charging and starting components. Because of their engine-related functions, these components are covered separately from chassis electrical devices such as the lights, instruments, etc. (which are included in Chapter 12).

Always observe the following precautions when working on the electrical systems:

a) *Be extremely careful when servicing engine electrical components. They're easily damaged if checked, connected or handled improperly.*
b) *Never leave the ignition switch on for long periods of time with the engine off.*
c) *Don't disconnect the battery cables while the engine is running.*
d) *Maintain correct polarity when connecting a battery cable from another vehicle during jump-starting.*
e) *Always disconnect the negative cable first and hook it up last or the battery may be shorted by the tool being used to loosen the cable clamps.*

It's also a good idea to review the safety-related information regarding the engine electrical systems located in the Safety first! Section near the front of this manual before beginning any operation included in this Chapter.

3.2 Unscrew the nuts (arrows), then pull the hold-down clamp off the battery

2 Battery - emergency jump starting

Refer to the *Booster battery (jump) starting* procedure at the front of this manual.

3 Battery - removal and installation

Refer to illustration 3.2
Caution: *Always disconnect the negative cable first and hook it up last or the battery may be shorted by the tool being used to loosen the cable clamps.*
1   Disconnect both cables from the battery terminals.
2   Remove the battery hold down clamp or strap **(see illustration)**.
3   Lift out the battery. Be careful - it's heavy.
4   While the battery is out, inspect the carrier (tray) for corrosion (see Chapter 1).
5   If you're replacing the battery, make sure you get one that's identical to the original, with the same dimensions, amperage rating, cold cranking rating, etc.
6   Installation is the reverse of removal.

4 Battery cables - check and replacement

1   Periodically inspect the entire length of each battery cable for damage, cracked or burned insulation and corrosion. Poor battery cable connections can cause starting problems and decreased engine performance.
2   Check the cable-to-terminal connections at the ends of the cables for cracks, loose wire strands and corrosion. The presence of white, fluffy deposits under the insulation at the cable terminal connection is a sign the cable is corroded and should be replaced. Check the terminals for distortion, missing mounting bolts and corrosion.
3   When removing the cables, always disconnect the negative cable first and hook it

up last or the battery may be shorted by the tool used to loosen the cable clamps. Even if only the positive cable is being replaced, be sure to disconnect the negative cable from the battery first (see Chapter 1 for further information regarding battery cable removal).
4   Disconnect the old cables from the battery, then trace each of them to their opposite ends and detach them from the starter solenoid and ground terminals. Note the routing of each cable to ensure correct installation.
5   If you're replacing either or both of the cables, take the old ones along when buying new cables. It's very important to replace the cables with identical parts. Cables have characteristics that make them easy to identify: positive cables are usually red, larger in cross-section and have a larger diameter battery post clamp; ground cables are usually black, smaller in cross-section and have a slightly smaller diameter clamp for the negative post.
6   Clean the threads of the solenoid or ground connection with a wire brush to remove corrosion. Apply a light coat of battery terminal corrosion inhibitor, or petroleum jelly, to the threads to prevent future corrosion.
7   Attach the cable to the solenoid or ground connection and tighten the mounting nut/bolt securely.
8   Before connecting a new cable to the battery, make sure it reaches the battery post without having to be stretched.
9   Connect the positive cable first, followed by the negative cable.

5 Ignition system - general information and precautions

Conventional electronic ignition system (1986 through 1998 models)

The ignition system on 1986 through 1998 models includes the battery, the ignition switch, the ignition coil, the distributor, the primary (low voltage) and secondary (high voltage) wiring circuits and the spark plugs. 1986 through 1989 Samurai and 1989 through 1991 Sidekick/Tracker systems use an "igniter" (ignition module) and a "generator" (also known as a signal generator or a pick-up coil), both of which are located inside the distributor. On these models, the generator is the timing device that turns the igniter on and off, which turns the primary side of the coil on and off. On 1992 through 1998 models, the Powertrain Control Module (PCM) controls switching of the igniter and the coil. These models use a Camshaft Position (CMP) sensor, instead of a generator, to tell the PCM when to turn the igniter on and off. On 1992 through 1995 models, the igniter is located outside the distributor, near the Powertrain Control Module (PCM). On 1996 through 1998 models, the ignition coil, the

igniter and the CMP sensor are all housed inside the distributor. If any one of these components fails on these later models, you must replace the distributor assembly.

Direct ignition (distributorless) system (1999 and later models)

The ignition system on 1999 and later models, which is known as a "direct ignition" or "distributorless" system, does not have a distributor. On these models, there are two ignition coils (SOHC engine) or four ignition coils (DOHC engine). On SOHC engines, the coils are located on the camshaft cover; on DOHC engines they're installed *in* the camshaft cover, right on top of each spark plug. On SOHC engines, each coil fires two spark plugs; on DOHC engines, each spark plug has its own coil. An igniter is built into each coil on all distributorless models. The igniter (or ignition module) is the device that toggles coil primary voltage on and off. The igniter is controlled by the Powertrain Control Module (PCM). The PCM uses the Camshaft Position (CMP) sensor, the Throttle Position (TP) sensor, the Engine Coolant Temperature (ECT) sensor and the Mass Air Flow (MAF) sensor to calculate the correct ignition timing. For more information on the PCM, the CMP sensor, the TP sensor and the ECT sensors, refer to Chapter 6.

When working on the ignition system, take the following precautions:

a) *Don't keep the ignition switch on for more than 10 seconds if the engine won't start.*
b) *Always connect a tachometer by following the manufacturer's instructions. Some tachometers may be incompatible with the ignition system. Consult a dealer service department before buying a tachometer.*
c) *Never ground the ignition coil terminals. Grounding the coil could result in damage to the igniter and/or the ignition coil.*
d) *Don't disconnect the battery when the engine is running.*
e) *Make sure the igniter is properly grounded.*

6 Ignition system - check

1986 through 1998 models

1   Attach an inductive timing light to each plug wire, one at a time, and crank the engine.

a) *If the light flashes, voltage is reaching the plug.*
b) *If the light doesn't flash, proceed to the next Step.*
2   Inspect the spark plug wire(s), distributor cap, rotor and spark plug(s) (see Chapter 1).
3   If the engine still won't start, check the ignition coil (see Section 7).

6.6 Use a calibrated spark tester to test for spark on 1999 and later models

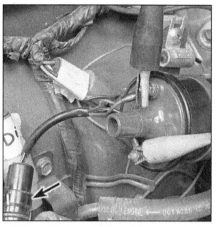

7.3a With the primary wiring electrical connector (arrow) unplugged, check the resistance across the primary terminals (early models)

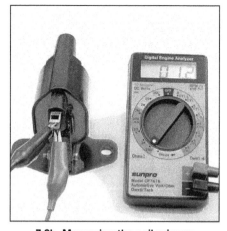

7.3b Measuring the coil primary resistance on later models (Sidekick/Tracker shown, Samurai similar) - to measure secondary resistance, measure between one of the primary terminals shown here and the secondary terminal (the terminal where the coil-to-distributor cap wire connects)

4   If the coil is functioning properly, proceed to check the igniter and generator (see Section 11).

1999 and later models

Refer to illustration 6.6
5   Remove the spark plug and inspect it (see Chapter 1).
6   If the spark plug is in good condition, install a calibrated spark tester in the spark plug cap and ground the tester to a good ground **(see illustration)**. (A spark tester can be obtained at most auto parts stores.) Crank the engine and verify that there is good spark at the tester.

  a) *If there is spark at the spark tester, voltage is reaching the plug.*
  b) *If there is no spark at the tester, proceed to the next Step.*

7   On SOHC engines, inspect the spark plug wires. (On DOHC engines, skip this Step.)
8   Check the ignition coil (see Section 7) and the igniter (see Section 11).

7   Ignition coil - check and replacement

Check

Note: *Inspect the ignition coil when the engine is cold.*
1   Disconnect the cable from the negative terminal of the battery. Inspect the area around the coil tower for carbon tracking, which looks like fine pencil-drawn lines. This indicates an electrical shorting condition - if any carbon tracking is found, the coil must be replaced and the secondary wiring should be checked for excessive resistance (see Chapter 1).

1986 through 1995 models

Refer to illustrations 7.3a, 7.3b and 7.4
2   Follow the primary wiring harness from the ignition coil up to the electrical connector, then unplug the connector. If you can't find the electrical connector, detach the wires right at the coil by removing the nuts from the

terminals. On later models disconnect the electrical connector at the coil. On all models, disconnect the high-tension lead from the coil tower.
3   Using an ohmmeter, measure the resistance across the primary terminals **(see illustrations)** and compare it to the primary resistance listed in this Chapter's Specifications. If it isn't within the specified range, replace the coil.
4   Now measure the secondary resistance of the coil. Insert a pointed metal tool into the coil tower and attach one of the ohmmeter leads to it **(see the accompanying illustration or illustration 7.3b or 7.3c)**. Connect the other lead to the positive primary terminal on the coil and note the indicated resistance. Compare your measurement to the coil secondary resistance listed in this Chapter's Specifications. If it isn't within the specified range, replace the coil.

7.4 When measuring the secondary coil resistance, insert a probe into the coil tower and connect the other lead to the positive (+) primary terminal (early model coil shown)

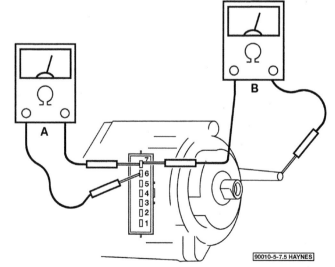

7.5 Ignition coil and igniter electrical connector terminal guide (1996 through 1998 models)

A   *Primary resistance check*
B   *Secondary resistance check*

7.10 On early models, the coil is attached to the firewall with two screws (arrows)

7.13 Unplug the electrical connector from the coil (1999 and later models)

7.14 Remove the coil retaining bolt, and then pull the coil straight up to remove it (1999 and later models)

1996 through 1998 models

Refer to illustration 7.5

5   Unplug the electrical connector from the distributor. To check the resistance on the primary side of the coil, touch the leads of an ohmmeter to the upper two terminals, on the distributor side of the connector **(see illustration)**. Compare your measurement to the primary resistance listed in this Chapter's Specifications. If the primary resistance is incorrect, replace the distributor.

6   To check the secondary resistance, remove the distributor cap screws and remove the cap. Then touch the leads of the ohmmeter to the upper terminal and the other lead to the high-tension tower **(see illustration 7.5)**. Compare your measurement to the coil secondary resistance listed in this Chapter's Specifications. If it isn't within the specified range, replace the distributor.

1999 and later models

7   At the time this book was revised to include 1999 and later models, there were no primary and secondary resistance specifications for the ignition coil available from the manufacturer. The only method prescribed by the manufacturer is to substitute a known good coil for a suspected defective unit. Since electrical components cannot be returned, we don't recommend this strategy. If the other ignition system components check out, but the ignition system is still malfunctioning, have the coils checked out by a dealer service department.

Replacement

1986 through 1995 models

Refer to illustration 7.10

8   Disconnect the cable from the negative battery terminal.

9   Label and disconnect the primary wires from the coil. Pull the high-tension lead out of the coil tower.

10   Remove the coil bracket-to-firewall screws **(see illustration)** and detach the coil.

11   To install the coil, reverse the removal

procedure. If the new coil doesn't come with a bracket, loosen the clamp screw on the old bracket and slide it off the coil. Slide the new coil into the bracket and tighten the clamp screw securely. When mounting the coil on the firewall, be sure to install the condenser, if equipped, under the upper mounting screw.

1996 through 1998 models

12   Replace the distributor (see Section 8). **Note:** *On 1996 through 1998 models, the ignition coil is an integral part of the distributor and cannot be replaced separately.*

1999 and later models

Refer to illustrations 7.13 and 7.14

13   Unplug the electrical connector from the coil **(see illustration)**.

14   Remove the coil retaining bolt **(see illustration)** and then pull the coil straight up to remove it.

15   Installation is the reverse of removal.

---

8   Distributor - removal and installation

Removal

Refer to illustrations 8.5a and 8.5b

1   Unplug the distributor primary lead wire (from the coil).

2   Unplug the electrical connector for the igniter. Follow the wires as they exit the distributor to find the connector.

3   Look for a raised "1" on the distributor cap. This marks the location for the number one cylinder spark plug wire terminal. If the cap doesn't have a mark for the number one terminal, locate the number one spark plug and trace the wire back to the terminal on the cap.

4   Remove the distributor cap (see Chapter 1) and turn the engine over until the rotor is pointing at the number one spark plug terminal (refer to the locating TDC procedure in Chapter 2).

5   On early models, make a mark on the

8.5a Before removing the distributor, use chalk or a felt-tip pen to make an alignment mark on the edge of the distributor base directly beneath the rotor tip - DO NOT use a lead pencil

edge of the distributor body directly below the rotor tip and in line with it **(see illustration)**. On all models, mark the distributor base and the gear case to ensure the distributor is reinstalled correctly **(see illustration)**.

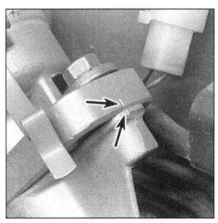

8.5b Mark the relationship of the distributor to the gear case as well

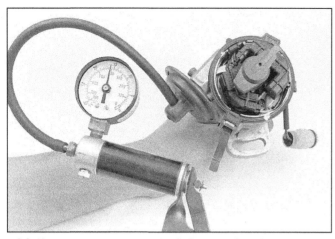

9.3 Use a vacuum pump to check the operation of the vacuum advance unit

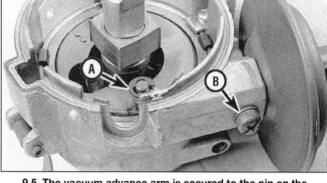

9.5 The vacuum advance arm is secured to the pin on the generator base plate with an E-clip (internal distributor components removed for clarity)

A   *E-clip*
B   *Advance mechanism mounting screw*

On 1996 models, also mark the distributor coupling position on the distributor housing.
6   Remove the distributor hold down bolt and clamp, then pull the distributor straight out to remove it. **Caution:** *DO NOT turn the crankshaft while the distributor is out of the engine - if you do, the alignment marks will be useless.*

Installation

Note: *If the crankshaft has been moved while the distributor is out, the number one piston must be repositioned at TDC. This can be done by feeling for compression pressure at the number one plug hole as the crankshaft is turned. Once compression is felt, align the ignition timing zero mark with the pointer.*

7   Insert the distributor into the engine in exactly the same position it was when removed.
8   To mesh the helical gears on the camshaft and distributor, it may be necessary to turn the rotor slightly. Recheck the alignment marks on the distributor base and gear case to verify the distributor is in the same position it was before removal. Also check the rotor to see if it's aligned with the mark on the edge of the distributor base.
9   Place the hold down clamp in position and loosely install the bolt.
10   Install the distributor cap.
11   Plug in the electrical connectors.
12   Reattach the spark plug wires to the plugs (if removed).
13   Connect the cable to the negative terminal of the battery.
14   Check the ignition timing (refer to Chapter 1) and tighten the distributor hold down bolt securely.

9   Vacuum advance unit - check and replacement (early models)

Check

Refer to illustration 9.3
1   Disconnect the cable from the negative

terminal of the battery.
2   Remove the distributor cap.
3   Disconnect the vacuum hose from the vacuum advance unit and connect a hand-held vacuum pump **(see illustration)**. Apply vacuum and watch the generator base plate - it should rotate counterclockwise slightly. The needle on the vacuum pump gauge should also remain steady, indicating the diaphragm inside the advance unit is in good condition.
4   If the generator base plate doesn't move and the vacuum diaphragm doesn't hold vacuum, replace the advance unit. If the generator base plate doesn't move but the advance unit does hold vacuum, disconnect the advance unit arm from the pin on the generator base plate and attempt to turn the plate. If it moves smoothly, replace the advance unit. If the plate is stuck or doesn't move smoothly, replace it (see Section 13).

Replacement

Refer to illustration 9.5
5   Using a small screwdriver or scribe, pry off the E-clip that retains the vacuum advance arm to the generator base plate **(see illustration)**. Lift up on the advance arm and turn the generator base plate to disconnect the arm from the plate.
6   Remove the screw that secures the advance unit to the distributor body **(see illustration 9.5)**. Slide the advance unit out of the distributor body, using a twisting motion if it's stuck.
7   Installation is the reverse of the removal procedure.

10   Centrifugal advance mechanism - check (early models)

1   Disconnect the cable from the negative terminal of the battery.
2   Remove the distributor cap.
3   Turn the rotor clockwise and see if it snaps back. It should turn smoothly, but you should be able to feel some spring resis-

tance. If the distributor has been removed, you'll have to hold the distributor drive gear.
4   If the rotor won't turn, snap back to its original position or if it's sticky, the distributor must be replaced (individual centrifugal advance parts aren't available separately, at least not at the time of this writing - check with a dealer parts department or a parts store before removing any parts).

11   Igniter and generator - check and replacement

1986 through 1989 Samurai and 1989 through 1991 Sidekick/Tracker

Refer to illustrations 11.3a and 11.3b
1   Remove the distributor cap and rotor and look at the relationship of the signal rotor to the igniter terminal. The terminal on the igniter should be pointing between two of the teeth on the signal rotor. If it isn't, turn the crankshaft until it is. Connect the positive lead of a voltmeter to the negative terminal on the coil. Connect the negative terminal of the voltmeter to a good ground. Disconnect the high-tension coil wire from the distributor and clip it to a good ground. Turn the ignition switch to the ON position, but don't crank the engine. Pass a standard screwdriver between the igniter terminal and a tooth on the signal rotor several times while looking at the voltmeter. The needle on the meter should fluctuate a little bit - about 0.5 to 1 volt - from normal battery voltage. If it does, the generator and igniter are working properly. If the needle doesn't fluctuate, remove the generator and the igniter and then check each of them to determine which one is defective.
2   Remove the distributor cap (see Chapter 1). Remove the distributor (see Section 8).
3   Remove the plastic dust covers from the igniter and generator. Remove the two screws that attach the igniter to the distributor body **(see illustration)**, push the rubber grommet on the wiring harness up and out of

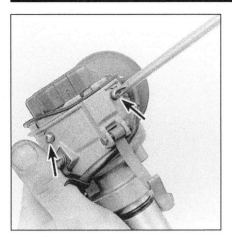

11.3a To detach the generator from the distributor body on early models, remove these two screws (arrows) (1986 through 1989 Samurai models)

11.3b To disconnect the igniter from the generator, remove the screws that secure the generator wires to the igniter; the red wire goes to the positive terminal and the white wire goes to the negative terminal (1986 through 1989 Samurai models)

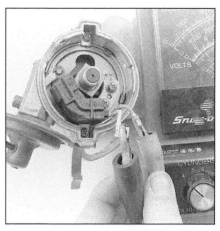

11.4 Measure the resistance of the generator; if it isn't as specified, replace it (1986 through 1989 Samurai models)

the groove, remove the screws that secure the generator leads to the igniter **(see illustration)** and then remove the igniter from the distributor. **Caution:** *Pay close attention to which generator lead is attached to which terminal. When reassembling these components, these leads must be reconnected to the same terminals of the igniter. Failure to do so could damage the generator and the igniter.*

Samurai models

Generator

Refer to illustrations 11.4 and 11.5

4   Connect the leads of an ohmmeter to the red and white wires of the generator **(see illustration)** and then measure the resistance. Compare your measurement to the generator resistance listed in this Chapter's Specifications. If the generator resistance is incorrect, replace the generator (see Step 5). If the generator resistance is correct, check the igniter (see Step 6).

5   To replace the generator, remove the two screws that secure it to the baseplate **(see illustration)**. After installing the new generator, and before tightening the generator retaining screws, adjust the signal rotor air gap (see Section 12). After the air gap has been adjusted, install the igniter. **Caution:** *Make sure that you don't reverse the red and white leads; doing so could damage the generator and the igniter.* Install the dust covers for the generator and igniter.

Igniter

Refer to illustration 11.6

6   Disconnect the red and white generator wires from the igniter and then remove the igniter from the distributor (see Step 3). Using a jumper cable, connect the positive terminal of a 12-volt battery to the igniter connector terminal for the black/white wire **(see illustration)**; using another jumper cable, hook up the negative battery terminal to a good ground on the igniter. **Caution:** *Don't reverse the jumper cables. Doing so might damage*

the igniter. Hook up a test light bulb to the positive battery lead and to the other igniter connector terminal (brown wire). Set an ohmmeter to the 1-to-10-ohm range, and then touch the negative lead of the ohmmeter to the igniter terminal for the red wire and the positive lead to the terminal for the white wire. **Caution:** *Don't reverse the ohmmeter leads. Doing so might damage the igniter.* The test bulb should come on if the igniter is functioning correctly. If the light doesn't come on, replace the igniter.

7   To install the igniter, reconnect the red and white generator wires to the two igniter terminals (white wire to the upper terminal; red wire to the lower terminal), and then install the two screws that secure the igniter

11.5 The generator is attached to the base plate by these two screws (1986 through 1989 Samurai models)

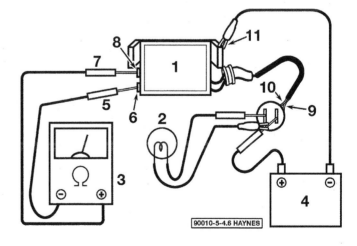

11.6 Test setup for testing the igniter (1986 through 1989 Samurai models)

1	Igniter	5	Negative ohmmeter lead
2	Test bulb	6	Red wire terminal
3	Ohmmeter	7	Positive ohmmeter lead
4	12-volt battery		

8	White wire terminal
9	Black/white wire
10	Brown wire
11	Ground

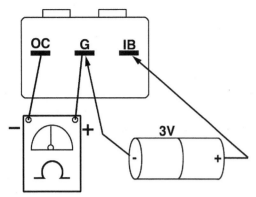

11.14 Checking the igniter on 1991 through 1995 Sidekick/Tracker models with two 1.5-volt batteries and an ohmmeter

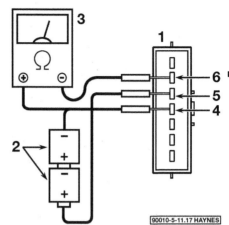

11.17 Checking the igniter on 1996 through 1998 Sidekick/Tracker models with two 1.5-volt batteries and an ohmmeter

1 *Distributor-side electrical connector*
2 *Two 1.5V batteries in series*
3 *Ohmmeter*
4 *Terminal No. 4*
5 *Terminal No. 5*
6 *Terminal No. 6*

to the distributor body **(see illustration 11.3a)**. Then push the grommet for the igniter wire harness into the cutout in the distributor body. Install the generator (see Step 5), check and - if necessary - adjust the air gap (see Section 12) and then install the distributor (see Section 8).

Sidekick models

Generator

8   Disconnect the high tension cable from the distributor cap and ground it on the engine. Remove the distributor cap and rotor (see Chapter 1). Verify that none of the signal rotor teeth are aligned with the generator. If any of the four signal rotor teeth are aligned with the generator, rotate the crankshaft slightly.
9   Connect a voltmeter between the negative terminal of the ignition coil and ground and verify that there's battery voltage available at the coil terminal. Insert the blade of a screwdriver between the signal rotor and the generator. Then repeatedly pull it up and down, removing it from between the rotor and generator and then putting it back in place. The voltmeter should indicate a fluctuating voltage that varies by about 0.5 to 1.0 volts as the screwdriver is inserted and removed. If it does indicate a slightly fluctuating voltage, the generator is functioning correctly. If the voltmeter doesn't indicate a slightly fluctuating voltage, replace the generator (see Step 5).

Igniter

10   There is no igniter test for these models. A defective igniter can only be identified by a process of elimination: If the rest of the ignition system appears to be functioning correctly, the igniter is probably defective. Of course, you can also purchase a new igniter and substitute it for a unit suspected to be defective. However, electrical components cannot be returned, so you could end up buying an igniter you don't need. The best way to verify a bad igniter is to have it checked by a dealer service department. To replace an igniter, refer to Step 7.

1990 and later Samurai and 1992 and later Sidekick/Tracker

11   These models have a computer-controlled ignition. They still use an igniter, but they don't use a generator. Instead, they use a Hall-effect-type device that senses the passing of a shutter-type signal rotor that is attached to the distributor shaft. The Hall-effect assembly senses the passing of each of the four shutters of the signal rotor (one for each engine cylinder) and sends a signal to the Powertrain Control Module (PCM), which switches the igniter on and off. On 1990 and 1991 Samurai models, this device is referred to as a "pick-up coil." On all 1992 and later models, it's referred to as a Camshaft Position (CMP) sensor. For more information about this system, refer to Section 5 in this Chapter and refer to "Camshaft Position (CMP) sensor - check and replacement" in Chapter 6.

Pick-up coil/Camshaft Position (CMP) sensor

12   Refer to "Camshaft Position (CMP) sensor - check and replacement" in Chapter 6.

Igniter

1991 through 1995 models

Refer to illustration 11.14
Note: *The following procedure applies to Sidekick and Tracker models only. There is no testing information available for 1990 and later Samurai models; take the vehicle to a dealer service department to have the igniter tested.*
13   Disconnect the electrical connector from the igniter, which is attached to the ignition coil and then remove the igniter from the coil bracket.
14   Connect two 1.5 volt batteries in series so that their total output voltage is around 3 volts. Connect the positive lead of an ohmmeter to the G terminal of the igniter and the negative lead to the OC terminal **(see illustration)**. Apply the 3 volts between the G and IB terminal and verify that the ohmmeter indi-

cates continuity. With no voltage applied, there should be no continuity. If the continuity is not as specified, replace the igniter.
15   To replace the igniter, disconnect the electrical connector, remove the mounting screws and lift the igniter from the coil bracket. Installation is the reverse of removal.

1996 through 1998 models

Refer to illustration 11.17
16   Disconnect the electrical connector from the distributor (the igniter is located inside the distributor on these models).
17   Connect two 1.5 volt batteries in series so that their total output voltage is around 3 volts. Connect the positive lead of an ohmmeter to the No. 4 terminal of the distributor and the negative lead to the No. 6 terminal **(see illustration)**. Apply the 3 volts to terminals 4 and 5 and verify that the ohmmeter indicates continuity. With no voltage applied, there should be no continuity. If the continuity is not as specified, replace the distributor (see Section 8). (The igniter is an integral part of the distributor on these models; it cannot be replaced separately.)

1999 and later models

18   At the time this book was revised to include 1999 and later models, there was no test for the igniters, which are an integral part of each ignition coil on these models. Testing an igniter is therefore impossible. The only method prescribed by the manufacturer is to substitute a known good coil/igniter unit for a suspected defective unit. Since electrical components cannot be returned, we don't recommend this strategy. If the other ignition system components check out, but the ignition system is still malfunctioning, have the coil/igniter units checked out by a dealer service department.

12   Signal rotor air gap - adjustment (early models)

Refer to illustrations 12.2 and 12.3
1   Any time you replace the igniter - or

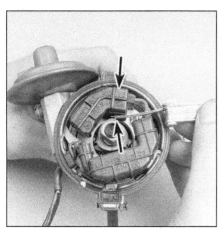

12.2 To measure the air gap, insert a non-magnetic feeler gauge between a tooth of the pole piece (lower arrow) and the pick-up coil (upper arrow)

remove it to get at the generator base plate - be sure to adjust the air gap before reinstalling the generator.

2   Remove the generator and loosen the igniter mounting screws. Place a BRASS feeler gauge of the specified thickness between one of the four projections on the

12.3 The air gap can be adjusted by loosening the two igniter screws, inserting a screwdriver into the slot in the igniter and turning the screwdriver to pry the igniter toward or away from the signal rotor

signal rotor and the igniter **(see illustration)**.

3   Gently pry the igniter toward the signal rotor until it's a snug - not tight - fit against the feeler gauge **(see illustration)**.

4   Tighten the igniter mounting screws.

5   Check the adjustment by noting the amount of drag on the feeler gauge when you pull it out of the gap between the signal rotor and igniter. You should feel a slight amount of drag. If you feel excessive drag on the gauge, the gap is probably too small. If you don't feel any drag on the gauge when you pull it out, the air gap is too large.

6   Install the generator.

13   Distributor - disassembly and reassembly (early models)

Refer to illustrations 13.5a, 13.5b, 13.5c and 13.6

Note: *On later models, the manufacturer does not recommend disassembling the distributor. If the distributor is damaged or the shaft is worn, replace the distributor as a unit.*

1   Disconnect the cable from the negative terminal of the battery.

2   Remove the distributor (see Section 8).

3   Remove the generator and igniter (see Section 11).

4   Remove the vacuum advance unit (see Section 9).

5   Remove the two screws that retain the generator base plate to the distributor body, and then lift out the plate **(see illustrations)**. Check the plate for smooth rotation - if it's stuck or doesn't turn smoothly, replace it.

6   To remove the distributor drive gear, grind off the ends of the set pin **(see illustration)**, then knock the pin out with a hammer and punch. The gear should now slide off the shaft. If it's stuck, a small gear puller can be used to remove it.

7   To install the gear, slide it onto the distributor shaft and align the set pin holes. Tap the new pin into place, then peen each end of the pin by hitting it with a hammer and punch while the other end is resting against a solid surface. This will prevent the pin from working its way out during engine operation.

8   Install the generator base plate - the four spring clips around the outer edge of the plate must line up with the four grooves in the

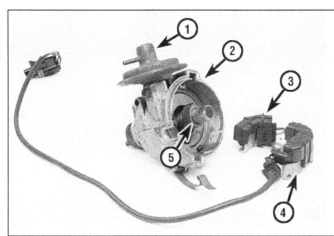

13.5a Disassembled view of early type distributor components

1   *Vacuum advance unit*
2   *Distributor housing*
3   *Pick-up coil*
4   *Module*
5   *Pole piece*

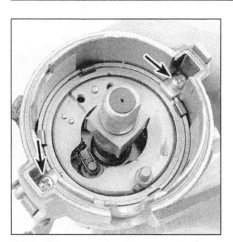

13.5b When removing the generator base plate screws, be careful not to lose the small hold down plates

13.5c When lifting the generator base plate out of the distributor body, be careful not to nick the signal rotor teeth

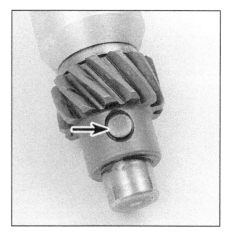

13.6 Before attempting to knock the drive gear set pin out, grind off the ends

distributor housing before the plate will drop down into position. Make sure the pin for the vacuum advance arm is located properly.

9   To reassemble the remainder of the distributor, reverse the disassembly sequence. Don't forget to adjust the signal rotor air gap following the procedure in Section 12.

14 Charging system - general information and precautions

The charging system includes the alternator, an internal voltage regulator, a charge indicator, the battery, a fusible link and the wiring between all the components. The charging system supplies electrical power for the ignition system, lights, radio, etc. The alternator is driven by a belt at the front of the engine.

The purpose of the voltage regulator is to limit the alternator's output to a preset value. This prevents power surges, circuit overloads, etc., during peak voltage output.

The fusible link is a short length of insulated wire integral with the engine compartment wiring harness. The link is four wire gauges smaller than the wire in the circuit it protects. Production fusible links and their identification flags are identified by the flag color. See Chapter 12 for additional information on the fusible links.

The charging system doesn't ordinarily require periodic maintenance. However, the drivebelt, battery and wires and connections should be inspected at the intervals outlined in Chapter 1.

The dashboard warning light should come on when the ignition key is turned to Start, then go off immediately. If it remains on, there's a malfunction in the charging system (see Section 15). Some vehicles are also equipped with a voltmeter. If the voltmeter indicates abnormally high or low voltage, check the charging system (see Section 15).

Be very careful when making electrical test connections on a vehicle equipped with an alternator and note the following:

a) *When reconnecting wires to the alternator from the battery, be sure to note the polarity.*
b) *Before using arc-welding equipment to repair any part of the vehicle, disconnect the wires from the alternator and the battery terminals.*
c) *Never start the engine with a battery charger connected.*
d) *Always disconnect both battery cables before using a battery charger.*
e) *The alternator is turned by an engine drivebelt, which could cause serious injury if hands, hair or clothes become entangled in it with the engine running.*
f) *Because the alternator is connected directly to the battery, it could arc or cause a fire if overloaded or shorted out.*
g) *Wrap a plastic bag over the alternator and secure it with rubber bands before steam cleaning the engine.*

h) *Never disconnect the battery while the engine is running.*

15 Charging system - check

Refer to illustration 15.6

1   If a malfunction occurs in the charging system, don't automatically assume the alternator is causing the problem. First check the following items:

a) *Check the drivebelt tension and condition (Chapter 1). Replace it if it's worn or deteriorated.*
b) *Make sure the alternator mounting and adjustment bolts are tight.*
c) *Inspect the alternator wiring harness and the connectors at the alternator. They must be in good condition and tight.*
d) *Check the fusible link (if equipped) located between the battery and the alternator. If it's burned, determine the cause, repair the circuit and replace the link (the battery won't charge and/or the accessories won't work if the fusible link blows). Sometimes a fusible link may look good, but still be bad. If in doubt, remove it and check it for continuity.*
e) *Start the engine and check the alternator for abnormal noises (a shrieking or squealing sound indicates a bad bearing).*
f) *Check the specific gravity of the battery electrolyte. If it's low, charge the battery (doesn't apply to maintenance free batteries).*
g) *Make sure the battery is fully charged (one bad cell in a battery can cause overcharging by the alternator).*
h) *Disconnect the battery cables (negative first, then positive). Inspect the battery posts and the cable clamps for corrosion. Clean them thoroughly if necessary (see Chapter 1). Reconnect the cable to the positive terminal.*
i) *With the key off, connect a test light between the negative battery post and the disconnected negative cable clamp.*

   1) *If the test light doesn't come on, reattach the clamp and proceed to Step 3.*
   2) *If the test light comes on, there's a short (drain) in the electrical system of the vehicle. The short must be repaired before the charging system can be checked.*
   3) *Disconnect the alternator wiring harness.*

      (a) *If the light goes out, the alternator is bad.*
      (b) *If the light stays on, pull each fuse until the light goes out (this will tell you which circuit is shorted).*

2   Using a voltmeter, check the battery voltage with the engine off. If should be approximately 12-volts.

3   Start the engine and check the battery voltage again. It should now be approxi-

15.6 If the charging voltage is low, insert a metal tool into the test hole in the rear of the alternator and take a voltage reading at the B+ terminal on the alternator - if the voltage rises, the regulator is faulty (alternator removed for clarity)

mately 14-to-15 volts.

4   Turn on the headlights. The voltage should drop, and then come back up, if the charging system is working properly.

5   If the voltage reading is more than the specified charging voltage, replace the voltage regulator (refer to Section 17). If the voltage is less, the alternator diode(s), stator or rectifier may be bad or the voltage regulator may be malfunctioning.

6   To isolate the problem component, insert a scribe or small screwdriver into the test hole in the rear of the alternator **(see illustration)**, then start the engine. Measure the output voltage at terminal B (the large threaded terminal on the alternator) - if the indicated voltage is higher than the standard charging value, replace the regulator.

7   If the indicated voltage is less than the standard charging value, the problem lies elsewhere in the alternator. It would be cheaper and easier to replace the alternator with a rebuilt unit rather than attempt to repair the problem.

16 Alternator - removal and installation

Refer to illustrations 16.3a, 16.3b and 16.3c

1   Detach the cable from the negative terminal of the battery.

2   Detach the electrical connectors from the alternator.

3   Loosen the alternator adjustment and pivot bolts and detach the drivebelt **(see illustrations)**.

4   Remove the bolts and separate the alternator from the engine.

5   If you're replacing the alternator, take the old one with you when purchasing a replacement unit. Make sure the new/rebuilt

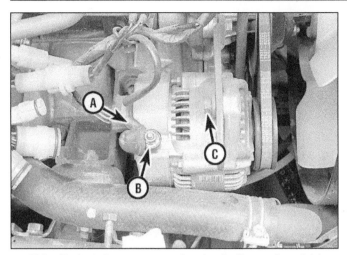

16.3a Typical alternator mounting details (Samurai models)

A   Electrical connector
B   B+ terminal
C   Adjusting bolt (pivot bolt on underside of alternator isn't visible in this photo)

16.3b Typical alternator mounting details (Sidekick models)

1   B+ (battery) terminal   3   Upper mounting bolt
2   Electrical connector

unit looks identical to the original. Look at the terminals - they should be the same in number, size and location as the terminals on the old alternator. Finally, look at the identification numbers - they'll be stamped into the housing or printed on a tag attached to the housing. Make sure the numbers are the same on both alternators.

6   Many new/rebuilt alternators DO NOT have a pulley installed, so you may have to switch the pulley from the old unit to the new/rebuilt one. When buying an alternator, find out the shop's policy regarding pulleys - some shops will perform this service free of charge.

7   Installation is the reverse of removal.

8   After the alternator is installed, adjust the drivebelt tension (see Chapter 1).

9   Check the charging voltage to verify proper operation of the alternator (see Section 14).

17 Starting system - general information and precautions

The sole function of the starting system is to turn over the engine quickly enough to allow it to start.

The starting system consists of the battery, the starter motor, the starter solenoid and the wires connecting them. The solenoid is mounted directly on the starter motor. The solenoid/starter motor assembly is installed on the lower part of the engine, bolted to the transmission bellhousing.

When the ignition key is turned to the Start position, the starter solenoid is actuated through the starter control circuit. The starter solenoid then connects the battery to the starter. The battery supplies the electrical energy to the starter motor, which does the actual work of cranking the engine.

The starter motor on a vehicle equipped with a manual transmission can only be operated when the clutch pedal is depressed; the starter on a vehicle equipped with an automatic transmission can only be operated when the transmission selector lever is in Park or Neutral.

Always observe the following precautions when working on the starting system:

a) Excessive cranking of the starter motor can overheat it and cause serious damage. Never operate the starter motor for more than 15 seconds at a time without pausing to allow it to cool for at least two minutes.

b) The starter is connected directly to the battery and could arc or cause a fire if mishandled, overloaded or shorted out.

c) Always detach the cable from the negative terminal of the battery before working on the starting system.

18 Starter motor - testing in vehicle

Note: Before diagnosing starter problems, make sure the battery is fully charged.

1   If the starter motor doesn't turn at all when the switch is operated, make sure the shift lever is in Neutral or Park (automatic transmission) or the clutch pedal is depressed (manual transmission).

2   Make sure the battery is charged and all cables, both at the battery and starter solenoid terminals, are clean and secure.

3   If the starter motor spins but the engine isn't cranking, the overrunning clutch in the starter motor is slipping and the starter motor must be replaced.

4   If, when the switch is actuated, the starter motor doesn't operate at all but the solenoid clicks, then the problem lies with either the battery, the main solenoid contacts

16.3c Lower alternator mounting bolt (arrow) (Sidekick models)

or the starter motor itself (or the engine is seized).

5   If the solenoid plunger can't be heard when the switch is actuated, the battery is bad, the fusible link is burned (the circuit is open) or the solenoid itself is defective.

6   To check the solenoid, connect a jumper lead between the battery (+) and the ignition switch wire terminal (the small terminal) on the solenoid. If the starter motor now operates, the solenoid is OK and the problem is in the ignition switch, neutral start switch or wiring.

7   If the starter motor still doesn't operate, remove the starter/solenoid assembly for disassembly, testing and repair.

8   If the starter motor cranks the engine at an abnormally slow speed, first make sure the battery is charged and all terminal connections are tight. If the engine is partially seized, or has the wrong viscosity oil in it, it will crank slowly.

9   Run the engine until normal operating

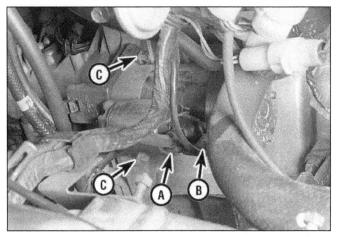

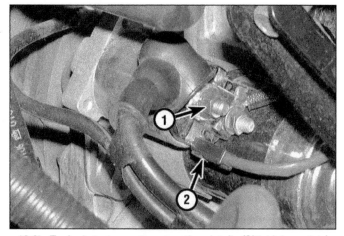

19.2a Typical starter motor mounting details (Samurai models)

A   Battery cable
B   From ignition switch (activates solenoid)
C   Starter motor mounting bolts

19.2b Typical starter motor mounting details (Sidekick models)

1   Battery cable
2   Magnetic switch (solenoid) connector

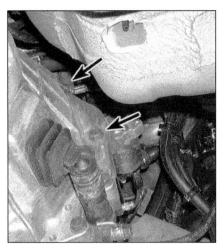

19.3 Typical starter mounting bolts (arrows); on some models, there is a nut on the other end of the lower bolt (Sidekick models)

20.3 To separate the solenoid from the starter motor, remove the nut and detach the lead (arrow) . . .

20.4 . . . then remove the solenoid mounting screws (arrows) and pull the solenoid straight off the starter flange

temperature is reached, then disconnect the coil wire from the distributor cap and ground it on the engine.

10   Connect a voltmeter positive lead to the positive battery post and connect the negative lead to the negative post.

11   Crank the engine and take the voltmeter readings as soon as a steady figure is indicated. Don't allow the starter motor to turn for more than 15 seconds at a time. A reading of 9 volts or more, with the starter motor turning at normal cranking speed, is normal. If the reading is 9 volts or more but the cranking speed is slow, the motor is faulty. If the reading is less than 9 volts and the cranking speed is slow, the solenoid contacts are probably burned, the starter motor is bad, the

battery is discharged or there's a bad connection.

19 Starter motor - removal and installation

Refer to illustrations 19.2a, 19.2b and 19.3

1   Detach the cable from the negative terminal of the battery.

2   Clearly label, then disconnect the wires from the terminals on the starter solenoid **(see illustrations)**.

3   Remove the mounting bolts **(see illustration)** and detach the starter.

4   Installation is the reverse of removal.

20 Starter solenoid - removal and installation

Refer to illustrations 20.3 and 20.4

1   Detach the cable from the negative terminal of the battery.

2   Remove the starter motor (see Section 19).

3   Disconnect the strap from the solenoid to the starter motor terminal **(see illustration)**.

4   Remove the screws that secure the solenoid to the starter motor **(see illustration)**.

5   Pull the solenoid off the starter body flange.

6   Installation is the reverse of removal.

Chapter 6
Emissions and engine control systems

Contents

Specifications

Carbureted models
PTC heater resistance .... 0.5 to 3.0 ohms

Models with throttle body injection
Throttle Position (TP) sensor resistance
Between terminals C and D
Throttle plate closed .... 0 to 500 ohms
Throttle plate wide open .... Infinity
Between terminals A and D .... 3.5 to 6.5 K-ohms
Between terminals B and D
Throttle plate closed .... 0 to 2 K-ohms
Throttle plate wide open .... 2 to 6.5 K-ohms
Idle Air Control (IAC) valve resistance (Sidekick/Tracker) ... 11 to 14 ohms (at 68 degrees F, 20 degrees C)
Idle Speed Control (ISC) solenoid valve resistance (Samurai) ... 30 to 33 ohms

Models with sequential multiport fuel injection
Crankshaft Position (CKP) sensor resistance
Resistance between CKP sensor terminals
SOHC engine .... 360 to 460 ohms
DOHC engine .... 484 to 656 ohms
Resistance between each terminal and ground (all engines) ... At least 1 M-ohm
EGR valve resistance
Between terminals A and B, C and B, F and E, and D and E ... 20 to 24 ohms (at 68 degrees F, 20 degrees C)
Between terminal B and valve body, and terminal E and valve body . Infinite
EVAP canister purge valve resistance ... 28 to 36 ohms (at 68 degrees F, 20 degrees C)
Idle Air Control (IAC) solenoid valve resistance
1992 through 1998 .... 11 to 14 ohms (at 68 degrees F, 20 degrees C)
1999 on .... 35 to 43 ohms
MAF sensor output voltage
Ignition key turned to ON, engine not running ... 1.0 to 1.6 volts
Engine running at idle .... 1.7 to 2.0 volts

Models with sequential multiport fuel injection (continued)

Oxygen sensor
  Heater resistance
    1992 through 1995
      California vehicles .... 11.7 to 15.5 ohms
      All other vehicles .... 3.7 to 5.1 ohms
    1996 through 2000
      Sensor No. 1, *with* warm-up three way catalyst (WU-TWC) ... 4.5 to 5.7 ohms
      Sensor No. 1 (upstream sensor), *without* WU-TWC............... 11.7 to 14.3 ohms
      Sensor No. 2 (downstream sensor) .... 11.7 to 14.3 ohms
    2001
      Sensor No. 1 (upstream sensor).. 5.0 to 6.4 ohms
      Sensor No. 2 (downstream sensor .... 11.7 to 14.3 ohms
Throttle Position (TP) sensor resistance
  Between terminals A and B
    Throttle lever-to-stop screw clearance 0.020 in (0.5 mm) .... 0 to 500 ohms
    Throttle lever-to-stop screw clearance 0.031 (0.8 mm) .... Infinity
  Between terminals A and D.. 3.5 to 6.5 K-ohms
  Between terminals A and C
    Throttle valve at idle.. 0.3 to 2 K-ohms
    Throttle plate wide open .... 2 to 6.5 K-ohms

Torque specifications   **Ft-lb** (unless otherwise indicated)

Camshaft Position (CMP) sensor retaining bolt
  SOHC engine .... 78 in-lbs
  DOHC engine
    CMP sensor retaining bolt .... 78 in-lbs
    CMP sensor case bolt .... 132 in-lbs
Crankshaft Position (CKP) sensor retaining bolt
  SOHC engine .... 90 in-lbs
  DOHC engine .... 54 in-lbs
Engine Coolant Temperature (ECT) sensor .... 132 in-lbs
Fuel tank pressure sensor bolts .... 14.4 in-lbs
Idle Air Control (IAC) valve retaining screws
  1992 through 1998 .... 43 in-lbs
  1999 on .... 30 in-lbs
Intake Air Temperature (IAT) sensor (pre-1999 models) .... 132 in-lbs
Oxygen sensor... 31 to 32
Throttle Position (TP) sensor screws .... 30 in-lbs
Vehicle Speed Sensor (VSS) retaining bolt (1999 on)............................ 54 in-lbs

---

1   General information

Refer to illustrations 1.1a, 1.1b and 1.7

To prevent pollution of the atmosphere from incompletely burned and evaporating gases, and to maintain good driveability and fuel economy, a number of emission control systems are incorporated.

All models are equipped with the following systems:

Catalytic converter
Evaporative Emission (EVAP) control system
Exhaust Gas Recirculation (EGR) system
Positive Crankcase Ventilation (PCV) system

Carbureted models are also equipped with the following systems:

Fuel feedback system
Hot Idle Compensator (HIC)
Deceleration mixture control system
Thermostatically Controlled Air Cleaner (TCAC)

Fuel-injected models are equipped with the following information sensors and systems:

Engine Coolant Temperature (ECT) sensor
Camshaft Position (CMP) sensor
Crankshaft Position (CKP) sensor
Fuel tank pressure sensor
Heated Oxygen Sensor (HO2S) (two sensors: one upstream, one downstream)
Intake Air Temperature (IAT) sensor
Manifold Absolute Pressure (MAP) sensor
Mass Air Flow (MAF) sensor
PTC heater system (early fuel-injected models with an automatic transmission)
Throttle Position (TP) sensor
Transmission Range sensor (if equipped with an automatic transmission)
Vehicle Speed Sensor (VSS)

Fuel-injected models are also equipped with the following switches and signals:

ABS control signal (if equipped with ABS)
Air conditioning signal (if equipped with air conditioning)
Battery voltage
Brake light switch (stop lamp switch) signal
Engine start signal
Fuel level signal
Heater blower fan switch signal
Lighting switch signal
Power steering pressure switch signal
Rear window defogger switch signal (if equipped with rear window defogger)

The engine management systems for fuel-injected engines use the same controls:

Air conditioning condenser fan control system (if equipped with air conditioning)
Air conditioning cut signal output (if equipped with air conditioning)

All of these systems are linked, directly or indirectly, to the computerized emission control system:

Fuel pump control system
Heater control system for heated oxygen sensors
Idle air control system
Ignition control system

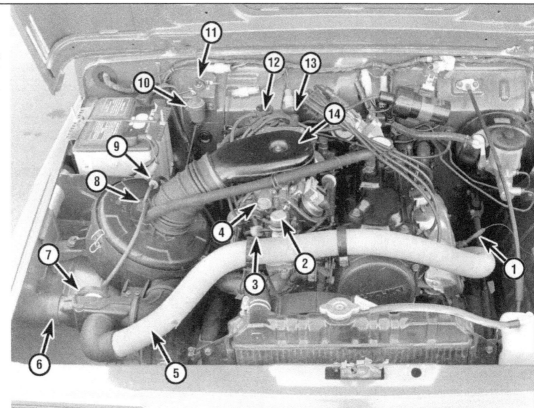

1.1a Typical emission control system component locations (carbureted Samurai model)

1 Oxygen sensor
2 Mixture Control Valve (MCV)
3 MCV jet
4 Vent solenoid valve
5 Warm air duct
6 Cool air duct
7 Air Control Actuator (ACA)
8 Thermo sensor
9 Thermo sensor check valve
10 High Altitude Compensator (HAC)
11 Engine compartment temperature sensor
12 EGR modulator
13 Three Way Switching Valve (TWSV)
14 Air intake case

1.1b Typical emission control system component location (fuel-injected Tracker model)

1 Mass Air Flow (MAF) sensor
2 Throttle Position (TP) sensor
3 Idle Air Control (IAC) valve
4 Manifold Differential Pressure (MDP) sensor
5 Crankcase breather hose
6 PCV hose
7 Camshaft Position (CMP) sensor

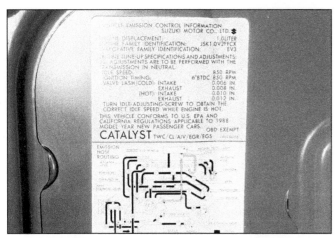

1.7 The Vehicle Emission Control Information (VECI) label contains tune-up specifications and vital information regarding the location of the emission control devices and vacuum hose routing

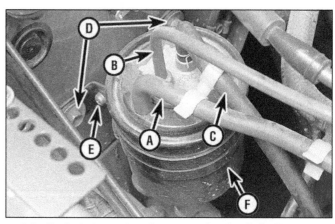

3.2a Typical charcoal canister mounting details (carbureted model shown; fuel-injected models similar)

A	Vent solenoid valve (port D)	E	Bracket clamp bolt
B	To intake manifold (port B)	F	Port C (on bottom of
C	To fuel tank port		canister, hidden from view)
D	Mounting bolts		

The Sections in this Chapter include general descriptions, checking procedures within the scope of the home mechanic and component replacement procedures (when possible) for each of the systems listed above.

Before assuming an emissions control system is malfunctioning, check the fuel and ignition systems carefully. The diagnosis of some emission control devices requires specialized tools, equipment and training. If checking and servicing become too difficult, or if a procedure is beyond your ability, consult a dealer service department. Remember, the most frequent cause of emissions problems is simply a loose or broken vacuum hose or wire, so always check the hose and wiring connections first.

This doesn't mean, however, that emission control systems are particularly difficult to maintain and repair. You can quickly and easily perform many checks and do most of the regular maintenance at home with common tune-up and hand tools. **Note:** *Because of a Federally mandated extended warranty which covers the emission control system components, check with a dealer service department about warranty coverage before working on any emissions-related systems. Once the warranty has expired, you may wish to perform some of the component checks and/or replacement procedures in this Chapter to save money.*

Pay close attention to any special precautions outlined in this Chapter. It should be noted that the illustrations of the various systems might not exactly match the system installed on your vehicle because of changes made by the manufacturer during production or from year-to-year.

A Vehicle Emissions Control Information label is located in the engine compartment **(see illustration)**. The label contains important emissions specifications and adjustment information, as well as a vacuum hose schematic with emissions components iden-

tified. When servicing the engine or emissions systems, the VECI label in your particular vehicle should always be checked for up-to-date information.

2 Positive Crankcase Ventilation (PCV) system

1 The Positive Crankcase Ventilation (PCV) system reduces hydrocarbon emissions by scavenging crankcase vapors. It does this by circulating fresh air from the air cleaner through the crankcase, where it mixes with blow-by gases and is then rerouted through a PCV valve to the intake manifold.

2 The main components of the PCV system are the PCV valve, a fresh air filtered inlet and the vacuum hoses connecting these two components with the engine.

3 To maintain idle quality, the PCV valve restricts the flow when the intake manifold vacuum is high. If abnormal operating condi-

tions (such as piston ring problems) arise, the system is designed to allow excessive amounts of blow-by gases to flow back through the crankcase vent tube into the air cleaner to be consumed by normal combustion.

4 Checking and replacement of the PCV valve is covered in Chapter 1.

3 Evaporative emission (EVAP) control system

General description

Refer to illustrations 3.2a, 3.2b, 3.4 and 3.5

1 The evaporative emission (EVAP) control system is designed to prevent hydrocarbons from being released into the atmosphere, by trapping and storing fuel vapor from the fuel tank, the carburetor or the fuel injection system.

2 The serviceable parts of the system include a charcoal filled canister, a two-way

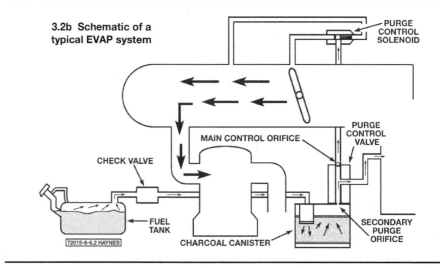

3.2b Schematic of a typical EVAP system

3.4 Location of the vent solenoid valve (arrow) on carbureted vehicles

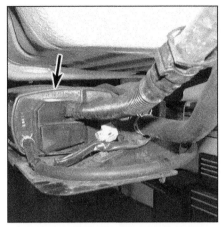

3.5 On 2.0L 2001 models equipped with an On-Board Refueling Vapor Recovery (ORVR) system, the canister (arrow) is located underneath the vehicle, on the right side, in front of the fuel tank

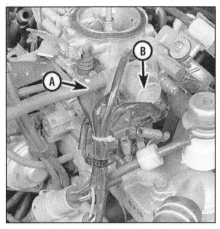

3.9 The hose from the vent solenoid valve goes directly to the canister

A   To canister
B   Vent solenoid valve

check valve (fuel-injected models), the fuel tank filler cap and the lines between the fuel tank and the rest of the system **(see illustrations)**.

3   Vapor trapped in the gas tank is vented through a line in the top of the tank. The vapor leaves the tank through the line and is routed to a carbon canister located in the engine compartment (non-ORVR vehicles) or under the vehicle (ORVR vehicles) where it's stored until the next time the engine is started.

4   The canister outlet is connected to an electrically actuated vent solenoid valve in the carburetor **(see illustration)**. The vent solenoid valve is normally closed. When the engine is started, the solenoid is energized by a signal from the PCM and allows intake vacuum to open the line between the canister and a port on the carburetor, which draws vapor stored in the canister through the carburetor and into the engine where it's burned. On fuel-injected models, the system operates continually - there is no solenoid valve.

5   On 2001 models equipped with the On-Board Refueling Vapor Recovery (ORVR) system, the charcoal canister is located under the right side of the vehicle, just ahead of the fuel tank **(see illustration)**.

Check

Warning: *When performing checks that require you to blow air into a hose or port, be extremely careful not to inhale or suck on the hose or port - fuel vapor is harmful.*

Charcoal canister

6   Except for the check valve(s), there are no moving parts and nothing to wear in the canister. Check for loose, missing, cracked or broken fittings and inspect the canister for cracks and other damage. If the canister is damaged, replace it (refer to Step 12). Never attempt to wash the canister.

7   On carbureted models, disconnect the cable from the negative terminal of the battery and detach the hoses from the canister. Cover port C and D with your fingers and blow into port A - air should come out of

port B **(see illustration 3.2a)**. When you blow into port B, air shouldn't come out any of the other ports. When air is blown into port C, it should come out all of the other ports. If the canister doesn't function as described, replace it (see Step 12).

8   On fuel-injected models, disconnect the negative battery cable and then disconnect the hoses from the canister. Blow into port A **(see illustration 3.2a)** - air should exit ports B and C. When air is blown into port B, no air should escape from the other ports. If the canister doesn't function as described, replace it (see Step 12).

Vent solenoid valve (carbureted models)

Refer to illustration 3.9

9   Pull the hose off the vent solenoid valve outlet on the carburetor **(see illustration)** and connect another length of hose to the port. With the engine off, blow air into the hose - the air should flow through the valve freely. The results of this check should be the same with the ignition switch in the Off and On positions.

10   Start the engine and let it idle. Blow into

the hose again - air should not pass through the valve.

11   If the vent solenoid valve doesn't operate as described, either the valve is defective or there's a problem with the electrical circuit or ECM, which will require diagnosis by a dealer service department to pinpoint.

EVAP canister purge valve (fuel-injected models)

Refer to illustration 3.12

12   To locate the EVAP canister purge valve, find the charcoal canister first, and then follow the hose from the canister to the air intake manifold. The EVAP canister purge valve **(see illustration)** is located on that hose somewhere between the EVAP canister and the intake manifold.

13   Unplug the electrical connector from the EVAP canister purge valve and measure the resistance between the two terminals. Compare your measurement to the purge valve resistance listed in this Chapter's Specifications. If the indicated resistance is outside the specified range, replace the purge valve.

14   Loosen the hose clamps and disconnect the two hoses from the purge valve. Let's call the hose coming from the charcoal canister

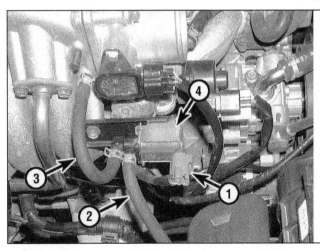

3.12 Typical EVAP canister purge valve (late-model fuel-injected vehicle shown)

1   Electrical connector
2   Hose from EVAP canister
3   Hose to intake manifold
4   EVAP canister purge valve

"hose A" and the hose going to the intake manifold "hose B." Blow into pipe A (the pipe for hose A); no air should come out pipe B (the pipe for hose B). Then, using jumper cables, hook up the battery to the purge valve electrical terminals. Again, blow into pipe A. This time, air should come out pipe B. If the canister purge valve doesn't operate as described, replace it.

Component replacement

Charcoal canister

15   Locate the canister. On carburetor equipped models it's mounted on the firewall, below the battery. On fuel-injected vehicles, it's mounted on the right inner fender panel. On 2001 2.0L models with the On-Board Refueling Vapor Recovery (ORVR) system, the canister is located underneath the vehicle, on the right side, in front of the fuel tank.
16   Disconnect the hoses from the canister and remove the mounting bolts **(see illustration 3.2a)**. Remove the canister from the engine compartment.
17   Installation is the reverse of removal.

EVAP canister purge valve

18   Unplug the electrical connector from the canister purge valve and then disconnect the two hoses from the valve **(see illustration 3.12)**. Remove the purge valve from its mounting bracket.
19   Installation is the reverse of removal.

All other components

20   Referring to the appropriate vacuum hose and vacuum valve schematics in this Section and on the VECI label in the vehicle, locate the component to be replaced.
21   Label the hoses and fittings, then detach the hoses and remove the component.
22   Installation is the reverse of removal.

---

4   Computerized emission control systems and the emissions warranty

---

The computerized emission control system (the electronic engine control system on fuel-injected models and the fuel feedback system on carburetor equipped models) monitors various engine operating conditions and alters the fuel/air mixture and engine idle speed to promote better fuel economy, improve driveability and reduce exhaust emissions.

Fortunately, all of these systems are protected by a Federally mandated, extended warranty (5 years/ 50,000 miles, whichever comes first, at the time this manual was written). This warranty covers all emission control components (EGR system, oxygen sensor, catalytic converter, fuel evaporative control system, PCV system, etc.). It also covers emissions-related parts like carburetor and fuel injection components, and major ignition system components. Refer to your owner's

6.3a The MCV clips into a bracket near the thermostat housing

manual (or supplement) to find out exactly what components and systems are covered by the Federal warranty. If you don't have an owner's manual or supplement, contact a dealer service department. Dealers are required by Federal law to provide a detailed list of the emission-related parts protected by the Federal warranty. We strongly discourage any attempt to test or repair the emission control system while it's under warranty.

---

5   Hot Idle Compensator (HIC) (carburetor equipped vehicles)

General description

1   The Hot Idle Compensator helps to maintain a stable idle speed during hot operation, when the mixture becomes rich under high intake manifold vacuum conditions. A bi-metallic valve, located in the top of the air intake case and connected to the intake manifold with a hose begins to open when the engine heat in the air intake case exceeds approximately 131-degrees F (55-degrees C). This creates a controlled vacuum leak, which leans out the fuel/air mixture and stabilizes the idle. It becomes fully open at 158-degrees F (70-degrees C).

Check

2   Unbolt the air intake case from the carburetor. Follow the hose from the HIC valve in the air intake case down to the intake manifold and disconnect it from the manifold. Attach a hand-held vacuum pump to the hose. With the engine cold, apply vacuum to the valve - the gauge on the pump should indicate vacuum and the needle should remain steady.
3   Plug the open port on the manifold, install the air intake case and start the engine, allowing it to warm up. When the temperature in the air intake case reaches 150-degrees F, operate the vacuum pump - the gauge should not show vacuum, indicating the HIC valve is open. If the valve doesn't operate as described, replace it.

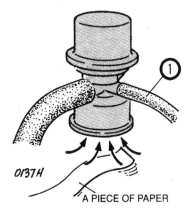

OI37H   A PIECE OF PAPER

6.3b Air should be drawn into the MCV when hose no. 1 is disconnected and reinstalled - it's normal for the engine to run rough or stall when performing this check

---

Component replacement

4   Remove the air intake case from the carburetor. Remove the screw from the under side of the case that retains the HIC valve and detach the valve. Installation is the reverse of the removal procedure.

---

6   Deceleration mixture control system (carburetor equipped vehicles)

General description

1   This system reduces the excessive HC and CO emissions that are generated during rapid deceleration conditions. It consists of a Mixture Control Valve (MCV), a jet and the necessary vacuum hoses.
2   The MCV contains a pressure balancing orifice and a check valve in the diaphragm. When manifold vacuum is constant, the valve remains closed. Under rapid deceleration, the valve opens and allows an additional amount of air into the intake manifold, which leans out the fuel/air mixture.

Check

Refer to illustrations 6.3a and 6.3b
3   Start the engine and allow it to reach normal operating temperature. Disconnect the small hose from the valve, then reconnect it. At the same time, hold a strip of paper close to the under side of the valve - the paper should be drawn up to the valve, indicating air is passing through it **(see illustrations)**.
4   Disconnect the hoses from the jet and blow into the gray side - air should pass through. If the valve is clogged, replace it - be sure to install it with the gray side toward the MCV.
5   If the valve doesn't operate as described in Step 3, check the hoses for cracks, holes and kinks, replacing them if

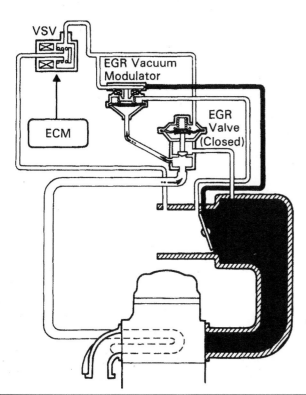

7.6 Check for movement of the EGR diaphragm with your finger, but be careful - the valve can get EXTREMELY hot!

7.1 Typical EGR system

7.11 Location of the EGR modulator on a carburetor equipped vehicle (arrow)

necessary. If the hoses are in good condition, replace the valve.

Component replacement

6   Disconnect the hoses from the valve and unclip the valve from the mount. Installation is the reverse of the removal procedure.

7   Exhaust Gas Recirculation (EGR) system

General description

Refer to illustration 7.1

1   The Exhaust Gas Recirculation (EGR) system **(see illustration)** is designed to reduce oxides of nitrogen in the engine exhaust. The EGR system recirculates a portion of the exhaust gas back into the intake manifold to be burned with the fuel/air mixture, which in turn brings down the temperature in the combustion chamber and reduces the oxides of nitrogen output.

2   The EGR valve is vacuum operated, controlled by the modulator. The modulator is operated by exhaust backpressure and allows the EGR valve to open wider under high-load conditions, but reduces or closes the EGR valve opening under light-load conditions.

3   Fuel-injected vehicles destined for the California market also employ a sub-EGR valve, which regulates the flow of exhaust gas to the main EGR valve. The sub-EGR valve is regulated by an PCM controlled Vacuum Switching Valve (VSV), much like the Three Way Solenoid Valve (TWSV) on carbu-

retor equipped models.

4   Under certain conditions, the PCM, Three Way Solenoid Valve (TWSV) or Vacuum Switching Valve (VSV) and the Bi-metal Vacuum Switching Valve (BVSV) can override the vacuum modulator (for instance, during low coolant temperatures when fifth gear is engaged and the High Altitude Compensator is turned on).

5   If the EGR system is malfunctioning and the following checks don't indicate a problem, have the system diagnosed by a dealer service department or a repair shop.

Check

1986 through 1998 models

EGR valve

Refer to illustration 7.6

Warning: *The EGR valve becomes very hot during engine operation - wear gloves when checking the valve to avoid burning your fingers. If this check is being performed on a carburetor equipped vehicle at an altitude of 4000 feet or higher, the High Altitude Compensator (HAC) electrical connector must be disconnected* **(see illustration 1.1** *at the beginning of this Chapter).*

6   The first check must be performed when the engine is cool (under 131-degrees F on carbureted models or 113-degrees F on fuel-injected models). Start the engine and rev it up a little while touching the valve diaphragm with your finger **(see illustration)** - the diaphragm shouldn't move.

7   Allow the engine to warm up to normal operating temperature and repeat the check - the diaphragm should now move when the

engine speed is increased.

8   If the diaphragm doesn't move, disconnect the vacuum hose to the valve and connect a hand-held vacuum pump to it. Apply vacuum - the diaphragm should move and the gauge on the pump should indicate vacuum. If it doesn't, replace the valve. If the valve works, proceed with the following checks.

9   If the vehicle is a fuel-injected California model, the sub-EGR diaphragm must be checked too. Look through the small inspection hole in the sub-EGR valve while revving the engine as in the EGR valve check - the diaphragm of the sub-EGR valve should move when the engine is warm, but not when it's cold. The sub-EGR valve can also be checked with a hand-held vacuum pump, just like the main EGR valve.

Hoses

Refer to illustration 7.11

10   Check all of the hoses associated with the EGR system for cracks, holes and signs of general deterioration, replacing them if necessary.

11   Mark the hoses at the modulator, then disconnect them. Unclip the modulator from the bracket **(see illustration)**.

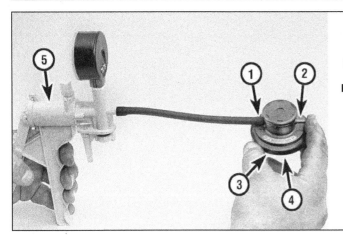

7.13 The EGR modulator should hold vacuum when it's applied to port P, port Q is plugged and air is blown into Port A

1   Port P
2   Port Q
3   Port A
4   Air
5   Vacuum pump gauge

7.17a The TWSV (sometimes called the Vacuum Switching Valve [VSV]) is fastened to a bracket on the rear of the carburetor

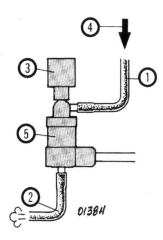

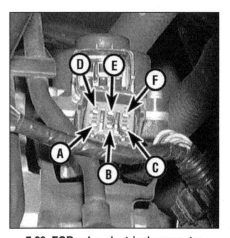

7.17b When the TWSV or VSV is not energized, air should pass through the valve when blown into hose 1 - when the valve is energized, air should come out of the filter

1   Hose to EGR valve
2   Hose to EGR modulator
3   Filter
4   Air
5   TWSV

7.19 On models with sequential multiport fuel injection, the EGR valve is located at the back of the cylinder head (DOHC engines, shown) or on the back of the air intake manifold plenum (SOHC engines); to unplug the electrical connector (arrow), release the locking spring clip and pull straight up

7.20 EGR valve electrical connector terminal guide (models with sequential multiport fuel injection)

EGR modulator

Refer to illustration 7.13

12   Plug one of the vacuum ports on the side of the valve and blow into the other port. Air should pass through the valve and come out the filter at the top of the valve.

13   Connect a vacuum pump to the modulator **(see illustration)**, plug the other vacuum port with your finger then blow into the port on the bottom. Operate the pump - the modulator should hold vacuum as long as air is being blown into the bottom port.

14   If the modulator doesn't operate as described, replace it.

Bi-metal Vacuum Switching Valve (BVSV)

15   With the engine cool, disconnect the hoses from the valve (located adjacent to the EGR vacuum modulator). Connect a length of hose to the top port on the valve and blow

into it - no air should come out of the bottom port.

16   Warm up the engine to normal operating temperature and repeat the check - air should now come out of the bottom port. If the valve doesn't check out as described, replace it (Step 23).

Three-way Solenoid Valve (TWSV) (carburetor equipped models) or Vacuum Switching Valve (VSV) (fuel-injected models)

Refer to illustrations 7.17a and 7.17b

17   Disconnect the hose from the EGR valve and the hose from the EGR modulator to the TWSV or VSV **(see illustrations)**. Blow into the hose that was connected to the EGR valve - air should pass through the valve and come out the other hose.

18   Unplug the electrical connector from the valve. Using jumper wires connected to the battery, energize the valve and blow into the hose again - air should come out of the filter now, not the hose. If the valve doesn't operate as described, replace it.

1999 and later models

EGR valve

Refer to illustrations 7.19 and 7.20

19   First, locate the EGR valve: On SOHC engines, it's located on the back of the air intake plenum; on DOHC engines, it's at the back of the cylinder head **(see illustration)**. Unplug the electrical connector from the EGR valve.

20   Using an ohmmeter, measure the resistance between terminals A and B, C and B, F and E, and D and E, and then verify that there is no continuity between terminal B and the EGR valve body or between terminal E and the valve body **(see illustration)**. Compare your measurements to the resistance listed in this Chapter's Specifications. If the resistance is incorrect, replace the EGR valve.

Component replacement

1986 through 1998 models

EGR valve

Refer to illustration 7.21

21   Unscrew the threaded fitting that con-

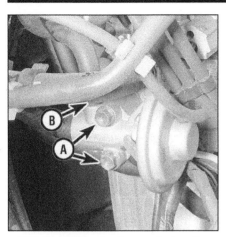

7.21 The EGR valve is mounted on the rear of the intake manifold

A   *Mounting bolts*
B   *EGR pipe fitting*

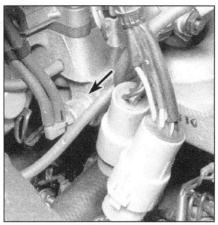

7.27 The BVSV on carburetor equipped vehicles is screwed into the intake manifold, below the carburetor

8.4 Pull the warm air duct off the air intake housing and feel the damper door - it should be shut with the engine off

nects the EGR pipe to the EGR valve, disconnect the vacuum hose and remove the two mounting bolts **(see illustration)**. Remove the valve and clean the mating surface on the intake manifold. If the passage in the manifold is clogged or restricted by exhaust deposits, clean it out before installing the valve.

22   Position the EGR valve and gasket against the intake manifold and install the bolts, but don't tighten them completely yet.

23   Start the EGR pipe fitting by hand to avoid cross-threading it. Tighten the two EGR valve bolts securely.

24   Tighten the EGR pipe fitting securely and connect the vacuum hose.

Bi-metal Vacuum Switching Valve (BVSV)

Refer to illustration 7.27

Warning: *This procedure must be done with the engine cool.*

25   Remove the radiator cap, squeeze the upper radiator hose, and then install the cap. This will create a slight vacuum in the cooling system and will allow removal of the BVSV without draining the coolant first.

26   Prepare the new valve by wrapping the threads with Teflon tape.

27   Disconnect the hoses from the valve, then unscrew it from the intake manifold **(see illustration)**.

28   Quickly install the new valve in the manifold (to avoid excessive coolant loss) and tighten it securely. Reconnect the hoses.

Three Way Solenoid Valve (TWSV) (carburetor equipped models) or Vacuum Switching Valve (VSV) (fuel-injected models)

29   Mark and disconnect the hoses from the TWSV, then unplug the electrical connector **(see illustrations 1.1a and 7.17a)**.

30   Remove the screws that retain the valve and detach it from the bracket.

31   Installation is the reverse of removal.

1999 and later models

EGR valve

32   Unplug the EGR valve electrical connector **(see illustration 7.19)**.

33   Remove the two EGR valve mounting bolts and then remove the EGR valve. Discard the old gasket.

34   Inspect the EGR valve, the valve seat and the operating rod for damage and wear. Clean the gasket mating surfaces of the EGR valve and its mounting flange.

35   Installation is the reverse of removal. Be sure to use a new gasket.

---

8   Thermostatically Controlled Air Cleaner (TCAC) (carburetor equipped vehicles)

General description

1   This system is designed to improve driveability and reduce emissions in cold weather, as well as to maintain a constant inlet air temperature during all driving conditions to improve fuel vaporization.

2   The incoming air can enter the air cleaner two ways - through the fresh air duct (cold air) or past the exhaust manifold cover, through a duct and into the air cleaner (warm air). A thermo sensor, located in the air filter, regulates vacuum to the Air Control Actuator (ACA) which operates a damper in the air intake housing. The ACA opens or closes the damper as necessary, mixing the required amounts of hot and cold air to come up with the desired temperature intake air charge.

3   An improperly functioning TCAC system can result in poor cold or hot driveability and excessive emissions.

Check

Air Control Actuator (ACA) and damper valve

Refer to illustrations 8.4 and 8.7

4   With the engine off, disconnect the

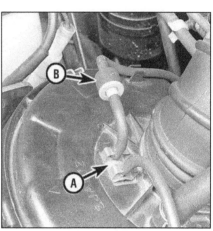

8.7 The TCAC thermo sensor is mounted in the air cleaner cover and retained by a spring clip

A   *Thermo sensor*   B   *Check valve*

warm air hose from the air intake housing and feel the damper valve **(see illustration)** - it should be shut (covering the warm air inlet duct).

5   Start the engine. If the air cleaner is cold, the damper door should now open up. Allow the engine to warm up to normal operating temperature - the door should begin to close, allowing cool air to pass into the air intake housing.

6   If the damper valve doesn't move as described above, disconnect the vacuum hose to the ACA and connect a hand-held vacuum pump to it. Apply vacuum and feel for the damper valve to open - if it doesn't, replace the ACA.

Thermo sensor

7   If the damper valve does open, check the thermo sensor. Disconnect the two hoses from the sensor, which is located on the air cleaner lid **(see illustration)**.

8   Measure the air temperature around the sensor and write it down. Attach a length of hose to one side of the sensor, cover the

other port with your finger and blow into the hose. If the recorded temperature is above 104-degrees F (40-degrees C), air should bleed out of the valve, indicating the valve is open.

9   If the temperature is below 77-degrees F (25-degrees C), no air should come out. If the thermo sensor doesn't work as described, replace it.

Check valve

Refer to illustration 8.10

10   Disconnect the hoses from the check valve, located between the thermo sensor and the intake manifold. Connect a vacuum pump to the orange side of the valve and apply vacuum **(see illustration)** - the needle on the gauge should indicate vacuum and the valve shouldn't leak.

11   Now connect the pump to the black side of the valve and apply vacuum - the valve should leak. If the valve doesn't work as described, replace it. **Note:** *When installing the valve, the orange side must connect to the hose coming from the thermo sensor.*

Component replacement

Air control actuator (ACA)

12   Pull the warm air duct out of the outlet on the air intake housing. Disconnect the vacuum hose from the ACA.

13   Rotate the ACA to align the retaining clip with the slot in the air intake housing, then pull the ACA out. Lift the ACA to disconnect the rod from the damper door.

14   When installing the ACA, position the rod with the bent end pointing down, push in on the damper door, then engage the rod with the hole in the damper door. A small mirror and flashlight will be helpful when doing this.

15   Clip the ACA into the hole, making sure the ends of the clip are securely fastened around the edges of the hole.

16   Reconnect the vacuum hose and the warm air duct.

Thermo sensor

17   Unplug the vacuum hoses from the sensor. Remove the top of the air cleaner, along with the air intake case, from the carburetor.

18   Pry the spring clip off the ports on the thermo sensor and remove the sensor from the under side of the air cleaner lid.

19   Installation is the reverse of the removal procedure.

9   PTC system

General description

Note: *This Section applies only to models with throttle body injection and an automatic transmission.*

1   The PTC system is designed to heat the fuel/air mixture to improve cold start performance. When the engine coolant is below 86-degrees F (30-degrees C), the PTC heater is

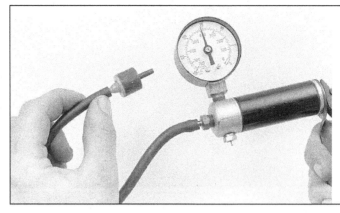

8.10 Check the TCAC check valve with a vacuum pump - when the pump is connected to the orange side of the valve, it should hold vacuum

energized for up to ten seconds.

2   The system consists of the PTC heater, located between the throttle body unit and the intake manifold, the PTC heater relay, the PCM and the coolant sensor.

3   An improperly functioning PTC could result in hard starting and/or stumbling and stalling when cold (no PTC operation) or sluggish performance and/or overheating when hot (PTC stays on).

Check

Note: *The engine must be cool for this check.*

4   Unplug the electrical connector from the PTC heater. Using an ohmmeter, measure the resistance between the PTC heater side of the electrical connector and ground, comparing your reading with the value listed in this Chapter's Specifications. If the resistance isn't as specified, replace the PTC heater.

5   Due to the fairly complex procedures for testing the rest of the system, all further diagnostic procedures should be left to a dealer service department or a repair shop.

Component replacement

6   Following the procedure in Chapter 4, remove the TBI unit from the intake manifold.

7   Unplug the PTC electrical connector and detach the heater from the intake manifold. Clean all traces of old gasket material off the manifold, TBI unit and PTC heater (if it's going to be reinstalled).

8   Installation is the reverse of the removal procedure.

10   Catalytic converter

Note: *Because of a Federally mandated extended warranty which covers emissions-related components such as the catalytic converter, check with a dealer service department before replacing the converter at your own expense.*

General description

1   The catalytic converter is an emission control device added to the exhaust system to reduce pollutants in the exhaust gas stream. There are two types of converters.

The conventional oxidation catalyst reduces the levels of hydrocarbon (HC) and carbon monoxide (CO). The three-way catalyst lowers the levels of oxides of nitrogen (NOx) as well as hydrocarbons (HC) and carbon monoxide (CO).

Check

2   The test equipment for a catalytic converter is expensive and highly sophisticated. If you suspect the converter is malfunctioning, take the vehicle to a dealer service department or authorized emissions inspection facility for diagnosis and repair.

3   Whenever the vehicle is raised for servicing of underbody components, check the converter for leaks, corrosion, dents and other damage. Check the welds/flange bolts that attach the front and rear ends of the converter to the exhaust system. If damage is discovered, the converter should be replaced.

4   Although catalytic converters don't break too often, they do become plugged. The easiest way to check for a restricted converter is to use a vacuum gauge to diagnose the effect of a blocked exhaust on intake vacuum.

a) *Connect a vacuum gauge to an intake manifold vacuum source.*

b) *Warm the engine to operating temperature, place the transaxle in Park and apply the parking brake.*

c) *Note and record the vacuum reading at idle.*

d) *Open the throttle until the engine speed is about 2000 rpm.*

e) *Release the throttle quickly and record the vacuum reading.*

f) *Perform the test three more times, recording the reading after each test.*

g) *If the reading after the fourth test is more than one in-Hg lower than the reading recorded at idle, the catalytic converter, muffler or exhaust pipes may be plugged or restricted.*

Component replacement

5   Because the converter is welded to the exhaust system, converter replacement requires removal of the exhaust pipe assembly (see Chapter 4). Take the vehicle, or the exhaust system, to a dealer service depart-

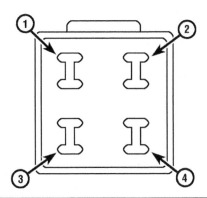

11.5 Obtain the codes on 1991 through 1995 Sidekick and Tracker models by using a jumper wire between the 2 and 3 terminals of the test connector located next to the battery

1 Duty check terminal
2 Diagnostic test terminal
3 Ground terminal
4 Test switch terminal

ment or a muffler shop.

6   If the converter is bolted to the exhaust system, refer to the exhaust system removal and installation section in Chapter 4.

11 Diagnosis system - general information and obtaining code output

Note: *The following trouble code charts apply to 1995 and earlier fuel-injected models only. To access the code information on 1996 and later models, it is necessary to use a Diagnostic Computer or hand-held SCAN tool. If you suspect a problem with the emission related components on a 1996 model, and do not have access to the proper scan tools, have the system tested by a dealer service department or other properly equipped repair facility.*

General information

1   The Powertrain Control Module (PCM) contains a built-in self-diagnosis system which detects and identifies malfunctions occurring in the engine management system. When the PCM detects a problem, three things happen: the Check Engine light comes on, the trouble is identified and a diagnostic code is recorded and stored. The PCM stores the failure code assigned to the specific problem area until the diagnosis system is canceled by removing the diagnostic fuse on models so equipped or by removing the tail lamp fuse from the fuse block (later models).

2   The Check Engine warning light, which is located on the instrument panel, comes on when the ignition switch is turned to On and the engine is not running. When the engine is started, the warning light should go out. If the light remains on, the diagnosis system has detected a malfunction in the system.

Obtaining diagnosis code output on OBD-I (1995 and earlier models)

Refer to illustration 11.5

3   To obtain an output of diagnostic codes, verify first that the battery voltage is above 11 volts, the throttle is fully closed, the transmission is in Neutral, the accessory switches are off and the engine is at normal operating temperature.

4   Turn the ignition switch to OFF. Do not start the engine. **Caution:** *The ignition key must be in the OFF position when disconnecting or reconnecting power to the ECM.*

5   On 1989 and 1990 Sidekick/Tracker models and all Samurai models, insert the spare fuse into the diagnostic terminal of the fuse block. On 1991 through 1995 Sidekick and Tracker models, connect a jumper wire between the number 2 and 3 terminals of the Check connector located adjacent to the battery **(see illustration)**.

6   Turn the ignition key ON. Read the diagnosis code as indicated by the number of flashes of the "Check Engine" light on the dash (see the accompanying chart). Normal system operation is indicated by Code 12 (no malfunctions) for all models. The "Check Engine" light displays a Code 12 by blinking the corresponding pattern one time only.

7   If there are any malfunctions in the system, their corresponding trouble codes are stored in computer memory and the light will blink the requisite number of times for the indicated trouble codes. If there's more than one trouble code in the memory, they'll be displayed in numerical order (from lowest to highest) with a pause interval between each one. After the code with the largest number flashes has been displayed, there will be another pause and then the sequence will begin all over again.

8   To ensure correct interpretation of the blinking "Check Engine" light, watch carefully for the interval between the end of one code and the beginning of the next (otherwise, you will become confused by the relative length of the pause between flashes and misinterpret the codes). The pause between codes is longer than the pause between digits of the same code.

9   To cancel stored trouble codes, remove the cable from the battery negative terminal.

10   If the diagnosis code is not canceled it will be stored by the PCM and appear with any new codes in the event of future trouble.

11   Should it become necessary to work on engine components requiring removal of the battery terminal, first check to see if a diagnostic code has been recorded.

Trouble code chart (OBD-I)

Trouble code	Circuit or symptom	Probable cause
Code 12 (1 flash, pause, 2 flashes)	Normal	This code will flash whenever the diagnostic terminal in the fuse block is activated, the ignition is switched to ON and there are no other codes stored in the ECM.
Code 13 (1 flash, pause, 3 flashes)	Oxygen sensor circuit (open circuit)	Check the wiring and connectors from the oxygen sensor. Replace the oxygen sensor.*
Code 14 (1 flash, pause, 4 flashes)	Coolant sensor circuit (low temperature)	Check all wiring and connectors associated with the coolant temperature sensor. Replace the coolant temperature sensor.
Code 15 (1 flash, pause, 5 flashes)	Coolant sensor circuit (high temperature)	If the engine is experiencing overheating problems, the problem must be rectified before continuing. Then check the wiring connections at the ECM.
Code 21 (2 flashes, pause, 1 flash)	Throttle Position Sensor (TPS)	Check the TPS adjustment and connection (see Chapter 4). Check the PCM connector. Replace the TPS*.
Code 22 (2 flashes, pause, 2 flashes)	Throttle Position Sensor (TPS)	Check the TPS adjustment and connection (see Chapter 4). Check the PCM connector. Replace the TPS*.

Trouble code chart (OBD-I)

Trouble code	Circuit or symptom	Probable cause
Code 23 (2 flashes, pause, 3 flashes)	Intake Air Temperature sensor	Check the IAT sensor, wiring and connectors for an open sensor circuit. Replace the IAT sensor*.
Code 24 (2 flashes, pause, 4 flashes)	Vehicle Speed Sensor (VSS)	A fault in this circuit should be indicated only when the vehicle is in motion. Check and repair the speedometer if it is functioning.
Code 25 (2 flashes, pause, 5 flashes)	Intake Air Temperature sensor	Check the IAT sensor, wiring and connectors for an open sensor circuit. Replace the IAT sensor.*
Code 31 (3 flashes, pause, 1 flash)	MAP sensor (TBI models)	Check the circuit for bare wire (stripped insulation) or damaged electrical connectors. Check the vacuum hose connection. Replace the MAP sensor if necessary.
Code 32 (3 flashes, pause, 2 flashes)	MAP sensor (TBI models)	Check the circuit for bare wire (stripped insulation) or damaged electrical connectors. Check the vacuum hose connection. Replace the MAP sensor if necessary.
Code 33 (3 flashes, pause, 3 flashes)	Mass Airflow (MAF) sensor (MPFI models)	Check the MAF sensor and electrical connections (see Chapter 4). Replace the MAF sensor, if necessary.
Code 34 (3 flashes, pause, 4 flashes)	Mass Airflow (MAF) sensor (MPFI models)	Check the MAF sensor and electrical connections (see Chapter 4). Replace the MAF sensor, if necessary.
Code 41 (4 flashes, pause, 1 flash)	Ignition signal	Inspect and repair any damaged electrical connectors and wire in the harness.
Code 42 (4 flashes, pause, 2 flashes)	Crank angle sensor (pick-up assembly)	Check for a faulty connector or circuit. Also check for an improper air gap in the distributor (see Chapter 5). Replace the sensor (pick-up assembly), if necessary.
Code 44 (4 flashes, pause, 4 flashes)	Throttle Position Sensor (TPS) idle switch circuit	Check the TPS and its adjustment, as described in Chapter 4. Replace the TPS, if necessary.
Code 45 (4 flashes, pause, 5 flashes)	Throttle Position Sensor (TPS) idle switch circuit	Check the TPS and its adjustment, as described in Chapter 4. Replace the TPS, if necessary.
Code 51 (5 flashes, pause, 1 flash)	Exhaust Gas Recirculation (EGR) system	Check for a faulty connector or circuit. Also check for high resistance in the EGR solenoid coil.
Code 53 (5 flashes, pause, 3 flashes)	Ground circuit	Poor ground connection at the PCM or engine block. Also, possibly a faulty ECM.

OBD II systems (1996 and later models)

Refer to illustration 11.16

12   For some years there has been a gradual process of making and enforcing a "universal" set of computer codes that would be applied by all auto manufacturers to their self-diagnostics. One of the first automotive applications of computers was self-diagnosis of system and component failures. While early on-board diagnostic computers simply lit a "CHECK ENGINE" light on the dash, present systems must monitor complex interactive emission control systems, and provide enough data to the technician to successfully isolate a malfunction.
13   The computer's role in self-diagnosing emission control problems has become so important that such computers are now required by Federal law. The requirements of the "first generation" system, nicknamed OBD I for On-Board Diagnostics, have been incorporated into 1992 through 1995 models. The purpose of On Board Diagnostics (OBD) is to ensure that emission related components and systems are functioning properly to reduce emission levels of several pollutants emitted by auto and truck engines. The first step is to detect that a malfunction has occurred which may cause increased emissions. The next step is for the system to notify the driver so that the vehicle can be serviced. The final step is to store enough information about the malfunction so that it can be identified and repaired.

14   The latest step has been the establishment of the OBD II system, which further defines emissions performance, and also regulates the code numbers and definitions.
15   The basic OBD II code is a letter followed by a four-digit number. Most manufacturers also have many additional codes that are *specific* to their vehicles. Although many of the self-diagnostic tests performed by early OBD systems are retained in OBD II systems, they are now more sensitive in detecting malfunctions. Where there may have been only one code for an oxygen sensor malfunction, there are now six codes that narrow down where the performance discrepancy is. OBD II systems perform many additional tests in areas not required under the earlier OBD. These include monitoring for

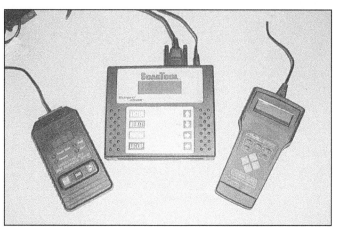

11.16 Scanners like the Actron Scantool and the AutoXray XP240 are powerful diagnostic aids - programmed with comprehensive diagnostic information, they can tell you just about anything you want to know about your engine management system, but they're expensive

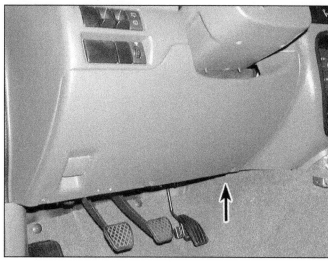

11.17 The 16-pin diagnostic connector is located under the dash on the driver's side on OBD-II models

engine misfires and detecting deterioration of the catalytic converter.

16 On the models covered by this manual, OBDII systems began in 1996. At first, a very expensive scan tool was required to read the codes. Now, however, the aftermarket has come up with inexpensive scan tools for the home mechanic to use **(see illustration)**. On these models, the scan tool is the only way to extract and clear trouble codes. **Note:** *To determine if your vehicle is OBD II or not, look at the VECI (Vehicle Emission Control Information) decal on the top of the radiator fan shroud. If it's OBD II, the decal will indicate "OBD II certified".*

Retrieving codes on OBD-II systems

Refer to illustration 11.17

17 Besides a standardized set of diagnos-

tic trouble codes, the Federal government's OBD-II program mandates a standard 16-pin diagnostic link connector (DLC) for all vehicles. The DLC is also referred to as a J1962 connector (a designation taken from the physical and electrical specification number assigned by the SAE). Besides its standard pin configuration, the J1962 must also provide power and ground circuits for scan tool hook-up. The SAE has also recommended locating the DLC or J1962 connector under the driver's side of the dash **(see illustration)**.

18 Now that all new vehicles have a standardized connector and a universal set of diagnostic trouble codes, the same scan tool can be used on any vehicle, and any home mechanic can access these codes with a relatively affordable generic scan tool.

19 All aftermarket generic scan tools include good documentation, so refer to the

manufacturer's hook-up instructions in your scan tool manual. Before plugging a scan tool into the DLC, inspect the condition of the DLC housing; make sure that all the wires are connected and that the contacts are fully seated in the housing. Inside the connector, make sure that there's no corrosion on the pins and that no pins are bent or damaged.

Clearing codes on OBD-II systems

20 After you've made the necessary repairs, you'll need to "clear" (erase) any stored trouble codes from the PCM. Most OBD-II systems require a scan tool to clear the codes. Refer to the manufacturer's documentation that comes with the aftermarket generic scan tool you're using. It will guide you through the correct procedure for clearing any codes stored in the PCM.

Trouble code chart (OBD-II)

Code	Description
P0101	Mass air flow circuit, range or performance problem
P0102	Mass air flow circuit, low input
P0103	Mass air flow circuit, high input
P0111	Intake air temperature circuit, range or performance problem
P0112	Intake air temperature circuit, low input
P0113	Intake air temperature circuit, high input
P0116	Engine coolant temperature circuit, range or performance problem
P0117	Engine coolant temperature circuit, low input
P0118	Engine coolant temperature circuit, high input
P0121	Throttle position circuit, performance problem
P0122	Throttle position circuit, low input
P0123	Throttle position circuit, high input
P0125	Insufficient coolant temperature for closed loop fuel control
P0131	O2 sensor circuit, low voltage (cylinder bank no. 1, sensor no. 1)
P0132	O2 sensor circuit, high voltage (cylinder bank no. 1, sensor no. 1)
P0133	O2 sensor circuit, slow response (cylinder bank no. 1, sensor no. 1)
P0134	O2 sensor circuit - no activity detected (cylinder bank no. 1, sensor no. 1)
P0135	O2 sensor heater circuit malfunction (cylinder bank no. 1, sensor no. 1)
P0136	O2 sensor circuit malfunction (cylinder bank no. 1, sensor no. 2)
P0141	O2 sensor heater circuit malfunction (cylinder bank no. 1, sensor no. 2)
P0171	System too lean (cylinder bank no. 1)
P0172	System too rich (cylinder bank no. 1)
P0300	Random/multiple cylinder misfire detected
P0301	Cylinder no. 1 misfire detected
P0302	Cylinder no. 2 misfire detected

Trouble code chart (OBD-II) (continued)

P0303	Cylinder no. 3 misfire detected		P0461	Fuel level sensor circuit, range or performance problem
P0304	Cylinder no. 4 misfire detected		P0463	Fuel level sensor circuit, high input
P0327	Knock sensor no. 1 circuit, low input (cylinder bank no. 1 or single sensor)		P0500	Vehicle speed sensor malfunction
			P0505	Idle control system malfunction
P0328	Knock sensor no. 1 circuit, high input (cylinder bank no. 1 or single sensor)		P0601	Internal control module, memory check sum error
P0335	Crankshaft position sensor circuit malfunction		P0705	Transmission range sensor, circuit malfunction (PRNDL input)
P0340	Camshaft position sensor circuit malfunction		P0720	Output speed sensor malfunction
P0400	Exhaust gas recirculation flow malfunction		P0741	Torque converter clutch (TCC) solenoid, circuit performance or stuck in OFF position
P0403	Exhaust gas recirculation circuit malfunction			
P0420	Catalyst system efficiency below threshold		P0743	Torque converter clutch (TCC) solenoid circuit, electrical problem
P0440	Evaporative emission control system malfunction			
P0443	Evaporative emission control system, EVAP canister purge valve circuit malfunction		P0751	Shift solenoid A (No. 1), performance problem or stuck in off position
P0450	Evaporative emission control system, pressure sensor malfunction		P0753	Shift solenoid A (No. 1), electrical problem
			P0756	Shift solenoid B (No. 2), performance problem or stuck in off position
P0451	Evaporative emission control system, pressure sensor range or performance problem			
			P0758	Shift solenoid B (No. 2), electrical problem
P0455	Evaporative emission control system, leak detected			

12 Camshaft Position (CMP) sensor - check and replacement

Description

1   The Camshaft Position (CMP) sensor is the timing device used by the PCM to calculate engine speed and the position of the pistons. The signal from the CMP sensor is one of the signals used by the PCM to control ignition and injector timing. On SOHC engines, the CMP sensor is bolted to the rear of the cylinder head. On DOHC engines, the CMP sensor is bolted to the sensor case, which houses the signal rotor. It's not necessary to remove the case for the CMP sensor on DOHC engines unless you're replacing or servicing the cylinder head. If you do remove the CMP sensor case on DOHC engines, be sure to adjust it (see Step 18).

1990 and later Samurai models and 1992 through 1998 Sidekick/Tracker models

2   On 1990 and later Samurai models, and on 1992 through 1998 Sidekick/Tracker models, the CMP sensor is located inside the distributor. (On 1990 through 1992 Samurai models, the CMP sensor is referred to by the manufacturer as a "Crank Angle Sensor, or a "CAS," but it is identical in form and function to a CMP sensor, so in this manual the CAS is also referred to as a CMP sensor.)

3   On these models, the CMP sensor consists of a signal rotor and a signal generator (which includes a magnet and a Hall effect element). The signal rotor is located on the distributor shaft, below the ignition rotor. The signal rotor has four "poles," one for each piston. As the signal rotor turns, each of these four poles causes the magnetic flux from the magnet to be applied to the Hall effect element. The Hall element generates a voltage signal in proportion to the magnetic flux. This wave-shaped signal is sent to the PCM, which uses it to calculate engine speed and the position of each piston, and to control ignition and injector timing. A problem in the CMP sensor circuit on 1996 and later models will be indicated by a P0340 diagnostic trouble code.

1999 and later Vitara/Tracker models

Refer to illustration 12.4

4   There is no distributor on 1999 and later Vitara/Tracker models. On these models, the CMP sensor is bolted to the back of the cylinder head **(see illustration)**. The CMP sensors used on SOHC and DOHC engines are slightly different: On SOHC engines, the signal rotor is an integral part of the camshaft and the CMP sensor is bolted to the rear of the head. On DOHC engines, the signal rotor is housed inside an adjustable case, which is bolted to the rear of the head. The exhaust camshaft drives the signal rotor via a pair of lugs that fit into a slot in the end of the cam. (Unless you're replacing or servicing the cylinder head on a DOHC engine, it's not necessary to remove the CMP sensor case.) A problem in the CMP sensor circuit will be indicated by a P0340 diagnostic trouble code.

Check

1990 and later Samurai models and 1992 through 1998 Sidekick/Tracker models

5   Remove the distributor cap and rotor (see Chapter 1). Also remove the plastic dust shield below the rotor (the shield protects the

12.4 A typical Camshaft Position (CMP) sensor (arrow) on 1999 and later Vitara/Tracker models (DOHC unit shown, SOHC unit in same location)

signal rotor and the CAS or CMP sensor from moisture and dirt).

6   Carefully inspect the CMP sensor and the electrical connector at the CMP sensor. Make sure that the electrical connector is securely plugged in and that the CMP sensor is securely installed in the distributor (see Chapter 5).

7   Unplug the electrical connector from the CMP sensor. The CMP sensor connector has three terminals: voltage supply (from the battery or from the PCM), voltage signal (to the PCM), and ground.

8   Hook up a voltmeter between the terminals for voltage supply and ground, on the harness side of the connector, and then turn the ignition to ON (don't start the engine). On Samurai models, there should be battery voltage; on Sidekick/Tracker models, there

12.10 To test the CMP sensor on 1999 and later models, unplug the electrical connector (upper arrow); to detach the CMP sensor, remove the sensor retaining bolt (lower arrow) (DOHC engine shown, SOHC engine similar)

should be 5 volts. There are a number of different wire colors and terminal configurations for the CAS/CMP sensor. The ground wire is usually black or black and green. Use a process of elimination to identify the terminal for supply voltage by grounding the voltmeter and then touching the other voltmeter lead to each of the other two terminals. Only one of them will give you voltage. If there is no voltage to the CMP sensor, check the wire harness back to the PCM. If the wiring back to the PCM is okay, have the PCM checked out by a dealer service department.

9   Make a note of which terminal is the ground terminal and which terminal is for voltage supply, and then reconnect the CMP sensor electrical connector. Now backprobe the third terminal, the one that sends the voltage signal to the PCM, with a straight pin, clip the positive lead of the voltmeter to the pin with an alligator clip and ground the other voltmeter lead. Make sure that the magnet between the signal rotor and the Hall element is free of any metal particles. Remove the spark plugs (see Chapter 1), and then slowly rotate the engine a couple of revolutions and watch the voltmeter. There are four poles on the signal rotor (one for each piston). When a pole piece on the signal rotor passes through the field of magnetic flux between the magnet and the Hall element, the magnetic flux field is "interrupted" (cut off), and the voltmeter should indicate battery voltage (Samurai models) or 5 volts (Sidekick/Tracker models). In between each pole passing through the magnetic flux field, i.e. when no pole is passing through the flux field, there should be zero voltage. If the CAS/CMP sensor doesn't operate as described, replace it.

1999 and later Vitara/Tracker models

Refer to illustration 12.10

10   Unplug the electrical connector from the CMP sensor **(see illustration)**. Hook up the

leads of a voltmeter to the terminals for the blue/black and the violet/red wires (the two outer terminals), on the harness side of the connector. Turn the ignition key to ON and measure the voltage. There should be 10 to 14 volts. If there isn't, inspect the wire harness between the CMP sensor and the PCM for a problem and repair as necessary.

11   Connect the voltmeter to the terminals for the yellow/blue wire (the center terminal) and the violet/red wire. With the ignition turned to ON, there should be 4 to 5 volts.

12   If the CMP sensor doesn't function as described, replace it.

Replacement

1990 and later Samurai models and 1992 through 1998 Sidekick/Tracker models

13   If the CMP sensor is defective, replace the distributor (see Chapter 5).

1999 and later Vitara/Tracker models

14   Disconnect the negative battery cable, and then unplug the electrical connector from the CMP sensor.

15   Remove the CMP sensor retaining bolt **(see illustration 12.10)**. Discard the old O-ring. If you're removing the CMP sensor to repair or replace the cylinder head on a DOHC engine, remove the CMP sensor case bolt and remove the case.

16   Installation is the reverse of removal. Be sure to use a new O-ring and coat it with clean engine oil.

17   Tighten the CMP sensor retaining bolt to the torque listed in this Chapter's Specifications and then reconnect the electrical connector.

18   If you removed the CMP sensor case on a DOHC engine, make sure that the lugs on the CMP sensor coupling fit into the slot on the end of the camshaft. The lugs are offset, so if they don't fit into the slot in the cam, rotate the CMP sensor shaft 180 degrees. When the CMP sensor case is flush with the cylinder head, install the CMP sensor retaining bolt and tighten it to the torque listed in this Chapter's Specifications.

---

13   Crankshaft Position (CKP) sensor - check and replacement

Description

1990 through 1992 Samurai models

1   On 1990 through 1992 Samurai models, the Crank Angle Sensor (CAS) is a Hall effect element located in the distributor. The CAS serves not only as an information sensor for the PCM (see Section 12) but as the "pick-up coil" for the ignition system (see Section 11 in Chapter 5). It consists of a generator (Hall effect element) and signal rotor. As the signal rotor turns, a square-wave variable voltage signal is generated by the Hall effect element.

This pulse signal is sent to the PCM, which uses it to calculate engine speed and piston position. The CAS signal is one of the signals used by the PCM to control ignition and injector timing. If a failure occurs, a code 42 will be set (the same code that's set for a problem in the CMP sensor circuit, because "CAS" is a name used for the device that is referred to as the CMP sensor on later models).

1996 and later models

2   On SOHC engines, the Crankshaft Position (CKP) sensor is located at the front of the engine, on the oil pan flange, below the crankshaft pulley. On DOHC engines, the CKP sensor is located at the back of the engine, above the rear crankshaft main seal. On these models, you must remove the transmission and the flywheel to replace the CKP sensor.

3   The CKP sensor is a magnetic pick-up coil. As the crankshaft rotates, the toothed timing belt pulley (SOHC engines) or the signal rotor on the crankshaft (DOHC engines) passes the sensor tip, which generates an AC voltage pulse at the sensor. The CKP sensor signal is one of the signals used by the PCM to calculate the crankshaft rotational speed, information it needs for monitoring misfires. A problem in the CKP sensor circuit will be indicated by a P0335 diagnostic trouble code.

Check

1990 through 1992 Samurai models

4   On 1990 through 1992 Samurai models, refer to the Crank Angle Sensor (CAS) check in Steps 5 through 9 in Section 12.

1996 and later models

5   Unplug the CKP sensor electrical connector and measure the resistance between the sensor terminals and between each sensor terminal and ground. (There are only two terminals in the electrical connector for the CKP sensor used on SOHC engines, but the connector for the CKP sensor on DOHC engines has *three* terminals. Ignore the terminal for the black/green wire; you want to measure the resistance across the terminals for the white/blue wire and for the orange/blue wire.) Compare your measurement to the CKP sensor resistance listed in this Chapter's Specifications. If the indicated sensor resistance is out of range, replace the CKP sensor. If the CKP sensor checks out okay, but problems persist, have the CKP sensor circuit checked by a dealer service department.

Replacement

1990 through 1992 Samurai models

6   Replace the distributor (see Chapter 5).

1996 and later models

7   On SOHC engines, raise the front of the vehicle and place it securely on jackstands. Unplug the electrical connector from the CKP sensor, remove the CKP sensor retaining bolt and remove the CKP sensor.

14.1   A typical Engine Coolant Temperature (ECT) sensor installation; on DOHC engines (shown), the ECT sensor is on the back of the cylinder head; on SOHC engines, it's on the intake manifold

A   *Electrical connector (release locking tang to remove)*
B   *Engine Coolant Temperature (ECT) sensor*

15.1   The fuel tank pressure sensor is bolted to the top of the fuel tank; to detach the sensor from the tank, remove these two bolts (arrows)

8   On DOHC engines, remove the transmission (see Chapter 7) and then remove the flywheel or driveplate (see Chapter 2). Unplug the electrical connector from the CKP sensor and then remove the CKP sensor retaining bolt.

9   Installation is the reverse of removal. Tighten the CKP sensor mounting bolt to the torque listed in this Chapter's Specifications.

14   Engine Coolant Temperature (ECT) sensor - check and replacement

General description

Refer to illustration 14.1

1   The Engine Coolant Temperature (ECT) sensor is a "thermistor," a resistor which changes its resistance from high to low as the temperature increases. On SOHC engines, the ECT sensor is located in the side of the intake manifold, right next to the Intake Air Temperature (IAT) sensor. On DOHC engines, the ECT sensor is located on the back of the engine **(see illustration)**. A failure in the ECT sensor circuit should set either a Code 14 or a Code 15 (OBD-I models) or a P0116, P0117 or P0118 (OBD-II models). These codes indicate a failure in the ECT sensor or in the ECT sensor circuit, so the appropriate solution to the problem will be either repair of a wire or replacement of the sensor.

Check

2   Unplug the electrical connector and then unscrew and remove the ECT sensor. Suspend the ECT sensor in a Pyrex, container, or some other container designed to withstand high temperature. Fill the container with water so that the business end of the ECT sensor is immersed in the water. Using a cooking thermometer, bring the water to a boil. As the water heats up, use an ohmmeter to measure the resistance across the sensor terminals, write down your measurements and then compare them to the accompanying ECT sensor temperature-to-resistance table. If the ECT temperature sensor resistance is incorrect, replace the ECT sensor.

ECT sensor temperature-to-resistance table

Temperature (degrees F)	1998 and earlier models Resistance (K-ohms)	1999 and later models Resistance (k-ohms)
32	5.21 to 6.37	5.74
68	2.21 to 2.69	2.28 to 2.61
104	1.14	1.15
140	0.58	0.584
176	0.29 to 0.35	0.303 to 0.326

Replacement

Warning: *Wait until the engine is completely cool before starting this procedure.*

3   Unplug the electrical connector from the ECT sensor and then unscrew the sensor from the intake manifold (SOHC engines) or from the head (DOHC engines). **Caution:** *Handle the coolant sensor with care. Damage to this sensor will affect the operation of the entire fuel system.*

4   Before installing the new sensor, wrap the threads with Teflon sealing tape to prevent leakage and thread corrosion.

5   Installation is the reverse of removal. Tighten the ECT sensor to the torque listed in this Chapter's Specifications.

6   Check the coolant level and add as necessary (see Chapter 1).

15   Fuel tank pressure sensor - check and replacement

General description

Refer to illustration 15.1

1   The fuel tank pressure sensor **(see illustration)**, which is located on top of the fuel tank, is part of the Evaporative Emission (EVAP) control system on OBD-II vehicles. When vapor pressure inside the fuel tank reaches a specified threshold, it pushes open the tank pressure control valve and migrates to the EVAP canister. However, if a malfunction in the system causes the fuel tank pressure to exceed a specified threshold (higher than ambient atmospheric pressure), the pressure sensor sends a voltage signal to the

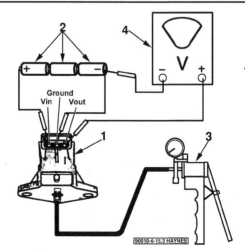

15.3   Fuel tank pressure sensor test setup

1   *Fuel tank pressure sensor*
2   *Three 1.5V batteries (4.5 to 5.0 volts total voltage)*
3   *Hand-held vacuum/pressure pump*
4   *Digital voltmeter*

15.5 After removing the fuel tank pressure sensor, inspect the air vent (upper arrow) and the pressure vent (middle arrow) for obstructions and clean them out if necessary with a wooden toothpick; also inspect the O-ring (lower arrow) for damage and deterioration

16.4 Unplug the electrical connector from the Intake Air Temperature (IAT) sensor, which is located in the air cleaner housing

16.5 To remove the IAT sensor from the air cleaner housing, pull it straight out of its grommet; make sure the grommet is neither damaged nor deteriorated before installing the IAT sensor again

PCM, which sets a diagnostic trouble code. A problem in the fuel tank pressure control system will be indicated by a P0450 or a P0451 diagnostic trouble code.

Check

Refer to illustration 15.3

2   Remove the fuel tank pressure sensor from the fuel tank (see Steps 4 through 6).
3   Using three 1.5V batteries in series (verify that the three batteries have at least 4.5 volts), connect the positive terminal of the batteries to the "Vin" (voltage in) terminal of the pressure sensor connector and connect the negative battery terminal to the connector Ground terminal **(see illustration)**. Then check the voltage between "Vout" (voltage out) and Ground. There should be about 2.5 volts between the "Vout" and Ground terminals. Next, hook up a hand-held vacuum/pressure pump and verify that the voltage decreases slightly when 6.6 kPa (50 mm Hg) vacuum is applied, then verify that the voltage increases slightly when 6.6 kPa (50 mm Hg) pressure is applied. If the fuel tank pressure sensor doesn't function as described, replace it.

Replacement

Refer to illustration 15.5

4   Remove the fuel tank (see Chapter 4).
5   Remove the fuel tank pressure sensor

bolts and remove the sensor **(see illustration)**.
6   Inspect the sensor air vent hole and the pressure passage for clogging. If either passage is clogged, carefully clean it out with a wooden toothpick. Also inspect the O-ring for damage and deterioration. Replace the O-ring if it's damaged or deteriorated.
7   Installation is the reverse of removal. Tighten the fuel tank pressure sensor bolts to the torque listed in this Chapter's Specifications.

---

16 Intake Air Temperature (IAT) sensor - check and replacement

General description

1   The Air Temperature Sensor (ATS), which is used on Samurai models with throttle body injection, is a thermistor which constantly measures the temperature of the air entering the intake manifold. The Intake Air Temperature (IAT) sensor, which is used on 1993 California and all 1994 and later models with sequential multiport fuel injection, is a thermistor which measures the temperature of the air entering the air cleaner housing. As air temperature varies, the PCM, which monitors the ATS/IAT sensor, adjusts the amount of fuel metered by the injector(s) in accordance with changes in the air temperature. A malfunction in the ATS circuit, or in a pre-OBD-II IAT sensor circuit, will set a diagnostic code 23 or code 25. A fault in the IAT sensor

circuit will set a P0111, a P0112 or a P0113 diagnostic trouble code.

Check

2   Remove the ATS or IAT sensor (see Steps 4 and 5).
3   Wipe off the ATS/IAT sensor and then suspend the sensor in a Pyrex, container, or some other container designed to withstand high temperature. Fill the container with water so that the business end of the ATS/IAT sensor is immersed in the water. Using a cooking thermometer, bring the water to a boil. As the water heats up, use an ohmmeter to measure the resistance across the sensor terminals, write down your measurements and then compare them to the accompanying ATS/IAT sensor temperature-to-resistance table. If the ATS/IAT sensor resistance is incorrect, replace the sensor.

Replacement

Refer to illustrations 16.4 and 16.5

4   Unplug the electrical connector from the ATS/IAT sensor **(see illustration)**.
5   On Samurai models with throttle body injection, unscrew the ATS from the intake manifold. On 1993 through 1995 models with sequential multiport fuel injection, unscrew the IAT sensor from the air cleaner housing. On 1996 and later models with sequential multiport fuel injection, simply pull the IAT sensor out of its grommet in the air cleaner housing **(see illustration)**.

ATS/IAT sensor temperature-to-resistance table

Temperature (degrees F)	Air Temperature Sensor (ATS) (Samurai models with throttle body injection) Resistance (K-ohms)	1992 through 1995 (pre-OBD-II) models and 1996 through 1998 (OBD-II) models with sequential multiport fuel injection Resistance (k-ohms)	1999 and later models with sequential multiport fuel injection (all OBD-II models) Resistance (k-ohms)
32	5.4 to 6.6	5.4 to 6.6	
68	2.28 to 2.87	2.28 to 2.87	2.09 to 2.81
104	1.06 to 1.36	1.06 to 1.36	
140	0.53 to 0.70	0.53 to 0.70	
176	0.29 to 0.39	0.29 to 0.39	0.322

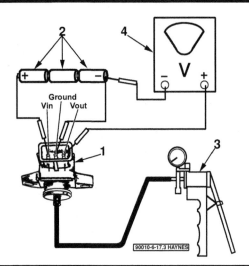

17.3 Manifold Differential Pressure (MDP)/Manifold Absolute Pressure (MAP) sensor test setup

1   *MDP/MAP sensor*
2   *Three new 1.5V batteries*
3   *Hand-held vacuum pump*
4   *Digital voltmeter*

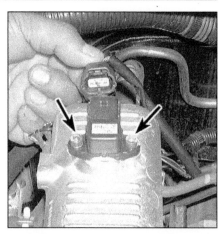

17.6 To remove the Manifold Differential Pressure (MDP) or Manifold Absolute Pressure (MAP) sensor, unplug the electrical connector and remove the mounting bolts (arrows); (1996 through 2001 MDP/MAP sensor shown; earlier sensor units similar, except that they have an external vacuum hose which must be disconnected for removal and testing)

6   If you're installing an ATS, be sure to use a new gasket and coat the threads of the sensor with Teflon, tape and then tighten the ATS securely. Reconnect the electrical connector to the ATS.

7   If you're installing an IAT sensor on a pre-1996 model, coat the threads of the sensor with Teflon, tape and then tighten the sensor to the torque listed in this Chapter's Specifications. If you're installing an IAT sensor on a 1996 or later model, make sure that the rubber grommet in the air cleaner housing is in good condition. If it isn't, replace it before installing the IAT sensor. Push the IAT sensor into the grommet until it's firmly seated and then reconnect the electrical connector.

---

17 Manifold Absolute Pressure (MAP)/Manifold Differential Pressure (MDP) sensor - check and replacement

General description

1   The Manifold Absolute Pressure (MAP) sensor (1993 through 1995 and 2001 models) or the Manifold Differential Pressure (MDP) sensor (1996 through 2000 models) monitors changes in pressure, caused by changes in engine load and speed, inside the intake manifold, and then converts this information into a variable voltage output to the PCM.

2   The PCM sends a 5-volt reference voltage to the MAP/MDP sensor. The PCM uses the altered returning signal to monitor changes in intake manifold pressure. The MAP/MDP sensor consists of a special semiconductor element that changes its resistance in proportion to changes in manifold pressure, thus converting each change in pressure into a change in voltage, and an electronic circuit that amplifies and corrects this changing voltage signal before sending it back to the PCM. The MAP/MDP sensor voltage output is one of the signals used by the PCM to control fuel delivery and ignition timing. The PCM also uses the MAP/MDP sen-

sor output signal to determine when the EGR valve is allowing recirculated exhaust gases to enter the intake manifold. Monitoring the changes in pressure in the intake manifold also enables the PCM to identify an excessive or inadequate EGR flow. A malfunction in the MAP/MDP sensor circuit on OBD-I vehicles will set a Code 31 or Code 32; a fault in the MAP/MDP sensor circuit on OBD-II vehicles will set a P1408 diagnostic trouble code (not all aftermarket generic scanners, however, will be able to identify this code).

Check

Refer to illustration 17.3

3   Remove the MAP/MDP sensor (see Steps 5 through 7).

4   Using three *new* 1.5V batteries in series, connect the positive terminal of the batteries to the "Vin" (voltage in) terminal of the MAP/MDP sensor connector and connect the negative battery terminal to the sensor connector Ground terminal **(see illustration)**. Then connect a hand-held vacuum pump to the vacuum pipe on the bottom of the sensor and, referring to the appropriate chart below, check the voltage between "Vout" (voltage out) and Ground and verify that the sensor output voltage decreases as vacuum is applied. If the MAP/MDP sensor output voltage doesn't fall within the specified voltage range for each corresponding applied vacuum, replace the sensor.

MAP sensor vacuum to voltage output (1993 through 1995 models)

Applied vacuum (mm Hg)	MAP sensor output voltage
760	3.6 to 4.4
733	3.5 to 4.2
707	3.4 to 4.1
682	3.2 to 4.0
658	3.1 to 3.8
634	3.0 to 3.7
611	2.9 to 3.6
589	2.8 to 3.4
567	2.7 to 3.3
546	2.6 to 3.2
526	2.5 to 3.1

MDP sensor vacuum to voltage (1996 through 2000 models)

Applied vacuum (mm Hg)	MAP sensor output voltage
200	2.40 to 4.40
250	2.13 to 4.13
300	1.86 to 3.86
350	1.59 to 3.59
400	1.32 to 3.32

MAP sensor vacuum to voltage (2001 models)

Applied vacuum (mm Hg)	MAP sensor output voltage
707 to 760	3.3 to 4.3
634 to 707	3.0 to 4.1
567 to 634	2.7 to 3.7
526 to 567	2.5 to 3.3

Replacement

Refer to illustration 17.6

5   Detach the vacuum hose from the sensor (if applicable). The MDP/MAP sensor unit on 1996 and later sensors doesn't have an external vacuum hose.

6   Unplug the electrical connector from the sensor **(see illustration)**.

7   Remove the sensor mounting bolts **(see illustration 17.6)** and then remove the sensor.

8   Installation is the reverse of removal.

---

18 Mass Air Flow (MAF) sensor - check and replacement

Description

1   The Mass Air Flow (MAF) sensor monitors the mass of the air being drawn into the

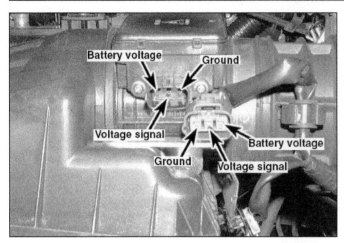

18.6 Mass Air Flow (MAF) sensor connector terminal guide

18.14 To detach the MAF sensor from the air cleaner housing on 1999 and later models, remove these two bolts (arrows)

engine and converts that information into a current signal that it sends to the PCM, which uses the signal to control various control devices. The MAF sensor consists of a heat resistor inside a "metering duct" (air passage) and a control circuit. The intake air passing over the heat resistor cools the resistor, and the control circuit regulates the temperature of the resistor by controlling the amount of current flowing through the resistor, which keeps it within a specified temperature range in relation to the ambient temperature. This "control value" is expressed as a current output signal, which the MAF sensor sends to the PCM.

2   A problem in the MAF sensor circuit will set a Code 33 or Code 34 (OBD-I vehicles), or a P0101, P0102 or P0103 diagnostic trouble code (OBD-II vehicles).

Check

Refer to illustration 18.6

3   Inspect the intake air duct from the MAF sensor to the throttle body. Any air leaks between the sensor and throttle body will cause the engine to run poorly.

4   Inspect the MAF sensor for cracks and other damage. Make sure the electrical connection is tight and clean.

5   If you suspect an intermittent malfunction in the MAF sensor, i.e. one that does not set a code, try tapping gently on the MAF sensor housing with a small tool while the engine is idling. If the engine runs poorly when the MAF sensor is tapped, the sensor is probably faulty.

6   Unplug the MAF sensor electrical connector **(see illustration)**. Connect the positive lead of a voltmeter to the battery voltage (B+) terminal of the harness side of the connector and connect the negative voltmeter lead to a good ground. Turn the ignition switch to ON and verify that battery voltage is available. If battery voltage is not available, inspect the wire harness between the connector and the PCM for an open or a bad connection. Turn the ignition switch to OFF.

7   Plug in the MAF sensor connector and

then, using a long straight pin, backprobe the connector terminal for the MAF sensor voltage signal to the PCM (the middle terminal of the connector). Clip your voltmeter positive lead to the straight pin and ground the negative voltmeter lead. Turn the ignition switch to ON and note the indicated voltage. Compare your measurement to the MAF sensor output voltage listed in this Chapter's Specifications.

8   Start the engine and verify that the MAF sensor signal voltage to the PCM is less than 5 volts (at idle, the MAF sensor signal voltage is typically from 1.7 to 2.0 volts). Finally, verify that the MAF sensor signal voltage rises as the engine speed increases.

9   If the MAF sensor doesn't function as described, the problem could be in the wire harness, or in the connector at the MAF sensor or the connector at the PCM. If the wire harness and the connectors are in good shape, the problem is either the MAF sensor (more likely) or in the PCM (less likely). If you're in any doubt, have the MAF sensor, the PCM and the circuit between them checked out by a dealer service department before condemning any particular component. MAF sensors and PCMs are expensive and cannot be returned once purchased.

Replacement

Refer to illustration 18.14

Caution: *When handling the MAF sensor, do NOT attempt to disassemble it, don't expose it to any shocks, don't blow it out with compressed air and don't put your finger or any tool inside the unit. Any of these things could damage the MAF sensor, which is more susceptible to rough treatment than other sensors.*

10   Disconnect the negative cable from the battery.

11   Remove the intake air duct from the MAF sensor.

12   Disconnect the electrical connector from the MAF sensor **(see illustration 18.6)**.

13   On 1998 and earlier models, disconnect and remove the upper half of the air cleaner housing and then detach the MAF sensor

from the upper half of the housing. Be sure to discard and replace the MAF sensor-to-air cleaner upper case seal.

14   On 1999 and later models, remove the MAF sensor mounting bolts **(see illustration)** and then remove the MAF sensor from the air cleaner housing. Inspect the MAF sensor seal for damage and deterioration. If it's damaged or deteriorated, replace it.

15   Installation is the reverse of removal.

---

19 Oxygen sensor - check and replacement

General description

Refer to illustration 19.3

1   The oxygen sensor is mounted in the exhaust system where it monitors the oxygen content of the exhaust gas stream and produces a voltage signal proportional to the amount of oxygen present in the exhaust gases. The oxygen in the exhaust reacts with the elements inside the oxygen sensor to produce a voltage output that varies from 0.1 volt (high oxygen, lean mixture) to 0.9 volt (low oxygen, rich mixture). The oxygen sensor (the pre-converter sensor on OBD-II vehicles - see Step 3) provides a feedback signal to the PCM that represents the amount of oxygen still remaining in the exhaust gas after combustion. The PCM monitors this voltage output continuously to determine the correct amount of fuel for the injectors to spray into the intake passages so that it can maintain the correct air/fuel ratio of 14.7:1 (14.7 parts air to 1 part fuel, by weight, not by volume). This ratio has been proven to be the ideal air/fuel mixture ratio for good performance and economy, and low emissions.

2   The oxygen sensor produces no voltage when it's below its normal operating temperature of about 600-degrees F. During this initial period before warm-up, the PCM operates in "open-loop" mode. In normal operation, when the oxygen sensor has been warmed by hot exhaust (about 2 minutes

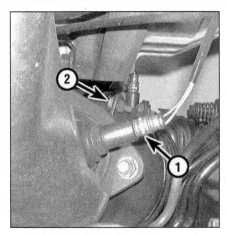

19.3 Typical OBD-II oxygen sensors (2001 model shown, 1996 and later models similar)

1   *Upstream (pre-converter) heated oxygen sensor*

2   *Downstream (post-converter) heated oxygen sensor (vehicle shown here is equipped with a Warm-Up Three-Way Catalyst, or WU-TWC, so the downstream sensor is located right below the WU-TWC; on vehicles without WU-TWC, the downstream sensor is located behind the regular catalytic converter)*

after start-up for an unheated sensor), the oxygen sensor will normally produce a signal voltage that varies between 1.0-volt (rich limit) and 0.1 volt (lean limit). Vehicles with sequential multiport fuel injection are equipped with heated oxygen sensors in order to reduce this two-minute warm-up period. Except for their resistor-type heater circuit, heated sensors function exactly the same way as unheated sensors.

3   There are *two* heated oxygen sensors on all OBD-II vehicles **(see illustration)**: The upstream, or pre-converter, oxygen sensor, which is located in the exhaust manifold right above the flange, functions just like the sensor in an OBD-I vehicle. The downstream, or post-converter, sensor, which is located *behind* the catalytic converter, enables the PCM to compare the oxygen content of the upstream exhaust gases with the oxygen content in the exhaust stream after it has been through the catalytic converter. The PCM uses this information to monitor the effectiveness of the reduction catalyst inside the converter, and to predict the failure of the converter before it completely fails. The post-converter oxygen sensor has no effect on PCM control of the air/fuel ratio, but it's identical to the pre-converter sensor in design and operation. However, a post-converter oxygen sensor produces a more slowly fluctuating voltage signal because the oxygen content is lower in the post-catalyst exhaust.

4   A problem in the oxygen sensor circuit will normally set a Code 13 (OBD-I vehicles) or a P0131, P0132, P0133, P0134, P0135, P0136, P0141, P0171 or P0172 (OBD-II vehi-

19.7 On models with carburetors, turn off the wide-open micro switch (arrow) by rotating the micro-switch lever clockwise (down)

cles). If the PCM stores some other diagnostic trouble code(s) along with a code related to the oxygen sensor code, try to diagnose and repair the problem that caused the other code(s) first. Sometimes, oxygen sensor codes are caused by malfunctions in other sensor circuits, and clearing up those codes first often eliminates the oxygen sensor code as well. When an oxygen sensor-related fault occurs, the PCM disregards the sensor signal voltage and reverts to open-loop fuel control.

Check

Caution: *The oxygen sensor is very sensitive to excessive circuit loads and circuit damage of any kind. For safest testing, disconnect the oxygen sensor connector, install jumper wires between the two connectors and connect your voltmeter to the jumper wires. If jumper wires aren't available, carefully backprobe the wires in the connector shell with suitable probes (such as T-pins). Do not puncture the oxygen sensor wires or try to backprobe the sensor itself. Use only a digital voltmeter to test an oxygen sensor (see* **Note** *below).*
Note: *Always check the oxygen sensor(s) with a high-impedance (10 M-ohm) digital voltmeter. This type of voltmeter has an inner resistance that is more than one M-ohm per volt. Other voltmeters are not sufficiently accurate to check an oxygen sensor.*
5   Before checking the oxygen sensor itself, verify that the following components are functioning correctly:

Make sure the air cleaner isn't clogged (see Chapter 1).
Make sure that there are no vacuum leaks anywhere in the intake system (which will cause the system to run excessively lean).
Inspect the spark plugs for contamination and make sure that they're correctly gapped (see Chapter 1).
If applicable, inspect the spark plug wires and caps and the coil high tension cable too. Look for cracks and deterioration (see Chapter 1).

Inspect the distributor cap and rotor.
Look for wear, cracks and carbon tracking (see Chapter 1).
Also, check the ignition timing, if applicable (see Chapter 1).
Check the engine compression.
Check any other components and/or systems that might affect the air/fuel mixture or combustion quality.

Often, when the above items are checked out and, if necessary, fixed, the "problem" in the oxygen sensor circuit disappears. If none of the above fails to bring the oxygen sensor circuit within its normal operating range, the sensor might be defective. The following tests will verify whether the sensor is good or bad.

Carbureted vehicles

Refer to illustration 19.7
6   Warm up the engine to normal operating temperature, then turn the engine off. Unplug the oxygen sensor electrical connector **(see illustration 48.2 in Chapter 1)** and connect the positive probe of the voltmeter to the sensor side of the connector and the negative lead to a good ground.
7   Start the engine again and, with the engine running between 1500 and 2000 rpm, turn off the wide-open micro switch by rotating the micro-switch lever down **(see illustration)** and then note the voltmeter reading. It should be about 0.8 volt. **Caution:** *Don't allow the oxygen sensor wire or either voltmeter lead to touch the exhaust pipe or manifold.*
8   With the engine running at 1000 to 1500 rpm, disconnect a vacuum hose from the intake manifold and note the voltmeter reading again. It should be about 0.2 volt.
9   Reconnect the vacuum hose to the intake manifold and reconnect the electrical connector for the oxygen sensor.
10   If the oxygen sensor doesn't function as described, replace it.

Vehicles with throttle body injection

11   Warm up the engine to normal operating temperature, then turn the engine off. Unplug the oxygen sensor electrical connector **(see illustration 48.2 in Chapter 1)** and connect the positive probe of the voltmeter to the sensor side of the connector and the negative lead to a good ground.
12   Start the engine again and run it at 2000 rpm for one minute while watching the voltmeter. The voltmeter reading should fluctuate repeatedly between just below and just above 0.45 volt. **Caution:** *Don't allow the oxygen sensor wire or either voltmeter lead to touch the exhaust pipe or manifold.*
13   If the voltmeter indicates a steady reading below 0.45 volt, maintain the engine speed at 2000 rpm for another minute. After one minute, disconnect the vacuum hose from the Manifold Absolute Pressure (MAP) sensor and recheck the voltmeter reading.
14   If the indicated voltage is still not 0.45 volt or more, replace the oxygen sensor.
15   If the voltage is now more than 0.45 volt, either the wire between the oxygen sensor

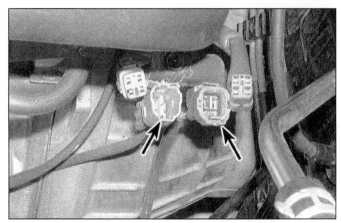

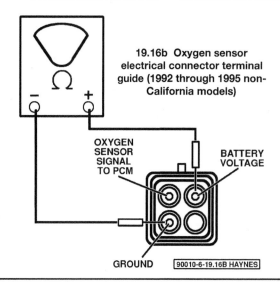

19.16b Oxygen sensor electrical connector terminal guide (1992 through 1995 non-California models)

19.16a Typical oxygen sensor electrical connectors (arrows) on OBD-II vehicle, located on left (driver's side) of engine, next to transmission bellhousing (2001 model shown, 1996 and later models similar)

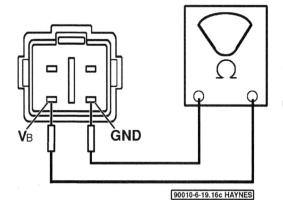

19.16c Oxygen sensor electrical connector terminal guide (1992 through 1995 California models; all 1996 and later models, both upstream and downstream sensor connectors)

and the PCM is open, there's a bad connection at the PCM, or the air/fuel mixture is too lean (go back and look for an air leak in the intake system).

Vehicles with sequential multiport fuel injection

Oxygen sensor heater

Refer to illustrations 19.16a, 19.16b and 19.16c

16   Unplug the oxygen sensor electrical connector **(see illustrations)** and measure the resistance between the two terminals for the sensor heater. Compare your measurement to the oxygen sensor heater resistance listed in this Chapter's Specifications. If the heater resistance is incorrect, replace the oxygen sensor.

Oxygen sensor

17   Plug in the oxygen sensor electrical connector and warm up the engine to its normal operating temperature, then turn it off. Using a T-pin, backprobe the sensor connector terminal for the sensor voltage signal to the PCM **(see illustration 19.16b)**. Connect the positive lead of a voltmeter to the T-pin and the negative lead to a good ground. Start the engine again and run it at 2000 rpm for one minute. After a minute, note the indicated voltage.

18   If the indicated voltage is zero, either your T-pin is not connecting to the signal terminal, or the oxygen sensor is defective. Check the T-pin connection and then repeat the test. If the indicated voltage is still zero, replace the oxygen sensor.

19   If the indicated voltage is steady and below 0.45 volt, run the engine at 2000 rpm for another minute and then watch the voltmeter while repeatedly racing the engine. Does the voltmeter now indicate at least 0.45 volt, or more, at least once? If it doesn't, replace the oxygen sensor. If it does, the air/fuel mixture is too lean. Inspect the MAF sensor (see Section 18) and the ECT sensor (see Section 14). If they're both functioning correctly, check the fuel pressure and inspect the injectors (see Chapter 4). If no problems

turn up during any of these checks, the PCM is probably defective. Have the PCM checked by a dealer service department.

20   If the indicated voltage is steady and above 0.45 volt, check the TP sensor (see Section 20), the MAF sensor (see Section 18) and the ECT sensor (see Section 14) and then, if all those devices check out, check the fuel pressure and the fuel injectors (see Chapter 4). If everything checks out, have the PCM checked by a dealer service department.

21   If the indicated voltage repeatedly fluctuates back and forth between just above and just below 0.45 volt, the heated oxygen sensor and its circuit are in good condition. If an intermittent condition persists, have the PCM checked out by a dealer service department.

Replacement

22   Disconnect the negative battery cable.
23   Trace the electrical lead from the sensor to the connector and unplug the connector.
24   Using a special oxygen sensor wrench (available at most auto parts stores), carefully unscrew and remove the sensor.
25   Coat the threads of the new sensor (or the old sensor, if you're planning to reuse it) with anti-seize compound and screw in the sensor. Tighten the sensor to the torque listed in this Chapter's Specifications.

20   Throttle Position (TP) sensor - check and replacement

Description

1   The Throttle Position (TP) sensor, which is located on the throttle body, on the end of the throttle valve shaft, monitors the throttle valve opening angle. The TP sensor is a potentiometer. A 5-volt reference voltage is sent to the TP sensor from the PCM. As the throttle valve opens, a "brush" (sweeper) moves across a printed resistor board, varying the output voltage accordingly. This output voltage signal is sent to the PCM, which uses this information to determine the throttle valve angle. The TP sensor signal is one of the signals used by the PCM to control the fuel injector pulse width (the interval during which the injectors are open), the Idle Air Control (IAC) valve, ignition timing, the EVAP canister purge valve and the EGR valve. A problem in the TP sensor circuit will set a Code 21 or Code 22 (OBD-I vehicles) or a P0122 or P0123 (OBD-II vehicles). When a trouble code is set, the PCM resorts to a preprogrammed default value for TP sensor input to allow you to get home. But performance will be seriously impaired.

Check and adjustment

1992 through 1998 models

Check

Refer to illustrations 20.3a and 20.3b

2   Disconnect the cable from the negative terminal of the battery.

3   Unplug the TPS electrical connector and, referring to the accompanying chart, check the resistance between the indicated terminals **(see illustrations)** when the throttle is fully open and fully closed. If the readings aren't as specified, adjust the throttle position sensor. If the correct readings cannot be obtained even after adjustment, then replace the sensor.

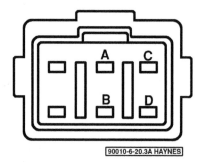

90010-6-20.3A HAYNES

20.3a Throttle Position (TP) sensor electrical connector terminal guide (models with throttle body injection)

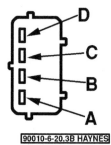

90010-6-20.3B HAYNES

20.3b Throttle Position (TP) sensor electrical connector terminal guide (1992 through 1998 models with sequential multiport fuel injection)

Throttle Position (TP) sensor resistance (1992 through 1998 models)

Resistance between terminals C and D	When clearance between throttle lever and stopper screw is 0.012 inch	0 to 500 ohms
Resistance between terminals C and D	When clearance between throttle lever and stopper screw is 0.020 inch	Infinite resistance
Resistance between terminals A and D		3.5 to 6.5 k-ohms
Resistance between terminals B and D	When throttle valve is at idle position	0.3 to 2 k-ohms
Resistance between terminals B and D	When throttle valve is fully open	2 to 6.5 k-ohms

Adjustment

Refer to illustration 20.5

4   On models with throttle body injection, disconnect the throttle opener vacuum hose from the Vacuum Switching Valve (VSV) and connect a hand-held vacuum pump in its place. Apply enough vacuum to move the throttle valve to the idle position.

5   Slightly loosen the TP sensor mounting screws **(see illustration)**. Referring to the accompanying chart, insert the specified feeler gauge between the throttle stop screw and the throttle lever. **Caution:** *Do NOT attempt to adjust or remove the throttle stop screw. It's precisely adjusted by the factory and will be very difficult to readjust if tampered with.*

Feeler gauge thickness for adjusting TP sensor

Samurai models	0.012 inch (0.3 mm)
Sidekick/Tracker models	0.026 inch (0.65 mm)

6   On models with throttle body injection, connect an ohmmeter between terminals C and D of the TP sensor electrical connector **(see illustration 20.3a)**. Loosen the TP sensor screws and turn the sensor clockwise until it stops. Now slowly turn the sensor counterclockwise until the ohmmeter changes from infinite to zero resistance. Tighten the screws to the torque listed in this Chapter's Specifications. Remove the feeler gauge.

7   On models with sequential multiport fuel injection, connect an ohmmeter between terminals A and B of the TP sensor electrical connector **(see illustration 20.3b)**. Loosen the TP sensor screws and turn the TP sensor

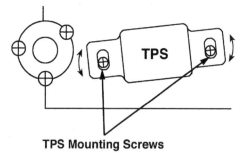

TPS Mounting Screws

20.5 On models with throttle body injection, turn the TP sensor clockwise until it stops, then turn it counterclockwise until the ohmmeter reading changes from infinity to zero (on models with sequential multiport fuel injection, it's just the opposite: turn the TP sensor *counterclockwise* until it stops, then turn it *clockwise* until the reading changes from infinity to zero)

counterclockwise until it stops. Then slowly turn the sensor clockwise until the ohmmeter changes from infinite to zero resistance. Tighten the screws the torque listed in this Chapter's Specifications. Remove the feeler gauge.

8   On models with sequential multiport fuel injection, verify that there is no continuity between terminals A and B when a 0.037-inch feeler gauge is inserted between the throttle stop screw and the throttle lever. Then verify that there is continuity between terminals A and B when a 0.020-inch feeler gauge is inserted between the throttle stop screw and the throttle lever. If either of these verification checks is unsatisfactory, the TP sensor is not adjusted correctly. Readjust and then recheck it again.

9   On models with throttle body injection, reconnect the throttle opener vacuum hose to the VSV.

10   Reconnect the TPS electrical connector and then reconnect the cable to the negative battery terminal.

1999 and later models

Refer to illustration 20.12

11   Disconnect the cable from the negative

terminal of the battery.

12   Unplug the TPS electrical connector and, referring to the accompanying chart, check the resistance between the indicated terminals **(see illustration)**. If the readings aren't as specified, replace the sensor.

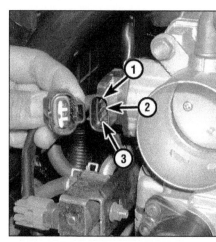

20.12 Throttle Position (TP) sensor electrical connector terminal guide (1999 and later models)

Throttle Position (TP) sensor resistance (1999 and later models)

Terminals	Resistance
Between terminals 1 and 3	4.0 to 6.0 K-ohms
Between terminals 1 and 2	0.02 to 6.0 K-ohms, in proportion to throttle valve opening

Replacement

Refer to illustration 20.15

13 Disconnect the negative battery cable.

14 Unplug the TP sensor electrical connector.

15 Remove the TP sensor screws **(see illustration)** and then remove the TP sensor.

16 When installing the TP sensor, make sure that the lugs on the throttle body lever engage their corresponding slots in the TP sensor pickup lever (the TP sensor cannot be installed flush with the throttle body unless the sensor pickup lever and the throttle body lever are fully engaged).

17 On models with throttle body injection, adjust the TP sensor as described above.

18 Install the TP sensor mounting screws and tighten them to the torque listed in this Chapter's Specifications.

19 Plug in the TP sensor electrical connector.

20 Reconnect the negative battery cable.

---

21 Vehicle Speed Sensor (VSS) - check and replacement

General description

Refer to illustrations 21.2

1 The Vehicle Speed Sensor (VSS) on 1992 through 1998 models is built into the speedometer. The VSS consists of a reed switch and magnet. As the magnet turns with the speedometer cable, its magnetic force causes the reed switch to turn on and off.

This pulsing voltage signal, which increases or decreases in proportion to the revolutions of the cable, is sent to the PCM, which uses it to calculate vehicle speed. If a failure occurs, a code 24 will be set.

2 The Vehicle Speed Sensor (VSS) on 1999 and later models is located on the side of the transmission (2WD models) **(see illustration)** or the transfer case (4WD models). The VSS consists of a Hall effect element, a magnet and a gear. As the sensor gear is turned by a drive gear on the transmission or transfer case output shaft, the signal voltage switches back and forth from high to low in a "square-wave" form. The frequency of this switching increases in proportion to vehicle speed. This signal is sent to the PCM, which uses it to control various devices. A P0500 diagnostic trouble code will be set by the PCM in the event that a problem occurs in the VSS circuit.

Check

1992 through 1998 models

Refer to illustrations 21.4a, 21.4b and 21.4c

Note: *Use a conventional ohmmeter with an analog display for this test. A digital ohmmeter is more difficult to read during a test of this type.*

3 Remove the instrument cluster (see Chapter 12).

4 Connect an ohmmeter to the RSW and GND terminals on Samurai models, or to the VSS and GND terminals on 1992 through 1995 Sidekick/Tracker models or to the G11-4 and G11-10 terminals on 1996 through 1998 Sidekick/Tracker models **(see illustrations)**.

20.15 To detach a Throttle Position (TP) sensor from the throttle body, remove the two screws (arrows)

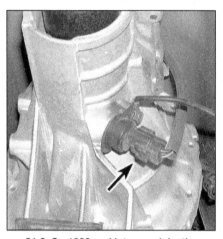

21.2 On 1999 and later models, the Vehicle Speed Sensor (VSS) (arrow) is located on the right side of the transmission (shown, on 2WD models) or on the transfer case (on 4WD models)

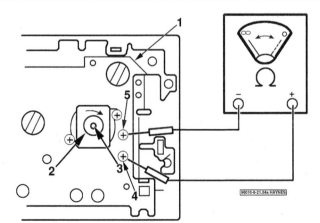

21.4a Vehicle Speed Sensor (VSS) test setup (Samurai models)

1 *Backside of instrument cluster*
2 *Speedometer cable lug-to-speedometer drive slot*
3 *Insert small slotted screwdriver into drive slot to activate VSS*
4 *RSW terminal*
5 *GND terminal*

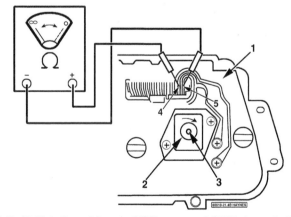

21.4b Vehicle Speed Sensor (VSS) test setup (1992 through 1995 Sidekick/Tracker models)

1 *Backside of instrument cluster*
2 *Speedometer cable lug-to-speedometer drive slot*
3 *Insert small slotted screwdriver into drive slot to activate VSS*
4 *VSS terminal*
5 *GND terminal*

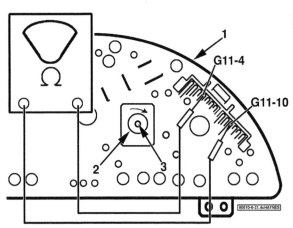

21.4c Vehicle Speed Sensor (VSS) test setup (1996 through 1998 Sidekick/Tracker models)

1   *Backside of instrument cluster*
2   *Speedometer cable lug-to-speedometer drive slot*
3   *Insert a small slotted screwdriver into drive slot and turn to activate VSS*

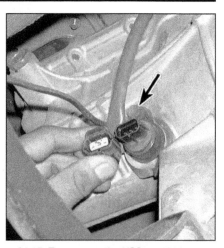

21.15 To remove the VSS on 1999 and later models, simply unplug the electrical connector and then remove the retaining bolt (arrow)

5   Insert the tip of a flat-bladed screwdriver in the speedometer drive mechanism, turn the speedometer with the screwdriver and watch the ohmmeter. The needle should fluctuate back and forth between zero resistance and infinite resistance four times for every revolution of the speedometer.
6   If the VSS doesn't operate as described, replace the instrument cluster (see Chapter 12).

1999 and later models

Note: *Use a conventional voltmeter with an analog display for this test; an analog voltmeter is easier to read during a test of this type.*
7   Raise the vehicle and place it securely on jackstands.
8   Release the parking brake lever, put the transmission shift lever in Neutral and the transfer case in 2H.
9   Using a T-pin, backprobe the VSS connector center terminal (the blue/yellow wire), which is the terminal for the VSS output signal to the PCM. Hook up the positive lead of a voltmeter to the T-pin and connect the negative lead to a good ground.
10   Turn the ignition key to ON (don't start the engine). Lock the left rear tire and slowly rotate the right rear wheel and watch the voltmeter. The needle should fluctuate between 0 to 1 and 8 to 14 volts for each revolution of the wheel.
11   If the VSS doesn't operate as described, replace it.

Replacement

1992 through 1998 models

12   Replace the instrument cluster (see Chapter 12). The VSS cannot be replaced separately.

1999 and later models

Refer to illustration 21.15
13   Disconnect the negative battery cable.
14   Raise the rear of the vehicle and place it securely on jackstands.
15   Unplug the electrical connector from the VSS **(see illustration)**.

16   Remove the VSS retaining bolt **(see illustration 21.15)**.
17   Remove the VSS and discard the old O-ring.
18   Installation is the reverse of removal. Be sure to use a new O-ring and tighten the VSS retaining bolt to the torque listed in this Chapter's Specifications.

22   Idle speed control system - check and component replacement

Description

1   A steady idle speed is critical to good fuel economy and low emission. But several factors can cause the idle speed to vary: the "load" (air conditioner, power steering, etc.) imposed on an idling engine, variations in atmospheric pressure, and even changes in the engine itself as it ages. Fuel-injected vehicles are equipped with idle speed control systems to keep the idle at its specified speed at all times. These systems also improve the starting performance of the engine, improve driveability during warm-up, and compensate the air/fuel mixture ratio during deceleration.

Idle Speed Control (ISC) solenoid valve (Samurai models with throttle body injection)

2   On Samurai models, the idle speed is controlled by the PCM through an output actuator known as the Idle Speed Control (ISC) solenoid valve. The ISC solenoid valve opens a bypass air passage in the throttle body when it's turned on by the PCM and closes the passage when it's turned off. The PCM can cycle the ISC solenoid valve on and off up to 12.5 times a second. The PCM controls bypass airflow by increasing and decreasing the on-time within each cycle. For example, when the engine is being cranked, the PCM keeps the ISC solenoid valve on for the maximum time rate within a cycle. When the vehicle is not moving, the throttle valve is

at the idle position and the engine is running, the PCM controls the bypass airflow by increasing or decreasing the on-time of the ISC solenoid valve to keep engine speed at a specified idle speed. When the air conditioning is turned on, a certain amount of bypass air is supplied by the air conditioning system Vacuum Switching Valve (VSV); the bypass air provided by the ISC solenoid valve is used for "fine tuning" the idle speed.

Idle Air Control (IAC) valve (Sidekick/Tracker models with throttle body injection and all models with sequential multiport fuel injection)

3   On Sidekick and Tracker models with throttle body injection, and on all models with sequential multiport fuel injection, the idle speed is controlled by the PCM through an output actuator known as the Idle Air Control (IAC) valve. The IAC valve regulates the bypass airflow by controlling the opening and closing of a bypass air passage in the throttle body. The IAC valve consists of a stepper motor and a rod-actuated valve. When the stepper motor is energized by the PCM, the valve position changes, increasing or reducing bypass airflow. This occurs at a rate of about 200 times a second.

Check

Idle Speed Control (ISC) solenoid valve (Samurai models with throttle body injection)

4   With the engine and the ignition turned off, unplug the ISC solenoid valve electrical connector. Connect an ohmmeter across the connector terminals, measure the resistance and then compare your measurement with the resistance listed in this Chapter's Specifications. If the resistance is out of range, replace the ISC solenoid valve. If the resistance is correct, proceed to the next check.

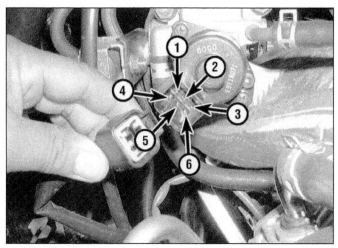

22.9 Idle Air Control (IAC) valve electrical connector terminal guide (1999 and later model shown, earlier models similar)

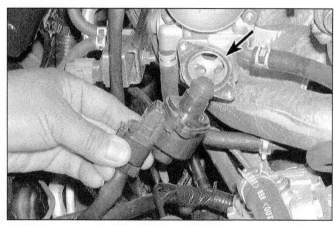

22.27 To remove the IAC valve from the throttle body on 1999 and later models, remove the two Allen screws and then pull out the IAC valve; be sure to discard the old O-ring (arrow) and replace it before installing the IAC valve

5   Plug in the ISC solenoid valve electrical connector. Warm up the engine to its normal operating temperature. With the engine still running, unplug the ISC valve connector and disconnect the hose between the air cleaner housing and the ISC valve (there are two hoses connected to the ISC valve; the hose from the air cleaner is the lower hose, the one that's closer to the electrical connector). Verify that no air is being drawn into this hose.
6   With the engine still running, use jumper cables to hook up the battery to the ISC solenoid valve connector terminals, and then verify that air *is* being drawn into the lower hose. If it isn't (or if air was being drawn into the lower hose before you energized the ISC valve), replace the valve.

Idle air control (IAC) valve (Sidekick/Tracker models with throttle body injection and all models with sequential multiport fuel injection)

1992 through 1998 models

7   Unplug the electrical connector from the IAC valve.
8   Using an ohmmeter, measure the resistance between each pair of terminals on the IAC valve side of the connector and then compare your measurement to the IAC valve resistance listed in this Chapter's Specifications. If the IAC valve resistance is out of range between any two terminals, replace the IAC valve.

1999 and later models
Refer to illustration 22.9
9   Unplug the electrical connector from the IAC valve **(see illustration)**.
10   Using an ohmmeter, measure the resistance between terminals 1 and 2, 3 and 2, 4 and 5, and 6 and 5. Compare your measurements to the IAC valve resistance listed in this Chapter's Specifications. If the IAC valve resistance is out of range for any terminal pair, replace the IAC valve.

Replacement
11   Disconnect the negative battery cable.

ISC solenoid valve (Samurai models with throttle body injection)
12   Disconnect the ISC solenoid valve electrical connector.
13   Clearly label and then disconnect the two large air hoses and the two smaller vacuum hoses from the ISC solenoid valve.
14   Remove the ISC solenoid valve.
15   Installation is the reverse of removal.

IAC valve (1992 through 1998 models)
Warning: *Perform the following procedure with the engine cold.*
16   Disconnect the negative battery cable.
17   Unplug the electrical connector from the IAC valve.
18   On models with throttle body injection, remove the EGR modulator from its bracket.
19   On models with sequential multiport fuel injection, remove the throttle cover.
20   Disconnect the air hose from the IAC valve.
21   Remove the radiator cap to relieve any residual pressure inside the cooling system.
22   Remove the coolant hoses from the IAC valve. Be prepared to mop up any spilled coolant.
23   Remove the two IAC valve retaining screws or bolts and remove the IAC valve from the throttle body.
24   Installation is the reverse of removal. Be sure to tighten the IAC valve retaining screws or bolts to the torque listed in this Chapter's Specifications.

IAC valve (1999 and later models)
Refer to illustration 22.27
25   Disconnect the negative battery cable.
26   Unplug the electrical connector from the IAC valve **(see illustration 22.9)**.
27   Remove the two IAC valve retaining screws and then remove the IAC valve **(see illustration)**. Discard the old O-ring.

28   Installation is the reverse of removal. Be sure to use a new O-ring and tighten the IAC valve retaining screws to the torque listed in this Chapter's Specifications.

---

23   Powertrain Control Module (PCM) - removal and installation

---

Samurai models
1   Disconnect the negative battery cable.
2   The PCM is located inside the right end of the dash. Remove the right kick panel for access (see Chapter 12).
3   Remove the fuel pump relay and the main relay from the PCM.
4   Unplug the two large multi-pin electrical connectors from the PCM.
5   Loosen the three PCM mounting screws and then remove the PCM.
6   Installation is the reverse of removal.

1992 through 1998 Sidekick/Tracker models
7   Disconnect the negative battery cable.
8   The PCM is located inside the left end of the dash. On 1992 through 1995 models, remove the driver's side speaker from the left end of the dash (see Chapter 12).
9   On 1992 through 1995 models, remove the PCM, bracket, fuse box and relays from the steering column holder and then unplug the electrical connectors from the PCM and remove the PCM from its mounting bracket.
10   On 1996 through 1998 models, unplug the electrical connectors from the PCM, remove the PCM mounting screws and then remove the PCM.
11   Installation is the reverse of removal.

1999 and later Vitara/Tracker models
Refer to illustrations 23.14 and 23.15
12   Disconnect the negative battery terminal.

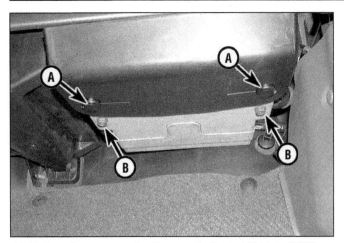

23.14 **To gain access to the Powertrain Control Module (PCM) on 1999 and later Vitara/Tracker models, remove the trim cover push fasteners (A) and the PCM mounting bolts (B)**

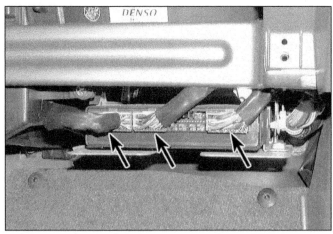

23.15 **Remove these three electrical connectors (arrows) and then pull out the PCM (1999 and later Vitara/Tracker models)**

13   Disable the air bag system (see Chapter 12).

14   Remove the PCM cover **(see illustration)**.

15   Unplug the PCM electrical connectors **(see illustration). Caution:** *Avoid static electricity damage to the Powertrain Control Module (PCM) by grounding yourself to the body of the vehicle before touching the PCM and by using a special anti-static pad to store the PCM on once it has been removed.*

16   Remove the PCM mounting bracket screws **(see illustration 23.14)**.

17   Remove the PCM.

18   Installation is the reverse of removal.

---

24   Emission maintenance reminder light

---

1   On 1989 and many later models, a "sensor" light will illuminate every 50,000, 80,000 and 100,000 miles. This will occur with a warm engine at 1500 - 2000 RPM. This indicates the PCM is in good condition and the oxygen sensor needs service.

2   When all the emission system services have been completed (see maintenance schedule), locate the red, white and blue 3-wire "sensor" light cancel switch attached to the steering column bracket. Slide the switch to the opposite position to reset the light.

3   Start vehicle and test drive vehicle to ensure light does not flash.

Chapter 7 Part A
Manual transmission

Contents

Specifications

Torque specifications

Ft-lbs (unless otherwise specified)

Oil filler, level and drain plugs	14 to 20
Transmission-to-engine bolts/nuts	
Samurai	16 to 25
Sidekick/X-90/Vitara/Tracker	51 to 72
Input shaft bearing retainer bolts	14 to 20
Shift lever retainer bolts (Samurai)	36 to 60 in-lbs

1 General information

All vehicles covered in this manual come equipped with either a 5-speed manual transmission or an automatic transmission. All information on the manual transmission is included in this Part of Chapter 7. Information on the automatic transmission can be found in Part B of this Chapter.

Due to the complexity, unavailability of replacement parts and the special tools necessary, internal repair by the home mechanic is not recommended. The information in this Chapter is limited to general information and removal and installation of the transmission.

Depending on the expense involved in having a faulty transmission overhauled, it may be a good idea to replace the unit with either a new or rebuilt one. Your local dealer or transmission shop should be able to supply you with information concerning cost, availability and exchange policy. Regardless of how you decide to remedy a transmission problem, you can still save a lot of money by removing and installing the unit yourself.

2 Oil seal replacement

Refer to illustrations 2.3, 2.4, 2.6, 2.12 and 2.13

Front oil seal

1 Remove the transmission as described in Section 6.
2 Remove the release bearing as described in Chapter 8.
3 Make a mark on the clutch release arm before removing it from the pivot shaft **(see illustration)**.

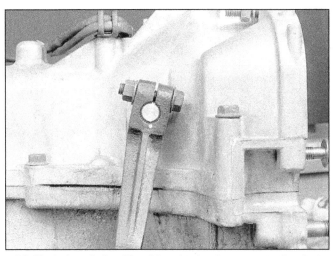

2.3 Mark the relationship of the clutch release arm to the pivot shaft to insure proper alignment during installation

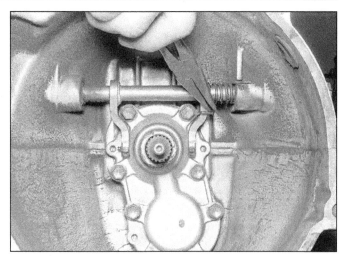

2.4 Use needle-nose pliers to release the end of the spring clip from the shaft

2.6 The transmission front oil seal is located in the front cover - to remove it, carefully pry it out with a screwdriver

4   Release the end of the return spring from the shaft on the inside of the bellhousing **(see illustration)**.
5   Remove the bolts that retain the input shaft bearing retainer to the case, then lift off the retainer.
6   Being careful not to nick or damage the bearing retainer, pry out the front oil seal **(see illustration)**.
7   Drive the new seal into the retainer using a seal driver or an appropriately-sized socket.
8   Apply a light coat of gear oil to the seal lips and the transmission input shaft, then reinstall the bearing retainer. Tighten the retainer bolts to the torque listed in this Chapter's Specifications.
9   Reinstall the remaining components in the reverse order of removal.

Rear oil seal - Samurai

10   Remove the driveshaft as described in Chapter 8.
11   On later models it may be necessary to remove the dust cover for access to the oil seal.

12   Being careful not to damage the output shaft of the transmission housing, use a seal remover to pry out the old seal **(see illustration)**.
13   Apply a coat of gear oil to the lips of the new seal and drive it into place using a seal driver or an appropriately-sized socket **(see illustration)**.
14   Reinstall the driveshaft.

---

3   Manual transmission shift lever (Samurai) - removal and installation

---

Refer to illustrations 3.3, 3.4, and 3.7
1   Place the shift lever in Neutral.
2   Remove the carpet from around the shifter area.
3   Remove the bolts that secure the shift lever boot and lift the boot off the floor **(see illustration)**.
4   Slide the small inner boot up toward the knob **(see illustration)**.

2.12 The rear oil seal can be removed by prying it out with a seal remover (Samurai)

2.13 A large socket and extension can be used to drive the seal into place - make sure the seal is installed squarely in the bore

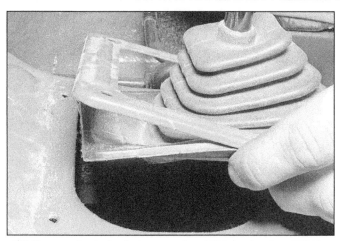

3.3 Remove the bolts that secure the shift lever boot and lift the boot off the floor

3.4 Pull the small inner boot up off the lever retainer

5   Remove the three bolts that secure the shift lever retainer to the transmission.

6   Pull the shift lever out of the case.

7   Check the lower portion of the shift lever for excessive wear **(see illustration)**. Also check the boot for damage.

8   Apply grease to the pivot portions and the seat. Install the shift lever into the case.

9   Tighten the shift lever retainer bolts to the torque listed in this Chapter's Specifications.

10   The remainder of the installation procedure is the reverse of removal.

4   Manual transmission shift lever (Sidekick/X-90/Vitara/Tracker) - removal and installation

Note: *The transfer case shift lever can also be removed using this procedure.*

1   Using an Allen wrench and a small Phillips screwdriver, remove the two screws at the front and the two clips at the rear of the console box and pull the console box up.

2   Remove the bolts that secure the console box bracket and lift up the boot cover.

3   Remove the boot clamp that attaches to the small inner boot then pull the boot up and away from the gear shift assembly.

4   With your fingers, push the gear shift control case cover down while simultaneously turning it counterclockwise. The shift lever can now be removed.

5   Check the lower portion of the shift lever for excessive wear and boot damage. Replace any worn parts.

6   Apply grease to the pivot areas and seat. Install the shift lever into the housing, push the control case cover down and turn it clockwise to lock it in place.

7   The remainder of installation is the reverse of the removal procedure.

5   Transmission mount - check and replacement

Refer to illustrations 5.2 and 5.3

1   Insert a large screwdriver or pry bar into the space between the transmission extension housing and the crossmember and try to

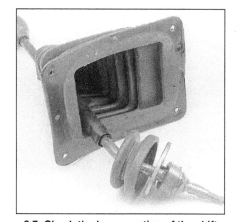

3.7 Check the lower portion of the shift lever for excessive wear

pry the transmission up slightly.

2   The transmission should not move away from the mount much at all **(see illustration)**.

3   To replace the mount, remove the bolts attaching the mount to the crossmember and the bolt attaching the mount to the transmission **(see illustration)**.

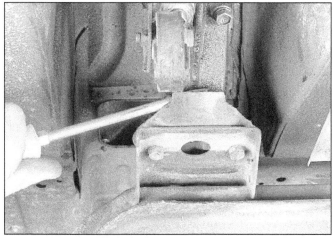

5.2 Using a screwdriver, try to pry the transmission up and away from the rubber mount - replace the mount if the play is excessive

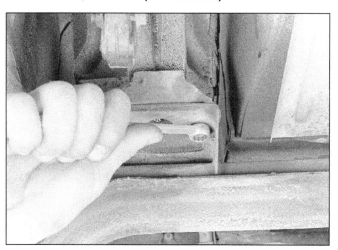

5.3 Remove the bolts that secure the transmission mount to the crossmember

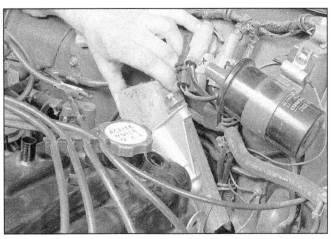

6.8 Insert a wood block behind the distributor housing to prevent damage to the housing and firewall

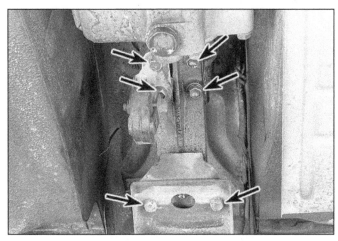

6.11 Remove the transmission rear support-to-crossmember nuts and bolts (Samurai)

4   Raise the transmission slightly with a jack and remove the mount.
5   Installation is the reverse of the removal procedure. Be sure to tighten the bolts securely.

6   Manual transmission - removal and installation

Samurai models

Removal

Refer to illustrations 6.8, 6.11 and 6.13
1   Disconnect the negative cable from the battery.
2   Drain the oil from the transmission and transfer case (see Chapter 1).
3   Remove the transfer case (see Chapter 7C).
4   Working inside the vehicle, remove the transmission shift lever (see Section 3).
5   Raise the vehicle and support it securely on jackstands.
6   Remove the exhaust system components as necessary for clearance (see Chapter 4).
7   Support the engine. This can be done from above with an engine hoist, or by placing a jack (with a wood block as an insulator)

under the engine oil pan. The engine should remain supported at all times while the transmission is out of the vehicle.
8   Remove the distributor (see Chapter 2) and place a wood block between the distributor housing and firewall to prevent damage to the housing and firewall **(see illustration)**.
9   Support the transmission with a jack - preferably a special jack made for this purpose. Safety chains will help steady the transmission on the jack.
10   Remove the starter (see Chapter 5).
11   Remove the rear transmission support-to-crossmember nuts and bolts **(see illustration)**.
12   Raise the transmission slightly and remove the crossmember.
13   Remove the front crossmember **(see illustration)**.
14   Remove the front and rear driveshafts (see Chapter 8).
15   Disconnect the clutch cable from the release arm.
16   Remove the transmission-to-engine bolts.
17   Make a final check that all wires and hoses have been disconnected from the transmission, then move the transmission and jack toward the rear of the vehicle until the transmission input shaft is clear of the

clutch pressure plate. Keep the transmission level as this is being done.
18   Once the input shaft is clear, lower the transmission and remove it from under the vehicle.
19   The clutch components can be inspected after removing them from the engine (see Chapter 8). In most cases, new clutch components should be routinely installed if the transmission is removed.

Installation

20   If removed, install the clutch components (see Chapter 8).
21   Position the transmission on the jack.
22   With the transmission secured to the jack as on removal, raise it into position and carefully slide it forward, engaging the input shaft with the clutch plate hub. Do not use excessive force to install the transmission - if the input shaft does not slide into place, readjust the angle of the transmission so it is level and/or turn the input shaft so the splines engage properly with the clutch hub.
23   Install the transmission-to-engine bolts. Tighten the bolts to the torque listed in this Chapter's Specifications.
24   Install the crossmembers and transmission support. Tighten all nuts and bolts securely.
25   Remove the jacks supporting the transmission and the engine.
26   Install the transfer case.
27   Install the various items removed previously, referring to Chapter 8 for the installation of the driveshaft, the installation and adjustment of the clutch cable and Chapter 4 for information regarding the exhaust system components.
28   Make a final check that all wires, hoses and the speedometer cable have been connected and that the transmission and transfer case has been filled with lubricant to the proper level (Chapter 1). Lower the vehicle.
29   Working inside the vehicle, install the shift lever (see Section 3).
30   Connect the negative battery cable. Road test the vehicle for proper operation and check for leakage.

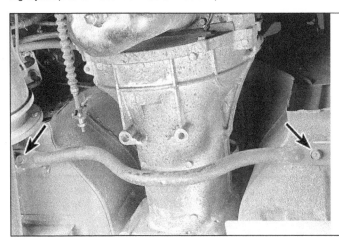

6.13 The front crossmember is attached by two bolts (Samurai)

Sidekick/X-90/Vitara/Tracker

Removal

31   Disconnect the negative cable from the battery.

32   Working inside the vehicle, remove the transfer case shift lever (see Section 4).

33   Remove the transmission shift lever (see Section 4).

34   Drain the lubricant from the transmission/transfer case unit (see Chapter 1).

35   Disconnect the breather hose from the clamp at the rear end of the cylinder head.

36   Disconnect the wiring harness clamp at the rear end of the intake manifold to free the wiring harness.

37   Follow Steps 5 through 19 of this Section **(see illustration 6.11b** for crossmember removal) **Note:** *Sidekick/X-90/Vitara/Tracker are equipped with a transmission/transfer case unit.*

Installation

38   Follow Steps 20 through 27 of this Section, ignoring the Steps which don't apply.

39   Connect the breather hose to the clamp at the rear end of the cylinder head.

40   Connect the wiring harness clamp at the rear end of the intake manifold to the wiring harness.

41   Make a final check that all wires, hoses and the speedometer cable have been connected and that the transmission/transfer case unit has been filled with lubricant to the proper level (see Chapter 1). Lower the vehicle.

42   Connect the negative battery cable. Road test the vehicle for proper operation and check for leakage.

---

7   Manual transmission overhaul - general information

Overhauling a manual transmission is a difficult job for the do-it-yourselfer. It involves the disassembly and reassembly of many small parts. Numerous clearances must be precisely measured and, if necessary, changed with select fit spacers and snaprings. As a result, if transmission problems arise, it can be removed and installed by a competent do-it-yourselfer, but overhaul should be left to a transmission repair shop. Rebuilt transmissions may be available - check with your dealer parts department and auto parts stores. At any rate, the time and money involved in an overhaul is almost sure to exceed the cost of a rebuilt unit.

Nevertheless, it's not impossible for an inexperienced mechanic to rebuild a transmission if the special tools are available and the job is done in a deliberate step-by-step manner so nothing is overlooked.

The tools necessary for an overhaul include internal and external snap-ring pliers, a bearing puller, a slide hammer, a set of pin punches, a dial indicator and possibly a hydraulic press. In addition, a large, sturdy workbench and a vise or transmission stand will be required.

During disassembly of the transmission, make careful notes of how each piece comes off, where it fits in relation to other pieces and what holds it in place.

Before taking the transmission apart for repair, it will help if you have some idea what area of the transmission is malfunctioning. Certain problems can be closely tied to specific areas in the transmission, which can make component examination and replacement easier. Refer to the Troubleshooting Section at the front of this manual for information regarding possible sources of trouble.

Notes

Chapter 7 Part B
Automatic transmission

Contents

Specifications

Torque specifications

Ft-lb (unless otherwise indicated)

Transmission-to-engine bolts and nuts	62
Torque converter-to-driveplate bolt	
3-speed	40
4-speed	48
Oil pan bolts	See Chapter 1

1 General information

All vehicles covered in this manual come equipped with either a 5-speed manual transmission or an automatic transmission. The automatic transmission is composed of a fully automatic three-speed or four-speed mechanism with a four element hydraulic torque converter including a torque converter clutch (TCC). All information on the automatic transmission is included in this Part of Chapter 7. Information for the manual transmission can be found in Part A of this Chapter.

Due to the complexity of the automatic transmissions covered in this manual and the need for specialized equipment to perform most service operations, this Chapter contains only general diagnosis, routine maintenance, adjustment and removal and installation procedures.

If the transmission requires major repair work, it should be left to a dealer service department or an automotive or transmission repair shop. You can, however, remove and install the transmission yourself and save the expense, even if the repair work is done by a transmission shop.

2 Diagnosis - general

Note: *Automatic transmission malfunctions may be caused by five general conditions: poor engine performance, improper adjustments, hydraulic malfunctions or mechanical malfunctions. Diagnosis of these problems should always begin with a check of the easily repaired items: fluid level and condition (see Chapter 1), shift linkage adjustment and throttle linkage adjustment. Next, perform a road test to determine if the problem has been corrected or if more diagnosis is necessary. If the problem persists after the preliminary tests and corrections are completed, additional diagnosis should be done by a dealer* service department or transmission repair shop. Refer to the Troubleshooting Section at the front of this manual for transmission problem diagnosis.

Preliminary checks

1 Drive the vehicle to warm the transmission to normal operating temperature.
2 Check the fluid level as described in Chapter 1:

a) If the fluid level is unusually low, add enough fluid to bring the level within the designated area of the dipstick, then check for external leaks.
b) If the fluid level is abnormally high, drain off the excess, then check the drained fluid for contamination by coolant. The presence of engine coolant in the automatic transmission fluid indicates that a failure has occurred in the internal radiator walls that separate the coolant from the transmission fluid (see Chapter 3).

c) If the fluid is foaming, drain it and refill the transmission then check for coolant in the fluid or a high fluid level.

3   Check the engine idle speed. **Note:** *If the engine is malfunctioning, do not proceed with the preliminary checks until it has been repaired and runs normally.*

4   Check the kickdown cable for freedom of movement. Adjust it if necessary (see Section 5). **Note:** *The kickdown cable may function properly when the engine is shut off and cold, but it may malfunction once the engine is hot. Check it cold and at normal engine operating temperature.*

5   Inspect the shift control cable (see Section 3). Make sure that it's properly adjusted and that the linkage operates smoothly.

Fluid leak diagnosis

6   Most fluid leaks are easy to locate visually. Repair usually consists of replacing a seal or gasket. If a leak is difficult to find, the following procedure may help.

7   Identify the fluid. Make sure it's transmission fluid and not engine oil or brake fluid (automatic transmission fluid is a deep red color).

8   Try to pinpoint the source of the leak. Drive the vehicle several miles, then park it over a large sheet of cardboard. After a minute or two, you should be able to locate the leak by determining the source of the fluid dripping onto the cardboard.

9   Make a careful visual inspection of the suspected component and the area immediately around it. Pay particular attention to gasket mating surfaces. A mirror is often helpful for finding leaks in areas that are hard to see.

10   If the leak still cannot be found, clean the suspected area thoroughly with a degreaser or solvent, then dry it.

11   Drive the vehicle for several miles at normal operating temperature and varying speeds. After driving the vehicle, visually inspect the suspected component again.

12   Once the leak has been located, the cause must be determined before it can be properly repaired. If a gasket is replaced but the sealing flange is bent, the new gasket will not stop the leak. The bent flange must be straightened.

13   Before attempting to repair a leak, check to make sure that the following conditions are corrected or they may cause another leak. **Note:** *Some of the following conditions cannot be fixed without highly specialized tools and expertise. Such problems must be referred to a transmission shop or a dealer service department.*

Gasket leaks

14   Check the pan periodically. Make sure the bolts are tight, no bolts are missing, the gasket is in good condition and the pan is flat (dents in the pan may indicate damage to the valve body inside).

15   If the pan gasket is leaking, the fluid level or the fluid pressure may be too high,

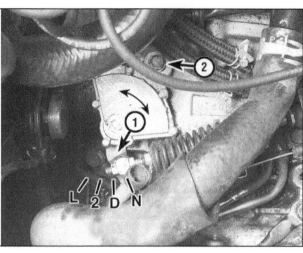

3.4 Rotate the switch until it clicks into Neutral before tightening the bolt

1   *Manual select lever*
2   *Bolt*

the vent may be plugged, the pan bolts may be too tight, the pan sealing flange may be warped, the sealing surface of the transmission housing may be damaged, the gasket may be damaged or the transmission casting may be cracked or porous. If sealant instead of a gasket has been used to form a seal between the pan and the transmission housing, it may be the wrong sealant.

Seal leaks

16   If a transmission seal is leaking, the fluid level or pressure may be too high, the vent may be plugged, the seal bore may be damaged, the seal itself may be damaged or improperly installed, the surface of the shaft protruding through the seal may be damaged or a loose bearing may be causing excessive shaft movement.

17   Make sure the dipstick tube seal is in good condition and the tube is properly seated. Periodically check the area around the speedometer gear or sensor for leakage. If transmission fluid is evident, check the O-ring for damage. Also inspect the side gear shaft oil seals for leakage.

Case leaks

18   If the case itself appears to be leaking, the casting is porous and will have to be repaired or replaced.

19   Make sure the oil cooler hose fittings are tight and in good condition.

Fluid comes out vent pipe or fill tube

20   If this condition occurs, the transmission is overfilled, there is coolant in the fluid, the case is porous, the dipstick is incorrect, the vent is plugged or the drain back holes are plugged.

---

3   Shift linkage - adjustment

Refer to illustration 3.4

1   This adjustment should not be considered routine and is not required unless there is wear in the shift cable or a new cable has

been installed.

2   Raise the front of the vehicle and support it securely on jack stands.

3   Locate the shift cable on the passenger side of the transmission. Clean the threads and nuts on the end of the cable.

4   Loosen the lock nut on the end of the shift cable **(see illustration)**. Turn the adjusting nut so it is closer to the rubber boot on the cable. Put the manual select lever (on the transmission) in the N position. **Note:** *The N position is the third from the lowest (L) position.*

5   Remove the screws at the front and the clips at the rear of the console box, then lift the console box up. Remove the four bolts that secure the manual selector to the floor and raise the unit.

6   Place the shift lever inside the vehicle to the N position also, and insert an appropriately-sized pin into the hole in the shift mechanism to hold the lever in place. **Note:** *The alignment hole is located on the side of the metal selector housing, toward the rear.*

7   Confirm the N position of the select lever on the transmission and turn the adjuster nut and the lock nut until they each contact the boss on the lever, without moving the lever in either direction. Tighten the nuts against the lever boss securely, at the same time, to ensure correct positioning of the select lever.

8   Make sure the vehicle only starts in Park and Neutral.

---

4   Shift cable - removal and installation

Removal

1   Remove the screws at the front and the clips at the rear of the console box. Lift the console box up.

2   Remove the four bolts that secure the manual selector to the floor and raise the unit.

3   Remove the cable end clip, washer and outer cable E-ring, then disconnect the cable from the selector assembly.

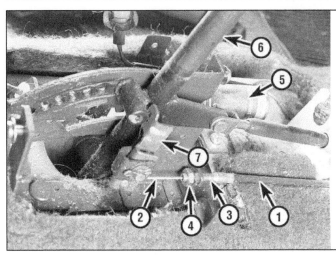

6.4 Backdrive cable adjustment details

1 Backdrive cable casing
2 Backdrive cable
3 Nut
4 Nut
5 Solenoid
6 Shift lever
7 Key release plate

4   Raise the vehicle and support it securely on jack stands. Remove the lock nut from the cable end at the transmission **(see illustration 3.4)**.
5   Pull down the manual select lever and disconnect the cable from the lever.
6   Remove the cable by taking the E-ring out of the shift cable bracket, then pull the cable from the vehicle.

Installation

7   Installation is the reverse of the removal procedure, but be sure to adjust the cable as described in the previous Section before installing the console box.
8   Confirm that the vehicle only starts in Park and Neutral.

5   Kickdown cable - removal, installation and adjustment

Removal and installation

1   Locate the kickdown cable on the throttle lever of the carburetor or T.B.I. unit. Push the plastic joint on the cable end off the lever pin.
2   Remove the cable by loosening the adjusting nut at the bracket and by pulling the cable through the slotted keyway of the bracket.
3   Remove the retaining plate from the cable bracket by unscrewing the bolt.
4   Clean the right side of the transmission where the cable is connected.
5   Using a screwdriver, pry the cable from the transmission case and disconnect the inner cable from the valve.
6   Install the cable by reversing the removal procedure. Be sure to route the cable clear of any hot or moving parts.

Adjustment

7   Adjust the cable by loosening the kickdown cable lock nut and adjusting nut.
8   Have an assistant depress the accelerator pedal to the floor, pull the kickdown cable in. On three-speed models, adjust the locknut-to-bracket clearance to 0.039 inch (1

mm) by turning the locknut. On four-speed models adjust the locknut-to-bracket clearance to 0.039 to 0.059 inch (0.8 to 1.5 mm) by turning the locknut. **Note:** *When adjusting the clearance, make sure the adjust nut does not contact with the bracket.*
9   Release the accelerator pedal and tighten the adjusting nut until it contacts the bracket, making sure the lock nut doesn't turn. To accomplish this, pull the adjusting nut and kickdown cable away from the lock tabs of the bracket while you tighten it.
10   With adjusting nut positioned against the bracket, tighten the lock nut securely.

6   Back drive system - adjustment

Refer to illustration 6.4
1   This system is a safety device to prevent the select lever from being released from the Park position without using the ignition key and to prevent the steering lock from engaging while driving. The system also prevents you from taking the key out of the ignition (which will prevent the steering wheel from locking) unless you have shifted the selector lever into Park.
2   Remove the center console box (see Chapter 11).
3   Turn the ignition switch to the ACC position.
4   Loosen the adjusting nut on the cable and remove the cable from the bracket but not from the backdrive cam **(see illustration)**.
5   Shift the selector to the P position and depress the push button all the way. Pull the cable tight and turn the adjusting nut so it contacts the bracket.
6   Release the push button and put a mark on the adjusting nut.
7   Tighten the adjusting nut one full rotation, then tighten the lock nut.
8   Check to make sure the key cannot be turned to the LOCK position (though it can be turned from ON to ACC) when the select lever is in any other position than Park. The select lever shouldn't be able to be shifted from the Park position to other positions

when key is in the LOCK position or when the key is not in the key slot.

7   Automatic transmission - removal and installation

Removal

1   Disconnect the negative cable from the battery.
2   Remove the transfer shift control lever (see Chapter 7 Part A, *Manual transmission shift lever - removal and installation*).
3   Disconnect the wiring harness couplers, breather hose clamp, kickdown cable and vacuum hoses.
4   Raise the vehicle and support it securely on jackstands.
5   Drain the transfer case fluid and transmission fluid (see Chapter 1), then reinstall the pan.
6   Remove the torque converter cover.
7   Mark the torque converter to the driveplate so they can be installed in the same position.
8   Remove the torque converter-to-driveplate bolts. Turn the crankshaft for access to each bolt. Turn the crankshaft in a clockwise direction only (as viewed from the front).
9   Remove the nut from the end of the select cable **(see illustration 3.4)** and the E-ring from the bracket to release the cable.
10   Remove the select cable bracket by removing its two bolts.
11   Loosen the clamps and disconnect the oil cooler hoses from the pipes. **Note:** *To avoid any leaking of transmission fluid, plug the open ends of the transmission cooler pipes and hoses.*
12   Remove the starter motor (see Chapter 5).
13   Remove the driveshafts (see Chapter 8).
14   Disconnect the speedometer cable.
15   On models so equipped, disconnect the vacuum hose from the modulator.
16   Remove any exhaust components which will interfere with transmission removal (see Chapter 4).
17   Support the engine with a jack. Use a block of wood under the oil pan to spread the load.
18   Support the transmission with a jack - preferably a jack made for this purpose. Safety chains will help steady the transmission on the jack.
19   Remove the rear mount to crossmember bolts and the crossmember-to-frame bolts.
20   Remove the two engine rear support-to-transmission extension housing bolts.
21   Raise the transmission enough to allow removal of the crossmember.
22   Remove the bolts securing the transmission to the engine.
23   Lower the transmission slightly.
24   Remove the transmission dipstick tube.
25   Move the transmission to the rear to disengage it from the engine block dowel pins and make sure the torque converter is detached from the driveplate. Secure the

torque converter to the transmission so it won't fall out during removal.

Installation

26 Prior to installation, make sure the torque converter hub is securely engaged in the pump.

27 With the transmission secured to the jack, raise it into position. Be sure to keep it level so the torque converter does not slide forward. Connect the transmission fluid cooler lines.

28 Turn the torque converter to line up the matchmark with the mark on the driveplate.

29 Move the transmission forward carefully until the dowel pins and the torque converter are engaged.

30 Install the transmission housing-to-engine bolts. Tighten them securely.

31 Install the torque converter-to-driveplate bolts. Tighten the bolts to the torque listed in this Chapter's Specifications.

32 Install the transmission mount cross-member bolts. Tighten the bolts and nuts securely.

33 Remove the jacks supporting the transmission and the engine.

34 Install the dipstick tube.

35 Install the starter motor (see Chapter 5).

36 Connect the vacuum hose(s) (if equipped).

37 Install the select cable into the select cable bracket. Install the bolts and the E-ring.

38 Plug in the transmission wire harness connectors.

39 Install the torque converter cover.

40 Install the driveshafts.

41 Connect the speedometer cable.

42 Adjust the shift linkage (see Section 3) and the kickdown cable (see Section 5).

43 Install any exhaust system components that were removed or disconnected.

44 Lower the vehicle.

45 Fill the transfer case and transmission with the specified fluid (see Chapter 1), run the engine and check for fluid leaks.

Chapter 7 Part C
Transfer case

Contents

Specifications

Torque specifications

	Ft-lb (unless otherwise indicated)
Transfer case-to-transmission nuts	20
Lubricant level and drain plugs	20
Shift lever case center bolt	60 in-lbs
Shift lever case bolts	15 to 20

1   General information

The transfer case is a device used on 4WD models that passes the power from the engine and transmission to the front and rear driveshafts.

This auxiliary transmission selects between four wheel drive (engaging the front and rear axles) and two wheel drive (rear axle), and between HIGH and LOW for four wheel drive.

2   Shift lever (Samurai models) - removal and installation

Refer to illustration 2.3
Note: *The transfer case shift lever removal and installation procedure for Sidekick and Tracker models can be found in Chapter 7 Part A (it's the same as the manual transmission shift lever procedure).*

Removal

1   Carefully pull the carpet away from the shift lever and remove the bolts that connect the boot to the floor.
2   Loosen the clamp and slide the boot up the transfer case shift lever.
3   Remove the retaining clip and the small inner boot, then twist the shift lever guide counterclockwise while pushing it down **(see illustration)**. This will unlock the lever from the transfer case.
4   Pull the shift lever out of the case.

Installation

5   Check the lower portion of the lever for excessive wear **(see illustration 3.7** in Chapter 7A) and also inspect the boot for damage. Replace parts as necessary.
6   Apply grease to the pivot portions and lever seat. Install the shift lever into the case, then push down on the guide clockwise to lock it in place.
7   The remainder of installation is the reverse of the removal procedure.

2.3 Push down on the lever guide while simultaneously twisting it counterclockwise

3   Transfer case - removal and installation

Samurai models
Removal
Refer to illustration 3.9
1   Disconnect the negative cable from the battery.
2   Raise the vehicle and support it securely on jackstands.
3   Drain the transfer case lubricant and the transmission lubricant (see Chapter 1).

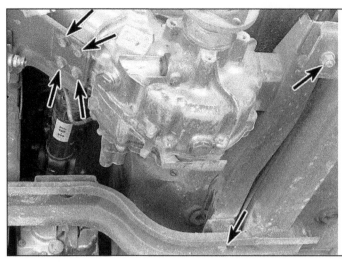

3.9 Remove the transfer case rear support-to-crossmember nuts and bolts (Samurai)

4   Disconnect the speedometer cable, shift lever, and the 4WD switch at the electrical connector.
5   Remove the driveshafts (see Chapter 8).
6   Remove the exhaust system components as necessary for clearance (see Chapter 4).
7   Support the transmission with a floor jack. The transmission must remain supported at all times while the transfer case is out of the vehicle. **Note:** *Place a wood block between the distributor housing and the firewall to prevent damage when the engine tilts.*
8   Support the transfer case with a floor jack. Safety chains will help steady the transfer case on the jack.
9   Remove the transfer case rear support-to-crossmember nuts and bolts **(see illustration)**.
10   Raise the transfer case slightly and remove the crossmember.
11   Make a final check that all wires and hoses have been disconnected from the transfer case, then move the transfer case and jack toward the rear of the vehicle.
12   Remove the intermediate shaft with the transfer case. The shaft slides out of the transmission housing.
13   Lower the transfer case and remove it from under the vehicle.

Installation

14   With the transfer case secured to the jack as on removal, raise it into position behind the transmission and carefully slide it forward, engaging the intermediate shaft with the transmission output shaft. Do not use excessive force to install the transfer case if the intermediate shaft does not slide into place. Readjust the angle so it is level and/or turn the intermediate shaft so the splines engage properly.
15   Install the crossmember and transmis-

sion support. Tighten the fasteners securely.
16   Remove the jacks supporting the transmission and transfer case.
17   Install the various items removed previously, referring to Chapter 8 for the installation of the driveshafts and Chapter 4 for information regarding the exhaust system components.
18   Make a final check that all wires, hoses, and the speedometer cable have been connected and that the transmission and transfer case have been filled with lubricant to the proper level (see Chapter 1).
19   Connect the negative battery cable. Road test the vehicle for proper operation and check for leakage.

Sidekick/X-90/Vitara and Tracker models

Removal

20   Disconnect the negative cable from the battery.
21   Remove the center console (see Chapter 11).
22   Remove the transmission shift lever (see Chapter 7A and 7B) and the transfer case shift lever (see Chapter 7A - its the same procedure as for the transmission shift lever).
23   Follow steps 2-10 of this Section.
24   Inside the passenger compartment, remove the bolts that secure the gear shift lever case.
25   Slide the breather hose clamp forward and remove the hose.
26   Remove the gear shift lever case and then remove the transfer case center bolt.
27   Remove the transfer case/engine rear mounting assembly.
28   Remove the transfer case-to-transmission bolts.
29   With the transfer case assembly supported with a jack (preferably a transmission

jack), slide it toward the rear of the vehicle and lower it.

Installation

30   With the transfer case secured to the jack, raise it into position behind the transmission and carefully slide it forward. Do not use excessive force to install the transfer case - make sure the splines are correctly aligned.
31   Install the transmission to transfer case bolts, tightening them to the torque listed in this Chapter's Specifications.
32   Follow steps 15-19 of this Section.

4   Transfer case overhaul - general information

Overhauling a transfer case is a difficult job for the do-it-yourselfer. It involves the disassembly and reassembly of many small parts. Numerous clearances must be precisely measured and, if necessary, changed with select fit spacers and snap-rings. As a result, if transfer case problems arise, it can be removed and installed by a competent do-it-yourselfer, but overhaul should be left to a transmission repair shop. Rebuilt transfer cases may be available - check with your dealer parts department and auto parts stores. At any rate, the time and money involved in an overhaul is almost sure to exceed the cost of a rebuilt unit.

Nevertheless, it's not impossible for an inexperienced mechanic to rebuild a transfer case if the special tools are available and the job is done in a deliberate step-by-step manner so nothing is overlooked.

The tools necessary for an overhaul include internal and external snap-ring pliers, a bearing puller, a slide hammer, a set of pin punches, a dial indicator and possibly a hydraulic press. In addition, a large, sturdy workbench and a vise or transmission stand will be required.

During disassembly of the transfer case, make careful notes of how each piece comes off, where it fits in relation to other pieces and what holds it in place. Noting how parts are installed when you remove them will make it much easier to get the transfer case back together.

Before taking the transfer case apart for repair, it will help if you have some idea what area of the transfer case is malfunctioning. Certain problems can be closely tied to specific areas in the transfer case, which can make component examination and replacement easier. Refer to the Troubleshooting Section at the front of this manual for information regarding possible sources of trouble.

Chapter 8
Clutch and drivetrain

Contents

Specifications

Torque specifications

	Ft-lbs (unless otherwise indicated)
Pressure plate-to-flywheel bolts	14 to 20
Driveshaft bolts and nuts	
Samurai	17 to 22
Sidekick/X-90/Vitara/Tracker	36 to 43
Driveshaft center support bolts	37
Freewheeling hub	
Manual locking	
Hub body bolts	15 to 22
Hub cover bolts	72 to 109 in-lbs
Automatic locking	
Hub body bolts	15 to 21
Kingpin bolts (Samurai)	15 to 22
Front wheel bearing nut (Samurai)	
Nut	96 to 132 in-lbs
Lock nut	43 to 65
Driveaxle flange bolt and nut (Sidekick/X-90/Vitara/Tracker)	29 to 43
Front axle housing mounting bolts (Sidekick/X-90/Vitara/Tracker)	
Left side	37
Right side	37
Rear	37
Front wheel bearing locknut (Sidekick/X-90/Vitara/Tracker)	
1992 and earlier (all models)	89 to 148
1993	
Sidekick	180
Tracker	155
1994 and later (all models	159
Brake backing plate bolts/nuts	14 to 20
Differential carrier bolts	
Samurai	20
Sidekick/X-90/Vitara/Tracker	41
Hydraulic clutch	
Master cylinder	
Mounting bolt	12
Fluid line flarenut	10
Release cylinder	
Mounting bolt	12
Fluid line bolt	17

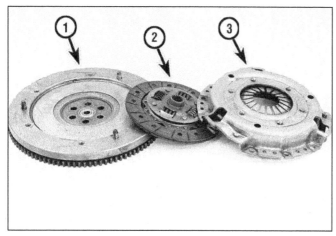

2.1 Exploded view of the clutch components

1 *Flywheel*	*3* *Pressure plate*
2 *Clutch disc*	

3.6 Mark the relationship of the pressure plate to the flywheel (arrow) (in case you are going to reuse the same pressure plate)

1   General information

The information in this Chapter deals with the components from the rear of the engine to the wheels, except for the transmission and transfer case, which are dealt with in the previous Chapter. For the purposes of this Chapter, these components are grouped into four categories; clutch, driveshaft, front axle and rear axle. Separate Sections within this Chapter offer general descriptions and checking procedures for components in each of the four groups.

Since nearly all the procedures covered in this Chapter involve working under the vehicle, make sure it's securely supported on sturdy jackstands or on a hoist where the vehicle can be easily raised and lowered.

2   Clutch - description and check

Refer to illustration 2.1

1   All vehicles with a manual transmission use a single dry plate, diaphragm spring type clutch **(see illustration)**. The clutch disc has a splined hub that allows it to slide along the input shaft splines of the transmission. Spring pressure exerted by the diaphragm in the pressure plate holds the clutch and pressure plate in contact.

2   The clutch release system is operated mechanically on 1998 and earlier models. The mechanical release system includes the clutch pedal, a clutch cable that actuates the clutch release lever and the release bearing. 1999 and later models use a clutch release system operated by hydraulic pressure. The hydraulic release system consists of the clutch pedal, a master cylinder and fluid reservoir, the hydraulic line and a hydraulic release cylinder.

3   When pressure is applied to the clutch pedal to release the clutch, pressure is exerted against the outer end of the clutch release lever by the cable or the hydraulic release cylinder. As the lever pivots the shaft fingers push against the release bearing. The bearing pushes against the fingers of the diaphragm spring of the pressure plate assembly, which in turn releases the clutch plate.

4   Terminology can be a problem when discussing the clutch components because common names are in some cases different from those used by the manufacturer. For example, the driven plate is also called the clutch plate or disc, and the clutch release bearing is sometimes called a throwout bearing.

5   Other than to replace components with obvious damage, some preliminary checks should be performed to diagnose clutch problems.

a) *To check "clutch spin down time," run the engine at normal idle speed with the transmission in Neutral (clutch pedal up - engaged). Disengage the clutch (pedal down), wait several seconds and shift the transmission into Reverse. No grinding noise should be heard. A grinding noise would most likely indicate a problem in the pressure plate or the clutch disc.*

b) *To check for complete clutch release, run the engine (with the parking brake applied to prevent movement) and hold the clutch pedal approximately 1/2-inch from the floor. Shift the transmission between 1st gear and Reverse several times. If the shift is rough, component failure is indicated.*

c) *Visually inspect the pivot bushing at the top of the clutch pedal to make sure there is no binding or excessive play.*

d) *A clutch pedal that is difficult to operate is most likely caused by a faulty clutch cable. Check the cable where it enters the housing for frayed wires, rust and other signs of corrosion. If it looks good, lubricate the cable with penetrating oil. If pedal operation improves, the cable is worn out and should be replaced.*

e) *Crawl under the vehicle and make sure the clutch release arm is solidly clamped to the release shaft.*

3   Clutch components - removal, inspection and installation

Warning: *Dust produced by clutch wear and deposited on clutch components may contain asbestos, which is hazardous to your health. DO NOT blow it out with compressed air and DO NOT inhale it. DO NOT use gasoline or petroleum-based solvents to remove the dust. Brake system cleaner should be used to flush the dust into a drain pan. After the clutch components are wiped clean with a rag, dispose of the contaminated rags and cleaner in a covered, marked container.*

Removal

Refer to illustration 3.6

1   Access to the clutch components is normally accomplished by removing the transmission, leaving the engine in the vehicle. If, of course, the engine is being removed for major overhaul, then check the clutch for wear and replace worn components as necessary. However, the relatively low cost of the clutch components compared to the time and trouble spent gaining access to them warrants their replacement anytime the engine or transmission is removed, unless they are new or in near perfect condition. The following procedures are based on the assumption the engine will stay in place.

2   Remove the clutch cable (see Section 6).

3   Referring to Chapter 7 Part A, remove the transmission from the vehicle. Support the engine while the transmission is out. Preferably, an engine hoist should be used to support it from above. However, if a jack is used underneath the engine, make sure a piece of wood is positioned between the jack and oil pan to spread the load. **Caution:** *The pickup for the oil pump is very close to the*

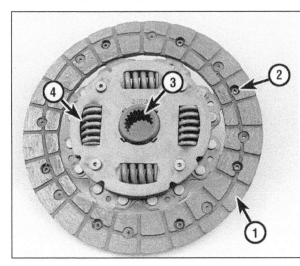

3.11 The clutch plate

1 **Lining** – *This will wear down in use*
2 **Rivets** – *These secure the lining and will damage the flywheel or pressure plate if allowed to contact the surfaces*
3 **Splined hub** – *the splines should not be worn and should slide smoothly on the input shaft splines*
4 **Springs** – *check for deformation*

There should be at least 1/16-inch of lining above the rivet heads. Check for loose rivets, distortion, cracks, broken springs and other obvious damage **(see illustration)**. As mentioned above, ordinarily the clutch disc is routinely replaced, so if in doubt about the condition, replace it with a new one.

12 The release bearing should also be replaced along with the clutch disc (see Section 4).

13 Check the machined surfaces and the diaphragm spring fingers of the pressure plate **(see illustration)**. If the surface is grooved or otherwise damaged, replace the pressure plate. Also check for obvious damage, distortion, cracking, etc. Light glazing can be removed with emery cloth or sandpaper. If a new pressure plate is required, new and factory-rebuilt units are available.

bottom of the oil pan. *If the pan is bent or distorted in any way, engine oil starvation could occur.*

4 The clutch fork and release bearing can remain attached to the housing for the time being.

5 To support the clutch disc during removal, install a clutch alignment tool through the clutch disc hub.

6 Carefully inspect the flywheel and pressure plate for indexing marks. The marks are usually an X, an O or a white letter. If they cannot be found, scribe marks yourself so the pressure plate and the flywheel will be in the same alignment during installation **(see illustration)**.

7 Turning each bolt only a little at a time, loosen the pressure plate-to-flywheel bolts. Work in a criss-cross pattern until all spring pressure is relieved. Then hold the pressure plate securely and completely remove the

bolts, followed by the pressure plate and clutch disc.

Inspection

Refer to illustrations 3.11 and 3.13

8 Ordinarily, when a problem occurs in the clutch, it can be attributed to wear of the clutch driven plate assembly (clutch disc). However, all components should be inspected at this time.

9 Inspect the flywheel for cracks, heat checking, grooves and other obvious defects. If the imperfections are slight, a machine shop can machine the surface flat and smooth, which is highly recommended regardless of the surface appearance. Refer to Chapter 2 for the flywheel removal and installation procedure.

10 Inspect the pilot bearing (see Section 5).

11 Inspect the lining on the clutch disc.

Installation

Refer to illustration 3.15

14 Before installation, clean the flywheel and pressure plate machined surfaces with lacquer thinner or acetone. It's important that no oil or grease is on these surfaces or the lining of the clutch disc. Handle the parts only with clean hands.

15 Position the clutch disc and pressure plate against the flywheel with the clutch held in place with an alignment tool **(see illustration)**. Make sure it's installed properly (most replacement clutch plates will be marked "flywheel side" or something similar - if not marked, install the clutch disc with the damper springs toward the transmission).

16 Tighten the pressure plate-to-flywheel bolts only finger tight, working around the pressure plate.

17 Center the clutch disc by ensuring the alignment tool extends through the splined hub and into the pilot bearing in the crankshaft. Wiggle the tool up, down or side-to-side as needed to bottom the tool in the pilot bearing. Tighten the pressure plate-to-

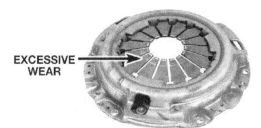

NORMAL FINGER WEAR

EXCESSIVE WEAR

EXCESSIVE FINGER WEAR   **BROKEN OR BENT FINGERS**

3.13 Replace the pressure plate if excessive wear is noted

3.15 Center the clutch disc in the pressure plate with an alignment tool before the bolts are tightened - a clutch alignment tool can be purchased at most auto parts stores and eliminates all guesswork when centering the clutch disc in the pressure plate

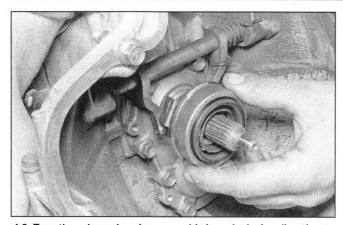

4.3 Turn the release bearing assembly in a clockwise direction to detach it from the release fork - when installing the bearing, make sure the pin on the fork is engaged with the slot in the release bearing

5.5 A slide hammer can be used for removing the pilot bearing

flywheel bolts a little at a time, working in a criss-cross pattern to prevent distorting the cover. After all the bolts are snug, tighten them to the specified torque. Remove the alignment tool.
18 Using high temperature grease, lubricate the inner groove of the release bearing (see Section 4). Also place grease on the release lever contact areas and the transmission input shaft bearing retainer.
19 Install the clutch release bearing as described in Section 4.
20 Install the transmission and all components removed previously. Tighten all fasteners to the proper torque specifications.

---

4 Clutch release bearing - removal, inspection and installation

Refer to illustration 4.3
Warning: *Dust produced by clutch wear and deposited on clutch components may contain asbestos, which is hazardous to your health. DO NOT blow it out with compressed air and DO NOT inhale it. DO NOT use gasoline or petroleum-based solvents to remove the dust. Brake system cleaner should be used to flush the dust into a drain pan. After the clutch components are wiped clean with a rag, dispose of the contaminated rags and cleaner in a covered, marked container.*

Removal

1 Disconnect the negative cable from the battery.
2 Remove the transmission (see Chapter 7).
3 Push the clutch release arm forward, then remove the bearing from the fork by turning the bearing assembly in a clockwise direction **(see illustration)**.

Inspection

4 Hold the center of the bearing and rotate the outer portion while applying pressure. If the bearing doesn't turn smoothly or if

it's noisy, replace it with a new one. Wipe the bearing with a clean rag and inspect it for damage, wear and cracks. Don't immerse the bearing in solvent - it's sealed for life and to do so would ruin it.

Installation

5 Using high temperature grease, lightly lubricate the release fork and the input shaft where they contact the bearing. Fill the inner groove of the bearing with the same grease.
6 Attach the release bearing to the clutch lever by turning counterclockwise. Make sure the pin on the fork is engaged with the slot in the release bearing ear.
7 Install the transmission.
8 The remainder of installation is the reverse of the removal procedure. Tighten all bolts to the specified torque.

---

5 Pilot bearing - inspection and replacement

Refer to illustration 5.5
1 The clutch pilot bearing is a roller type bearing which is pressed into the rear of the crankshaft. It is greased at the factory and does not require additional lubrication. Its primary purpose is to support the front of the transmission input shaft. The pilot bearing should be inspected whenever the clutch components are removed from the engine. Due to its inaccessibility, if you are in doubt as to its condition, replace it with a new one.
Note: *If the engine has been removed from the vehicle, disregard the following steps which do not apply.*
2 Remove the transmission (see Chapter 7 Part A).
3 Remove the clutch components (see Section 3).
4 Inspect for any excessive wear, scoring, lack of grease, dryness or obvious damage. If any of these conditions are noted, the bearing should be replaced. A flashlight will be helpful to direct light into the recess.

5 Removal can be accomplished with a special puller, available at most auto parts stores **(see illustration)**.
6 If a puller is not available, try using a hooked tool (position the hook behind the bearing.) Tap the tool outward with a hammer. If there's nothing to tap on, clamp vise-grip pliers securely to the tool and tap on them.
7 To install the new bearing, lightly lubricate the outside surface with lithium-based grease, then drive it into the recess with a soft-face hammer.
8 Install the clutch components, transmission and all other components removed previously, tightening all fasteners properly.

---

6 Clutch cable - replacement

Refer to illustrations 6.2 and 6.5

Removal

1 Disconnect the negative cable from the battery.
2 Remove the clutch cable joint nut and loosen the cable casing nuts **(see illustration)**.

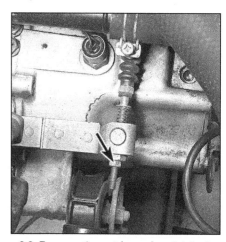

6.2 Remove the nut (arrow) and detach the clutch cable from the release arm

6.5 Remove the two bolts to disconnect the clutch cable from the firewall

7.3 If possible, use a flare-nut wrench when unscrewing the fluid line fitting to avoid rounding it off

3   Remove the cable from the release arm and the cable bracket.
4   Follow the clutch cable up to the firewall and remove any clutch cable clamps.
5   Remove the two bolts holding the cable to the firewall **(see illustration).**
6   Disconnect the clutch cable hook at the clutch pedal and remove the cable from the vehicle.

Installation

7   Apply grease to the cable hook and the joint pin.
8   Position the cable in the vehicle and hook the cable end to the clutch pedal. Install the two cable casing-to-firewall bolts.
9   Install the clutch cable into the cable bracket and the release arm.
10   Refer to Chapter 1 for the clutch cable and pedal adjustments.
11   Connect negative battery cable.
12   Check for proper clutch operation.

7   Clutch master cylinder - removal and installation

Removal

Refer to illustrations 7.3 and 7.4
1   Disconnect the negative cable from the battery.
2   In the engine compartment, use a syringe to remove the fluid from the reservoir.
3   Disconnect the hydraulic line at the clutch master cylinder. If available, use a flare-nut wrench on the fitting, which will prevent the fitting from being rounded off **(see illustration).** Have rags handy as some fluid will be lost as the line is removed. **Caution:** *Don't allow brake fluid to come into contact with paint, as it will damage the finish.*
4   Remove the lower bolt securing the master cylinder to the firewall **(see illustration).**
5   Under the dashboard, disconnect the pushrod from the top of the clutch pedal. It's held in place with a clevis pin.

6   From under the dashboard remove the nut that secures the master cylinder to the firewall. Remove the master cylinder, again being careful not to spill any of the fluid.

Installation

7   Position the master cylinder on the fire-wall, installing the mounting nuts finger-tight.
8   Connect the hydraulic line to the master cylinder, moving the cylinder slightly as necessary to thread the fitting properly into the bore. Don't cross-thread the fitting as it's installed.
9   Tighten the mounting nut(s) and the hydraulic line fitting securely.
10   Connect the pushrod to the clutch pedal.
11   Fill the clutch master cylinder reservoir with brake fluid and bleed the clutch system (see Section 9).
12   Check the clutch pedal height and freeplay and adjust if necessary, following the procedure in Chapter 1.

8   Clutch release cylinder - removal and installation

Removal

Refer to illustration 8.3
1   Disconnect the negative cable from the battery.
2   Raise the vehicle and support it securely on jackstands. Have a small can and rags handy, as some fluid will be spilled as the cylinder hydraulic line is removed.
3   Disconnect the hydraulic line from the release cylinder by unscrewing the bolt then remove the mounting bolts and separate the release cylinder from the bellhousing **(see illustration).**

Installation

4   Connect the hydraulic line to the release cylinder. Install the release cylinder on the clutch housing, making sure the pushrod is seated in the release fork pocket. Tighten the

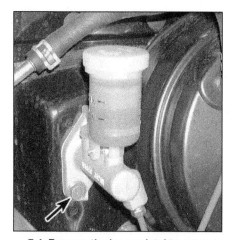

7.4 Remove the lower clutch master cylinder retaining bolt (arrow)

bolts to the torque listed in this Chapter's Specifications.
5   Tighten the hydraulic line fitting bolt securely.
6   Fill the clutch master cylinder with brake fluid.

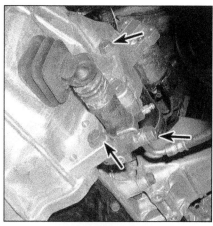

8.3 Remove the release cylinder hydraulic line bolt, followed by the two retaining bolts (arrows)

7   Bleed the system (see Section 9).
8   Lower the vehicle and connect the negative battery cable.

9   Clutch hydraulic system - bleeding

1   The hydraulic system should be bled of all air whenever any part of the system has been removed or if the fluid level has been allowed to fall so low that air has been drawn into the master cylinder. The procedure is very similar to bleeding a brake system.
2   Fill the master cylinder with new brake fluid conforming to DOT 3 specifications. **Caution:** *Do not re-use any of the fluid coming from the system during the bleeding operation or use fluid which has been inside an open container for an extended period of time.*
3   Raise the vehicle and place it securely on jackstands to gain access to the release cylinder, which is located on the left side of the clutch housing.
4   Remove the dust cap that fits over the bleeder valve and push a length of plastic hose over the valve. Place the other end of the hose into a clear container with about two inches of brake fluid in it. The hose end must be submerged in the fluid.
5   Have an assistant depress the clutch pedal and hold it. Open the bleeder valve on the release cylinder, allowing fluid to flow through the hose. Close the bleeder valve when fluid stops flowing from the hose. Once closed, have your assistant release the pedal.
6   Continue this process until all air is evacuated from the system, indicated by a full, solid stream of fluid being ejected from the bleeder valve each time and no air bubbles in the hose or container. Keep a close watch on the fluid level inside the clutch master cylinder reservoir; if the level drops too low, air will be sucked back into the system and the process will have to be started all over again.
7   Install the dust cap and lower the vehicle. Check carefully for proper operation before placing the vehicle in normal service.

10   Clutch pedal - removal and installation

Removal

1   Loosen the pedal shaft clamp bolt.
2   Remove the clutch pedal shaft arm.
3   Remove the clutch pedal spring while sliding the clutch pedal out.

Installation

4   Connect the clutch pedal spring while sliding the clutch pedal into position through the shaft bushing and the pedal bracket.
5   Install the clutch pedal shaft arm.
6   Tighten the clamp bolt securely.

11   Clutch start switch - removal, installation and adjustment

Removal

1   Apply the parking brake firmly and put the transmission in neutral.
2   Locate the clutch start switch next to the clutch pedal bracket under the dash. Disconnect the electrical connector.
3   Loosen the lock nut and unscrew the clutch start switch.

Installation

4   Installation is the reverse of removal.

Adjustment

5   Depress the clutch pedal all the way and bring it back up two to three inches off the floor.
6   With an ohmmeter connected to the switch, slowly screw the switch in until there is continuity.
7   Tighten the lock nut and connect the electrical connector.
8   Depress the clutch pedal all the way to the floor and check clearance B. If the clearance is not correct, the switch may be damaged.

12   Driveshafts, differentials and axles - general information

Four different driveshaft assemblies are used on the vehicles covered in this manual. Samurai models use three drive shafts and Sidekick/X-90/Vitara and Tracker models use only two (or one on 2WD models). On the Samurai, a short driveshaft is used to transfer power from the transmission to the transfer case. On all models, a front (4WD models) and rear driveshaft transmit power from the transfer case to the front and rear axles. Since the transfer case on Sidekick/X-90/-Vitara and Tracker and models is mounted to the rear of the transmission, the short driveshaft is not necessary. Later model longwheel base 2WD models use a two-piece driveshaft with a center bearing support.
    All universal joints are of the solid type and can be replaced separate from the driveshaft.
    The driveshafts are finely balanced during production and whenever they are removed or disassembled, they must be reassembled and reinstalled in the exact manner and positions they were originally in, to avoid excessive vibration.
    The rear axle on all models (and the front axle on Samurai models) is of the semi-floating type, which is held in proper alignment with the body by the suspension.
    Mounted in the center of the axle is the differential, which transfers the turning force of the driveshaft to the axleshafts.
    The rear wheel bearing provides support for the axle outer ends. At their inner ends the axles are splined to fit into the splines in the differential gears. The outer end of the front axles are supported by bearings located in the spindles.
    Because of the complexity and critical nature of the differential adjustments, as well as the special equipment needed to perform the operations, we recommend any disassembly of the differential be done by a dealer service department or other qualified repair facility.

13   Driveshaft - removal and installation

Refer to illustrations 13.3 and 13.7

Removal

1   Disconnect the negative cable from the battery.
2   Raise the vehicle and support it securely on jackstands. Place the transmission in Neutral with the parking brake off.
3   Using a scribe, paint or a hammer and punch, place marks on the driveshaft and the differential flange (and transfer case flange on Samurai models) in line with each other **(see illustration)**. This is to make sure the driveshaft is reinstalled in the same position to preserve the balance.
4   Remove the rear universal joint bolts and nuts. Turn the driveshaft (or tires) as necessary to bring the bolts into the most accessible position. On 2WD models with two-piece driveshafts, remove the center bearing support bolts.
5   Lower the rear of the driveshaft and slide the front out of the transmission. On some models it will be necessary to unbolt the driveshaft from the transmission.
6   To prevent loss of fluid and protect against contamination while the driveshaft is out, wrap a plastic bag over the transmission housing and hold it in place with a rubber band (Sidekick/X-90/Vitara and Tracker models only).

13.3   Before removing the driveshaft, mark the relationship of the driveshaft yoke to the differential flange

13.7 On Samurai models, if the splined portion of the driveshaft slips out of the yoke end during removal, be sure to align the match marks

Installation

7   Remove the plastic bag from the transmission and wipe the area clean. Inspect the oil seal carefully. Procedures for replacement of this seal can be found in Chapter 7. On Samurai models if the driveshaft slips apart, be sure to realign match marks when reinstalling (see illustration).
8   Slide the front of the driveshaft into the transmission. On 2WD models with two-piece driveshafts, raise the center bearing support into position making sure it is aligned at 90° to the driveshaft centerlines and install the retaining bolts.
9   Raise the rear of the driveshaft into position, checking to be sure the marks are in alignment. If not, turn the rear wheels to match the pinion flange and the driveshaft.
10   Tighten the bolts and nuts to the specified torque.

14   Universal joints - replacement

Refer to illustrations 14.2, 14.4 and 14.9
Note: A press or large vise will be required for this procedure. It may be a good idea to take the driveshaft to a repair or machine shop where; the universal joints can be replaced for you, normally at a reasonable charge.
1   Remove the driveshaft as outlined in the previous Section.
2   Using a small pair of pliers, remove the snap-rings from the spider (see illustration).
3   Supporting the driveshaft, place it in position on either an arbor press or on a workbench equipped with a vise.
4   Place a piece of pipe or a large socket with the same inside diameter over one of the bearing caps. Position a socket which is of slightly smaller diameter than the cap on the opposite bearing cap (see illustration) and use the vise or press to force the cap out (inside the pipe or large socket), stopping just before it comes completely out of the yoke. Use the vise or large pliers to work the cap the rest of the way out.

14.2 Use snap-ring pliers to remove the snap-rings

5   Transfer the sockets to the other side and press the opposite bearing cap out in the same manner.
6   Pack the new universal joint bearings with grease. Ordinarily, specific instructions for lubrication will be included with the universal joint servicing kit and should be followed carefully.
7   Position the spider in the yoke and partially install one bearing cap in the yoke. If the replacement spider is equipped with a grease fitting, be sure it's offset in the proper direction (toward the driveshaft).
8   Start the spider into the bearing cap and then partially install the other cap. Align the spider and press the bearing caps into position, being careful not to damage the dust seals.
9   Install the snap-rings. If difficulty is encountered in seating the snap-rings, strike the driveshaft yoke sharply with a hammer. This will spring the yoke ears slightly and allow the snap-rings to seat in the groove (see illustration).
10   Install the grease fitting and fill the joint with grease. Be careful not to overfill the joint, as this could blow out the grease seals.
11   Install the driveshaft. Tighten the flange bolts and nuts to the specified torque.

14.9 To relieve stress produced by pressing the bearing caps into the yokes, strike the yoke in the area shown

14.4 To press the universal joint out of the driveshaft, set it up in a vise with the small socket (on the left) pushing the joint and bearing cap into the large socket

15   Front axleshaft and steering knuckle (Samurai models) - removal, inspection and installation

Refer to illustrations 15.6, 15.8, 15.12 and 15.14

Removal

1   Loosen the wheel lug nuts, raise the front of the vehicle and support it securely on jackstands. Apply the parking brake. Remove the wheel.
2   Remove the disc brake caliper and carrier and hang them out of the way with a piece of wire (see Chapter 9).
3   Remove the brake disc (see Chapter 9).
4   Remove the freewheeling hub or automatic hub (see Section 24).
5   Remove the front wheel bearings and hub (see Section 25).
6   Loosen the upper and lower kingpin bolts but don't remove them at this time. Mark the upper and lower kingpins so they can be installed in their original positions (see illustration).

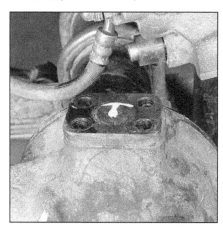

15.6 Before removing the kingpins, be sure to mark them so they will be reinstalled in their original positions

15.8 Mark the spindle so it can be reinstalled in its original position

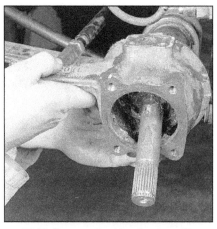

15.12 Remove the steering knuckle carefully, because the lower kingpin bearing could fall out

15.14 To remove the axleshaft, simply pull it out

7   Remove the disc brake dust cover.
8   Mark the relationship of the spindle to the steering knuckle and unbolt it **(see illustration)**.
9   Separate the tie-rod end from the steering knuckle (see Chapter 10).
10   Remove the steering knuckle oil seal (see Chapter 1).
11   Remove the kingpins, noting the locations of any shims that may be present between the kingpins and the steering knuckle.
12   Remove the steering knuckle, taking care not to drop the kingpin bearing **(see illustration)**.
13   Drain the front differential lubricant (see Chapter 1).
14   Remove the axleshaft by pulling it straight out of the housing **(see illustration)**.

Inspection

15   Check all parts for cracking, distortion, dents, deformed splines or any other type of damage and replace as necessary.

Installation

16   Installation is the reverse of removal, but before installing the axleshaft, pack the constant velocity joint with chassis grease. Also, be sure to lubricate the kingpins and the kingpin bearing with the same grease.
17   Tighten all fasteners to the torque listed in this Chapter's Specifications.
18   Fill the front differential with the recommended lubricant (see Chapter 1).

---

16   Front axleshaft oil seal (Samurai models) - removal and installation

---

Refer to illustrations 16.2 and 16.3

1   Remove the front axleshaft (see Section 15).
2   Remove the seal with a seal remover tool or a small pry bar **(see illustration)**.
3   Lubricate the seal lip, then drive the seal

16.2 Using a small prybar, pry the axle seal out of the housing

into position with a seal installation tool, a piece of pipe, or socket with an outside diameter slightly smaller than the outside diameter of the seal **(see illustration)**.
4   Install the axleshaft (see Section 15).

---

17   Driveaxle (Sidekick/X-90/Vitara/ Tracker 4WD models) only) - removal and installation

---

Refer to illustration 17.10

Removal

1   Loosen the lug nuts, raise the front of the vehicle and support it securely on jackstands. Remove the front wheel.
2   Drain the differential lubricant (see Chapter 1).
3   Remove the locking hub (see Section 24).
4   Remove the driveaxle snap-ring with a pair of snap-ring pliers.
5   Disconnect the stabilizer bar from the lower arm (see Chapter 10).

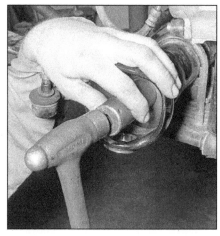

16.3 A socket with an outside diameter slightly smaller than the outside diameter of the seal can be used to install the seal

6   Remove the tie-rod ends (see Chapter 10).
7   Remove the caliper bolts and suspend the caliper with a piece of wire (see Chapter 9).
8   Loosen, but do not remove, the steering knuckle balljoint nut (see Chapter 10). Support the lower arm with a jack and remove the balljoint nut.
9   Lower the jack slowly and swing the steering knuckle clear of the lower arm.

Right side

10   Pry the driveaxle out of the differential **(see illustration)**.
11   Remove the driveaxle from the hub, being careful not to damage the axle boots during removal.

Left side

12   Unbolt the three driveaxle bolts and remove the driveaxle, being careful not to damage the axle boots during removal. Removal of the differential side gear shaft can be accomplished by prying it out, using the technique described in Step 10.

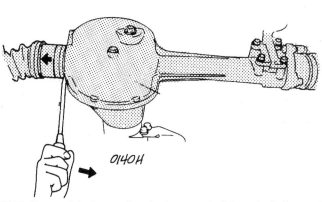

17.10 Be careful when prying the inner end of the axleshaft out of the differential (the seal is very close to the edge of the joint and could be damaged if the prybar is inserted too far)

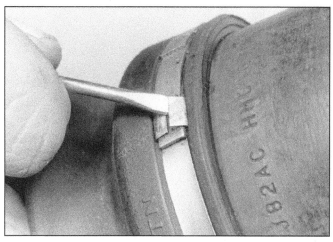

19.3 Pry the boot clamp retaining tabs up with a small screwdriver and slide the clamps off the boot

Installation

13 Inspect all parts for damage, warpage, cracks, breakage and corrosion. Replace parts as necessary.
14 Installation is the reverse of the removal procedure. Tighten all fasteners to the torque values listed in this Chapter's Specifications.

18 Front axle housing - removal and installation

Removal

1 Loosen the front wheel lug nuts, raise the front of the vehicle and support it securely on jackstands positioned under the frame rails. Remove the front wheels.
2 On 4WD models, mark the relationship of the front driveshaft to the front differential pinion shaft yoke, then disconnect the driveshaft from the yoke (see Section 13). With a piece of wire, suspend the driveshaft up out of the way. **Note:** *On Sidekick/X-90/Vitara and Tracker models, don't pull the driveshaft out of the transfer case unless you drain the transfer case lubricant first.*

Samurai models

3 Unbolt the front brake calipers and hang them out of the way with a piece of wire - don't let the calipers hang by the brake hose (see Chapter 9).
4 Remove the brake pads, caliper holder and the brake discs (see Chapter 9).
5 Remove the stabilizer bar (See Chapter 10).
6 Disconnect the tie-rod ends from the steering knuckles (see Chapter 10). Position them out the way and hang them with pieces of wire from the underbody.
7 Position a hydraulic jack under the differential. If two jacks are available, place one under the right side axle tube to balance the assembly.
8 Unbolt the lower ends of the shock

absorbers from the axle.
9 Remove the nuts from the leaf spring U-bolts (see Chapter 10).
10 Remove the leaf spring front mounting bolts and loosen the rear shackle bolts. Allow the leaf springs to swing down.
11 Slowly lower the assembly down and out from under the vehicle.

Sidekick/X-90/Vitara/Tracker 4WD models

12 Remove the differential breather hose.
13 Remove the 4 bolts from the left hand mounting bracket and the 3 bolts and nuts from the driveaxle flange. Separate the driveaxle from the flange and hang it from the underbody with a piece of wire.
14 Remove the 2 bolts from the differential rear mounting bracket.
15 Place a jack under the differential housing - a transmission jack is preferred, as the unit can be secured to the jack with a chain.
16 Remove the 3 right side mounting bolts.
17 Using a pry bar, pry the right side driveaxle from the housing. As this is done, the axle housing will move to the left considerably, so make sure the jack is allowed to roll with it so it doesn't fall off. Also, don't let the driveaxle fall as it comes out of the housing - support it with a piece of wire.
18 Slowly lower the jack and remove the housing from the vehicle.

Installation

19 Installation is the reverse of the removal procedure. On Samurai models, make sure the leaf springs are positioned properly on the axle housing (see Chapter 10). Tighten all fasteners the torque values listed in this Chapter's Specifications.

19 Driveaxle boot (Sidekick/X-90/Vitara/Tracker 4WD models only) - replacement

Note: *Some auto parts stores carry "split" type replacement boots, which can be installed without removing the driveaxle from the vehicle. This is a convenient alternative; however, it's recommended that the driveaxle be removed and the CV joints cleaned to ensure that the joint is free from contaminants such as moisture and dirt, which will accelerate CV joint wear.*

Inner CV joint and boot

Disassembly

Refer to illustrations 19.3, 19.4, 19.5, 19.6 and 19.7

1 Remove the driveaxle from the vehicle (see Section 17).
2 Mount the driveaxle in a vise. The jaws of the vise should be lined with wood or rags to prevent damage to the axleshaft.
3 Pry the boot clamp retaining tabs up with a small screwdriver and slide the clamps off the boot **(see illustration)**.
4 Slide the boot back on the axleshaft and pry the wire ring ball retainer from the outer race **(see illustration)**.

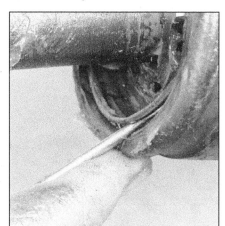

19.4 Pry the wire ring ball retainer out of the outer race

19.5 Slide the outer race off the inner bearing assembly

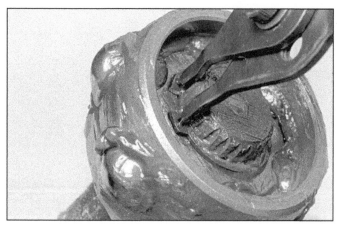

19.6 Remove the snap-ring from the end of the axle

19.7 Apply match marks to the bearing to identify which side faces out during reassembly

19.10 Wrap the splined area of the axle with tape to prevent damage to the boot when installing it

19.13 Pack the inner bearing assembly full of CV joint grease (also note that the larger diameter side, or "bulge", is facing the axleshaft end)

5   Pull the outer race off the inner bearing assembly **(see illustration)**.

6   Remove the snap-ring from the groove in the axleshaft with a pair of snap-ring pliers **(see illustration)**.

7   Mark the inner bearing assembly to ensure that it is reassembled with the correct side facing out **(see illustration)**.

8   Slide the inner bearing assembly off the axleshaft.

Inspection

9   Clean the components with solvent to remove all traces of grease. Inspect the cage, balls and races for pitting, score marks, cracks and other signs of wear and damage. Shiny, polished spots are normal and will not adversely affect CV joint performance.

Reassembly

Refer to illustrations 19.10, 19.13, 19.16, 19.17a and 19.17b

10   Wrap the axleshaft splines with tape to avoid damaging the boot. Slide the small boot clamp and boot onto the axleshaft, then remove the tape **(see illustration)**.

11   Install the inner bearing assembly on the axleshaft with the larger diameter side or

"bulge" of the cage (and the previously applied mark) facing the axleshaft end.

12   Install the snap-ring in the groove. Make sure it's completely seated by pushing on the inner bearing assembly.

13   Fill the outer race and boot with the specified type and quantity of CV joint grease (normally included with the new boot kit). Pack the inner bearing assembly with grease, by hand, until grease is worked completely into the assembly **(see illustration)**.

14   Slide the outer race down onto the inner race and install the wire ring retainer.

15   Wipe any excess grease from the axle boot groove on the outer race. Seat the small diameter of the boot in the recessed area on the axleshaft. Push the other end of the boot onto the outer race.

16   Equalize the pressure in the boot by inserting a dull screwdriver between the boot and the outer race **(see illustration)**. Don't damage the boot with the tool.

17   Install the boot clamps **(see illustrations)**.

18   If you are working on the right side driveaxle, install a new circlip on the inner CV joint stub axle.

19   Install the driveaxle as described in Section 14.

Outer CV joint and boot

Disassembly

20   Following Steps 1 through 8, remove the inner CV joint from the axleshaft.

21   Remove the outer CV joint boot clamps, using the technique described in Step 3. Slide the boot off the axleshaft.

19.16 Equalize the pressure inside the boot by inserting a small, dull screwdriver between the boot and the outer race

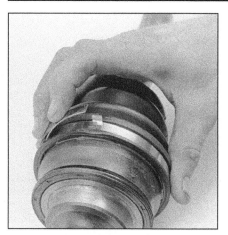

19.17a To install the new clamps, bend the tang down . . .

19.17b . . . fold the tabs over to hold it in place

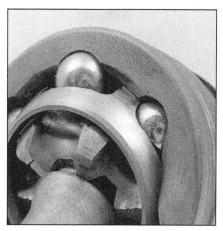

19.23 After the old grease has been rinsed away and the solvent has been blown out with compressed air, rotate the outer joint housing through its full range of motion and inspect the bearing surfaces for wear and damage - if any of the balls, the race or the cage look damaged, replace the driveaxle and outer joint assembly

Inspection

Refer to illustration 19.23

22 Thoroughly wash the inner and outer CV joints in clean solvent and blow them dry with compressed air, if available. **Note:** *Because the outer joint cannot be disassembled, it is difficult to wash away all the old grease and to rid the bearing of solvent once it's clean. But it is imperative that the job be done thoroughly, so take your time and do it right.*

23 Bend the outer CV joint housing at an angle to the driveaxle to expose the bearings, inner race and cage **(see illustration)**. Inspect the bearing surfaces for signs of wear. If the bearings are damaged or worn, replace the driveaxle.

Reassembly

24 Slide the new outer boot onto the driveaxle. It's a good idea to wrap vinyl tape around the spline of shaft to prevent damage to the boot **(see illustration 19.10)**. When the boot is in position, add the specified amount of grease (included in the boot replacement kit) to the outer joint and the boot (pack the joint with as much grease as it will hold and put the rest into the boot). Slide the boot on the rest of the way and install the new clamps **(see illustrations 19.17a and 19.17b)**.

25 Proceed to clean and install the inner CV joint and boot by following Steps 9 through 18, then install the driveaxle as outlined in Section 17.

---

20 Differential carrier assembly - removal and installation

Removal

All models

1 Loosen the front or rear wheel lug nuts and raise the front or rear of the vehicle, supporting it securely on jackstands. Remove the front or rear wheels.

2 Drain the differential lubricant (See Chapter 1).

3 Remove the front or rear axleshafts or driveaxles and side gear shaft (see Sec-

tions 15, 17 and 21).

4 Remove the driveshaft hang it up out of the way with a piece of wire (see Section 13).

Samurai models (front and rear)

5 Remove the 8 carrier-to-differential housing bolts.

6 Place a jack under the carrier, pull the carrier out of the housing and slowly lower the assembly.

Sidekick/X-90/Vitara/Tracker models

Front

7 The front differential carrier must be removed as a complete unit with the axle housing (see Section 18).

Rear

8 Place a jack under the differential, remove the upper arm-to-carrier bolts and lower the rear axle down onto a pair of jackstands or large blocks of wood.

9 Remove the 8 differential bolts and, with the jack, lower the carrier assembly.

Installation (all models)

10 Clean the mating surfaces of the differential carrier and the rear axle housing and apply RTV sealant to them.

11 Installation is the reverse of removal.

---

21 Rear axleshaft and bearing assembly - removal and installation

Refer to illustration 21.8

Note: *A slide hammer and an adapter may be needed to perform this procedure.*

Removal

1 Loosen the rear wheel lug nuts, raise the rear of the vehicle and support it securely on jackstands. Remove the wheels.

2 Remove the brake drum (see Chapter 1).

3 Drain the differential lubricant (see Chapter 1).

Samurai

4 Disconnect the parking brake cable from the rear brake assembly and backing plate (see Chapter 9).

5 Unscrew the brake line fitting from the wheel cylinder - don't pull the line away from the cylinder, as it may become kinked. When the axle and brake assembly are pulled from the housing, plug the line.

6 Remove the brake backing plate nuts from the axle housing.

7 Pull the axle and brake assembly away from the axle housing. If necessary, attach a slide hammer to the axle flange to accomplish this.

8 Slide the axle out of the housing, being careful not to damage the seal or the splines on the inner end **(see illustration)**.

21.8 After breaking the axleshaft loose, detach the puller and slide the axle assembly out of the housing, being careful not to damage the axle seal or splines (Samurai)

23.2 Pry out the seal, being careful not to damage the axle housing

23.4 To install the axle seal, use a large socket, piece of pipe or seal installation tool

24.2 Before removing the hub cover, apply match marks

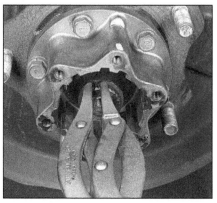

24.4 Using snap-ring pliers, remove the snap-ring from the end of the axle

Sidekick/X-90/Vitara/Tracker models

9   Remove the brake backing plate/wheel bearing retainer nuts from the axle housing.
10   Check to make sure there is enough clearance between the bearing retainer and the parking brake lever to allow the bearing retainer to pass through. If there isn't, pry the parking brake lever back while performing the next Step.
11   Remove the axleshaft from the housing. If it won't come out, attach a slide hammer to the axle flange and pull the axle from the housing, being careful not to pull the brake backing plate along with the axleshaft.

Inspection

12   If the axle bearing must be replaced, take the assembly to a dealer service department or repair shop to have the old bearing removed and a new one pressed on. If a new bearing is installed, be sure to also replace the axleshaft oil seal (see Section 26).

Installation

13   Wipe the bearing bore in the axle housing clean. Apply a thin coat of grease to the outer surface of the bearing.
14   Smear the lips of the axleshaft seal with grease, then guide the axle straight into the axle housing, being careful not to damage the seal.
15   Installation is the reverse of the removal procedure. Be sure to tighten the brake backing plate/bearing retainer nuts to the torque listed in this Chapter's Specifications. On Samurai models, bleed the brakes as described in Chapter 9.

---

22 Rear axle housing - removal and installation

Removal

1   Loosen the rear wheel lug nuts, raise the vehicle and support it securely on jackstands placed underneath the frame. Remove the wheels.

2   Support the rear axle assembly with a floor jack placed underneath the differential.
3   Disconnect the driveshaft from the differential pinion shaft yoke and hang the rear of the driveshaft from the underbody with a piece of wire (see Section 13).
4   Disconnect the parking brake cables from the levers on the backing plates (Samurai) or from the equalizer (Sidekick/X-90/-Viyara/Tracker) (see Chapter 9).
5   Disconnect the flexible brake hose from the junction block on the rear axle housing. Plug the end of the hose or wrap a plastic bag tightly around it to prevent excessive fluid loss and contamination.

Samurai

6   Remove the shock absorber lower mounting nuts and compress the shocks to get them out of the way (see Chapter 13).
7   Remove the U-bolt nuts from the leaf spring plates (see Chapter 10).
8   Raise the rear axle assembly slightly, then unbolt the springs from the shackles (see Chapter 10) and lower the rear ends of the springs to the floor.
9   Lower the jack and move the axle assembly out from under the vehicle.

Sidekick/X-90/Vitara/Tracker and Tracker models

10   Disconnect the breather hose from the axle housing.
11   Unbolt the upper arm from the differential housing (see illustration 20.8).
12   Loosen the trailing rod nuts, but don't remove them yet.
13   Remove the coil springs (see Chapter 10).
14   Remove the trailing rod nuts and bolts. Lower the jack and move the axle assembly out from under the vehicle.

Installation (all models)

15   Installation is the reverse of the removal procedure. Be sure to tighten the fasteners torque values listed in this Chapter's Specifications and the Chapter 10 Specifications, where applicable.

---

23 Rear axle oil seal - replacement

Refer to illustrations 23.2 and 23.4
1   Remove the axleshaft (see Section 21).
2   Remove the axle seal with a pry bar or seal removal tool (see illustration).
3   Wipe the seal bore in the axle housing clean, then apply a coat of grease to the seal lips.
4   Install the seal with a large socket or piece of pipe (see illustration).
5   Install the axleshaft.

---

24 Freewheel hub - removal and installation

1   Raise the front of the vehicle and place it securely on jackstands.

Manual locking hubs

Refer to illustrations 24.2 and 24.4
2   Set the hub cover to the Free position and apply match marks between the cover and hub body (see illustration).
3   Remove the hub cover mounting bolts and pull off the cover along with the clutch.
4   Using snap-ring pliers, remove the snap-ring from the end of the axle (see illustration).

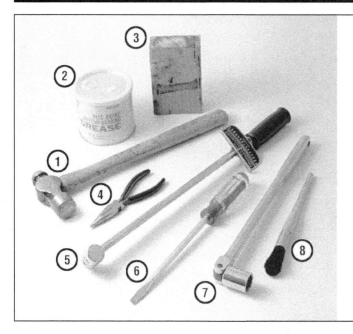

25.1 Tools and materials needed for front wheel bearing maintenance

1   **Hammer** - A common hammer will do just fine
2   **Grease** - High-temperature grease that is formulated specially for front wheel bearings should be used
3   **Wood block** - If you have a scrap piece of 2x4, it can be used to drive the new seal into the hub
4   **Needle-nose pliers** - Used to straighten and remove the cotter pin in the spindle
5   **Torque wrench** - This is very important in this procedure; if the bearing is too tight, the wheel won't turn freely - if it's too loose, the wheel will "wobble" on the spindle. Either way, it could mean extensive damage
6   **Screwdriver** - Used to remove the seal from the hub (a long screwdriver is preferred)
7   **Socket/breaker bar** - Needed to loosen the nut on the spindle if it's extremely tight
8   **Brush** - Together with some clean solvent, this will be used to remove old grease from the hub and spindle

25.7a With a hammer and chisel, bend back the locking tabs

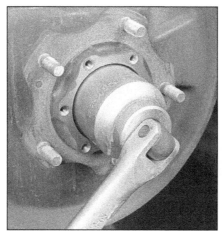

25.7b Loosen the front wheel bearing lock nut with a large socket

25.8 Remove the wheel bearing lock nut, lock washer, wheel bearing nut and spacer

5   Remove the mounting nuts from the freewheel hub body.
6   Pull the freewheel hub body from the wheel hub.
7   Installation is the reverse of the removal procedure. Use new gaskets and apply multi-purpose grease to the inner hub splines. The control handle should be set to the Free position and the cover should be attached to the body with the match marks aligned.
8   If match marks were not made or were wiped off during cleaning, align the cover to hub body so the two stopper nails are fitted freely into the wide slots in the hub body.

Automatic locking hubs

9   Remove the front hub cover.
10   Remove the circlip with a pair of snap-ring **pliers (see illustration 24.4)**.
11   Remove the bolts and remove the drive flange.
12   Installation is the reverse of removal.

25   Front wheel bearing check, repack and adjustment

Check

Refer to illustration 25.1

1   In most cases the front wheel bearings will not need servicing until the brake pads are changed. However, the bearings should be checked whenever the front of the vehicle is raised for any reason. Several items, including a torque wrench and special grease, are required for this procedure **(see illustration)**.
2   With the vehicle securely supported on jackstands, spin each wheel and check for noise, rolling resistance and free play.
3   Grasp the top of each tire with one hand and the bottom with the other. Move the wheel in-and-out on the spindle. If there's any noticeable movement, the bearings

should be checked and then repacked with grease or replaced if necessary.
4   Remove the wheel.
5   Remove the brake caliper and disc (see Chapter 9). Hang the caliper out of the way with a piece of wire.
6   Remove the freewheeling hub assembly (4WD models) (see Section 24).

Samurai models

Refer to illustrations 25.7a, 25.7b, 25.8 and 25.11

7   Straighten the bent ends of the lock washer, then remove the wheel bearing lock nut **(see illustrations)**.
8   Remove the wheel bearing nut and spacer from the end of the spindle **(see illustration)**.
9   Pull the hub assembly out slightly, then push it back into its original position. This should force the outer bearing off the spindle enough so it can be removed.

25.11 Use a screwdriver to pry the grease seal out of the hub

10   Pull the hub off the spindle.
11   Use a screwdriver to pry the seal out of the rear of the hub **(see illustration)**. As this is done, note how the seal is installed.
12   Remove the inner wheel bearing from the hub.

Sidekick/X-90/Vitara/Tracker models

13   Remove the front wheel bearing lock plate by removing the four screws.
14   Remove the front wheel bearing lock nut and thrust washer.
15   Remove the hub assembly from the spindle.
16   Use a screwdriver to pry out the inner wheel bearing seal.
17   Using a pair of snap-ring pliers, remove the wheel bearing snap-ring followed by the bearing assembly. It may be necessary to have the bearing assembly pressed from the hub.

Repack and adjustment

Samurai models

Refer to illustration 25.20
18   Use solvent to remove all traces of the old grease from the bearings, hub and spindle. A small brush may prove helpful; however make sure no bristles from the brush break off inside the bearing. Allow the parts to air dry.
19   Carefully inspect the bearings for

cracks, heat discoloration, worn rollers, etc. Check the bearing races inside the hub for wear and damage. If the bearing races are defective, the hubs should be taken to a machine shop with the facilities to remove the old races and press new ones in. Note that the bearings and races come as matched sets and old bearings should never be installed on new races.
20   Use high-temperature front wheel bearing grease to pack the bearings. Work the grease completely into the bearings, forcing it between the rollers, cone and cage from the back side **(see illustration)**.
21   Apply a thin coat of grease to the spindle at the outer bearing seat, inner bearing seat, shoulder and seal seat.
22   Put a small quantity of grease inboard of each bearing race inside the hub. Using your finger, form a dam at these points to provide extra grease availability and to keep thinned grease from flowing out of the bearing.
23   Place the grease-packed inner bearing into the rear of the hub.
24   Place a new seal over the inner bearing and tap the seal evenly into place with a hammer and block of wood or a large socket until it's flush with the hub.
25   Carefully place the hub assembly onto the spindle and push the grease-packed outer bearing into position.
26   Install the spacer and wheel bearing nut.
27   Tighten the nut to 57 ft-lbs while spinning the hub by hand. Spin the hub in a forward direction a few more times to seat the bearings and remove any grease or burrs which could cause excessive bearing play later.
28   Loosen the spindle nut until it's just loose, no more.
29   Using a torque wrench, tighten the nut to the torque listed in this Chapter's Specifications. Install the lock washer and the wheel bearing lock nut.
30   Tighten the lock nut to the torque listed in this Chapter's Specifications.
31   Bend the ends of the lock washer up until they're flat against the lock nut.

Sidekick/X-90/Vitara/Tracker models

Note: *1993 and later models use a sealed cartridge type wheel bearing which is replaced as a unit. These bearings require no adjustment as the preload is set by the lock-*

25.20 Work grease into the bearing rollers by pressing it against the palm of your hand

nut. Therefore it is important that the locknut is tightened to the torque listed in this Chapter's Specifications.
32   Lubricate the wheel bearing assembly with high temperature wheel bearing grease. Install the bearing into the hub, followed by the snap-ring.
33   Install the hub seal and wheel bearing seal and coat the lips of the seals with wheel bearing grease.
34   Pack the area behind the seal lips with wheel bearing grease, and apply a light coat of the same grease to the spindle.
35   Install the hub assembly on the spindle.
36   Install the spindle thrust washer.
37   Install the wheel bearing nut and tighten it to the torque listed in this Chapter's Specifications.
38   Install the bearing lock plate and screws. If the screw holes don't line up, tighten the wheel bearing nut a little more until they are aligned Installation (all models).
39   Install the brake disc and caliper (see Chapter 9).
40   Install the freewheeling hub (4WD models) (see Section 24).
41   Install the tire/wheel assembly on the hub and tighten the lug nuts to the torque listed in the Chapter 1 Specifications.
42   Grasp the top and bottom of the tire and check the bearings in the manner described earlier in this Section.
43   Lower the vehicle.

Chapter 9 Brakes

Contents

Specifications

General

Brake fluid type	See Chapter 1
Brake light switch plunger clearance	1/64 to 1/32-inch (.5 to 1 mm)

Disc brakes

Disc thickness*	
Two-door models	
Standard	0.394 in (10.0 mm)
Minimum	0.315 in (8.0 mm)
Four-door models	
Standard	0.670 in (17.0 mm)
Minimum	0.590 in (15.0 mm)
Disc runout (maximum)	0.006 in (0.15 mm)
Brake pad minimum thickness	See Chapter 1

* Refer to marks cast in the disc (they supersede information printed here)

Drum brakes

Drum diameter*	
Two-door models	
Standard	8.66 in (220 mm)
Service limit	8.74 in (222 mm)
Four-door models	
Standard	10.0 in (254 mm)
Service limit	10.07 in (256 mm)
Minimum brake lining thickness	See Chapter 1

* Refer to marks cast in the drum (they supersede information printed here)

Parking brake lever travel

Samurai	3 to 8 clicks
Sidekick/X-90/Vitara and Tracker	7 to 9 clicks

Brake light switch clearance
1995 and earlier models .... 0.02 to 0.04 inch
1996 and later models .... 0.06 to 0.08 inch

Power brake booster
Booster-to-master cylinder clearance
    Without ABS... 0.010 to 0.020 in (0.25 to 0.50 mm)
    With ABS.. 0.006 to 0.014 in (0.14 to 0.35 mm)
Booster pushrod length
    1998 and earlier models... 4.94 to 4.98 in (125.5 to 126.5 mm)
    1999 and later models.. 4.31 to 4.35 in (109.5 to 110.5 mm)

Torque specifications   Ft-lbs (unless otherwise indicated)
Brake bleeder screw... 96 in-lbs
Power brake booster mounting nuts .... 84 to 144 in-lbs
Master cylinder mounting nuts .... 84 to 144 in-lbs
Caliper mounting bolts .... 18 to 20
Caliper carrier bolts .... 51 to 72
Brake hose-to-caliper inlet fitting bolt .... 14 to 18
Wheel cylinder mounting bolts .... 72 to 108 in-lbs

1   General information

The vehicles covered by this manual are equipped with hydraulically operated front and rear brake systems. The front brakes are disc type and the rear brakes are drum type. Both the front and rear brakes are self-adjusting. The front disc brakes automatically compensate for pad wear, while the rear drum brakes incorporate an adjustment mechanism which is activated as the brakes are applied when the vehicle is driven in reverse.

Hydraulic system

The hydraulic system consists of two separate circuits. The master cylinder has separate reservoirs for the two circuits and in the event of a leak or failure in one hydraulic circuit, the other circuit will remain operative. A visual warning of circuit failure or air in the system is given by a warning light activated by displacement of the piston in the pressure differential switch portion of the combination valve from its normal "in balance" position.

Proportioning and bypass valve (Samurai only)

A proportioning and bypass valve, located in the engine compartment below the master cylinder, consists of three sections providing the following functions. The metering section limits pressure to the front brakes until a predetermined front input pressure is reached and until the rear brakes are activated. There is no restriction at inlet pressures below 3 psi, allowing pressure equalization during non-braking periods. The proportioning section proportions outlet pressure to the rear brakes after a predetermined rear input pressure has been reached, preventing early rear wheel lock-up under heavy brake loads. The valve is also designed to assure full pressure to one brake system should the other system fail.

Load sensing proportioning valve (LSPV) - Sidekick/X-90/ Vitara and Tracker

The LSPV is included within the hydraulic circuit which connects the master cylinder and the rear brakes. The LSPV controls the hydraulic pressure applied to the rear brakes according to the loaded state of the vehicle, preventing the rear wheels from locking prematurely.

If pressure drops in the front brake hydraulic circuit, the LSPV releases this control over the rear brake hydraulic pressure. This allows maximum braking effort at the rear if the front brakes fail.

Power brake booster

The power brake booster, utilizing engine manifold vacuum and atmospheric pressure to provide assistance to the hydraulically operated brakes, is mounted on the firewall in the engine compartment.

Parking brake

The parking brake operates the rear brakes only, through cable actuation. It's activated by a lever located between the front seats.

Service

After completing any operation involving disassembly of any part of the brake system, always test drive the vehicle to check for proper braking performance before resuming normal driving. When testing the brakes, perform the tests on a clean, dry flat surface. Conditions other than these can lead to inaccurate test results.

Test the brakes at various speeds with both light and heavy pedal pressure. The vehicle should stop evenly without pulling to one side or the other. Avoid locking the brakes because this slides the tires and diminishes braking efficiency and control of the vehicle.

Tires, vehicle load and front-end alignment are factors that also affect braking performance.

2   Disc brake pads (Samurai) - replacement

Refer to illustrations 2.5 and 2.6a through 2.6h

Warning: *Disc brake pads must be replaced on both front wheels at the same time - never replace the pads on only one wheel. Also, the dust created by the brake system may contain asbestos, which is harmful to your health. Never blow it out with compressed air and don't inhale any of it. An approved filtering mask should be worn when working on the brakes. Do not, under any circumstances, use petroleum-based solvents to clean brake parts. Use brake cleaner or denatured alcohol only!*

1   Remove the cover from the brake fluid reservoir.
2   Loosen the wheel lug nuts, raise the front of the vehicle and support it securely on jackstands.
3   Remove the front wheels. Work on one brake assembly at a time, using the assembled brake for reference if necessary.
4   Inspect the brake disc carefully as outlined in Section 5. If machining is necessary, follow the information in that Section to remove the disc, at which time the pads can be removed from the calipers as well.
5   Push the piston back into the bore to provide room for the new brake pads. A C-clamp can be used to accomplish this **(see illustration)**. As the piston is depressed to the bottom of the caliper bore, the fluid in the master cylinder will rise. Make sure it doesn't overflow. If necessary, siphon off some of the fluid.
6   Follow **the accompanying illustrations, beginning with 2.6a,** for the actual pad removal procedure. Be sure to stay in

2.5 Using a large C-clamp, push the piston back into the caliper bore - note that one end of the clamp is on the flat area on the backside of the caliper and the other end (screw end) is pressing against the outer brake pad

2.6a Before removing the caliper, wash off all traces of brake dust with brake system cleaner

2.6b Using a screwdriver, remove the anti-rattle clip

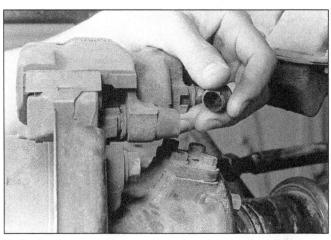

2.6c Remove the rubber caps to gain access to the caliper bolts

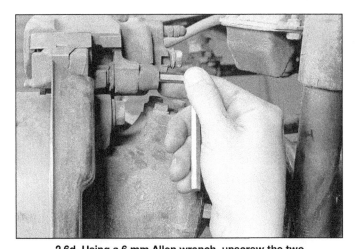

2.6d Using a 6 mm Allen wrench, unscrew the two caliper mounting bolts

2.6e Using a screwdriver, remove the pad protector clips and lift off the caliper

order and read the caption under each illustration. Before installing the new brake pads, it's a good idea to apply anti-squeal compound to the backing plates. Follow the manufacturer's instructions.

7   When reinstalling the caliper, be sure to tighten the mounting bolts to the specified torque. After the job has been completed, firmly depress the brake pedal a few times to bring the pads into contact with the disc,

then check the brake fluid level.
8   Check for fluid leakage and make sure the brakes operate normally before driving in traffic.

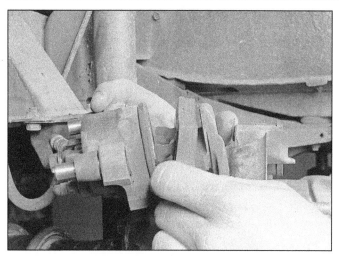

2.6f Pull the inner pad straight out, disengaging the retainer spring from the caliper piston

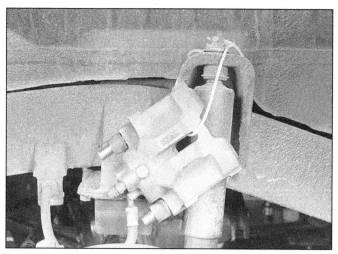

2.6g Hang the caliper with a piece of wire - DON'T let it hang by the brake hose

2.6h Slide the outer brake pad out of the caliper holder - as you do this, note how it fits into the frame so you can install the new one the same way

3.6a To detach the caliper from the support bracket, remove these two bolts (wrench on lower bolt, arrow on upper bolt)

3.6b Remove the brake pads - inner shown

3.6c Apply anti-squeal compound to the outside of the anti-squeal shim (outer pad) and a small amount to the inner pad where it will contact the piston - reinstall the caliper in the reverse order of removal

4.8 With the caliper padded to catch the piston, use compressed air to force the piston out of its bore - make sure your hands and fingers are not between the piston and caliper frame!

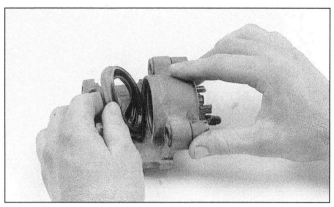

4.9 Remove the dust boot from the caliper

3 Disc brake pads - replacement (Sidekick/X-90/Vitara and Tracker)

Refer to illustrations 3.6a through 3.6c
Warning: *Disc brake pads must be replaced on both front wheels at the same time - never replace the pads on only one wheel. Also, the dust created by the brake system may contain asbestos, which is harmful to your health. Never blow it out with compressed air and don't inhale any of it. An approved filtering mask should be worn when working on the brakes. Do not, under any circumstances, use petroleum-based solvents to clean brake parts. Use brake cleaner or denatured alcohol only!*

1 Remove the cover from the brake fluid reservoir.
2 Loosen the wheel lug nuts, raise the front of the vehicle and support it securely on jackstands.
3 Remove the front wheels. Work on one brake assembly at a time, using the assembled brake for reference if necessary.
4 Inspect the brake disc carefully as outlined in Section 5. If machining is necessary, follow the information in that Section to remove the disc, at which time the pads can be removed from the calipers as well.
5 Push the piston back into the bore to provide room for the new brake pads. A C-clamp can be used to accomplish this **(see illustration 2.5).** As the piston is depressed to the bottom of the caliper bore, the fluid in the master cylinder will rise. Make sure it doesn't overflow. If necessary, siphon off some of the fluid.
6 Before removing the caliper, wash off all traces of brake dust with brake system cleaner **(see illustration 2.6a).** Follow the **accompanying illustrations, beginning with 3.6a** for the actual pad removal procedure. Be sure to stay in order and read the caption under each illustration. Before installing the new brake pads, it's a good idea to apply anti-squeal compound to the backing plates. Follow the manufacturer's instructions.
7 When reinstalling the caliper, be sure to tighten the mounting bolts to the specified

torque. After the job has been completed, firmly depress the brake pedal a few times to bring the pads into contact with the disc, then check the brake fluid.
8 Check for fluid leakage and make sure the brakes operate normally before driving in traffic.

4 Disc brake caliper - removal, overhaul and installation

Warning: *Dust created by the brake system may contain asbestos which is harmful to your health. Never blow it out with compressed air and don't inhale any of it. An approved filtering mask should be worn when working on the brakes. Do not, under any circumstances, use petroleum-based solvents to clean brake parts. Use brake cleaner or denatured alcohol only!*
Note: *If an overhaul is indicated (usually because of fluid leakage) explore all options before beginning the job. New and factory rebuilt calipers are available on an exchange basis, which makes this job quite easy. If it's decided to rebuild the calipers, make sure a rebuild kit is available before proceeding. Always rebuild the calipers in pairs - never rebuild just one of them.*

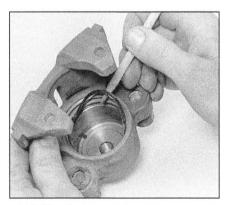

4.10 The piston seal should be removed with a plastic or wooden tool to avoid damage to the bore and seal groove - a pencil will do the job

Removal

1 Remove the cover from the brake fluid reservoir, siphon off two-thirds of the fluid into a container and discard it.
2 Loosen the wheel lug nuts, raise the front of the vehicle and support it securely on jackstands. Remove the front wheels.
3 Bottom the piston in the caliper bore **(see illustration 2.5).**
4 **Note:** *Do not remove the brake hose from the caliper if you are only removing the caliper.* Remove the brake hose inlet fitting bolt and detach the hose. Have a rag handy to catch spilled fluid and wrap a plastic bag tightly around the end of the hose to prevent fluid loss and contamination.
5 Remove the two mounting bolts and detach the caliper from the vehicle (refer to Section 2 or 3 if necessary).

Overhaul

Refer to illustrations 4.8, 4.9, 4.10, 4.11, 4.15, 4.17 and 4.18
6 Refer to Section 2 and remove the brake pads from the caliper.
7 Clean the exterior of the caliper with brake cleaner or denatured alcohol. Never use gasoline, kerosene or petroleum-based cleaning solvents. Place the caliper on a clean workbench.
8 Position a wooden block or several shop rags in the caliper as a cushion, then use compressed air to remove the piston from the caliper **(see illustration).** Use only enough air pressure to ease the piston out of the bore. If the piston is blown out, even with the cushion in place, it may be damaged.
Warning: *Never place your fingers in front of the piston in an attempt to catch or protect it when applying compressed air, as serious injury could occur.*
9 Carefully remove the dust boot out of the caliper bore **(see illustration).**
10 Using a wood or plastic tool, remove the piston seal from the groove in the caliper bore **(see illustration).** Metal tools may cause bore damage.
11 Remove the caliper bleeder screw, then remove and discard the bushings from the

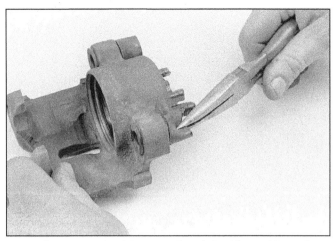

4.11 Grab the ends of the mounting pin bushings with needle-nose pliers and, using a twisting motion, push them through the caliper ears

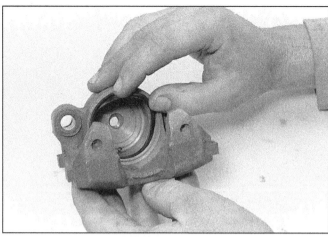

4.15 Push the new seal into the groove with your fingers, then check to see that it is not twisted or kinked

caliper ears. Discard all rubber parts **(see illustration)**.

12   Clean the remaining parts with brake system cleaner or denatured alcohol then blow them dry with compressed air.

13   Carefully examine the piston for nicks and burrs and loss of plating. If surface defects are present, the parts must be replaced.

14   Check the caliper bore in a similar way. Light polishing with crocus cloth is permissible to remove light corrosion and stains. Discard the mounting bolts if they're corroded or damaged.

15   When assembling, lubricate the piston bores and seal with clean brake fluid. Position the seal in the caliper bore groove **(see illustration)**.

16   Lubricate the piston with clean brake fluid, then install a new boot in the piston groove with the fold toward the open end of the piston.

17   Insert the piston squarely into the caliper bore, then apply force to the piston and move it about half way down **(see illustration)**.

18   Install the lip of the dust boot in the groove on the piston **(see illustration)**. Make sure the boot is recessed evenly below the caliper face. Push the piston the rest of the way down.

19   Install the bleeder screw.

20   Lubricate the new bushings with silicone grease and install them in the mounting bolt holes.

Installation

21   Inspect the mounting bolts for excessive corrosion.

22   Place the caliper in position over the rotor and mounting bracket, install the bolts and tighten them to the specified torque.

23   Install the brake hose and inlet fitting bolt, using new copper washers, then tighten the bolt to the specified torque.

24   If the line was disconnected, be sure to bleed the brakes (Section 11).

25   Install the wheels and lower the vehicle.

26 After the job has been completed, firmly depress the brake pedal a few times to bring the pads into contact with the disc.

27   Check brake operation before driving the vehicle in traffic.

5   Brake disc - inspection, removal and installation

Refer to illustrations 5.3, 5.4a, 5.4b 5.5, 5.6 and 5.7

Inspection

1   Loosen the wheel lug nuts, raise the vehicle and support it securely on jackstands. Remove the wheel and install two lug nuts to hold the disc in place.

2   Remove the brake caliper as outlined in Section 4. It's not necessary to disconnect the brake hose. After removing the caliper bolts, suspend the caliper out of the way with a piece of wire.

3   Visually inspect the disc surface for score marks and other damage. Light scratches and shallow grooves are normal after use and may not always be detrimental to brake operation, but deep score marks - over 0.015-inch (0.38 mm) - require disc removal and refinishing by an automotive machine shop. Be sure to check both sides of the disc **(see illustration)**. If pulsating has

4.17 Position the piston squarely and push it straight into the caliper

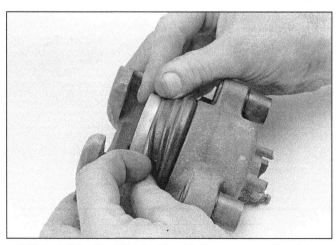

4.18 Install the lip of the dust boot in the groove on the piston

5.3 The brake pads on this vehicle were obviously neglected, as they wore down to the rivets and cut deep grooves into the disc - wear this severe will require replacement of the disc

5.4a To check disc runout, mount a dial indicator as shown and rotate the disc

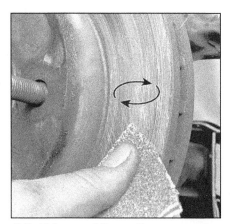

5.4b Using a swirling motion, remove the glaze from the disc surface with sandpaper or emery cloth

5.5 Use a micrometer to measure disc thickness

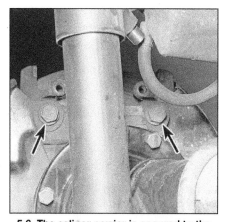

5.6 The caliper carrier is secured to the steering knuckle with two bolts (arrows)

been noticed during application of the brakes, suspect disc runout. Be sure to check the wheel bearings to make sure they're properly adjusted.

5.7 To help free the disc, thread bolts of the appropriate size into the two holes provided in the disc - alternate between the bolts, turning them a little at a time, until the disc is free

4   To check disc runout, place a dial indicator at a point about 1/2-inch from the outer edge of the disc **(see illustration)**. Set the indicator to zero and turn the disc. The indicator reading should not exceed the specified allowable runout limit. If it does, the disc should be refinished by an automotive machine shop. **Note:** *Professionals recommend resurfacing of brake discs regardless of the dial indicator reading (to produce a smooth, flat surface that will eliminate brake pedal pulsations and other undesirable symptoms related to questionable discs). At the very least, if you elect not to have the discs resurfaced, deglaze them with sandpaper or emery cloth (use a swirling motion to ensure a nondirectional finish)* **(see illustration)**.

5   The disc must not be machined to a thickness less than the minimum thickness cast into the inside of the disc. The disc thickness can be checked with a micrometer **(see illustration)**. If the minimum thickness is not cast into the disc, refer to this Chapter's Specifications.

Removal

6   Remove the two caliper carrier-to-steer-

ing knuckle bolts and remove the carrier **(see illustration)**.

7   Remove the two lug nuts which were put on to hold the disc in place and remove the disc from the hub. If the disc is stuck to the hub and won't come off, thread two bolts into the holes provided **(see illustration)**.

Installation

8   Place the disc in position over the threaded studs.

9   Install the caliper carrier and tighten the bolts to the torque listed in this Chapter's Specifications.

10   Install the caliper and brake pad assembly over the disc and position it on the steering knuckle (refer to Section 4 for the caliper installation procedure, if necessary). Tighten the caliper bolts to the torque listed in this Chapter's Specifications.

11   Install the wheel, then lower the vehicle to the ground. Depress the brake pedal a few times to bring the brake pads into contact with the disc. Bleeding of the system will not be necessary unless the brake hose was disconnected from the caliper. Check the operation of the brakes carefully before placing the vehicle into normal service.

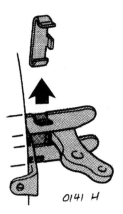

6.4a To retract the brake shoes on Samurai models, disconnect the parking brake cable (Section 14), then remove the stopper plate, as shown

6.4b Before removing anything, clean the brake assembly with brake cleaner and allow it to dry - position a drain pan under the brake to catch the fluid and residue - DO NOT USE COMPRESSED AIR TO BLOW THE DUST FROM THE PARTS!

6   Drum brake shoes - replacement

Refer to illustrations 6.4a through 6.4u
Warning: *Drum brake shoes must be replaced on both wheels at the same time - never replace the shoes on only one wheel. Also, the dust created by the brake system*

6.4c Using a pair of pliers, unhook the long return spring from the front brake shoe . . .

may contain asbestos, which is harmful to your health. Never blow it out with compressed air and don't inhale any of it. An approved filtering mask should be worn when working on the brakes. Do not, under any circumstances, use petroleum-based solvents to clean brake parts. Use brake cleaner or denatured alcohol only!
Caution: *Whenever the brake shoes are replaced, the retractor and hold-down springs should also be replaced. Due to the continuous heating/cooling cycle that the springs are subjected to, they lose their tension over a period of time and may allow the shoes to drag on the drum and wear at a much faster rate than normal. When replacing the rear brake shoes, use only high quality nationally recognized brand-name parts.*

1   Loosen the wheel lug nuts, raise the rear of the vehicle and support it securely on jackstands. Block the front wheels to keep the vehicle from rolling.
2   Release the parking brake.
3   Remove the wheel. **Note:** *All four rear brake shoes must be replaced at the same time, but to avoid mixing up parts, work on only one brake assembly at a time.*
4   Follow **the accompanying illustrations (6.4a through 6.4u)** for the inspection and replacement of the brake shoes. Be sure to

stay in order and read the caption under each illustration. **Note:** *If the brake drum cannot be easily pulled off the axle and shoe assembly (see Chapter 1), make sure that the parking brake is completely released, then apply some penetrating oil at the hub-to-drum joint. Allow the oil to soak in and try to pull the drum off. If the drum still cannot be pulled off, the brake shoes will have to be retracted. On Samurai models, to retract the brake shoes, disconnect the parking brake cable (Section 14) then remove the parking brake lever stopper plate* **(see illustration 6.4a)**. *On Sidekick/X90/Vitara and Tracker models, to retract the brake shoes, pull the shoe hold down pin about 1/4-inch. The drum should now come off.*
5   Before reinstalling the drum it should be checked for cracks, score marks, deep scratches and hard spots, which will appear as small discolored areas. If the hard spots cannot be removed with fine emery cloth or if any of the other conditions listed above exist, the drum must be taken to an automotive machine shop to have it turned. **Note:** *Professionals recommend resurfacing the drums whenever a brake job is done. Resurfacing will eliminate the possibility of out-of-round drums. If the drums are worn so much that they can't be resurfaced without exceeding*

6.4d . . . then remove the short return spring

6.4e On the rear shoe, using a pair of pliers, push in on the hold-down spring, turn it 90-degrees and release it

6.4f Unhook the rear shoe from the brake strut rod (arrow)

6.4g Unhook the shoe return spring from the front brake shoe

6.4h On Sidekick/X-90/Vitara and Tracker models, grasp the end of the parking brake cable with a pair of pliers and pull it out of the parking brake lever

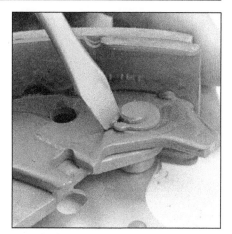

6.4i On Sidekick/X-90/Vitara and Tracker models, pry the C-washer apart and remove it to separate the parking brake lever and adjusting lever from the rear shoe

6.4j Using a pair of pliers, remove the hold-down spring on the front shoe

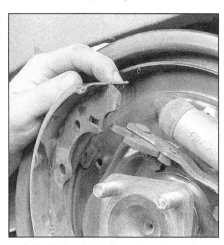

6.4k Unhook the front shoe from the brake strut

6.4l Lubricate the brake shoe contact areas with high-temperature grease

the maximum allowable diameter (stamped into the drum), then new ones will be required. At the very least, if you elect not to have the drums resurfaced, remove the glaz-

ing from the surface with medium-grit emery cloth using a swirling motion.
6   Install the brake drum on the axle flange.
7   Mount the wheel, install the lug nuts,

then lower the vehicle.
8   Make a number of forward and reverse stops to adjust the brakes until satisfactory pedal action is obtained.

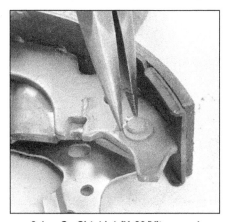

6.4m On Sidekick/X-90/Vitara and Tracker models, assemble the parking brake lever and adjuster lever on the new rear shoe and crimp the C washer closed with a pair of pliers

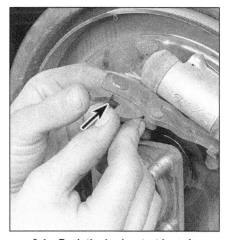

6.4n Push the brake strut lever in about halfway

6.4o Hook the new front shoe on the brake strut and install the hold-down spring

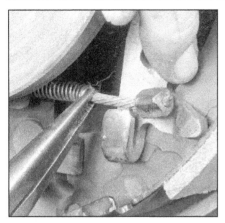

6.4p On Sidekick/X-90/Vitara and Tracker models, pull the parking brake cable spring back and hold it there with a pair of pliers, then place the cable into the hooked end of the parking brake lever

6.4q Connect the shoe return spring to both shoes and make sure the spring and shoe is installed behind the anchor plate

6.4r Hook the rear shoe to the brake strut and install the hold-down spring - hook the long and short return springs to both shoes

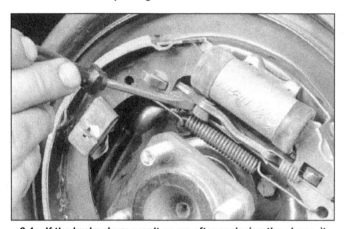

6.4s If the brake drum won't go on after replacing the shoes, it may be necessary to retract the shoes - put a screwdriver between the two parts of the ratchet and push in - spring pressure will retract the shoes

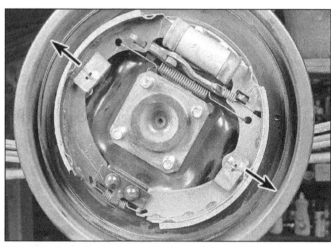

6.4t Wiggle the brake assembly (arrows) to make sure it's seated properly against the backing plate

9   Check brake operation before driving the vehicle in traffic.

7   Wheel cylinder - removal, overhaul and installation

Note: *If an overhaul is indicated (usually because of fluid leakage or sticky operation) explore all options before beginning the job.* New wheel cylinders are available, which makes this job quite easy. If it's decided to rebuild the wheel cylinder, make sure that a rebuild kit is available before proceeding. Never overhaul only one wheel cylinder - always rebuild both of them at the same time.

Removal

Refer to illustration 7.4

1   Raise the rear of the vehicle and support it securely on jackstands. Block the front wheels to keep the vehicle from rolling.
2   Remove the brake shoe assembly (Section 6).

3   Remove all dirt and foreign material from around the wheel cylinder.
4   Unscrew the bleeder screw and the brake line fitting **(see illustration)**. Don't pull

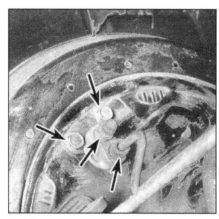

7.4 Remove the brake bleeder screw and the brake line fitting, then remove the two wheel cylinder bolts (arrows)

the brake line away from the wheel cylinder.
5   Remove the wheel cylinder mounting bolts.
6   Detach the wheel cylinder from the brake backing plate and place it on a clean workbench. Immediately plug the brake line to prevent fluid loss and contamination. **Note:** *If the brake shoe linings are contaminated with brake fluid, install new brake shoes.*

Overhaul

Refer to illustration 7.7

7   Remove the cups, pistons, boots and spring assembly from the wheel cylinder body **(see illustration)**.
8   Clean the wheel cylinder with brake fluid, denatured alcohol or brake system cleaner. **Warning:** *Do not, under any circumstances, use petroleum based solvents to clean brake parts!*
9   Use compressed air to remove excess fluid from the wheel cylinder and to blow out the passages.
10   Check the cylinder bore for corrosion

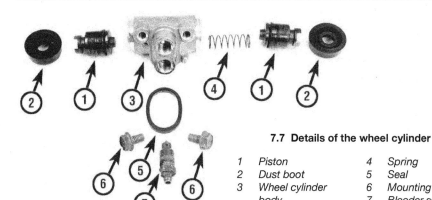

7.7 Details of the wheel cylinder

1	Piston	4	Spring
2	Dust boot	5	Seal
3	Wheel cylinder body	6	Mounting bolts
		7	Bleeder screw

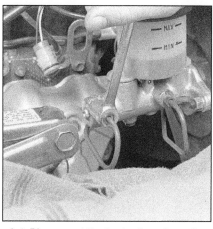

8.4 Disconnect the brake lines from the master cylinder - a flare nut wrench should be used

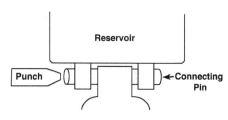

8.8b On Sidekick/X-90/Vitara and Tracker models, use a punch and hammer to knock out the connecting pin

8.6 After disconnecting the electrical connector and the brake lines, remove the two master cylinder mounting nuts (arrows)

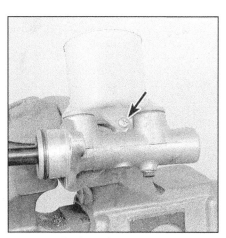

8.8a On Samurai models, remove the reservoir mounting screw (arrow), then pull the reservoir out of the grommets

and score marks. Crocus cloth can be used to remove light corrosion and stains, but the cylinder must be replaced with a new one if the defects cannot be removed easily, or if the bore is scored.

11   Lubricate the new cups with brake fluid.

12   Assemble the wheel cylinder components. Make sure the cup lips face in **(see illustration 7.7)**.

Installation

13   Place the wheel cylinder in position and install the bolts. Tighten them to the torque listed in this Chapter's Specifications.

14   Connect the bleeder screw and the brake line and tighten the fitting. Install the brake shoe assembly.

15   Bleed the brakes (Section 11).

16   Check brake operation before driving the vehicle in traffic.

8   Master cylinder - removal, overhaul and installation

Refer to illustrations 8.4, 8.6, 8.8a, 8.8b, 8.9, 8.10, 8.11a, 8.11b and 8.11c

Note: *Before deciding to overhaul the master cylinder, check on the availability and cost of*

a new or factory rebuilt unit and also the availability of a rebuild kit.

Removal

1   The master cylinder is located in the engine compartment, mounted to the power brake booster.

2   Remove as much fluid as you can from the reservoir with a syringe.

3   Place rags under the fluid fittings and prepare caps or plastic bags to cover the ends of the lines once they are disconnected. **Caution:** *Brake fluid will damage paint. Cover all body parts and be careful not to spill fluid during this procedure.*

4   Loosen the tube nuts at the ends of the brake lines where they enter the master cylinder **(see illustration)**. If possible, use a flare-nut wrench because it wraps securely around the flats of the nuts and prevents rounding them off.

5   Pull the brake lines slightly away from the master cylinder and plug the ends to prevent contamination.

6   Disconnect the electrical connector at the master cylinder, then remove the two nuts attaching the master cylinder to the power booster **(see illustration)**. Pull the master cylinder off the studs and out of the engine compartment. Again, be careful not to spill the fluid as this is done.

Overhaul

7   Before attempting the overhaul of the master cylinder, obtain the proper rebuild kit, which will contain the necessary replacement parts and also any instructions which may be specific to your model.

8   Inspect the reservoir grommets for indications of leakage near the base of the reservoir. Remove the reservoir **(see illustrations)**.

9   Place the cylinder in a vise and use a punch or Phillips screwdriver to depress the pistons until they bottom against the other end of the master cylinder **(see illustration)**. Hold the pistons in this position and remove the stop screw on the side of the master cylinder.

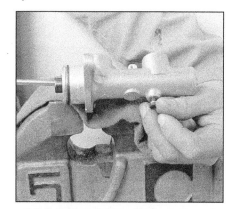

8.9 Push the pistons all the way in and remove the piston stop screw

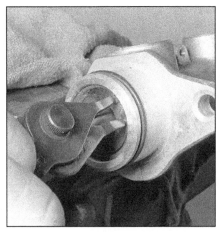

8.10 While still holding the pistons in, use snap-ring pliers to remove the snap-ring

8.11a Remove the primary piston assembly from the master cylinder bore

8.11b To remove the secondary piston assembly, tap the master cylinder firmly against a block of wood

10   Carefully remove the snap-ring at the end of the master cylinder **(see illustration)**.

11   The internal components can now be removed from the cylinder bore **(see illustrations)**. Make a note of the proper order of the components so they can be returned to their original locations. **Note:** *The two springs are of different tension, so pay particular attention to their order. Also, do not disassemble the piston components - they are serviced as an assembly.*

12   Carefully inspect the bore of the master cylinder. Any deep scoring or other damage will mean a new master cylinder is required.

13   Replace all parts included in the rebuild kit, following any instructions in the kit. Clean all reused parts with clean brake fluid or denatured alcohol. Do not use any petroleum-based cleaners. During assembly, lubricate all parts liberally with clean brake fluid.

14   Push the assembled components into the bore, bottoming them against the end of the master cylinder, then install the stop screw. Be sure to install a new gasket on the stop screw.

15   Install the new snap-ring, making sure it is seated properly in the groove.

16   Before installing the master cylinder it should be bench bled. Because it will be necessary to apply pressure to the master cylinder piston and, at the same time, control flow from the brake line outlets, it is recommended that the master cylinder be mounted in a vise, with the jaws of the vise clamping on the mounting flange. The master cylinder is aluminum, so be careful not to damage it.

17   Insert threaded plugs into the brake line outlet holes and snug them down so that there will be no air leakage past them, but not so tight that they cannot be easily loosened.

18   Fill the reservoir with brake fluid of the recommended type (see Chapter 1).

19   Remove one plug and push the piston assembly into the master cylinder bore to expel the air from the master cylinder. A large Phillips screwdriver can be used to push on the piston assembly.

20   To prevent air from being drawn back into the master cylinder the plug must be replaced and snugged down before releasing the pressure on the piston assembly.

21   Repeat the procedure until only brake fluid is expelled from the brake line outlet hole. When only brake fluid is expelled, repeat the procedure with the other outlet hole and plug. Be sure to keep the master cylinder reservoir filled with brake fluid to prevent the introduction of air into the system.

22   Since high pressure is not involved in the bench bleeding procedure, an alternative to the removal and replacement of the plugs with each stroke of the piston assembly is available. Before pushing in on the piston assembly, remove the plug as described in Step 19. Before releasing the piston, however, instead of replacing the plug, simply put your finger tightly over the hole to keep air from being drawn back into the master cylinder. Wait several seconds for brake fluid to be drawn from the reservoir into the piston bore, then depress the piston again, removing your finger as brake fluid is expelled. Be sure to put your finger back over the hole each time before releasing the piston, and when the bleeding procedure is complete for that outlet, replace the plug and snug it before going on to the other port.

Installation

23   Install the master cylinder over the studs on the power brake booster and tighten the attaching nuts only finger tight at this time.

24   Thread the brake line fittings into the master cylinder. Since the master cylinder is still a bit loose, it can be moved slightly for the fittings to thread in easily. Do not strip the threads as the fittings are tightened.

25   Tighten the mounting nuts to the torque listed in this Chapter's Specifications, then tighten the fittings securely.

26   Fill the master cylinder reservoir with fluid, then bleed the master cylinder (only if the cylinder has not been bench bled) and the brake system as described in Section 11. To bleed the cylinder on the vehicle, have an assistant pump the brake pedal several times and then hold the pedal to the floor. Loosen the fitting nut to allow air and fluid to escape. Repeat this procedure on both fittings until

8.11c Exploded view of the master cylinder

1   *Snap-ring*
2   *Piston stopper*
3   *Piston stopper seals*
4   *Primary piston*
5   *Piston cup*
6   *Secondary piston pressure cup*
7   *Piston cup*
8   *Secondary piston*
9   *Secondary piston return spring seat*
10   *Spring*
11   *Secondary piston stopper bolt*
12   *Master cylinder body*

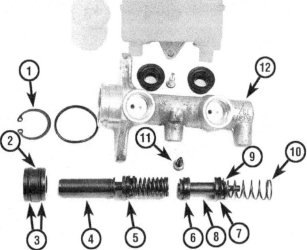

the fluid is clear of air bubbles. Test the operation of the brake system carefully before placing the vehicle into normal service.

9 Load sensing proportioning valve (LSPV) - removal and installation (Sidekick/X-90/Vitara and Tracker)

Note: *Special tools, test equipment and skills are required to diagnose problems with the LSPV. The home mechanic should not attempt such diagnosis; take the vehicle to a dealer service department or other properly equipped shop. If the LSPV is found to be faulty, the home mechanic can replace it using the procedure below. The LSPV must be replaced as a unit; do not disassemble it.*

1   Remove as much brake fluid as you can from the master cylinder reservoir with a syringe.
2   Raise the rear of the vehicle and support it securely on jackstands. Block the front wheels to keep the vehicle from rolling.
3   Remove the brake lines from the LSPV.
4   Remove the LSPV assembly, along with the load sensing spring and stay, as a unit.
5   Before installing the new LSPV, apply multi-purpose grease to the upper and lower joint of the load sensing spring.
6   Installation is the reverse of removal.
7   Fill the master cylinder reservoir with brake fluid and bleed the air from the brake system (see Section 11).

10 Brake hoses and lines - inspection and replacement

Inspection

1   About every six months, with the vehicle raised and supported securely on jackstands, the rubber hoses which connect the steel brake lines with the front and rear brake assemblies should be inspected for cracks, chafing of the outer cover, leaks, blisters and other damage. These are important and vulnerable parts of the brake system and inspection should be complete. A light and mirror will be helpful for a thorough check. If a hose exhibits any of the above conditions, replace it with a new one.

Replacement

Front and rear brake hose

Refer to illustrations 10.2 and 10.3

2   Disconnect the brake line from the hose fitting on the frame bracket, being careful not to bend the frame bracket or brake line **(see illustration)**.
3   Use a pair of pliers or a screwdriver to remove the U-clip from the female fitting at the frame bracket, then detach the hose from the bracket **(see illustration)**.
4   If you are replacing a front brake hose, remove the union bolt from the fitting at the

10.2 On Samurai models, use a flare-nut wrench to prevent rounding off the corners of the nut

caliper end of the hose, then separate the hose from the caliper. Note that there are two copper sealing washers on either side of the fitting - they should be replaced with new ones upon installation.
5   If you are replacing the rear brake hose, disconnect it at the axle end the same way you disconnected it at the frame end.
6   If you are installing a front hose, pass the caliper fitting end through the bracket, then connect the fitting to the caliper with the union bolt and copper washers. Make sure that the locating lug on the fitting is engaged with the hole in the caliper, then tighten the fitting to the specified torque.
7   To attach a front hose to the frame bracket, or to attach either end of the rear hose, install the female fitting of the hose in the bracket. It will fit the bracket in only one position. Be careful not to twist the hose.
8   Install the U-clip retaining the female fitting to the bracket.
9   Attach the brake line to the hose fitting.
10   When the brake hose installation is complete, there should be no kinks in the hose. Make sure the hose doesn't contact any part of the suspension. Check this by turning the wheels to the extreme left and right positions. If the hose makes contact, remove it and correct the installation as necessary. Bleed the system (Section 11).

Metal brake lines

11   When replacing brake lines be sure to use the correct parts. Don't use copper tubing for any brake system components. Purchase steel brake lines from a dealer or auto parts store.
12   Prefabricated brake line, with the tube ends already flared and fittings installed, is available at auto parts stores and dealers. These lines are also bent to the proper shapes.
13   When installing the new line make sure it's securely supported in the brackets and has plenty of clearance between moving or hot components.
14   After installation, check the master

10.3 Using a pair of pliers or a screwdriver, pry the U-clip off the hose

cylinder fluid level and add fluid as necessary. Bleed the brake system as outlined in the next Section and test the brakes carefully before driving the vehicle in traffic.

11 Brake system bleeding

Refer to illustration 11.8

Warning: *Wear eye protection when bleeding the brake system. If the fluid comes in contact with your eyes, immediately rinse them with water and seek medical attention.*

Note: *Bleeding the hydraulic system is necessary to remove any air that manages to find its way into the system when it's been opened during removal and installation of a hose, line, caliper or master cylinder.*

1   It will probably be necessary to bleed the system at all four brakes if air has entered the system due to low fluid level, or if the brake lines have been disconnected at the master cylinder.
2   If a brake line was disconnected only at a wheel, then only that caliper or wheel cylinder must be bled.
3   If a brake line is disconnected at a fitting located between the master cylinder and any of the brakes, that part of the system served by the disconnected line must be bled.
4   Remove any residual vacuum from the brake power booster by applying the brake several times with the engine off.
5   Remove the master cylinder reservoir cover and fill the reservoir with brake fluid. Reinstall the cover. **Note:** *Check the fluid level often during the bleeding operation and add fluid as necessary to prevent the fluid level from falling low enough to allow air bubbles into the master cylinder.*
6   Have an assistant on hand, as well as a supply of new brake fluid, a clear container partially filled with clean brake fluid, a length of 3/16-inch plastic, rubber or vinyl tubing to fit over the bleeder valve and a wrench to open and close the bleeder valve.
7   Beginning at the left rear wheel, loosen the bleeder valve slightly, then tighten it to a point where it is snug but can still be loos-

11.8 When bleeding the brakes, a hose is connected to the bleeder screw at the caliper or wheel cylinder and then submerged in brake fluid - air will be seen as bubbles in the tube and container (all air must be expelled before moving to the next wheel)

ened quickly and easily.

8   Place one end of the tubing over the bleeder valve and submerge the other end in brake fluid in the container **(see illustration)**.

9   Have the assistant pump the brakes slowly a few times to get pressure in the system, then hold the pedal firmly depressed.

10   While the pedal is held depressed, open the bleeder valve just enough to allow a flow of fluid to leave the valve. Watch for air bubbles to exit the submerged end of the tube. When the fluid flow slows after a couple of seconds, close the valve and have your assistant release the pedal.

11   Repeat Steps 9 and 10 until no more air is seen leaving the tube, then tighten the bleeder valve and proceed to the right rear wheel on Samurai models or the load sensing proportioning valve on Sidekick/X-90/Vitara and Tracker models (see Section 9). Perform the same procedure, then repeat the procedure at the right front wheel and the left front wheel, in that order. Be sure to check the fluid in the master cylinder reservoir frequently.

12   Never use old brake fluid. It contains moisture which can boil, rendering the brakes useless.

13   Refill the master cylinder with fluid at the end of the operation.

14   Check the operation of the brakes. The pedal should feel solid when depressed, with no sponginess. If necessary, repeat the entire process. **Warning:** *Do not operate the vehicle if there is in doubt about the effectiveness of the brake system.*

---

12   Power brake booster - check, removal and installation

Refer to illustrations 12.7, 12.12 and 12.13

Operating check

1   Depress the brake pedal several times

12.7 Remove the cotter pin and slide out the clevis pin - the booster is fastened to the firewall by four nuts, three of which are visible in this photo (arrows)

with the engine off and make sure that there is no change in the pedal reserve distance.

2   Depress the pedal and start the engine. If the pedal goes down slightly, operation is normal.

Airtightness check

3   Start the engine and turn it off after one or two minutes. Depress the brake pedal several times slowly. If the pedal goes down farther the first time but gradually rises after the second or third depression, the booster is airtight.

4   Depress the brake pedal while the engine is running, then stop the engine with the pedal depressed. If there is no change in the pedal reserve travel after holding the pedal for 30 seconds, the booster is airtight.

Removal

5   Power brake booster units should not be disassembled. They require special tools not normally found in most service stations or shops. They are fairly complex and because of their critical relationship to brake performance it is best to replace a defective booster unit with a new or rebuilt one.

6   To remove the booster, first remove the brake master cylinder as described in Section 8.

7   Locate the pushrod clevis connecting the booster to the brake pedal **(see illustration)**. This is accessible from the interior in front of the driver's seat.

8   Remove the cotter pin from the clevis pin with pliers and pull out the pin.

9   Disconnect the hose leading from the engine to the booster. Be careful not to damage the hose when removing it from the booster fitting.

10   Remove the four nuts and washers holding the brake booster to the firewall. You may need a light to see these, as they are up under the dash area **(see illustration 12.7)**.

11   Slide the booster straight out from the

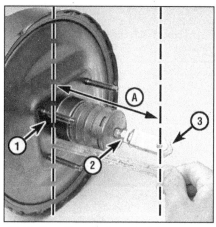

12.12 To adjust the length of the booster pushrod, loosen the locknut and turn the pushrod in or out, as necessary, to achieve the specified setting

    A   Pushrod length (see this Chapter's Specifications)
    1   Gasket
    2   Pushrod locknut
    3   Pushrod clevis

firewall until the studs clear the holes and pull the booster, brackets and gaskets from the engine compartment area.

Installation

12   Installation procedures are basically the reverse of those for removal. Check the length of the booster pushrod before installing the booster **(see illustration)**. Compare the length with this Chapter's Specifications. Tighten the pushrod locknut securely. Tighten the booster mounting nuts to the torque listed in this Chapter's Specifications.

13   If the power booster unit is being replaced, the clearance between the master cylinder piston and the pushrod in the vacuum booster must be measured. Using a depth micrometer or vernier calipers, mea-

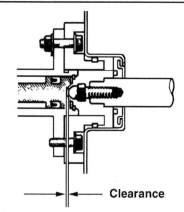

12.13 The booster pushrod-to-master cylinder clearance must be as specified - if there is interference between the two, the brakes may drag; if there is too much clearance, there will be excessive brake pedal travel

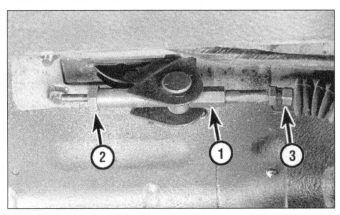

13.4 On Samurai models, loosen the stopper nut (1) and turn the adjusting nut (2) while holding the hold nut (3) with a wrench to prevent the inner cable from twisting - tighten or loosen until the parking brake lever travel is as specified

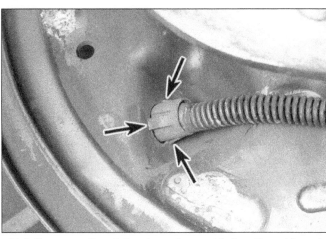

14.4 Compress the retainer tangs (arrows) to free the cable and housing from the backing plate

sure the distance from the seat (recessed area) in the master cylinder to the master cylinder mounting flange. Next, apply a vacuum of 20 in-Hg to the booster (using a hand vacuum pump) and measure the distance from the end of the vacuum booster pushrod to the mounting face of the booster (including gasket) where the master cylinder mounting flange seats. Subtract the two measurements to get the clearance (see illustration). If the clearance is more or less than specified, turn the adjusting screw on the end of the power booster pushrod until the clearance is within the specified range.

14   After the final installation of the master cylinder and brake hoses and lines, the brake pedal height and free play must be adjusted and the system must be bled. See the appropriate Sections of this Chapter for the procedures.

13 Parking brake - adjustment

Refer to illustration 13.4

1   The adjustment of the parking brake, often overlooked or put off by many motorists, is actually a fairly critical adjustment. If the parking brake cables are too slack, the brake won't hold the vehicle on an incline - if they're too tight, the brakes may drag, causing them to wear prematurely. Another detrimental side effect of a tightly adjusted parking brake cable is the restriction of the automatic adjuster assembly on the rear drum brakes, which will not allow them to function properly.

2   The first step in adjusting slack parking brake cables is to ensure the correct adjustment of the rear drum brakes. This can be accomplished by making a series of forward and reverse stops (approximately 10 of them), which will bring the brake shoes into proper relationship with the brake drums.

3   Raise the rear of the vehicle and support it securely on jackstands.

4   On Samurai models, loosen the stopper nut and turn the adjusting nut while holding the hold nut (see illustration).

5   On Sidekick/X-90/Vitara and Tracker models, loosen or tighten the self-locking nut.

6   Release the parking brake and apply it, making sure it travels within specifications. If it travels too far, tighten the equalizer locknut a little more. If the travel is less than the Specifications, the locknut will have to be loosened.

7   After the parking brake has been properly adjusted, place the handle in the released position and rotate the rear wheels, making sure the brakes don't drag.

8   Lower the vehicle and test the operation of the parking brake on an incline.

14 Parking brake cables - replacement

1   Release the parking brake. On Sidekick/X-90/Vitara and Tracker models loosen the rear wheel lug nuts. Raise the rear of the vehicle and support it securely on jackstands.

Sidekick/X-90/Vitara and Tracker models

Refer to illustration 14.4

2   Remove the rear wheel and brake drum. Loosen the self-locking nut fully.

3   Following the procedure in Section 6, remove the brake shoes, then disconnect the parking brake cable end from the parking brake lever.

4   Compress the cable housing retainer tangs at the brake backing plate (see illustration) and push the cable and housing through the backing plate.

5   Remove the self-locking nut and disconnect the cable at the lever, then pry cable out of the slot in the frame bracket.

Samurai

Refer to illustrations 14.6a and 14.6b

6   With a pair of pliers, remove the parking brake lever return spring, then remove the clip and pin (see illustrations).

7   Loosen the stopper nut and remove the adjusting nut while holding the hold nut (see illustration 13.4).

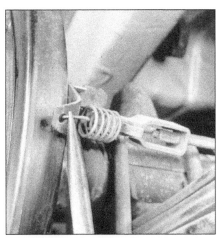

14.6a With a pair of pliers, remove the parking brake lever return spring . . .

14.6b . . . then remove the parking brake cable clip and pin

All models

8   Installation is the reverse of the removal procedure. Be sure to adjust the parking brake as described in Section 13.

---

15   Brake light switch - removal, installation and adjustment

Refer to illustration 15.4

1   Remove the lower dash panel trim under the steering column (Chapter 11).
2   Locate the brake light switch, which is mounted at the top of the brake pedal support and unplug it.
3   Unscrew the locknut on the switch and unscrew the switch from its bracket.
4   Installation is the reverse of removal. To adjust the switch, loosen the locknut and turn the switch in or out until the plunger clearance is within the range listed in this Chapter's Specifications **(see illustration)**. Tighten the locknut when the correct adjustment is obtained.

---

16   Anti-lock brake system - general information

The Anti-lock brake system is designed to maintain vehicle maneuverability, directional stability and optimum deceleration under severe braking conditions on most road surfaces. It does so by monitoring the rotational speed of the wheels and controlling the brake line pressure during braking. This prevents the wheels from locking up prematurely.

Two types of systems are used: Rear Wheel Anti-lock (RAWL) and Four Wheel Anti-lock (ABS). RAWL only controls lockup on the rear wheels, whereas ABS prevents lockup on all four wheels.

Actuator assembly

The actuator assembly includes the master cylinder and control valve(s) that consists of a dump valve an isolation valve. The valve operates by changing brake fluid pressure in response to signals from the control module (RAWL) or the electronic brake control module (ABS).

Control module (RAWL) or electronic control module (ABS VI)

The control module (RAWL) models is mounted behind the driver side of the instrument panel adjacent to the fuse block. The electronic brake control module (ABS) is mounted on the passenger side of the brake pedal brace. The function of the module is to accept and process information received from the speed sensors and brake light switch to control the hydraulic line pressure, avoiding wheel lock up. The module also constantly monitors the system, even under normal driving conditions, to find faults with the system.

If a problem develops within the system, the BRAKE warning light will glow on the dashboard. A diagnostic code will be stored, which when retrieved by a service technician, will indicate the problem area or component.

Speed sensor(s)

The single speed sensor on RAWL models is located in the rear differential carrier. On ABS models, there is a speed sensor on each front wheel and a single sensor in the rear differential carrier. The speed sensor(s) sends a signal to the control module indicating the rotational speed of the rear and/or front wheels.

Brake light switch

The brake light switch (see Section 15) signals the control module when the driver

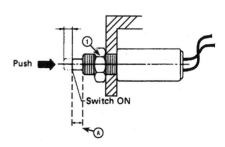

15.4 To adjust the brake light switch, loosen the locknut (1) and turn switch in or out until the clearance (A) is as specified

---

steps on the brake pedal. Without this signal the anti-lock system won't work.

Diagnosis and repair

If the BRAKE warning light on the dashboard comes on and stays on, make sure the parking brake is not applied and there's no problem with the brake hydraulic system. If neither of these is the cause the system is probably malfunctioning. Although a special electronic tester is necessary to properly diagnose the system, the home mechanic can perform a few preliminary checks before taking the vehicle to a dealer service department which is equipped with this tester.

a) *Make sure the brakes, calipers and wheel cylinders are in good condition.*
b) *Check the electrical connectors at the control module assembly.*
c) *Check the fuses.*
d) *Follow the wiring harness to the speed sensors and brake light switch and make sure all connections are secure and the wiring isn't damaged.*

If the above preliminary checks don't rectify the problem, the vehicle should be diagnosed by a dealer service department.

Chapter 10
Suspension and steering systems

Contents

Specifications

Torque specifications

Front suspension

	Ft-lbs
Samurai	
Leaf spring U-bolt nuts	44 to 58
Leaf spring shackle pin nut	22 to 40
Leaf spring eye-to-frame bolt	33 to 50
Sidekick/X-90/Vitara and Tracker	
1998 and earlier	
Front strut/shock absorber upper mounting nuts	14 to 22
Front strut/shock absorber-to-steering knuckle nuts	58 to 75
Lower control arm	
Front nut	50 to 75
Rear nut	65 to 100
Balljoint stud nut	32 to 50
Balljoint-to-lower arm nuts	50 to 75
Spindle-to-steering knuckle bolts	29 to 43
1999 and later	
Support brace bolts	36
Front strut/shock absorber upper mounting nuts	40
Front strut/shock absorber-to-steering knuckle nuts	69
Lower control arm	
Front nut	61
Rear nut	92
Balljoint stud nut	44
Spindle-to-steering knuckle bolts	36

Torque specifications

Ft-lbs

Rear suspension

Samurai .... See Front suspension

Sidekick/X-90/Vitara and Tracker

1998 and earlier

Upper arm rear balljoint boss-to-axle bolts .... 29 to 43

Upper arm-to-chassis bolts .... 58 to 72

Proportioning stay bolts 17

Trailing rod nuts .... 58 to 72

Shock absorber nuts 21

1999 and later

Lateral rod nuts 62

Upper arm bolts 65

Trailing rod nuts 65

Shock absorber nuts

Upper 21

Lower 61

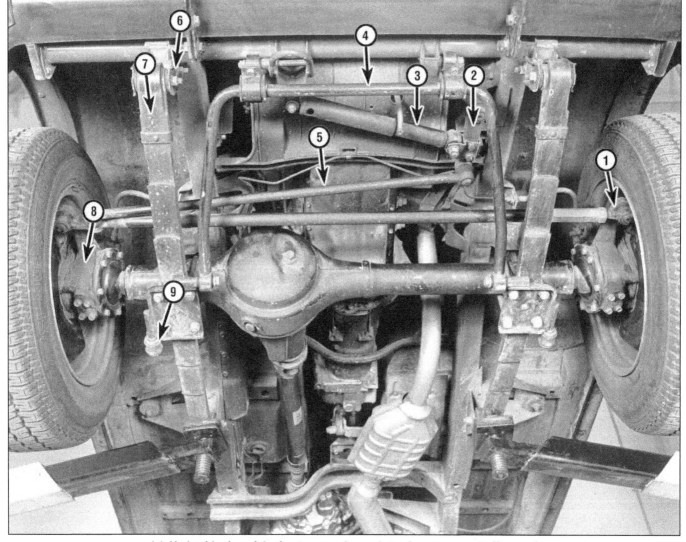

1.1 Underside view of the front suspension and steering components (Samurai)

1	Tie-rod end	4	Stabilizer bar	7	Leaf spring
2	Pitman arm	5	Drag rod	8	Steering knuckle
3	Steering damper	6	Spring shackle	9	Shock absorber

Steering

1998 and earlier (recirculating ball-type)
Steering wheel nut... 18 to 28
Steering gear mounting bolts.. 51 to 65
Intermediate shaft pinch bolt .... 14 to 22
Pitman arm-to-steering gear .... 101 to 129
Tie-rod end ballstud nut.. 22 to 40
Steering shaft rubber joint bolts.. 11 to 18
Pitman arm nut .... 101 to 129

Samurai
Drag rod nut .... 22 to 50

Sidekick and Tracker
Center link nut .... 22 to 50
Idler arm nut .... 50 to 72

1999 and later (rack-and-pinion-type)
Steering wheel nut... 23.5
Steering gear mounting bolts.. 40
Intermediate shaft pinch bolt 20
Tie-rod end ballstud nut.. 31

Wheel lug nuts .... See Chapter 1

1.2 Underside view of rear suspension (Samurai)

1   Leaf spring   2   Spring shackle   3   Shock absorber   4   Differential housing

1.3 Underside view of the front suspension and steering components (Sidekick/X-90/Vitara/Tracker)

1 Rack and pinion steering gear	2 Front suspension lower arm	3 Tie-rod end 4 Balljoint	5 Stabilizer bar 6 Spring 7 Shock absorber location

1 General information

Refer to illustrations 1.1, 1.2, 1.3 and 1.4

Samurai models covered by this manual utilize a solid front axle, suspended by two leaf springs. A dual-action shock absorber is mounted on each side. The steering knuckles pivot on kingpins and a stabilizer bar controls body roll **(see illustration)**.

The rear axle on Samurai models is also suspended by two leaf springs and two dual-action telescopic shock absorbers **(see illustration)**. A stabilizer bar is installed on most models.

Sidekick/X-90/Vitara and Tracker models employ a strut/shock absorber design front suspension **(see illustration)**. Coil springs are mounted between the lower arm and the frame. The rear suspension on 1998 and earlier models uses coil springs, shock absorbers and a solid axle, located by lower trailing rods and an upper control arm con-

nected to the differential housing by a balljoint. 1999 and later models use a five-link system with coil springs, shock absorbers and solid axle located by two upper and two lower trailing rods and a lateral rod between the differential housing and the right side chassis frame **(see illustration)**.

Steering is either manual or power assisted. On 1998 and earlier models, a recirculating ball-type steering gearbox transmits the turning force through the steering linkage to the steering knuckle. A steering damper is mounted between the frame and the Pitman arm to reduce unwanted bump steer. On 1999 and later models a rack-and-pinion steering gear located in front of the engine actuates the tie-rods, which are attached to the steering knuckles. On all models an intermediate shaft connects the steering gear to the steering column.

Frequently, when working on the suspension or steering system components, you may come across fasteners that seem

impossible to loosen. These fasteners on the underside of the vehicle are continually subjected to water, road grime, mud, etc., and can become rusted or "frozen", making them extremely difficult to remove. In order to unscrew these stubborn fasteners without damaging them (or other components), be sure to use lots of penetrating oil and allow it to soak in for a while. Using a wire brush to clean exposed threads will also ease removal of the nut or bolt and prevent damage to the threads. Sometimes a sharp blow with a hammer and punch is effective in breaking the bond between nut and bolt threads, but care must be taken to prevent the punch from slipping off the fastener and ruining the threads. Heating the stuck fastener and surrounding area with a torch sometimes helps too, but isn't recommended because of the obvious dangers associated with fire. Long breaker bars and extension, or "cheater," pipes will increase leverage, but never use an extension pipe on a ratchet - the ratcheting

1.4 Underside view of rear suspension (Sidekick/X-90/Vitara/Tracker)

| 1 | Lateral rod | 3 | Lower control arm | 5 | Differential housing |
| 2 | Shock absorber | 4 | Upper control arm | 6 | Coil spring location |

mechanism could be damaged. Sometimes, turning the nut or bolt in the tightening (clockwise) direction first will help to break it loose. Fasteners that require drastic measures to unscrew should always be replaced with new ones.

Since most of the procedures that are dealt with in this chapter involve jacking up the vehicle and working underneath it, a good pair of jackstands will be needed. A hydraulic floor jack is the preferred type of jack to lift

the vehicle, and it can also be used to support certain components during various operation. **Warning:** *Never, under any circumstances, rely on a jack to support the vehicle while working on it. Whenever any of the suspension or steering fasteners are loosened or removed they must be inspected and, if necessary, be replaced with new ones of the same part number or of original equipment quality and design. Torque specifications must be followed for proper reassembly and component retention. Never attempt to heat or straighten any suspension or steering component. Instead, replace any bent or damaged part with a new one.*

---

2   Front stabilizer bar - removal and installation

Removal

1   Apply the parking brake. Raise the front of the vehicle and support it securely on jackstands.

Samurai

Refer to illustrations 2.2 and 2.3

2   Remove the stabilizer bar-to-leaf spring plate bolts, **(see illustration)**.

3   Remove the stabilizer bar bracket bolts

and detach the bar from the vehicle **(see illustration)**.

4   Pull the brackets off the stabilizer bar and inspect the bushings for cracks, hardness and other signs of deterioration. If the bushings are damaged, replace them.

2.2 Remove the stabilizer bar-to-leaf spring plate bolt (arrow) to detach the bar

2.3 The stabilizer bar is attached to the frame with two brackets like this - remove the bolts (arrows) to detach the bar - the rubber bushings should be replaced if they are hard, cracked or otherwise deformed

2.5 Make sure the stabilizer bar bushing and clamp positions are marked before removal so they can easily be centered during installation

2.6 Remove the nut (arrow) retaining the stabilizer bar to the lower arm

3.2 When removing the front shock absorber, remove the lock-nut first then the stem nut - it may be necessary to hold the stem with an open-end wrench or locking pliers to prevent it from turning

Sidekick and Tracker models

Refer to illustrations 2.5 and 2.6

5   Before removing the stabilizer bar, check to see if any alignment paint marks are present where the bushings support the bar. If none are visible, apply them with a marking pen before removal (this will help center the bar when installing it) **(see illustration)**.

6   Remove the stabilizer ball joint from the front suspension lower arms noting how the spacers, washers and bushings are positioned.

7   Remove the stabilizer bar bracket bolts and detach the bar from the vehicle.

Installation

Samurai models

8   Position the stabilizer bar bushings on the bar with the slits facing the top of the vehicle.

9   Push the brackets over the bushings and raise the bar up to the frame. Install the bracket bolts but don't tighten them completely at this time.

10   Install the stabilizer bar-to-lower control

3.3 The lower end of the shock absorber is connected to the front axle housing by a nut and washer

arm bolts, washers, spacers and rubber bushings and tighten the nuts securely.

11   Tighten the bracket bolts.

Sidekick/X-90/Vitara/Tracker models

12   Position the stabilizer bar bushings on the bar with the slits facing the front of the vehicle. **Note:** *Align the bushings with the alignment marks on the bar.*

13   Push the brackets over the bushings and raise the bar up to the frame. Install the bracket bolts but do not tighten them completely at this time.

14   Install the stabilizer ball joint.

---

3   Front shock absorber (Samurai models) - removal and installation

---

Refer to illustrations 3.2 and 3.3

Removal

1   Loosen the wheel lug nuts, raise the vehicle and support it securely on jackstands. Apply the parking brake. Remove the wheel.

2   Remove the upper shock absorber lock nut, then remove the upper shock absorber stem nut **(see illustration)**. Use an open end wrench to keep the stem from turning. If the nut won't loosen because of rust, squirt some penetrating oil on the stem threads and allow it to soak in for awhile. It may be necessary to keep the stem from turning with a pair of locking pliers, since the flats provided for a wrench are quite small.

3   Remove the lower shock mount nut **(see illustration)** and remove the shock absorber. Remove the washers and the rubber grommets from the top of the shock absorber.

Installation

4   Extend the new shock absorber as far as possible. Position a new washer and rubber grommet on the stem and guide the shock up and into the upper mount.

5   Install the upper rubber grommet and

washer and wiggle the stem back-and-forth to ensure that the grommets are centered in the mount. Tighten the stem nut securely, then install the lock nut.

6   Install the lower mounting nut and tighten them securely.

---

4   Front leaf spring (Samurai models) - removal and installation

---

Refer to illustrations 4.4 and 4.6

Warning: *Whenever any of the suspension or steering fasteners are loosened or removed they must be inspected and, if necessary, replaced with new ones of the same part number or of original equipment quality and design. Torque specifications must be followed for proper reassembly and component retention.*

Removal

1   Loosen the front wheel lug nuts, raise the front of the vehicle and support it securely on jackstands. Remove the wheel.

2   Support the axle assembly with a floor jack positioned underneath the differential. Raise the axle just enough to take the spring pressure off of the shock absorbers.

3   Disconnect the shock absorber from the axle bracket (see Section 3).

4   Support the axle, then unscrew the U-bolt nuts **(see illustration)**. Remove the spring plate.

5   Remove the spring eye-to-frame bracket bolt.

6   Remove the spring-to-shackle bolt and remove the spring from the vehicle **(see illustration)**.

Installation

7   Installation is the reverse of the removal procedure. Be sure to tighten the spring mounting bolts and the spring plate U-bolt

4.4 The front axle must be supported before removing the stabilizer bolt (1), front shock absorber nut (2) and the four U-bolt nuts (3)

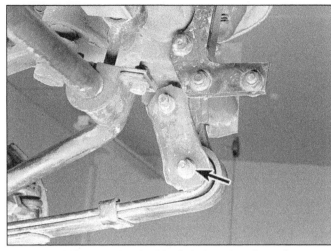

4.6 Remove the shackle pin bolt and nut (arrow)

nuts to the specified torque. **Note:** *The vehicle must be standing at normal ride height before tightening the front and rear mounting bolts.*

---

5 Front strut/shock absorber assembly (Sidekick/X-90/ Vitara/Tracker models) - removal and installation

Refer to illustrations 5.4, 5.5a, 5.5b and 5.6
Note: *The shock absorbers are not serviceable and must be replaced as complete assemblies.*

Removal

1   Loosen the front wheel lug nuts, raise the vehicle and support it securely on jackstands. Remove the wheel.
2   Place a floor jack under the lower control arm and raise it slightly. The jack must remain in this position throughout the entire procedure.

3   On later models, remove the bolt and ABS speed sensor harness from the strut.
4   Remove the clip securing the brake hose to the strut/shock absorber **(see illustration)**.
5   On later models, unscrew the three bolts and detach the strut tower bar. Unscrew the three upper mount-to-strut tower retaining nuts **(see illustrations)**.
6   Remove the strut-to-spindle nuts and bolts **(see illustration)**.
7   Separate the strut/shock absorber assembly from the spindle and remove it from the vehicle. Be careful not to overextend the driveaxle on 4WD models.

Installation

8   Guide the assembly into position in the wheel well, pushing the upper mount studs through the holes in the strut tower. Install the three nuts and tighten them to the specified torque.
9   Insert the spindle into the lower mounting flange of the strut/shock assembly and install the two bolts from the front side. Install

5.4 Using a pair of pliers, remove the clip - when attaching the clip, make sure not to twist the brake hose

the nuts and tighten them to the specified torque.
10   Remove the jack from under the lower

5.5a On later models, remove the bolts (arrows) and detach the strut tower bar for access to the shock tower nuts

5.5b The upper end of the front strut/shock absorber assembly is fastened to the shock tower with three support nuts (arrows)

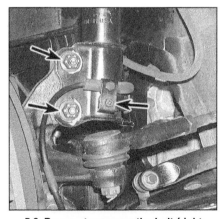

5.6 Be sure to remove the bolt (right arrow) and detach the ABS harness before removing the strut-to-spindle bolts (left arrows) - it may be necessary to drive out the bolts with a hammer and punch

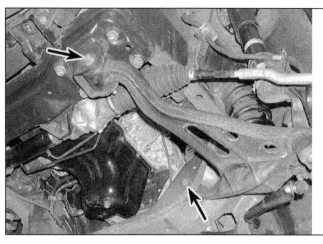

7.3 Remove the nuts and bolts (arrows) attaching the inner end of the lower suspension arm to the frame

control arm and install the brake hose clip without twisting the brake hose.
11   Install the wheel, lower the vehicle and tighten the lug nuts to the torque specified in Chapter 1.

6   Front coil spring (Sidekick/X-90/ Vitara/Tracker models) - removal and installation

Warning: *The following procedure is potentially dangerous if the proper safety precautions are not taken.*

Removal

1   Loosen the wheel lug nuts on the side to be dismantled. Raise the vehicle, support it securely on jackstands and remove the wheel.
2   Remove the locking hub on 4WD models or end cap (2WD) (see Chapter 8).
3   Remove the front driveaxle snap-ring and washer (see Chapter 8).
4   Remove the brake caliper and hang it out of the way with a piece of wire (see Chapter 9).
5   Remove the brake disc (see Chapter 9).
6   Disconnect the stabilizer bar joint from the lower control arm (see Section 2).
7   Remove the tie-rod end from the steering knuckle (see Section 18).
8   Use a jack to support the lower suspension arm.
9   Remove the steering knuckle (see Section 8).
10   Slowly lower the jack until the coil spring is completely extended, then remove the coil spring.

Installation

Note: *The upper and lower diameters of the coil spring are different.*
11   Guide the small diameter end of the coil spring up into the upper pocket and place the larger diameter end in the spring seat area of the control arm.
12   The remainder of the installation procedure is the reverse of removal. Tighten all fasteners to the specified torque.

7   Front suspension lower arm (Sidekick/X-90/Vitara/Tracker models) - removal and installation

Refer to illustration 7.3
1   Loosen the wheel lug nuts, raise the vehicle and support it securely on jackstands. Remove the wheel.
2   Remove the coil spring (see Section 6).
3   Remove the suspension arm-to-frame bolts **(see illustration)**.
4   Installation is the reverse of removal. The vehicle should be sitting at normal ride height before tightening the arm-to-frame bolts to the torque listed in this Chapter's Specifications.

8   Steering knuckle (Sidekick/X-90/ Vitara/Tracker models) - removal and installation

Warning: *Whenever any of the suspension or steering fasteners are loosened or removed they must be inspected and, if necessary, replaced with new ones of the same part number or of original equipment quality and design. Torque specifications must be followed for proper reassembly and component retention. Dust created by the brake system may contain asbestos, which is harmful to your health. Never blow it out with compressed air and don't inhale any of it. Do not, under any circumstances, use petroleum-based solvents to clean brake parts. Use brake cleaner or denatured alcohol only.*

Removal

1   Loosen the wheel lug nuts, raise the vehicle and support it securely on jackstands. Remove the wheel. On ABS-equipped models, remove the speed sensor and wiring harness clamp. Remove the brake caliper and support it with a piece of wire as described in Chapter 9.
2   Support the lower suspension arm with a floor jack.
3   Remove the brake disc (see Chapter 9)

and the front hub assembly (see Chapter 8).
4   Remove the 4WD model driveaxle snapring and washer (see Chapter 8).
5   Remove the brake dust cover (see Chapter 9).
6   With a soft face hammer, tap the wheel spindle off.
7   Remove the strut/shock absorber-to-steering knuckle bolts/nuts (see Section 5).
8   Separate the tie-rod from the steering knuckle arm as outlined in Section 18.
9   Separate the balljoint from the steering knuckle (see Section 9).
10   Remove the steering knuckle assembly from the strut/shock, balljoint and driveaxle. If the steering knuckle will not break loose from the balljoint, it may be necessary to hit the steering knuckle with a hammer. If the 4WD driveaxle sticks in the hub splines, push it from the hub with a puller tool. Support the end of the driveaxle with a piece of wire to prevent damage to the inner CV joint.

Installation

11   Guide the knuckle assembly into position, inserting the driveaxle into the hub.
12   Push the knuckle into the shock flange and install the bolts, but don't tighten them yet.
13   Insert the balljoint stud into the steering knuckle hole and install the nut, but don't tighten it yet.
14   Attach the tie-rod to the steering knuckle arm as described in Section 18. Tighten the strut bolt nuts, the balljoint nut and the tie-rod nut to the specified torque values.
15   Attach the wheel spindle to the steering knuckle using silicone sealant and lubricant.
16   Install the brake dust cover, tightening the fasteners to the torque listed in the Chapter 9 Specifications.
17   Install the front hub assembly and brake disc.
18   Install the 4WD driveaxle snap-ring and washer.
19   Install the caliper as outlined in Chapter 9.
20   Remove the floor jack supporting the suspension arm.
21   Install the wheel and lug nuts.
22   Lower the vehicle and tighten the lug nuts to the torque listed in the Chapter 9 Specifications.

9   Balljoints (Sidekick/X-90/ Vitara/Tracker models) - check and replacement

Warning: *Whenever any of the suspension or steering fasteners are loosened or removed, they must be inspected and, if necessary, replaced with new ones of the same part number or of original equipment quality and design. Torque specifications must be followed for proper reassembly and component retention.*

10.2a The lower end of the rear shock absorber on Samurai models mounts to a stud on the leaf spring plate and is retained by a nut and washer

10.2b The rear shock absorber lower end on Sidekick/X-90/ Vitara/Tracker models mounts to a bracket on the rear axle and is retained by a nut and washer

Check

1   Raise the vehicle and support it securely on jackstands.
2   Visually inspect the rubber boot for cuts, tears or leaking grease. If any of these conditions are noticed, the balljoint should be replaced.
3   Place a large pry bar under the balljoint and attempt to push the balljoint up. Next, position the pry bar between the steering knuckle and the lower arm and apply downward pressure. If any movement is seen or felt during either of these checks, a worn out balljoint is indicated.
4   Have an assistant grasp the tire at the top and bottom and shake the top of the tire in an in-and-out motion. Touch the balljoint stud nut. If any looseness is felt, suspect a worn out balljoint stud or a widened hole in the steering knuckle boss. If the latter problem exists, the steering knuckle should be replaced as well as the balljoint. On 1999 and later models the balljoint is not replaceable. If the ball joint is faulty or damaged the lower suspension arm assembly must be replaced with a new one (Section 7).

Replacement

1998 and earlier models

5   Loosen the wheel lug nuts, raise the vehicle and support it securely on jackstands. Remove the wheel.
6   Support the lower suspension arm with a floor jack.
7   Remove the steering knuckle (see Section 8).
8   Unscrew the three balljoint-to-lower arm bolts and remove the balljoint.
9   To install the balljoint, position it on the lower arm and install the three bolts to the specified torque.
10   Install the steering knuckle (see Section 8) and remove the floor jack.
11   The remainder of the installation procedure is the reverse of removal.
12   Install the wheel and lug nuts. Lower the vehicle and tighten the lug nuts to the specified torque.

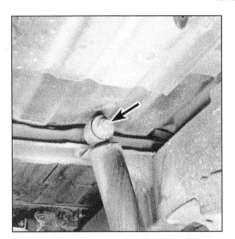

10.3a The upper end of the rear shock absorber on Samurai models mounts to a stud on the frame with a nut and washer (arrow)

---

10   Rear shock absorber - removal and installation

Refer to illustrations 10.2a, 10.2b, 10.3a and 10.3b
Warning: *Whenever any of the suspension or steering fasteners are loosened or removed, they must be inspected and, if necessary, replaced with new ones of the same part number or of original equipment quality and design.*

Removal

1   Loosen the rear wheel lug nuts, raise the rear of the vehicle and support it securely on jackstands. Block the front wheels and remove the rear wheel(s).
2   Position a floor jack under the rear axle housing and raise it just enough to take some of the spring pressure off the shock absorber. On Samurai models, remove the lower nut and washer **(see illustration)**. On Sidekick/X-90/Vitara and Tracker models, remove the shock absorber lower mounting bolt **(see illustration)**.

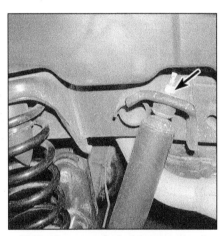

10.3b The upper end of the rear shock absorber on Sidekick/X-90/Vitara/Tracker models mounts to a bracket on the frame with a nut and washer

3   On Samurai models, remove the upper mounting nut and washer **(see illustration)** and detach the shock absorber **(see illustration)**. On Sidekick/X-90/Vitara/Tracker models, remove the shock absorber lock nut, then remove the shock absorber nut and detach the shock absorber.

Installation

4   Installation is the reverse of the removal procedure. Be sure to tighten the fasteners securely.

---

11   Rear leaf spring (Samurai models) - removal and installation

---

Warning: *Whenever any of the suspension or steering fasteners are loosened or removed they must be inspected and, if necessary, replaced with new ones of the same part number or of original equipment quality and design. Torque specifications must be followed for proper reassembly and component retention.*

12.2 Remove the lower trailing rod-to-axle nut and bolt (arrow)

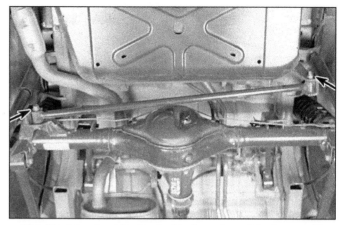

12.4 The lateral rod is attached to the axle and the frame with bolts and nuts (arrows)

Removal

1   Loosen the rear wheel lug nuts, raise the rear of the vehicle and support it securely on jackstands. Remove the wheel.
2   Support the rear axle assembly with a floor jack positioned underneath the differential. Raise the axle just enough to take the spring pressure off of the shock absorbers.
3   Disconnect the shock absorber from the axle bracket (see Section 10).
4   Support the axle, then unscrew the U-bolt nuts **(refer to illustration 4.5)**. Re-move the spring plate.
5   Remove the spring eye-to-frame bracket bolt.
6   Remove the spring-to-shackle bolt and remove the spring from the vehicle.

Installation

7   Installation is the reverse of the removal procedure. Be sure to tighten the spring mounting bolts and the spring plate U-bolt nuts to the specified torque. **Note:** *The vehicle must be standing at normal ride height before tightening the front and rear mounting bolts.*

---

12   Rear suspension trailing rod and lateral rod (Sidekick/ X-90/Vitara/Tracker models) - removal and installation

Removal

Refer to illustrations 12.2 and 12.4

1   Loosen the wheel lug nuts, raise the vehicle and support it securely on jackstands. Remove the wheel(s) and place a floor jack under the rear axle to support it when the trailing rod(s) and lateral rod are removed.
2   Remove the lower rod rear attaching nut and bolt from the axle bracket, then unbolt the parking brake cable guide **(see illustration)**.
3   Remove the front attaching bolt and nut and remove the lower rod from the vehicle.
4   Remove the retaining bolts and detach the lateral rod from the axle and frame brack-

ets **(see illustration)**.

Installation

5   Place the trailing rod(s) and the lateral rod in their mounting brackets and install the bolts and nuts (don't tighten them yet).
6   Install the parking brake cable guide on the trailing rod.
7   Remove the floor jack supporting the rear axle.
8   Install the wheel and lug nuts and lower the vehicle. Tighten the lug nuts to the torque listed in the Chapter 1 Specifications.
9   With the vehicle off the jackstands and in a non-loaded condition, tighten the trailing rod front and rear bolts and nuts and the lateral rod bolts to the specified torque.

---

13   Rear suspension upper arm (Sidekick/X-90/Vitara/Tracker models) - removal and installation

1998 and earlier models

Removal

1   Raise the rear of the vehicle and support it securely on jackstands placed beneath the frame rails. Block the front wheels.
2   Remove the proportioning valve stay from the rear suspension upper arm (see Chapter 9).
3   Position a jack under the differential and raise it slightly.
4   Remove the ball joint boss from the differential carrier.
5   Remove the upper arm-to-frame pivot bolts and nuts and remove the arm from the vehicle.

Installation

6   Position the leading end of the suspension arm in the frame bracket. Install the pivot bolts and nuts with the bolts installed from the inside, but don't fully tighten the nuts at this time.
7   Place the other end of the arm over the

differential carrier. It may be necessary to jack up the rear axle to align the holes. Install the balljoint boss bolts and tighten them to the specified torque.
8   Install the proportioning valve stay.
9   Remove the floor jack supporting the differential housing.
10   Lower the vehicle and tighten the upper arm-to-frame pivot bolts to the specified torque.

1999 and later models

Removal

Refer to illustration 13.14

11   Raise the rear of the vehicle and support it securely on jackstands placed beneath the frame rails. Block the front wheels.
12   On ABS-equipped models, detach the speed sensor harness clamp on the left side arm.
13   Position a jack under the differential and raise it slightly.
14   Note the directions in which the bolts are installed then remove the upper arm-to-frame pivot bolts and nuts from each end and remove the arm(s) from the vehicle **(see illustration)**.

13.14 Note the direction of installation of the upper rod-to-bracket bolts (arrows)

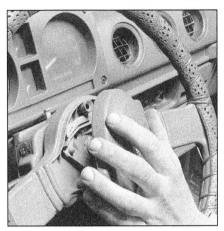

15.2a To remove the horn pad, pull it straight off the wheel (non-airbag models)

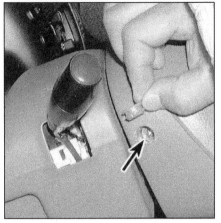

15.2b On airbag equipped models, pry out the screw cover, then loosen the airbag retaining screws (arrow) - there is one screw on each side

15.2c Lift the airbag module off

Installation

15   Place the front ends of the suspension arms in the frame brackets. Install the pivot bolts and nuts with the bolts installed in their original directions, but don't fully tighten the nuts at this time.

16   Place the other ends of the arms in the axle brackets. It may be necessary to jack up the rear axle to align the holes. Install the pivot bolts and nuts and tighten them to the specified torque.

17   Install the ABS harness (if equipped) in the left arm.

18   Remove the floor jack supporting the differential housing.

19   Lower the vehicle and tighten the pivot bolts to the specified torque.

14   Rear coil spring (Sidekick/ X-90/Vitara/Tracker models) - removal and installation

Removal

1   Loosen the wheel lug nuts, raise the rear of the vehicle and support it on jackstands. Remove the wheel and place a floor jack

under the rear axle to support it.

2   Remove the shock absorber lower mounting bolt.

3   Lower the floor jack slowly until the coil spring is fully extended, then remove the spring from the vehicle.

Installation

4   Guide the spring into position. Make sure the end of the spring rests in the stepped area of the spring seat.

5   The remainder of installation is the reverse of the removal procedure.

15   Steering wheel - removal and installation

Refer to illustrations 15.2a, 15.2b, 15.2c, 15.2d, 15.3, 15.4a and 15.4b

Warning: *1996 and later models are equipped with an airbag system. Always disable the airbags prior to working in the vicinity of the steering column, instrument panel or airbag components to avoid the possibility of accidental deployment of the airbags, which could cause personal injury (see Chapter 12).*

1   Disconnect the cable from the negative battery terminal. Place the front wheels in the straight-ahead position and turn the ignition switch to the Lock position. Disable the airbag (see Section 17 in Chapter 12).

2   On non-airbag equipped earlier models, detach the horn pad from the steering wheel **(see illustration)**. On 1997 and earlier models equipped with airbags, remove the screw cover from the right side of the steering wheel and disconnect the wiring connectors. Remove the airbag module retaining screws (one on each side of the steering wheel) and remove the airbag module **(see illustrations)**. Place the airbag module in a safe place with the trim cover facing up. **Warning:** *Be extremely careful when handling a live airbag module, carry the module with the trim cover pointed away from your body. Never place the module on any surface with the trim cover facing down or serious injury could occur in the event of an accidental deployment.*

3   Remove the steering wheel retaining nut and mark the relationship of the steering shaft to the hub to ensure steering wheel alignment **(see illustration)**.

4   Use a puller to detach the steering wheel from the shaft **(see illustrations)**. Don't hammer on the steering wheel or shaft to dislodge the wheel.

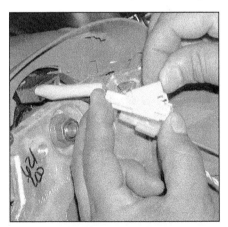

15.2d Unplug the airbag harness connector

15.3 Before removing the steering wheel, check to see if any relationship marks exist (arrows) - if not, use a sharp scribe or white paint to make your own marks

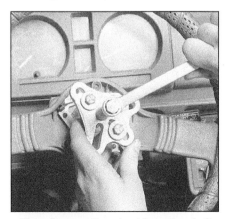

15.4a Remove the wheel from the shaft with a puller - DO NOT HAMMER ON THE SHAFT!

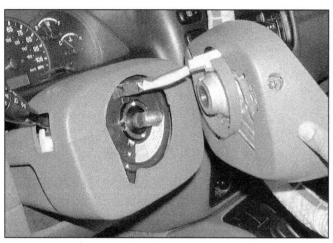

15.4b Unplug the electrical connector and remove the steering wheel

16.2 Mark the upper end of the intermediate shaft and universal joint then mark the universal joint to steering shaft (Samurai)

5   To install the steering wheel, align the mark on the steering wheel hub with the mark on the shaft. On 1996 and later models, align the two grooves on the back of the steering wheel with the lugs on the airbag coil. Install the steering wheel and tighten the nut to the torque listed in this Chapter's Specifications.

6   The remainder of installation is the reverse of removal.

16   Intermediate shaft - removal and installation

Refer to illustration 16.2

Warning 1: *On models equipped with airbags, DO NOT rotate the steering shaft while the intermediate shaft is removed from the vehicle or damage to the airbag coil will occur. To prevent the shaft from turning, turn the ignition key to the Lock position before beginning work, or wrap the seat belt around the steering wheel and clip the seat belt buckle into place.*

Warning 2: *Whenever any of the suspension or steering fasteners are loosened or removed they must be inspected and if necessary, replaced with new ones of the same part number or of original equipment quality and design. Torque specifications must be followed for proper reassembly and component retention. Never attempt to heat, straighten or weld any suspension or steering component. Instead, replace any bent or damaged part with a new one.*

1   Turn the front wheels to the straight ahead position.

2   Using white paint, place alignment marks on the upper universal joint, the steering shaft, the lower universal joint or lower flexible coupling and the steering gear input shaft **(see illustration).**

3   Remove the upper and lower universal joint pinch bolts.

4   Pry the intermediate shaft out of the steering shaft universal joint with a large screwdriver, then pull the shaft from the steering gearbox.

5   Installation is the reverse of the removal procedure. Be sure to align the marks and tighten the pinch bolts to the specified torque.

17   Steering gear - removal and installation

Warning 1: *On models equipped with airbags, DO NOT rotate the steering shaft while the steering gear is removed from the vehicle or damage to the airbag coil will occur. To prevent the shaft from turning, turn the ignition key to the Lock position before beginning work, or wrap the seat belt around the steering wheel and clip the seat belt buckle into place.*

Warning 2: *Whenever any of the suspension or steering fasteners are loosened or removed they must be inspected and if necessary, replaced with new ones of the same part number or of original equipment quality and design. Torque specifications must be followed for proper reassembly and component retention. Never attempt to heat, straighten or weld any suspension or steering component. Instead, replace any bent or damaged part with a new one.*

Note: *1998 and earlier models are equipped with recirculating ball-type steering while 1999 and later models use rack-and-pinion steering.*

Recirculating ball-type steering

Refer to illustrations 17.5 and 17.6

Removal

1   Raise the front of the vehicle and support it securely on jackstands. Apply the parking brake.

2   Position a drain pan under the steering gear (power steering only). Remove the hoses/lines and cap the ends to prevent excessive fluid loss and contamination.

3   Mark the relationship of the lower intermediate shaft universal joint to the steering gear input shaft. Remove the lower intermediate shaft pinch bolt.

4   On Samurai models, mark the relationship of the Pitman arm to the shaft so it can be installed in the same position. Remove the nut and washer. On Sidekick and Tracker models, disconnect the center link from the Pitman arm (see Section 18).

5   Remove the Pitman arm from the shaft with a two-jaw puller **(see illustration).**

17.5 Use a two-jaw puller to separate the pitman arm from the steering gear shaft

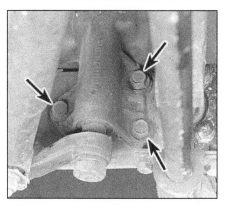

17.6 The recirculating ball-type steering gear is mounted to the frame rail with three bolts (arrows)

6   Support the steering gear and remove the mounting bolts **(see illustration)**. Lower the unit, separate the intermediate shaft from the steering gear input shaft and remove the steering gear from the vehicle.

Installation

7   Raise the steering gear into position and connect the intermediate shaft, aligning the marks.
8   Install the mounting bolts and washers and tighten them to the specified torque.
9   On Samurai models, slide the Pitman arm onto the shaft. Make sure the marks are aligned. Install the washer and nut and tighten the nut to the specified torque. On Sidekick and Tracker models, install the center link to the Pitman arm and tighten the nut to the specified torque.
10   Install the lower intermediate shaft pinch bolt and tighten it to the specified torque.
11   Connect the power steering hoses/lines to the steering gear and fill the power steering pump reservoir with the recommended fluid (see Chapter 1).
12   Lower the vehicle and bleed the steering system as outlined in Section 20.

Models with rack-and-pinion steering

Refer to illustrations 17.14 and 17.18

Removal

13   Loosen the front wheel lug nuts, raise the front of the vehicle and support it securely on jackstands. Apply the parking brake and remove the front wheels.
14   Mark the relationship of the intermediate shaft coupler to the steering gear input shaft and remove the pinch bolt **(see illustration)**.
15   Detach the tie-rod ends from the steering arms (Section 18).
16   Position a drain pan under the steering gear. Using a flare-nut wrench (if available) to unscrew the power steering pressure and return lines from the steering gear **(see illustration 17.14)**.
17   Unscrew the power steering pressure and return line fittings from the steering gear. Cap the ends to prevent fluid loss and the entry of contaminants.
18   Unscrew the four mounting bolts, detach the steering rack unit and carefully lower it from the vehicle **(see illustration)**.

Installation

19   Installation is the reverse of the removal procedure, with the following points:

a) *Tighten all fasteners to the torque values listed in this Chapter's Specifications.*
b) *Tighten the power steering pressure and return line fittings securely.*
c) *Add power steering fluid to the pump reservoir to bring it up to the desired level (see Chapter 1).*
d) *Lower the vehicle and tighten the lug nuts to the torque listed in the Chapter 1 Specifications.*
e) *Bleed the power steering system (see Section 20).*
f) *Have the wheel alignment checked and if necessary, adjusted.*

18   Steering linkage - inspection, removal and installation

Warning: *Whenever any of the suspension or steering fasteners are loosened or removed they must be inspected and if necessary, replaced with new ones of the same part number or of original equipment quality and design. Torque specifications must be followed for proper reassembly and component retention. Never attempt to heat, straighten or weld any suspension or steering component. Instead, replace any bent or damaged part with a new one.*
Caution: *DO NOT use a "pickle fork" type balljoint separator - it may damage the balljoint seals.*

Recirculating ball-type steering (1998 and earlier models)

Inspection

1   The steering linkage steering connects the steering gear to the front wheels and keeps the wheels in proper relation to each other. The linkage consists of the Pitman arm that is fastened to the steering gear shaft. On Samurai models, the Pitman arm moves the drag rod back-and-forth. The back-and-forth motion of the drag rod is transmitted to the steering knuckles through a tie-rod assembly. On Sidekick and Tracker models, the Pitman arm moves the center link back-and-forth. The center link is supported on the other end by an idler arm. The back-and-forth motion of the center link is transmitted to the steering knuckles through a pair of tie-rod assemblies.
2   Set the wheels in the straight ahead position and lock the steering wheel.
3   Raise one side of the vehicle until the tire is approximately 1-inch off the ground.
4   Mount a dial indicator with the needle resting on the outside edge of the wheel. Grasp the front and rear of the tire and using

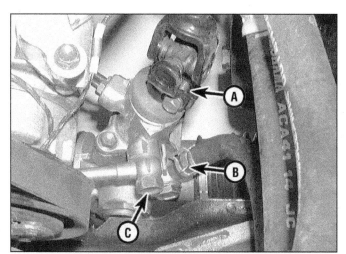

17.14 The intermediate shaft pinch bolt (A) is located next to the power steering gear housing fluid connections for inlet (B) and pressure (C)

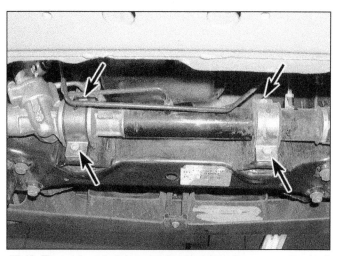

17.18 The rack-and-pinion steering housing clamps are attached with four bolts (arrows)

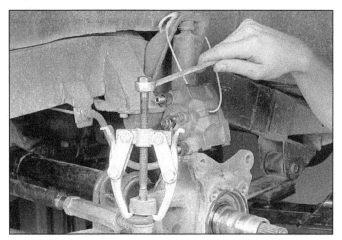

18.9 Use a two-jaw puller to detach the tie-rod end from the steering knuckle

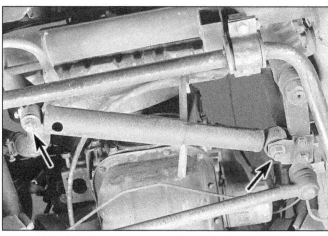

18.33 Remove the nuts and bolts (arrows) to remove the steering damper

light pressure, wiggle the wheel back-and-forth and note the dial indicator reading. If the play in the steering system is excessive, inspect each steering linkage pivot point and ball stud for looseness and replace parts if necessary.

5   On Sidekick and Tracker models, raise the vehicle and support it on jackstands. Push up, then pull down on the center link of the idler arm, exerting a force of approximately 25 pounds each way. Measure the total distance the end of the arm travels. If the play is excessive, replace the idler arm.

6   Check for torn ball stud boots, frozen joints and bent or damaged linkage components.

Removal and installation

Refer to illustrations 18.9 and 18.33

Tie-rod

7   Loosen the wheel lug nuts, raise the vehicle and support it securely on jackstands. Apply the parking brake. Remove the wheel.

8   Remove the cotter pin and loosen, but do not remove, the castellated nut from the ball stud.

9   Using a two jaw puller, separate the tie-rod end from the steering knuckle **(see illustration)**. Remove the castellated nut and pull the tie-rod end from the knuckle.

10   If the tie-rod or tie-rod end must be replaced, measure the distance from the end of the rod end connector to the center of the ball stud and record it. Loosen the lock nut and unscrew the tie-rod end.

11   Lubricate the threaded portion of the tie-rod end with chassis grease. Screw the new tie-rod end into the adjuster tube and adjust the distance from the tube to the ball stud to the previously measured dimension. The number of threads showing on the tie-rod and tie-rod end should be equal within three threads. Don't tighten the lock nut yet.

12   To install the tie-rod, insert the tie-rod end ball stud into the center link or drag rod until it's seated. Install the nut and tighten it to the specified torque. If the ball stud spins

when attempting to tighten the nut, force it into the tapered hole with a large pair of pliers.

13   Connect the tie-rod end to the steering knuckle and install the castellated nut. Tighten the nut to the specified torque and install a new cotter pin. If necessary, tighten the nut slightly to align a slot in the nut with the hole in the ball stud.

14   Tighten the lock nuts on the tie-rod connector.

15   Install the wheel and lug nuts, lower the vehicle and tighten the lug nuts to the specified torque. Drive the vehicle to an alignment shop to have the front end alignment checked and, if necessary, adjusted.

Idler arm (Sidekick and Tracker)

16   Raise the vehicle and support it securely on jackstands. Apply the parking brake.

17   Loosen but do not remove the idler arm-to-center link nut.

18   Separate the idler arm from the center link with a two jaw puller. Remove the nut.

19   Remove the idler arm-to-frame bolts.

20   To install the idler arm, position it on the frame and install the bolts, tightening them to the specified torque.

21   Insert the idler arm ball stud into the center link and install the nut. Tighten the nut to the specified torque. If the ball stud spins when attempting to tighten the nut, force it into the tapered hole with a large pair of pliers.

Drag rod (Samurai)

22   Raise the vehicle and support it securely on jackstands. Apply the parking brake.

23   Separate the tie-rod end from the drag rod.

24   Separate the drag rod from the Pitman arm.

25   Installation is the reverse of the removal procedure. If the ball studs spin when attempting to tighten the nuts, force them into the tapered holes with a large pair of pliers. Be sure to tighten all of the nuts to the specified torque.

Center link (Sidekick and Tracker)

26   Raise the front of the vehicle and support it securely on jackstands. Apply the parking brake.

27   Loosen, but do not remove, the nut securing the center link to the tie-rod. Separate the joint with a two jaw puller then remove the nut.

28   Separate the center link from the Pitman arm.

29   Separate the center link from the idler arm.

30   Installation is the reverse of the removal procedure. If the ball studs spin when attempting to tighten the nuts, force them into the tapered holes with a large pair of pliers. Be sure to tighten all of the nuts to the specified torque.

Pitman arm

31   Refer to Section 17 of this Chapter for the Pitman arm removal procedure.

Steering damper (Samurai)

32   Raise the front of the vehicle and support it securely on jackstands.

33   Unbolt the damper from the damper stay **(see illustration)**.

34   Remove the damper to frame nut and remove the damper from the vehicle.

35   Installation is the reverse of the removal procedure.

Rack-and-pinion steering (1999 and later models)

Inspection

Refer to illustration 18.41

36   The steering linkage steering connects the steering gear to the front wheels and keeps the wheels in proper relation to each other. The linkage consists of the steering rack housing which is mounted to the chassis with an internal rack that is connected to the steering arms and a pinion fastened to the steering gear shaft that moves the rack back-and-forth. The back-and-forth motion of the rack is transmitted to the steering knuckles

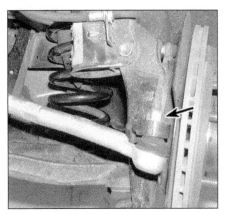

18.41 Remove the tie rod end nut (arrow)

through the tie-rod ends. Looseness in the steering can be caused by wear in the steering shaft intermediate coupler, loose retaining bolts, the steering gear and the tie rod ends.
37   Loosen the wheel lug nuts, raise the vehicle and support it securely on jackstands. Apply the parking brake. Remove the wheel.
38   Crawl under the vehicle and check for loose rack-and-pinion housing clamp bolts. Inspect the steering rack boots for damage and signs of leaking grease or power steering fluid.
39   Have a helper turn the steering wheel from side-to-side to check for loose tie-rod ends. Check the tie-rod end boots for cracks or leaking grease. If they are loose or damaged replace the tie-rod ends with new ones.

Tie-rod end replacement

40   Loosen the tie-rod end jam nut then mark the position of the tie-rod end.
41   Remove the nut attaching tie-rod end to the steering knuckle **(see illustration)**. Using a two jaw puller, separate the tie-rod end from the steering knuckle **(see illustration 18.9)**.
42   Unscrew the tie-rod end.
43   Screw the new tie-rod end onto the tie-rod, threading it in to the mark made during disassembly. Don't tighten the lock nut yet.
44   Insert the tie-rod end in the steering knuckle, install the nut and tighten it to the specified torque.
45   Tighten the tie-rod jam nut securely.

46   Install the wheel and lug nuts, lower the vehicle and tighten the lug nuts to the specified torque. Drive the vehicle to an alignment shop to have the front end alignment checked and, if necessary, adjusted.

19 Power steering pump (Sidekick/X-90/Vitara/Tracker models) - removal and installation

Refer to illustration 19.7

Removal

1   Disconnect the cable from the negative terminal of the battery.
2   Place a drain pan under the power steering pump. Remove the drivebelt (see Chapter 1).
3   Loosen the power steering pressure hose union bolt and let the fluid drain out, then remove the hose.
4   Disconnect the power steering suction hose from the power steering fluid reservoir.
5   Disconnect the power steering pressure switch electrical connector from the pump.
6   Remove the engine oil filter (see Chapter 2).
7   Remove the power steering pump mounting and adjusting bolts **(see illustration)**.
8   Remove the pump from the vehicle, taking care not to spill fluid on the painted surfaces.

Installation

9   Installation is the reverse of removal.
10   Fill the power steering reservoir with the recommended fluid and bleed the system following the procedure described in the next Section.

20 Power steering system - bleeding

1   Following any operation in which the power steering fluid lines have been disconnected, the power steering system must be bled to remove all air and obtain proper steering performance.
2   With the front wheels in the straight ahead position, check the power steering fluid level and, if low, add fluid until it reaches the Cold mark on the dipstick.
3   Start the engine and allow it to run at fast idle. Recheck the fluid level and add more if necessary to reach the Cold mark on the dipstick.
4   Bleed the system by turning the wheels from side-to-side, without hitting the stops. This will work the air out of the system. Keep the reservoir full of fluid as this is done.
5   When the air is worked out of the system, return the wheels to the straight ahead position and leave the vehicle running for several more minutes before shutting it off.
6   Road test the vehicle to be sure the steering system is functioning normally and noise free.
7   Recheck the fluid level to be sure it is up to the Hot mark on the dipstick while the engine is at normal operating temperature. Add fluid if necessary (see Chapter 1).

21 Wheels and tires - general information

Refer to illustration 21.1

All vehicles covered by this manual are equipped with metric-sized fiberglass or steel belted radial tires **(see illustration)**. Use of other size or type of tires may affect the ride

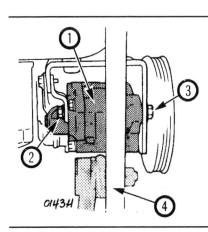

19.7 Power steering pump mounting details (typical Sidekick and Tracker models)

1 *Power steering pump*
2 *Mounting bolt*
3 *Power steering pump adjusting bolt*
4 *Center link*

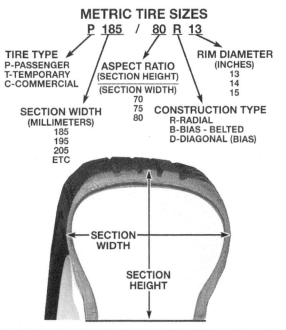

21.1 Metric tire size code

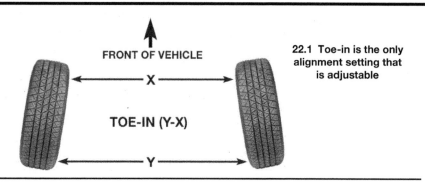

FRONT OF VEHICLE

TOE-IN (Y-X)

22.1 Toe-in is the only alignment setting that is adjustable

and handling of the vehicle. Don't mix different types of tires, such as radials and bias belted, on the same vehicle as handling may be seriously affected. It's recommended that tires be replaced in pairs on the same axle, but if only one tire is being replaced, be sure it's the same size, structure and tread design as the other.

Because tire pressure has a substantial effect on handling and wear, the pressure on all tires should be checked at least once a month or before any extended trips (see Chapter 1).

Wheels must be replaced if they are bent, dented, leak air, have elongated bolt holes, are heavily rusted, out of vertical symmetry or if the lug nuts won't stay tight.

Wheel repairs that use welding or peening are not recommended.

Tire and wheel balance is important to the overall handling, braking and performance of the vehicle. Unbalanced wheels can adversely affect handling and ride characteristics as well as tire life. Whenever a tire is installed on a wheel, the tire and wheel should be balanced by a shop with the proper equipment.

22   Front end alignment - general information

Refer to illustration 22.1

A front end alignment refers to the adjustments made to the front wheels so they are in proper angular relationship to the suspension and the ground. Front wheels that are out of proper alignment not only affect steering control, but also increase tire wear. The only front end adjustment possible on these vehicles is toe-in **(see illustration)**.

Getting the proper front wheel alignment is a very exacting process, one in which complicated and expensive machines are necessary to perform the job properly. Because of this, you should have a technician with the proper equipment perform these tasks. We will, however, use this space to give you a basic idea of what is involved with front end alignment so you can better understand the process and deal intelligently with the shop that does the work.

Toe-in is the turning in of the front wheels. The purpose of a toe specification is to ensure parallel rolling of the front wheels. In a vehicle with zero toe-in, the distance between the front edges of the wheels will be the same as the distance between the rear edges of the wheels. The actual amount of toe-in is normally only a fraction of an inch. Toe-in adjustment is controlled by the tie-rod end position on the inner tie-rod. Incorrect toe-in will cause the tires to wear improperly by making them scrub against the road surface.

Chapter 11 Body

Contents

1 General information

The vehicles covered in this manual have a separate frame and body.

Certain components are particularly vulnerable to accident damage and can be unbolted and repaired or replaced. Among these parts are the body moldings, bumpers, hood, doors and all glass.

Only general body maintenance practices and body panel repair procedures within the scope of the do-it-yourselfer are included in this Chapter.

2 Body - maintenance

1   The condition of your vehicle's body is very important, because the resale value depends a great deal on it. It's much more difficult to repair a neglected or damaged body than it is to repair mechanical components. The hidden areas of the body, such as the wheel wells, the frame and the engine compartment, are equally important, although they don't require as frequent attention as the rest of the body.
2   Once a year, or every 12,000 miles, it's a good idea to have the underside of the body steam cleaned. All traces of dirt and oil will be removed and the area can then be inspected carefully for rust, damaged brake lines, frayed electrical wires, damaged cables and

other problems. The front suspension components should be greased after completion of this job.
3   At the same time, clean the engine and the engine compartment with a steam cleaner or water soluble degreaser.
4   The wheel wells should be given close attention, since undercoating can peel away and stones and dirt thrown up by the tires can cause the paint to chip and flake, allowing rust to set in. If rust is found, clean down to the bare metal and apply an anti-rust paint.
5   The body should be washed about once a week. Wet the vehicle thoroughly to soften the dirt, then wash it down with a soft sponge and plenty of clean soapy water. If the surplus dirt is not washed off very carefully, it can wear down the paint.
6   Spots of tar or asphalt thrown up from the road should be removed with a cloth soaked in solvent.
7   Once every six months, wax the body and chrome trim. If a chrome cleaner is used to remove rust from any of the vehicle's plated parts, remember that the cleaner also removes part of the chrome, so use it sparingly.

3 Upholstery and carpets - maintenance

1   Every three months remove the carpets or mats and clean the interior of the vehicle

(more frequently if necessary). Vacuum the upholstery and carpets to remove loose dirt and dust.
2   Leather upholstery requires special care. Stains should be removed with warm water and a very mild soap solution. Use a clean, damp cloth to remove the soap, then wipe again with a dry cloth. Never use alcohol, gasoline, nail polish remover or thinner to clean leather upholstery.
3   After cleaning, regularly treat leather upholstery with a leather wax. Never use car wax on leather upholstery.
4   In areas where the interior of the vehicle is subject to bright sunlight, cover leather seats with a sheet if the vehicle is to be left out for any length of time.

4 Body repair - minor damage

See photo sequence

Repair of minor scratches

1   If the scratch is superficial and does not penetrate to the metal of the body, repair is very simple. Lightly rub the scratched area with a fine rubbing compound to remove loose paint and built up wax. Rinse the area with clean water.
2   Apply touch-up paint to the scratch, using a small brush. Continue to apply thin layers of paint until the surface of the paint in the scratch is level with the surrounding

paint. Allow the new paint at least two weeks to harden, then blend it into the surrounding paint by rubbing with a very fine rubbing compound. Finally, apply a coat of wax to the scratch area.

3   If the scratch has penetrated the paint and exposed the metal of the body, causing the metal to rust, a different repair technique is required. Remove all loose rust from the bottom of the scratch with a pocket knife, then apply rust inhibiting paint to prevent the formation of rust in the future. Using a rubber or nylon applicator, coat the scratched area with glaze-type filler. If required, the filler can be mixed with thinner to provide a very thin paste, which is ideal for filling narrow scratches. Before the glaze filler in the scratch hardens, wrap a piece of smooth cotton cloth around the tip of a finger. Dip the cloth in thinner and then quickly wipe it along the surface of the scratch. This will ensure that the surface of the filler is slightly hollow. The scratch can now be painted over as described earlier in this section.

Repair of dents

4   When repairing dents, the first job is to pull the dent out until the affected area is as close as possible to its original shape. There is no point in trying to restore the original shape completely as the metal in the damaged area will have stretched on impact and cannot be restored to its original contours. It is better to bring the level of the dent up to a point which is about 1/8-inch below the level of the surrounding metal. In cases where the dent is very shallow, it is not worth trying to pull it out at all.

5   If the back side of the dent is accessible, it can be hammered out gently from behind using a soft-face hammer. While doing this, hold a block of wood firmly against the opposite side of the metal to absorb the hammer blows and prevent the metal from being stretched.

6   If the dent is in a section of the body which has double layers, or some other factor makes it inaccessible from behind, a different technique is required. Drill several small holes through the metal inside the damaged area, particularly in the deeper sections. Screw long, self tapping screws into the holes just enough for them to get a good grip in the metal. Now the dent can be pulled out by pulling on the protruding heads of the screws with locking pliers.

7   The next stage of repair is the removal of paint from the damaged area and from an inch or so of the surrounding metal. This is easily done with a wire brush or sanding disk in a drill motor, although it can be done just as effectively by hand with sandpaper. To complete the preparation for filling, score the surface of the bare metal with a screwdriver or the tang of a file or drill small holes in the affected area. This will provide a good grip for the filler material. To complete the repair, see the Section on filling and painting.

Repair of rust holes or gashes

8   Remove all paint from the affected area and from an inch or so of the surrounding metal using a sanding disk or wire brush mounted in a drill motor. If these are not available, a few sheets of sandpaper will do the job just as effectively.

9   With the paint removed, you will be able to determine the severity of the corrosion and decide whether to replace the whole panel, if possible, or repair the affected area. New body panels are not as expensive as most people think and it is often quicker to install a new panel than to repair large areas of rust.

10   Remove all trim pieces from the affected area except those which will act as a guide to the original shape of the damaged body, such as headlight shells, etc. Using metal snips or a hacksaw blade, remove all loose metal and any other metal that is badly affected by rust. Hammer the edges of the hole inward to create a slight depression for the filler material.

11   Wire brush the affected area to remove the powdery rust from the surface of the metal. If the back of the rusted area is accessible, treat it with rust inhibiting paint.

12   Before filling is done, block the hole in some way. This can be done with sheet metal riveted or screwed into place, or by stuffing the hole with wire mesh.

13   Once the hole is blocked off, the affected area can be filled and painted. See the following subsection on filling and painting.

Filling and painting

14   Many types of body fillers are available, but generally speaking, body repair kits which contain filler paste and a tube of resin hardener are best for this type of repair work. A wide, flexible plastic or nylon applicator will be necessary for imparting a smooth and contoured finish to the surface of the filler material. Mix up a small amount of filler on a clean piece of wood or cardboard (use the hardener sparingly). Follow the manufacturer's instructions on the package, otherwise the filler will set incorrectly.

15   Using the applicator, apply the filler paste to the prepared area. Draw the applicator across the surface of the filler to achieve the desired contour and to level the filler surface. As soon as a contour that approximates the original one is achieved, stop working the paste. If you continue, the paste will begin to stick to the applicator. Continue to add thin layers of paste at 20-minute intervals until the level of the filler is just above the surrounding metal.

16   Once the filler has hardened, the excess can be removed with a body file. From then on, progressively finer grades of sandpaper should be used, starting with a 180-grit paper and finishing with 600-grit wet-or-dry paper. Always wrap the sandpaper around a flat rubber or wooden block, otherwise the surface of the filler will not be completely flat. During the sanding of the filler surface, the wet-or-

dry paper should be periodically rinsed in water. This will ensure that a very smooth finish is produced in the final stage.

17   At this point, the repair area should be surrounded by a ring of bare metal, which in turn should be encircled by the finely feathered edge of good paint. Rinse the repair area with clean water until all of the dust produced by the sanding operation is gone.

18   Spray the entire area with a light coat of primer. This will reveal any imperfections in the surface of the filler. Repair the imperfections with fresh filler paste or glaze filler and once more smooth the surface with sandpaper. Repeat this spray-and-repair procedure until you are satisfied that the surface of the filler and the feathered edge of the paint are perfect. Rinse the area with clean water and allow it to dry completely.

19   The repair area is now ready for painting. Spray painting must be carried out in a warm, dry, windless and dust free atmosphere. These conditions can be created if you have access to a large indoor work area, but if you are forced to work in the open, you will have to pick the day very carefully. If you are working indoors, dousing the floor in the work area with water will help settle the dust which would otherwise be in the air. If the repair area is confined to one body panel, mask off the surrounding panels. This will help minimize the effects of a slight mismatch in paint color. Trim pieces such as chrome strips, door handles, etc., will also need to be masked off or removed. Use masking tape and several thicknesses of newspaper for the masking operations.

20   Before spraying, shake the paint can thoroughly, then spray a test area until the spray painting technique is mastered. Cover the repair area with a thick coat of primer. The thickness should be built up using several thin layers of primer rather than one thick one. Using 600-grit wet-or-dry sandpaper, rub down the surface of the primer until it is very smooth. While doing this, the work area should be thoroughly rinsed with water and the wet-or-dry sandpaper periodically rinsed as well. Allow the primer to dry before spraying additional coats.

21   Spray on the top coat, again building up the thickness by using several thin layers of paint. Begin spraying in the center of the repair area and then, using a circular motion, work out until the whole repair area and about two inches of the surrounding original paint is covered. Remove all masking material 10 to 15 minutes after spraying on the final coat of paint. Allow the new paint at least two weeks to harden, then use a very fine rubbing compound to blend the edges of the new paint into the existing paint. Finally, apply a coat of wax.

---

5   Body repair - major damage

1   Major damage must be repaired by an auto body shop specifically equipped to repair major damage. These shops have the

8.1 Remove the mounting screws and pull off the headlight guards

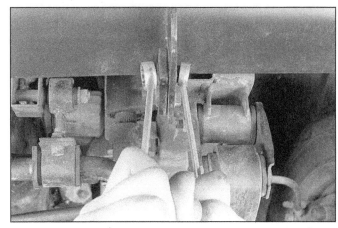

8.2 Unscrew the lower mounting bolts and remove the grille guard (if equipped)

specialized equipment required to do the job properly.

2   If the damage is extensive, the body must be checked for proper alignment or the vehicle's handling characteristics may be adversely affected and other components may wear at an accelerated rate.

3   Due to the fact that all of the major body components (hood, fenders, etc.) are separate and replaceable units, any seriously damaged components should be replaced rather than repaired. Sometimes the components can be found in a wrecking yard that specializes in used vehicle components, often at considerable savings over the cost of new parts.

6   Hinges and locks - maintenance

Once every 3000 miles, or every three months, the hinges and latch assemblies on the doors, hood and tailgate should be given a few drops of light oil or lock lubricant. The door latch strikers should also be lubricated with a thin coat of grease to reduce wear and ensure free movement. Lubricate the door and tailgate locks with spray-on graphite lubricant.

7   Fixed glass - replacement

Replacement of the windshield and fixed glass requires the use of special fast-setting adhesive/caulk materials and some specialized tools and techniques. These operations should be left to a dealer service department or a shop specializing in glass work.

8   Radiator grille - removal and installation

Refer to illustration 8.1, 8.2, 8.3, 8.4 and 8.5
Note: *The radiator grille on some earlier Tracker and Sidekick models is not removable.*

Samurai

1   If the vehicle is equipped with optional headlight guards (Samurai models), remove the screws and pull off the guards **(see illustration)**. **Note:** *The screws have backing plates located inside the fender. Reach inside the fender to hold them so they won't fall when you remove the screws.*

2   If the vehicle is equipped with an optional grille guard, remove the bolts under the bumper and remove the guard **(see**

illustration).

3   Remove the screws that attach the grille to the body.

4   Pry the grille away from the fender to free the locating pins **(see illustration)**.

5   If equipped, remove the mesh insert from the body **(see illustration)**.

Sidekick/X-90/Vitara/Tracker

1998 and earlier models

6   Open the hood.

7   Remove the screw at the top center of the grille.

8   Pull the grille assembly outward, detach the two retainers at the lower corners and lift it from the vehicle.

1999 and later models

9   Pry out the four plastic clips along the top of the grille fascia assembly.

10   Detach the grille assembly by pulling it straight out.

9   Hood - removal, installation and adjustment

Refer to illustrations 9.2 and 9.10
Note: *The hood is heavy and somewhat awk-*

8.4 Carefully pry the grille away from the fender to free the locating pins (arrows) - be careful not to break the grille!

8.5 Remove the mesh insert from the body

These photos illustrate a method of repairing simple dents. They are intended to supplement *Body repair - minor damage* in this Chapter and should not be used as the sole instructions for body repair on these vehicles.

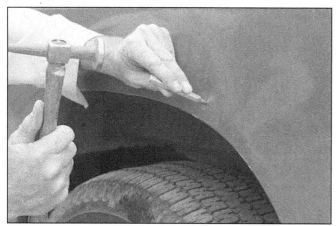

1 If you can't access the backside of the body panel to hammer out the dent, pull it out with a slide-hammer-type dent puller. In the deepest portion of the dent or along the crease line, drill or punch hole(s) at least one inch apart . . .

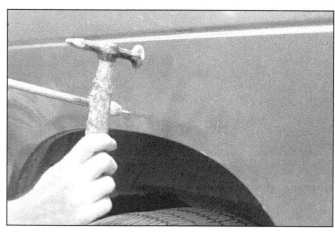

2 . . . then screw the slide-hammer into the hole and operate it. Tap with a hammer near the edge of the dent to help 'pop' the metal back to its original shape. When you're finished, the dent area should be close to its original contour and about 1/8-inch below the surface of the surrounding metal

3 Using coarse-grit sandpaper, remove the paint down to the bare metal. Hand sanding works fine, but the disc sander shown here makes the job faster. Use finer (about 320-grit) sandpaper to feather-edge the paint at least one inch around the dent area

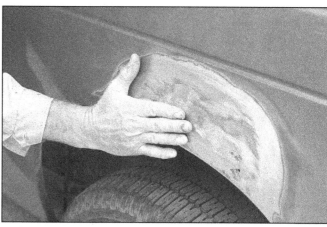

4 When the paint is removed, touch will probably be more helpful than sight for telling if the metal is straight. Hammer down the high spots or raise the low spots as necessary. Clean the repair area with wax/silicone remover

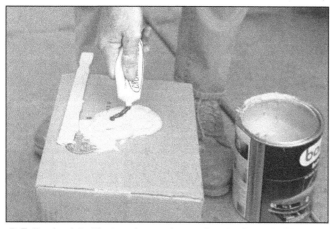

5 Following label instructions, mix up a batch of plastic filler and hardener. The ratio of filler to hardener is critical, and, if you mix it incorrectly, it will either not cure properly or cure too quickly (you won't have time to file and sand it into shape)

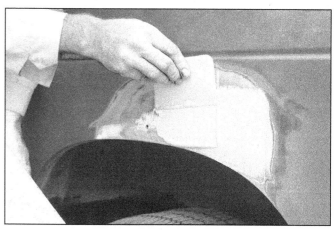

6 Working quickly so the filler doesn't harden, use a plastic applicator to press the body filler firmly into the metal, assuring it bonds completely. Work the filler until it matches the original contour and is slightly above the surrounding metal

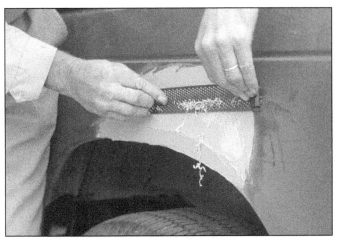

7 Let the filler harden until you can just dent it with your fingernail. Use a body file or Surform tool (shown here) to rough-shape the filler

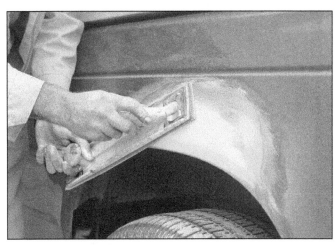

8 Use coarse-grit sandpaper and a sanding board or block to work the filler down until it's smooth and even. Work down to finer grits of sandpaper - always using a board or block - ending up with 360 or 400 grit

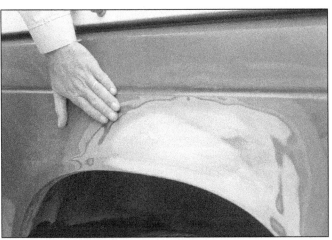

9 You shouldn't be able to feel any ridge at the transition from the filler to the bare metal or from the bare metal to the old paint. As soon as the repair is flat and uniform, remove the dust and mask off the adjacent panels or trim pieces

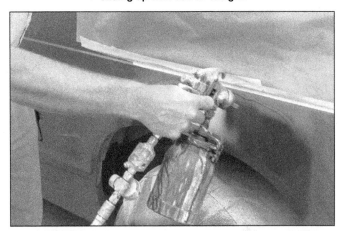

10 Apply several layers of primer to the area. Don't spray the primer on too heavy, so it sags or runs, and make sure each coat is dry before you spray on the next one. A professional-type spray gun is being used here, but aerosol spray primer is available inexpensively from auto parts stores

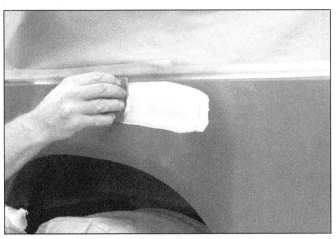

11 The primer will help reveal imperfections or scratches. Fill these with glazing compound. Follow the label instructions and sand it with 360 or 400-grit sandpaper until it's smooth. Repeat the glazing, sanding and respraying until the primer reveals a perfectly smooth surface

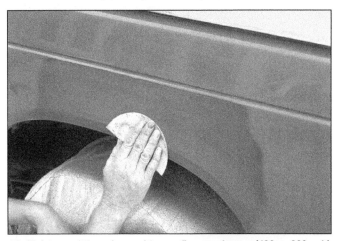

12 Finish sand the primer with very fine sandpaper (400 or 600-grit) to remove the primer overspray. Clean the area with water and allow it to dry. Use a tack rag to remove any dust, then apply the finish coat. Don't attempt to rub out or wax the repair area until the paint has dried completely (at least two weeks)

ward to remove and install - at least two people should perform this procedure. On earlier
models there is no provision for adjustment of
the hood latch so the hood alignment can
only be adjusted on later models.

1 Use blankets or pads to cover the cowl
area of the body and the fenders. This will
protect the body and paint as the hood is
lifted off.
2 Scribe or paint alignment marks around
the hinges to insure proper alignment during
installation **(see illustration)**.
3 Disconnect any cables or wire harnesses which will interfere with removal.
4 Have an assistant support the weight of
the hood. Remove the hinge-to-hood bolts.
5 Lift off the hood.
6 Installation is the reverse of removal.

Adjustment

7 Fore-and-aft and side-to-side adjustment of the hood is done by moving the hood
in relation to the hinge flanges after loosening
the bolts **(see illustration)**.
8 Scribe or trace a line around the entire
hinge plate.
9 Loosen the bolts and move the hood
into correct alignment. Move it only a little at

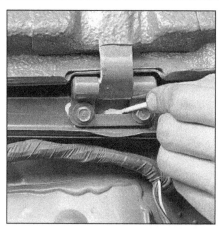

**9.2 Scribe or paint marks around the
hinges to insure proper alignment
on installation**

a time. Tighten the hinge bolts and carefully
lower the hood to check the position.
10 If necessary after installation, the entire
hood latch assembly can be adjusted up-
and-down as well as from side-to-side on the
radiator support so the hood closes securely
and flush with the fenders. Scribe a line or

**9.10 Loosen the hood latch bolts (arrows),
then move the latch as necessary to adjust
the hood-closed position**

mark around the hood latch mounting bolts
to provide a reference point, then loosen
them and reposition the latch assembly, as
necessary **(see illustration)**. Following
adjustment, retighten the mounting bolts.
11 Finally, adjust the hood bumpers on the
radiator support so the hood, when closed, is
flush with the fenders.
12 The hood latch assembly, as well as the
hinges, should be periodically lubricated with
white, lithium-base grease to prevent binding
and wear.

10 Bumper cover and bumper -
removal and installation

Front bumper cover

Refer to illustrations 10.3a and 10.3b
1 Apply the parking brake, raise the vehicle and support it securely on jackstands.
2 Remove the parking light screws.
3 Remove the bolts and screws securing
the bottom and sides of the bumper cover
(see illustrations).
4 Detach the cover and remove it from the
vehicle.
5 Installation is the reverse of removal.
Make sure the tabs on the back of the

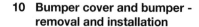

10.3a Remove the bolts (arrows) along the lower edge of the bumper cover

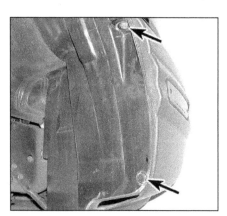

**10.3b Remove the bumper cover screws
(arrows) in the wheelwell**

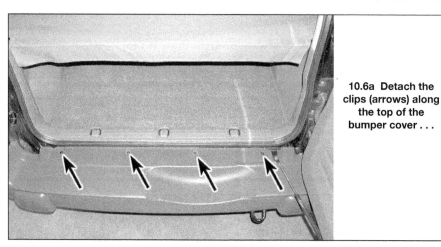

**10.6a Detach the
clips (arrows) along
the top of the
bumper cover . . .**

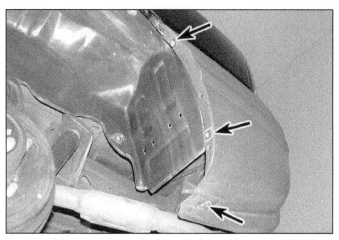

10.6b . . . and the clips (arrows) in each wheelwell

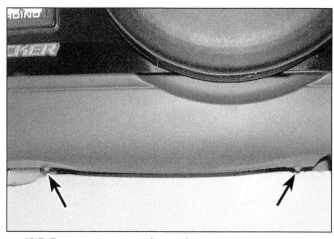

10.7 Remove the screws (arrows) on the bottom of the rear bumper cover

bumper cover fit into the corresponding clips on the body before attaching the bolts and screws. An assistant would be helpful at this point.

Rear bumper cover

Refer to illustrations 10.6a, 10.6b and 6.7

6   Open the tailgate and detach the remove the clips or screws securing the upper edge of the bumper cover, at the front edges in the wheel wells and along the bottom **(see illustrations)**.

7   Working under the vehicle, the screws securing the lower edge of the bumper cover **(see illustration)**. Pull the bumper cover out and away from the vehicle and remove it. On some models it will be necessary to remove the screws attaching the carpet sill and remove the sill so the cover can be pulled away.

8   Installation is the reverse of removal.

Bumper

Refer to illustrations 10.11 and 10.12

9   Detach the bumper cover (if equipped).

10   On earlier models, disconnect the turn signal electrical connectors located under the hood, near each headlight assembly.

11   Support the bumper with a jack or jackstand. Alternatively, have an assistant support the bumper as the bolts are removed **(see illustration)**.

12   Remove the retaining bolts located under the wheelwells **(see illustration)** and detach the bumper.

13   Installation is the reverse of removal.

14   Tighten the retaining bolts securely.

15   Install the bumper cover and any other components that were removed.

11   Seat belt check

1   Check the seat belts, buckles, latch plates and guide loops for obvious damage and signs of wear.

2   Check that the seat belt reminder light

10.11 Support the bumper with a jack - place a wood block between the jack head and bumper (Samurai)

comes on when the key is turned to the Run or Start positions. A chime should also sound.

3   The seat belts are designed to lock up during a sudden stop or impact, yet allow free movement during normal driving. Check that the retractors return the belt against your chest while driving and rewind the belt fully when the buckle is unlatched.

4   If any of the above checks reveal problems with the seat belt system, replace parts as necessary.

12   Door trim panel - removal and installation

Refer to illustrations 12.2a, 12.2b, 12.3 and 12.6

1   Disconnect the negative cable from the battery.

2   Remove all door trim panel retaining screws and door pull/armrest assemblies **(see illustration)**. On later models, remove

10.12 Remove the bumper retaining bolts located under the wheelwells (Samurai)

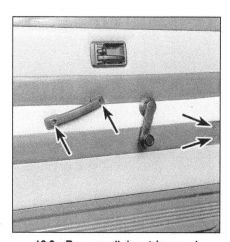

12.2a Remove all door trim panel retaining screws and the armrest

12.2b On later models, remove the door handle (arrow)

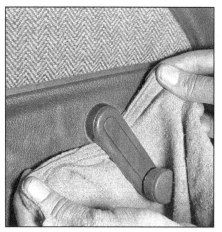

12.3 Work a cloth up behind the regulator handle and move it back-and-forth

12.6 Remove the plastic watershield, being careful not to tear it

the door handle retaining screw and detach the handle assembly **(see illustration)**.

3   Remove the window crank. Pop out the snap ring using a cloth **(see illustration)**.

4   Insert a putty knife between the trim panel and the door and disengage the retaining clips. Work around the outer edge until the panel is free.

5   Once all of the clips are disengaged, detach the trim panel, unplug any wire harness connectors and remove the trim panel from the vehicle.

6   For access to the inner door, carefully peel back the plastic watershield **(see illustration)**.

7   Prior to installation of the door panel, be sure to reinstall any clips in the panel which may have come out during the removal procedure and remain in the door itself.

8   Plug in the wire harness connectors and place the panel in position in the door. Press the door panel into place until the clips are seated and install the armrest/door pulls. Install the window crank handle.

---

13 Door - removal, installation and adjustment

Refer to illustration 13.4

1   Remove the door trim panel (see Section 12). Disconnect any wire harness connectors and push them through the door opening so they won't interfere with door removal.

2   Place a jack or jackstand under the door or have an assistant on hand to support it when the hinge bolts are removed. **Note:** *If a jack or jackstand is used, place a rag between it and the door to protect the door's painted surfaces.*

3   Scribe around the door hinges.

4   Remove the hinge-to-door bolts and carefully lift off the door **(see illustration)**.

5   Installation is the reverse of removal.

6   Following installation of the door, check

the alignment and adjust it if necessary as follows:

a) *Up-and-down and forward-and-backward adjustments are made by loosening the hinge-to-body bolts and moving the door as necessary.*

b) *The door lock striker can also be adjusted both up-and-down and side-to-side to provide positive engagement with the lock mechanism. This is done by loosening the mounting screws and moving the striker as necessary.*

c) *If the striker is too far to the front or rear to properly engage the door latch, the striker may be adjusted by adding or removing shims from beneath it.*

---

14 Tailgate - removal, installation and adjustment (hardtop models only)

Removal

Refer to illustration 14.8

1   Remove the spare tire assembly.

2   Remove the trim panel by pushing the

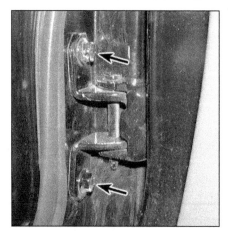

13.4 With the door supported, remove the hinge bolts (arrows)

plastic stud retainers in the center to release them.

3   Remove the plastic watershield. Be careful not to tear it.

4   Disconnect the electrical connectors for the rear window defogger and wiper and also disconnect the license plate lamp.

5   To ensure proper adjustment when the tailgate is reinstalled, use a scribe or a marker to mark around the tailgate hinge plates.

6   Disconnect the support strut from the tailgate.

7   Have an assistant support the tailgate. Alternatively, support the tailgate using a wood block and a floor jack. Place a towel or cloth between the wood and the tailgate.

8   Remove the upper and lower hinge bolts from the tailgate **(refer to illustration)**.

9   Remove the tailgate.

Installation

10   Have an assistant help you lift the tailgate into position. Alternatively, lift it with a floor jack and a wood block.

11   Install the tailgate following the reverse order of the above procedure. To ensure proper adjustment, be sure to align the marks you made on removal.

14.8 Support the tailgate and remove the hinge bolts (arrows)

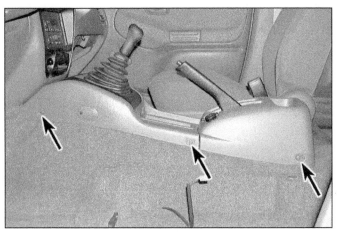

15.3 Remove the plastic retainers (arrows) on each side of the center console

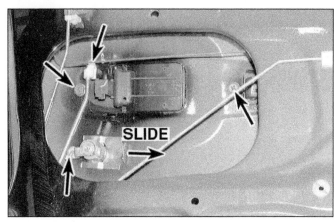

16.3 Detach the control rods, remove the door handle retaining screws and detach the lock cylinder retaining clip by sliding it in the direction shown with a screwdriver (arrows)

Adjustment

12   After installation, close the tailgate and check that it is in proper alignment with the surrounding body panels. Adjustments to the tailgate are made by moving the position of the hinge bolts in their slots. To adjust, loosen the hinge bolts, reposition the tailgate the desired amount and retighten the bolts.

13   The engagement of the tailgate can be adjusted by loosening the lock striker screws, repositioning the striker on the body and retightening the screws. If the striker is too far to the left or right to properly engage the latch on the tailgate, add or remove shims from beneath the striker. **Note:** *Do not adjust the latch.*

15 Center console - removal and installation (Sidekick/X-90/Vitara and Tracker)

Warning: *1996 and later models are equipped with an airbag system. Always disable the airbags prior to working in the vicinity of the steering column, instrument panel or airbag components to avoid the possibility of accidental deployment of the airbags, which*

could cause personal injury (see Chapter 12). Refer to illustration 15.3

1   Remove the two Phillips screws on the side of the console, then remove the two clips at the rear and pull off the rear console cover. To remove the clips, push in the center pins first.

2   On earlier models, use a 3 mm hex drive to remove the two screws at the rear of the console.

3   On later models, remove the three plastic retainers on each side by pushing in the center pins **(see illustration)**. Lift off the console.

4   Installation is the reverse of removal.

16 Door latch, lock cylinder and handles - removal and installation

Refer to illustrations 16.3 and 16.7

1   Remove the door trim panel as described in Section 12.

2   Remove the plastic watershield, taking care not to tear it **(see illustration 12.6)**.

3   Disengage the control rods from the connections at the door outside handle lock

cylinder by prying gently with a screwdriver **(see illustration)**.

4   Remove the screw(s) that retain the exterior handle assembly and lift it out.

5   Disengage the door lock rod from the door lock assembly.

6   Remove the retaining clip from the door lock cylinder and remove the cylinder.

7   Remove the door latch assembly mounting screws, located at the end of the door, then detach the control rods and lift out the latch assembly **(see illustration)**.

8   Installation is the reverse of removal. **Note:** *During installation, apply grease to the sliding surface of all levers and springs.*

17 Door window glass and regulator - removal and installation

Refer to illustrations 17.2, 17.3a, 17.3b, 17.4a and 17.4b

1   Remove the door trim panel and watershield (see Section 12).

2   Using a curved tool, remove the two molding pieces located on the top of the door **(see illustration)**.

3   Remove the glass mounting screws and

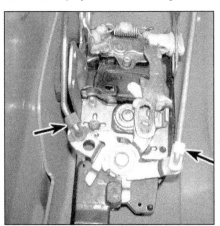

16.7 Detach the two control rods from the latch assembly (arrows)

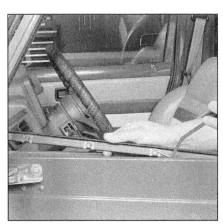

17.2 Locate the molding clips and pry them up to remove the molding pieces (Samurai)

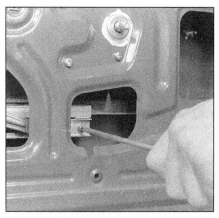

17.3a With the window rolled down, remove the two mounting screws (Samurai)

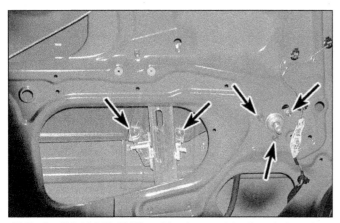

17.3b On Sidekick/X-90/Vitara/Tracker models, the window glass is held in place by two screws and the regulator is retained by three screws (arrows)

17.3c Remove the glass by carefully pulling it up - tilt it so it clears the channel

pull the glass out the top of the door **(see illustrations)**.

4   Remove the regulator mounting screws and remove the regulator through the large access hole **(see illustrations)**. On Sidekick/X-90/Vitara and Tracker models, **refer to illustration 17.3b** for the regulator nut locations.

5   Prior to installing the regulator and guide channel, apply a light coat of lithium-based grease to all their sliding surfaces.

6   Installation is the reverse of the removal procedure. Following installation of the window, roll it up completely and use the following procedure to bring it into correct alignment.

a) *The tilt of the window can be adjusted by loosening the regulator retaining screws and moving the glass so that its upper edge is parallel with the upper edge of the door. Following adjustment, retighten the screws.*

b) *To adjust the window in the fore-and-aft position, loosen the regulator retaining screws, then adjust the glass so that its rear edge and upper rear corner are seated firmly in the rubber of the door frame. Following adjustment, retighten the retaining screws.*

18 Outside mirror - removal and installation

1   Remove the mirror bezel.
2   Remove mirror attaching screws.
3   Remove mirror assembly.
4   Installation is the reverse of the above procedure.

19 Instrument panel - removal and installation

Warning: *1996 and later models are equipped with an airbag system. Always disable the airbags prior to working in the vicinity of the steering column, instrument panel or airbag components to avoid the possibility of accidental deployment of the airbags, which could cause personal injury (see Chapter 12).*

Samurai

1   Disconnect the cable from the negative battery terminal.
2   Remove the steering wheel (see Chapter 10).
3   Remove the radio (see Chapter 12), the cigar lighter and the ashtray.

4   Detach the fuse panel and hood opening cable from the instrument panel.
5   Remove the cables from the heater control panel and remove the panel.
6   Detach the defroster hoses.
7   Disconnect the electrical connectors from the instrument panel and switches. Disconnect the speedometer cable from the speedometer.
8   Disconnect any remaining wiring harness retainers connected to the instrument panel.
9   Remove the instrument panel retaining screws and remove the panel from the vehicle.
10   Installation is the reverse of removal.

Sidekick/X-90/Vitara/Tracker

Refer to illustrations 19.13a, 19.13b, 19.16, 19.20a, 19.20b and 19.20c

11   Disconnect the cable from the negative battery terminal. On models equipped with airbags, disable the airbag modules (see Chapter 12).
12   Remove the steering wheel (see Chapter 10).
13   Remove the lower steering column trim panel and the steering column covers **(see illustrations)**.

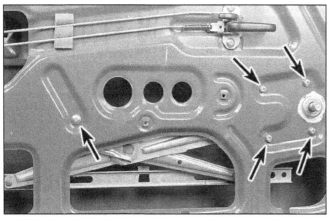

17.4a Locations of the regulator mounting screws (Samurai)

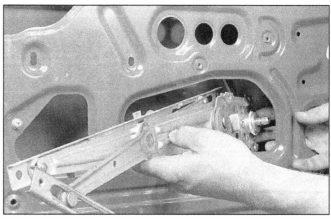

17.4b The regulator can be removed through the large access hole in the door

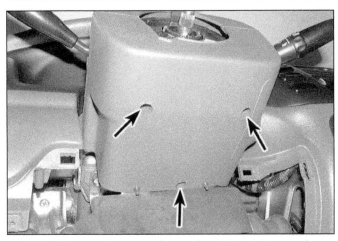

19.13a Remove the screws (arrows) and detach the steering column covers

19.13b The lower steering column cover on later Sidekick/X-90/Vitara/Tracker models is held in place by three screws (arrows)

14   Remove the turn signal/combination switch from the steering column (see Chapter 12).
15   Remove the instrument cluster (see Chapter 12).
16   Remove the center trim bezel, the radio and the heater/air conditioning control panel **(see illustration)**.
17   Remove the glove compartment **(see illustration)**. On 1996 and later models, remove the passenger airbag module from the instrument panel and store it in a safe place with the trim cover facing up. **Warning:** *Be extremely careful when handling a live airbag module, carry the module with the trim cover pointed away from your body. Never place the module on any surface with the trim cover facing down or serious injury could occur in the event of an accidental deployment.*
18   Remove the instrument panel retaining screws.
19   On 1995 and earlier models, pry off the screw covers along the top of the instrument panel and remove the screws underneath. Remove the grab handle attaching nuts and remove the handle. Remove the screws along

19.16 Remove the three center trim bezel retaining screws (arrows)

the bottom right and left sides of the instrument panel.
20   On 1996 and later models. Pry off the screw covers at each end of the instrument panel and remove the screws. Pry off the speaker grilles and remove the screws underneath. Remove the screws retaining the

19.17 Remove the glove compartment by pushing in on the sides and lowering the glove compartment assembly for access to the retaining screws

instrument panel support beam to the body. Remove the screws along the bottom right and left sides of the instrument panel. Remove the screws along the top of the instrument panel **(see illustrations)**.
21   Pull the instrument panel out slightly, detach the defroster ducts, and disconnect

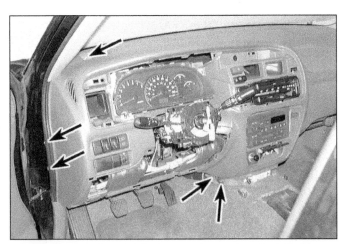

19.20a Remove the retaining screws (arrows) on the left side . . .

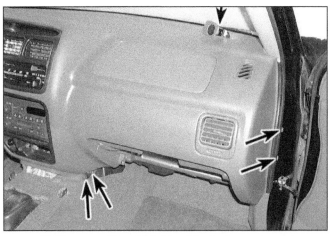

19.20b . . . and the right side of the instrument panel

any remaining electrical connectors or cables from the panel. Disconnect the wiring harness clamps and detach the wiring harness from the instrument panel.

22   Remove the instrument panel from the vehicle.

23   Installation is the reverse of removal. On 1996 and later models, be sure to center the airbag coil when installing the combination switch/airbag coil (see Chapter 12).

___

20   Cowl cover - removal and installation

___

Refer to illustration 20.2

1   Remove the hood (Section 9)

2   Remove the retaining nuts and screws and detach the cowl cover from the vehicle.

3   Installation is the reverse of removal.

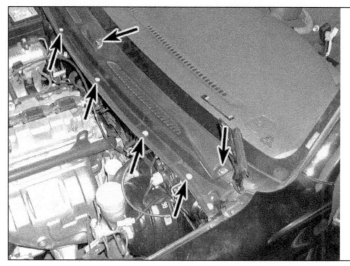

20.2 Remove the cowl cover retaining screws and nuts (arrows)

Chapter 12
Chassis electrical system

Contents

1 General information

The electrical system is a 12-volt, negative ground type. Power for the lights and all electrical accessories is supplied by a lead/acid-type battery that is charged by the alternator.

This Chapter covers repair and service procedures for the various electrical components not associated with the engine. Information on the battery, alternator, distributor and starter motor can be found in Chapter 5.

It should be noted that when portions of the electrical system are serviced, the negative battery cable should be disconnected from the battery to prevent electrical shorts and/or fires.

2 Electrical troubleshooting - general information

Refer to illustrations 2.5a, 2.5b, 2.6, 2.9 and 2.15

A typical electrical circuit consists of an electrical component, any switches, relays, motors, fuses, fusible links or circuit breakers related to that component and the wiring and connectors that link the component to both the battery and the chassis. To help you pinpoint an electrical circuit problem, wiring diagrams are included at the end of this Chapter.

Before tackling any troublesome electrical circuit, first study the appropriate wiring diagrams to get a complete understanding of what makes up that individual circuit. Trouble spots, for instance, can often be narrowed down by noting if other components related to the circuit are operating properly. If several components or circuits fail at one time, chances are the problem is in a fuse or ground connection, because several circuits are often routed through the same fuse and ground connections.

Electrical problems usually stem from simple causes, such as loose or corroded connections, a blown fuse, a melted fusible link or a failed relay. Visually inspect the condition of all fuses, wires and connections in a problem circuit before troubleshooting the circuit.

If test equipment and instruments are going to be utilized, use the diagrams to plan

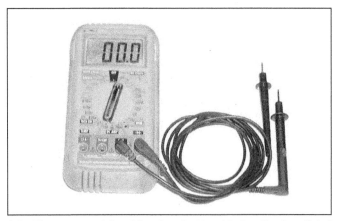

2.5a The most useful tool for electrical troubleshooting is a digital multimeter that can check volts, amps, and test continuity

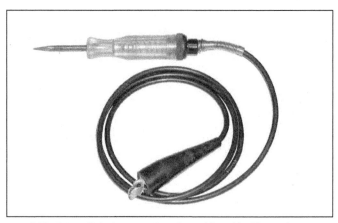

2.5b A test light is a very handy tool for checking voltage

ahead of time where you will make the necessary connections in order to accurately pinpoint the trouble spot.

The basic tools needed for electrical troubleshooting include a circuit tester or voltmeter (a 12-volt bulb with a set of test leads can also be used), a continuity tester, which includes a bulb, battery and set of test leads, and a jumper wire, preferably with a circuit breaker incorporated, which can be used to bypass electrical components **(see illustrations)**. Before attempting to locate a problem with test instruments, use the wiring diagram(s) to decide where to make the connections.

Voltage checks

Voltage checks should be performed if a circuit is not functioning properly. Connect one lead of a circuit tester to either the negative battery terminal or a known good ground. Connect the other lead to a connector in the circuit being tested, preferably nearest to the battery or fuse **(see illustration)**. If the bulb of the tester lights, voltage is present, which means that the part of the circuit between the connector and the battery is problem free.

Continue checking the rest of the circuit in the same fashion. When you reach a point at which no voltage is present, the problem lies between that point and the last test point with voltage. Most of the time the problem can be traced to a loose connection. **Note:** *Keep in mind that some circuits receive voltage only when the ignition key is in the Accessory or Run position.*

Finding a short

One method of finding shorts in a circuit is to remove the fuse and connect a test light or voltmeter in place of the fuse terminals. There should be no voltage present in the circuit. Move the wiring harness from side-to-side while watching the test light. If the bulb goes on, there is a short to ground somewhere in that area, probably where the insulation has rubbed through. The same test can be performed on each component in the circuit, even a switch.

Ground check

Perform a ground test to check whether a component is properly grounded. Disconnect the battery and connect one lead of a continuity tester or multimeter (set to the ohms scale), to a known good ground. Connect the other lead to the wire or ground connection being tested. If the resistance is low (less than 5 ohms), the ground is good. If the bulb on a self-powered test light does not go on, the ground is not good.

Continuity check

A continuity check is done to determine if there are any breaks in a circuit - if it is passing electricity properly. With the circuit off (no power in the circuit), a self-powered continuity tester or multimeter can be used to check the circuit. Connect the test leads to both ends of the circuit (or to the "power" end and a good ground), and if the test light comes on the circuit is passing current properly **(see illustration)**. If the resistance is low (less than 5 ohms), there is continuity; if the reading is 10,000 ohms or higher, there is a break somewhere in the circuit. The same procedure can be used to test a switch, by connecting the continuity tester to the switch terminals. With the switch turned On, the test light should come on (or low resistance should be indicated on a meter).

2.6 In use, a basic test light's lead is clipped to a known good ground, then the pointed probe can test connectors, wires or electrical sockets - if the bulb lights, the part being tested has battery voltage

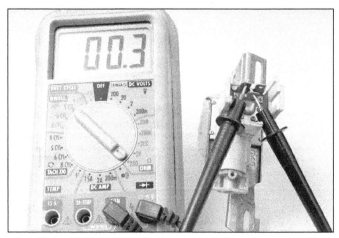

2.9 With a multimeter set to the ohms scale, resistance can be checked across two terminals - when checking for continuity, a low reading indicates continuity, a high reading indicates lack of continuity

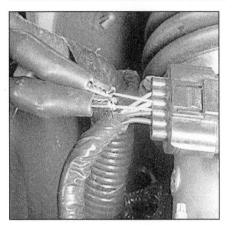

2.15 To backprobe a connector, insert a small, sharp probe (such as a straight-pin) into the back of the connector alongside the desired wire until it contacts the metal terminal inside; connect your meter leads to the probes - this allows you to test a functioning circuit

Finding an open circuit

When diagnosing for possible open circuits, it is often difficult to locate them by sight because the connectors hide oxidation or terminal misalignment. Merely wiggling a connector on a sensor or in the wiring harness may correct the open circuit condition. Remember this when an open circuit is indicated when troubleshooting a circuit. Intermittent problems may also be caused by oxidized or loose connections.

Electrical troubleshooting is simple if you keep in mind that all electrical circuits are basically electricity running from the battery, through the wires, switches, relays, fuses and fusible links to each electrical component (light bulb, motor, etc.) and to ground, from which it is passed back to the battery. Any electrical problem is an interruption in the flow of electricity to and from the battery.

Connectors

Most electrical connections on these vehicles are made with multiwire plastic con-

3.1a The fuse box on Samurai models is located under the left side of the dash, on the firewall

nectors. The mating halves of many connectors are secured with locking clips molded into the plastic connector shells. The mating halves of large connectors, such as some of those under the instrument panel, are held together by a bolt through the center of the connector.

To separate a connector with locking clips, use a small screwdriver to pry the clips apart carefully, then separate the connector halves. Pull only on the shell, never pull on the wiring harness as you may damage the individual wires and terminals inside the connectors. Look at the connector closely before trying to separate the halves. Often the locking clips are engaged in a way that is not immediately clear. Additionally, many connectors have more than one set of clips.

Each pair of connector terminals has a male half and a female half. When you look at the end view of a connector in a diagram, be sure to understand whether the view shows the harness side or the component side of the connector. Connector halves are mirror images of each other, and a terminal shown on the right side end-view of one half will be on the left side end-view of the other half.

It is often necessary to take circuit voltage measurements with a connector con-

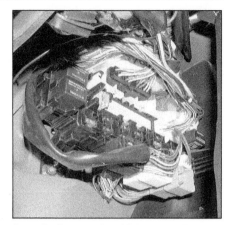

3.1b On Sidekick/X-90/Vitara and Tracker models, the interior fuse/relay panel is located under the left side of the instrument panel on the kick panel

nected. Whenever possible, carefully insert a small straight pin (not your meter probe) into the rear of the connector shell to contact the terminal inside, then clip your meter lead to the pin. This kind of connection is called "backprobing" (see illustration). When inserting a test probe into a terminal, be careful not to distort the terminal opening. Doing so can lead to a poor connection and corrosion at that terminal later. Using the small straight pin instead of a meter probe results in less chance of deforming the terminal connector.

3 Fuses - general information

Refer to illustrations 3.1a, 3.1b, 3.1c and 3.3

The electrical circuits of the vehicle are protected by a combination of fuses, circuit breakers and fusible links. The interior fuse/relay block is located under the instrument panel on the left side of the dashboard, while the main fuse/relay panel is in the engine compartment (see illustrations).

Each of the fuses is designed to protect a specific circuit, and the various circuits are identified on the fuse panel itself.

3.1c On Sidekick/X-90/Vitara and Tracker models the main fuse/relay box is located in the engine compartment - the cover has a legend identifying the fuses and relays

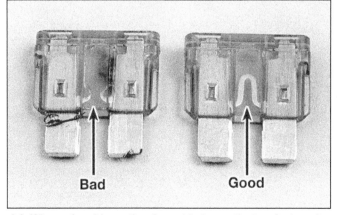

3.3 When a fuse blows, the element between the terminals melts - the fuse on the left is blown, the fuse on the right is good

4.2 A fusible link protects the wire harness from overload and possible damage (Samurai)

6.1 The turn signal and hazard flasher relay is located in or near the interior fuse box (Samurai model shown)

Miniaturized fuses are employed in the fuse block. These compact fuses, with blade terminal design, allow fingertip removal and replacement. If an electrical component fails, always check the fuse first. The easiest way to check fuses is with a test light. Check for power at the exposed terminal tips of each fuse. If power is available on one side of the fuse but not the other, the fuse is blown. A blown fuse can also be confirmed by visually inspecting it **(see illustration)**.

Be sure to replace blown fuses with the correct type. Fuses of different ratings are physically interchangeable, but only fuses of the proper rating should be used. Replacing a fuse with one of a higher or lower value than specified is not recommended. Each electrical circuit needs a specific amount of protection. The amperage value of each fuse is molded into the fuse body.

If the replacement fuse immediately fails, don't replace it again until the cause of the problem is isolated and corrected. In most cases, the cause will be a short circuit in the wiring caused by a broken or deteriorated wire.

4   Fusible links - general information

Refer to illustration 4.2

Some circuits are protected by fusible links. The links are used in circuits which are not ordinarily fused, such as the ignition circuit.

Although the fusible links appear to be a heavier gauge than the wire they are protecting, the appearance is due to the thick insulation. All fusible links are several wire gauges smaller than the wire they are designed to protect **(see illustration)**. **Note:** *The fusible link on later models is similar to a large fuse, located in the underhood fuse box.*

Fusible links cannot be repaired, but a new link of the same size wire can be put in its place. The procedure is as follows:

a) *Disconnect the negative cable from the battery.*

b) *Disconnect the fusible link from the wiring harness.*
c) *Cut the damaged fusible link out of the wiring just behind the connector.*
d) *Strip the insulation back approximately 1/2-inch.*
e) *Position the connector on the new fusible link and crimp it into place.*
f) *Use rosin core solder at each end of the new link to obtain a good solder joint.*
g) *Use plenty of electrical tape around the soldered joint. No wires should be exposed.*
h) *Connect the battery ground cable. Test the circuit for proper operation.*

5   Relays - general information

Several electrical accessories in the vehicle use relays to transmit the electrical signal to the component. If the relay is defective, that component will not operate properly.

The various relays are grouped together in several locations.

If a faulty relay is suspected, it can be removed and tested by a dealer service department or a repair shop. Defective relays must be replaced as a unit.

6   Turn signal and hazard flashers - check and replacement

Refer to illustration 6.1

Turn signal flasher

1   The turn signal flasher, a small canister or box-shaped unit located in or near the fuse block **(see illustration)**, flashes the turn signals. On later models the turn signal and hazard flashers are contained in one unit.
2   When the flasher unit is functioning properly, an audible click can be heard during its operation. If the turn signals fail on one side or the other and the flasher unit does not make its characteristic clicking sound a faulty turn signal bulb is indicated.
3   If both turn signals fail to blink, the problem may be due to a blown fuse, a faulty

flasher unit, a broken switch or a loose or open connection. If a quick check of the fuse box indicates that the turn signal fuse has blown, check the wiring for a short before installing a new fuse.
4   To replace the flasher, simply pull it out of the fuse block or wiring harness.
5   Make sure that the replacement unit is identical to the original. Compare the old one to the new one before installing it.
6   Installation is the reverse of removal.

Hazard flasher

7   The hazard flasher, which is integral with the turn signal flasher on later models, flashes all four turn signals simultaneously when activated.
8   The hazard flasher is checked in a fashion similar to the turn signal flasher (see Steps 2 and 3).
9   To replace the hazard flasher, pull it from the back of fuse block.
10   Make sure the replacement unit is identical to the one it replaces. Compare the old one to the new one before installing it.
11   Installation is the reverse of removal.

7   Ignition switch - removal and installation

Refer to illustrations 7.2, 7.3 and 7.4
Warning: *1996 and later models are equipped with an airbag system. Always disable the airbags prior to working in the vicinity of the steering column, instrument panel or airbag components to avoid the possibility of accidental deployment of the airbags, which could cause personal injury (see Section 17).*

Removal

1   Disconnect the negative cable at the battery. Remove the steering wheel (see Chapter 10).
2   Remove the steering column covers **(see illustration)**.
3   Locate the shear bolts that attach the ignition lock onto the steering column **(see illustration)** and remove the shear bolts.

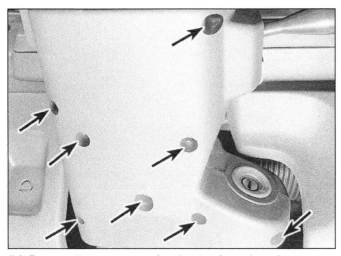

7.2 Remove the screws securing the steering column lower cover and detach the cover from the steering column (Samurai model shown)

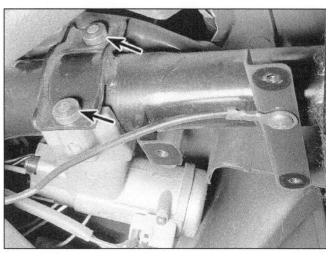

7.3 The ignition switch/lock is fastened to the steering column with two shear bolts (arrows)

Note: *There are several methods used to remove these specialized locking bolts. The most common method is to use a sharp chisel and notch the head and turn the bolt counterclockwise while tapping the chisel with a hammer* **(see illustration).** *If that method does not work, use a medium size drill bit and an easy-out.*

4   Trace the switch wires to the connector and unplug them **(see illustration).**

Installation

5   Installation is the reverse of the removal procedure. **Note:** *Use new shear bolts and tighten them until the heads break off, as an anti-theft measure.*

8   Turn signal/combination switch - removal and installation

Refer to illustrations 8.4, 8.5a, 8.5b and 8.6
Warning: *1996 and later models are equipped with an airbag system. Always disable the airbags prior to working in the vicinity of the steering column, instrument panel or airbag components to avoid the possibility of*

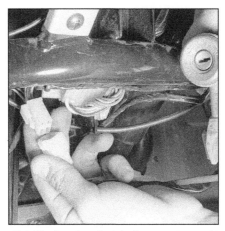

7.4 Unplug the electrical connector and remove the lock from the steering column (Samurai model shown)

accidental deployment of the airbags, which could cause personal injury (see Section 17).
1   Disconnect the negative cable at the battery.
2   Remove the steering column lower cover **(see illustration 7.2).**

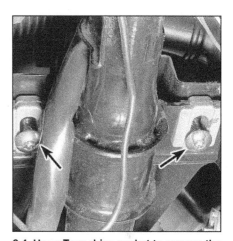

8.4 Use a Torx-drive socket to remove the steering column bolts

3   Remove the steering wheel (see Chapter 10).
4   Lower the steering column by removing the two Torx-head bolts located under the dash **(see illustration).**
5   Remove the screws that hold the switch to the column **(see illustrations).**

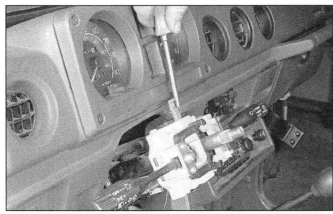

8.5a Remove the two upper and the two lower combination switch mounting screws (Samurai model shown)

8.5b Sidekick/X-90/Vitara and Tracker combination switch retaining screws (arrows)

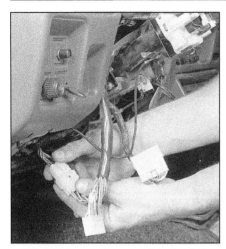

8.6 Disconnect the turn signal switch electrical connectors (Samurai models)

9.3 Loosen only the screws that retain the bulb (arrows) - do not disturb the adjustment screws (Samurai)

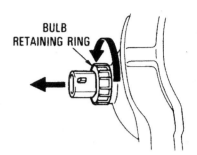

9.9a Turn the bulb retaining ring counterclockwise . . .

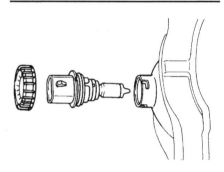

9.9b then pull the bulb and holder out of the housing

6   Trace the combination switch wires to the connector under the instrument panel, disconnect them and remove the switch assembly **(see illustration)**. **Note:** *The switch is a combination turn signal, dimmer and wiper/washer switch.*

7   On models equipped with airbags, center the airbag coil as follows: **Caution:** *Failure to center the airbag coil before installing the combination switch/coil assembly will result in damage to the coil.*

    a) *Position the front wheels pointing straight-ahead.*

    b) *Slowly rotate the coil hub counterclockwise until it reaches its stop.*

    c) *Rotate the coil clockwise 2-1/2 turns and align the arrows on the coil hub and body.*

8   Installation is the reverse of the removal procedure.

9   Headlights - removal and installation

1   Disconnect the negative cable from the battery.

Samurai sealed-beam type

Refer to illustration 9.3

2   Remove the front grille assembly (see Chapter 11).

3   Remove the headlight retainer screws, taking care not to disturb the adjustment screws **(see illustration)**. **Note:** *On models with round headlights, simply loosen the screws enough to turn the retaining ring.*

4   Remove the retainer and pull the headlight out sufficiently to allow the connector to be unplugged.

5   Remove the headlight.

6   To install, plug the connector securely into the headlight, place the headlight in position and install the retainer and screws. Tighten the screws securely.

7   Install the front grille assembly.

Sidekick/X-90/Vitara and Tracker bulb type

Refer to illustrations 9.9a and 9.9b

Warning: *The halogen gas filled bulbs used on these models are under pressure and may shatter if the surface is scratched or the bulb is dropped. Wear eye protection and handle the bulbs carefully, grasping only the base whenever possible. Do not touch the surface of the bulb with your fingers because the oil from you skin could cause it to overheat and fail prematurely. If you do touch the bulb surface, clean it with rubbing alcohol.*

8   Open the hood. Unplug the electrical connector.

9   Reach behind the headlight assembly. On earlier models, grasp the bulb retaining ring and turn it counterclockwise to remove it **(see illustrations)**. On later models, detach the spring clip. On all models, lift the holder assembly out for access to the bulb.

10   Push in and rotate the bulb counterclockwise to remove it.

11   Insert the new bulb into the holder and turn it clockwise to seat it in the holder.

12   Install the bulb holder in the headlight assembly.

10   Headlights - adjustment

Refer to illustration 10.2

Note: *The headlights must be aimed correctly. If adjusted incorrectly they could blind the driver of an oncoming vehicle and cause a serious accident or seriously reduce your ability to see the road. The headlights should be checked for proper aim every 12 months and any time a new headlight is installed or front end body work is performed. It should be emphasized that the following procedure is only an interim step which will provide temporary adjustment until the headlights can be adjusted by a properly equipped shop.*

1   Headlights have two spring loaded adjusting screws, one on the top controlling up-and-down movement and one on the side controlling left-and-right movement.

2   There are several methods of adjusting the headlights. The simplest method requires a blank wall 25 feet in front of the vehicle and a level floor **(see illustration)**.

3   Position masking tape vertically on the wall in reference to the vehicle centerline and the centerlines of both headlights.

4   Position a horizontal tape line in reference to the centerline of all the headlights. **Note:** *It may be easier to position the tape on the wall with the vehicle parked only a few inches away.*

5   Adjustment should be made with the vehicle sitting level, the gas tank half-full and no unusually heavy load in the vehicle.

6   Starting with the low beam adjustment, position the high intensity zone so it is two inches below the horizontal line and two inches to the side of the headlight vertical line, away from oncoming traffic. Adjustment is made by turning the top adjusting screw clockwise to raise the beam and counterclockwise to lower the beam. The adjusting screw on the side should be used in the same manner to move the beam left or right.

7   With the high beams on, the high intensity zone should be vertically centered with the exact center just below the horizontal line. **Note:** *It may not be possible to position the headlight aim exactly for both high and low beams. If a compromise must be made, keep in mind that the low beams are the most used and have the greatest effect on driver safety.*

8   Have the headlights adjusted by a

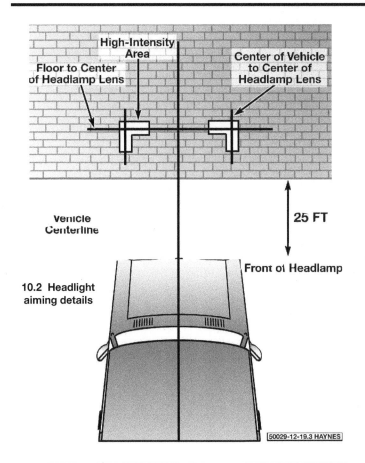

10.2 Headlight aiming details

11.2 Remove the retaining plate nuts with a deep socket (Samurai)

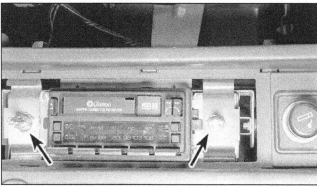

11.4 Unscrew the nuts on the front brace (see arrows) and remove the radio from the back (Samurai)

dealer service department or service station at the earliest opportunity.

11 Radio and speakers - removal and installation

1 Disconnect the cable from the negative terminal of the battery.

Radio

Samurai

Refer to illustrations 11.2 and 11.4
2 Pull off the radio knobs and remove the retaining plate nuts with a deep socket **(see illustration)**.
3 Remove the bottom retaining screws and allow the radio/console unit to drop slightly.
4 Unscrew the nuts on the front brace and remove the radio from the back **(see illustration)**.
5 Installation is the reverse of the removal.

1998 and earlier Sidekick/Tracker

Radio
6 Remove the rear retaining bracket screw.
7 Remove the four pins from the four corners of the radio front.
8 Disconnect the electrical connectors and antenna lead from the radio.
9 Installation is the reverse of removal.

Front speaker
10 Remove the speaker grilles from the instrument panel by removing the two screws.
11 Remove the screws from the speaker.
12 Disconnect the electrical connector from the speaker.
13 Installation is the reverse of removal.

Rear speaker
14 Remove the interior trim side panel.
15 Remove the speaker securing screws.
16 Disconnect the electrical connector from the speaker.
17 Installation is the reverse of removal.

1999 and later Sidekick/X-90/Vitara and Tracker

Radio
18 Remove the three center trim bezel retaining screws **(see illustration 19.16 in Chapter 11)**.
19 Remove the four retaining screws at each corner of the radio and pull it out for access to the electrical connector.
20 Unplug the connector and remove the radio.
21 Installation is the reverse of removal.

Speakers
22 Locate the speaker.
23 Rear speakers are accessible after opening the liftgate and detaching the trim

panel. On front speakers it will be necessary to remove the door trim panel (Chapter 11).
24 Remove the retaining screws, withdraw the speaker sufficiently to allow the connector to be unplugged and remove the speaker from the vehicle.
25 Installation is the reverse of removal.

12 Bulb replacement

Front end

Turn signal/parking and side marker lights
1 The turn signal/parking light bulb can be replaced after opening the hood. Grasp the bulb and rotate the bulb holder counterclockwise and replace the bulb by pushing it in and rotating it counterclockwise. On earlier models replace the side marker light bulb by pushing the light housing toward the rear of the vehicle to detach it from the clip. On later models, remove the two screws then detach the side marker housing. On all models, replace the bulb by pushing it in and rotating it counterclockwise.

Interior

Dome/map light
2 Detach the lens by prying carefully then pull down it down and remove it. Replace the bulb by pulling it straight out. On spotlight-

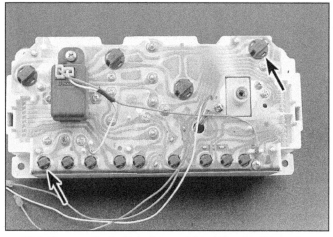

12.4 Remove the holders (arrows) from the circuit board or instrument cluster for access to the bulbs

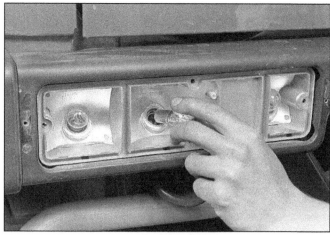

12.6 Push the bulb in and twist it counterclockwise to remove it - when installing the bulb, make sure the pins on the bulb line up with the proper slots in the socket (Samurai)

13.2 Remove the dashboard upper mounting screws and pull the dashboard away from the firewall (Samurai)

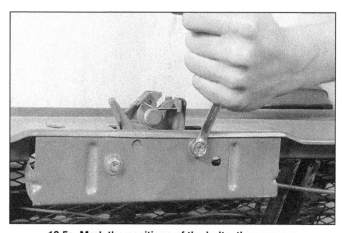

13.5a Mark the positions of the bolts, then remove the latch (Samurai)

type lamps, pry off the lens off for access, remove the screws and lower the housing. Remove the bulb holder by rotating it counterclockwise and replace the bulb by pulling it straight out of the holder.

Instrument panel lights

Refer to illustrations 12.4

3   The instrument panel bulbs can be replaced after removal of the instrument cluster (Section 14).

4   Rotate the holder, withdraw it from the panel then remove the bulb from the holder **(see illustration)**.

Rear end

Rear lights

Refer to illustration 12.6

5   To replace the rear side marker light bulb on earlier models push the light housing toward the rear of the vehicle and detach it from the clip. On later models, remove the two screws, then detach the side marker housing. On all models, replace the bulb by pushing it in and twisting it counterclockwise.

6   On Samurai models, remove the screws

and take off the taillight lens. To replace a bulb, push it in and twist it counterclockwise **(see illustration)**. When installing, make sure the tabs snap into position when the housing is pressed into place. On Sidekick/X-90/ Vitara and Tracker models, remove the screws and detach the taillight housing for access to the tail, back up and parking light bulbs. Rotate the bulb holder counterclockwise to remove it.

---

13   Windshield wiper motor - removal and installation

Samurai

Refer to illustrations 13.2, 13.5a, 13.5b, 13.6, 13.7 and 13.9

1   Disconnect the negative cable at the battery.

2   Remove the plastic molding covers **(see illustration)** on the top of the dashboard to expose the dashboard mounting bolts.

3   Remove the dashboard mounting bolts and pull the dashboard out slightly.

4   Open the glove box and remove the glove box mounting screws.

5   Mark the hood latch bolts with a scribe or paint. Remove the hood latch and disconnect the cable from it **(see illustrations)**.

6   Pull the glove compartment with the

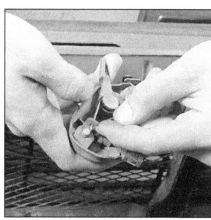

13.5b To disconnect the cable, pull it up and slide it through the slot in the latch (Samurai)

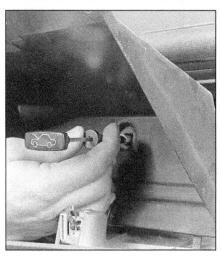

13.6 Remove the glove compartment with the hood latch cable (Samurai)

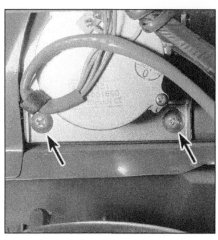

13.7 Remove the two lower windshield wiper motor mounting screws from inside the glove box compartment (arrows) (Samurai)

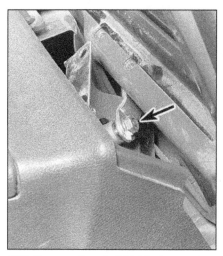

13.9 Remove the windshield wiper motor-to-linkage nut (Samurai)

hood latch cable out of the dashboard **(see illustration)**.

7   The windshield wiper motor is located under the dash on the passenger side, in the corner near the windshield. Remove the two lower mounting screws through the glove compartment opening **(see illustration)**.

8   Remove the two upper mounting bolts from above the dashboard.

9   Carefully pull the windshield wiper motor out from the firewall and remove the nut that attaches the wiper motor arm to the linkage **(see illustration)**. Disconnect the electrical connectors and remove the motor through the glove box compartment.

10   Installation is the reverse of the removal.

Sidekick/X-90/Vitara and Tracker

Refer to illustration 13.11

11   The windshield wiper motor is located

on the drivers side of the engine compartment, on the firewall. Disconnect the electrical connector, remove the mounting bolts and pull the motor out partially **(see illustration)**.

12   Remove the linkage nut, detach the windshield wiper motor and lift it out of the engine compartment.

13   Installation is the reverse of the removal.

14   Instrument cluster - removal and installation

Refer to illustration 14.7

Warning: *1996 and later models are equipped with an airbag system. Always disable the airbags prior to working in the vicinity of the steering column, instrument panel or airbag components to avoid the possibility of accidental deployment of the airbags, which could cause personal injury (see Section 17).*

Samurai

Refer to illustration 14.7

1   Disconnect the negative cable at the battery.

2   Remove the steering wheel (see Chapter 10).

3   Remove the steering column covers (see Section 8).

4   Remove the Torx drive bolts that support the steering column and lower the steering column (see Section 8).

5   Remove the instrument cluster cover screws and remove the cover.

6   Disconnect the speedometer cable from the rear of the cluster.

7   Disconnect the wire harness connector **(see illustration)**.

8   Remove the instrument cluster.

9   Check the printed circuit board for any flaws or breaks before replacing any of the combination meter's components.

10   Installation is the reverse of removal.

13.11 Unplug the electrical connector and remove the mounting bolts (arrows), then pull the motor out to partially expose the linkage nut (Sidekick/X-90/Vitara and Tracker)

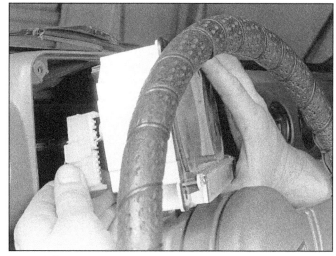

14.7 Press the locking tab on the electrical connector then unplug the connector from the instrument cluster (Samurai)

14.12 Remove the instrument cluster retaining screws (arrows) (Sidekick/X-90/Vitara and Tracker)

14.13 Remove the instrument cluster retaining screws (arrows) (Sidekick/X-90/Vitara and Tracker)

Sidekick/X-90/Vitara and Tracker

Refer to illustrations 14.12 and 14.13

11   Disconnect the battery negative cable.

12   Remove the screws and pull the bezel out **(see illustration)**.

13   Remove the cluster retaining screws **(see illustration)**.

14   Disconnect the speedometer cable.

15   Pull the cluster out, disconnect the wiring and remove it from the instrument panel.

16   Installation is the reverse of removal.

15   Cruise control - description and check

The cruise control system maintains vehicle speed with a vacuum actuated servo motor located in the engine compartment, which is connected to the throttle linkage by a cable. The system consists of a cruise control module, servo motor, clutch switch, brake switch, transmission range switch (automatic transmission) and vehicle speed sensor.

Because of the complexity of the cruise control system and the special tools and techniques for diagnosis, repair should be left to a dealer service department or other repair shop. However, it is possible for the home mechanic to make simple checks of the wiring for minor faults which can be easily repaired. These include:

a) *Inspect the cruise control switches for broken wires and loose connections.*

b) *Check the cruise control fuse.*

16   Speedometer cable - replacement

Refer to illustration 16.2

1   Disconnect the negative cable from the battery.

2   Disconnect the speedometer cable from

the transfer case **(see illustration)** by removing the retaining bolt.

3   Detach the cable from the routing clips in the engine compartment and pull it up to provide enough slack to allow disconnection from the speedometer.

4   Remove the instrument cluster screws, pull the cluster out and disconnect the speedometer cable from the back of the cluster.

5   Remove the cable from the vehicle.

6   Prior to installation, lubricate the speedometer end of the cable with spray-on speedometer cable lubricant (available at auto parts stores).

7   Installation is the reverse of removal.

17   Airbag system - general information

Warning: *1996 and later models are equipped with driver and passenger airbags. Airbag system components are located in the steering wheel, steering column, instrument panel and center console. The airbags could accidentally deploy if any of the system com-*

16.2 Remove the retaining bolt and pull the speedometer cable out of the transfer case

ponents or wiring harnesses are disturbed, so be extremely careful when working in these areas and don't disturb any airbag system components or wiring. You could be injured if an airbag accidentally deploys, or the airbag might not deploy correctly in a collision if any components or wiring in the system have been disturbed. The yellow wires and connectors routed through the instrument panel and center console are for this system. Do not use electrical test equipment on these yellow wires or tamper with them in any way while working in their vicinity.

Description

1   The models covered by this manual are equipped with a Supplemental Inflatable Restraint (SIR) system, more commonly known as an airbag system. The SIR system is designed to protect the driver and passenger from serious injury in the event of a head-on or frontal collision.

2   The SIR system consists of two airbags; one located in the center of the steering wheel, and another located in the top of the dashboard, above the glove box and the Sensing and Diagnostic Module, located under the center console.

Airbag modules

3   The airbag module contains a housing incorporating a cushion (airbag) and inflator unit. The inflator is mounted on the back of the housing over a hole through which gas is expelled, inflating the bag almost instantaneously when an electrical signal is sent from the system. The igniter in the airbag converts the electrical signal to heat, causing a chemical reaction producing nitrogen gas, which inflates the bag. The driver's airbag uses a coil assembly to transmit the electrical signal to the inflator regardless of steering wheel position. The coil assembly is mounted on the steering column under the steering wheel.

Sensing and Diagnostic Module (SDM)

4   The sensing circuitry is contained in the SDM. If a frontal crash of sufficient force is

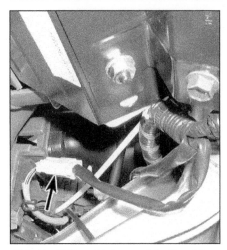

17.9 On 1998 and later models, the airbag fuse is located in the wiring harness inline fuse holder (arrow)

detected, the circuitry allows current to flow to the airbags, inflating them. The SDM also contains a back-up power supply to deploy the airbags in the event battery power is lost during a collision.

5   The SDM contains an on-board microprocessor which monitors the operation of the system. It performs a diagnostic check of the system every time the vehicle is started. If the system is operating properly, the AIRBAG warning light will blink on and off seven times. If there is a fault in the system, the light will remain on and the module will store fault codes indicating the nature of the fault. If the AIRBAG warning light remains on after starting, or comes on while driving, the vehicle should be taken to your dealer immediately for service.

Operation

6   For the airbags to deploy, an impact of sufficient G force must occur within 30-degrees of the vehicle centerline. When this condition occurs, the circuit to the airbag

inflators are closed and the airbags inflate. If the battery is destroyed by the impact, or is too low to power the inflators, a back-up power supply inside the SDM supplies current to the airbags.

Self-diagnosis system

7   A self-diagnosis circuit in the module displays a light when the ignition switch is turned to the On position. If the system is operating normally, the light should go out after seven flashes. If the light doesn't come on, or doesn't go out after seven flashes, or if it comes on while you're driving the vehicle, there's a malfunction in the SIR system. Have it inspected and repaired as soon as possible. Do not attempt to troubleshoot or service the SIR system yourself. Even a small mistake could cause the SIR system to malfunction when you need it.

Servicing components near the SIR system

8   Nevertheless, there are times when you need to remove the steering wheel, radio or service other components on or near the instrument panel. At these times, you'll be working around components and wiring harnesses for the SIR system. SIR system wiring is easy to identify; they're all covered by a bright yellow conduit. Do not unplug the connectors for the SIR system wiring, except to disable the system. And do not use electrical test equipment on the SIR system wiring. ALWAYS DISABLE THE SIR SYSTEM BEFORE WORKING NEAR THE SIR SYSTEM COMPONENTS OR RELATED WIRING.

Disabling the SIR system

Refer to illustration 17.9

9   Turn the steering wheel to the straight ahead position, place the ignition switch in Lock and remove the key. Remove the AIRBAG fuse from the fuse block (1996 and 1997 models) or from the inline fuse holder (1998 and later models) **(see illustration)**.

10   Unplug the yellow Connector Position Assurance (CPA) connectors at the airbag modules as described in the following steps.

Driver's side airbag

11   Remove the cover from the left side of the steering wheel (1996 and 1997 models). Unplug the yellow Connector Position Assurance (CPA) steering column harness connector.

Passenger's side airbag

12   Remove the glove compartment (Chapter 11).

13   Unplug the CPA electrical connector from the passenger inflator module under the right side of the dash.

Enabling the SIR system

14   After you've disabled the airbag and performed the necessary service, plug in the steering column (driver's side) and passenger side CPA connectors. Reinstall the steering wheel cover and the glove box.

15   Install the AIRBAG fuse.

---

18   Wiring diagrams - general information

Since it isn't possible to include all wiring diagrams for every model and year covered by this manual, the following diagrams are those that are typical and most commonly needed.

Prior to troubleshooting any circuit, check the fuse and circuit breakers (if equipped) to make sure they're in good condition. Make sure the battery is properly charged and check the cable connections (see Chapter 1).

When checking a circuit, make sure that all connectors are clean, with no broken or loose terminals. When unplugging a connector, do not pull on the wires. Pull only on the connector housings themselves.

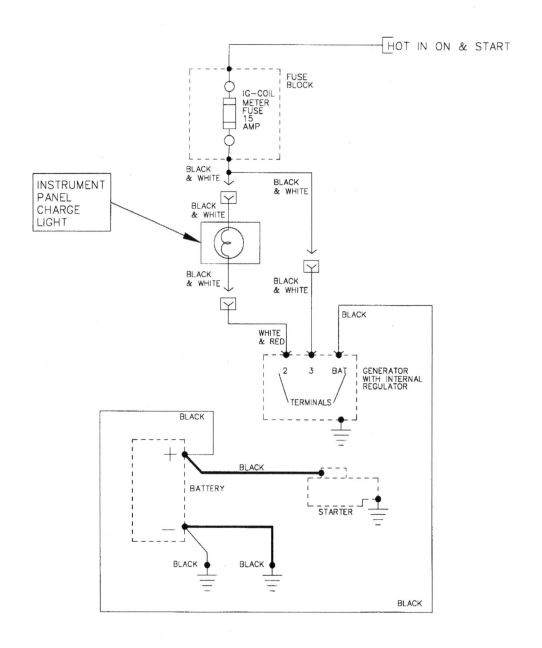

Typical charging system - 1998 and earlier Sidekick/Tracker

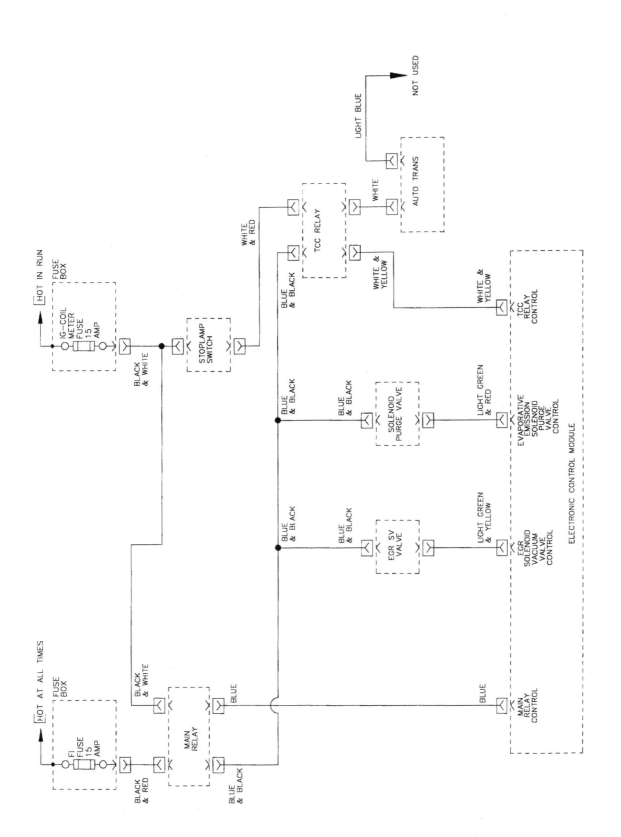

Typical emission and engine control system - 1998 and earlier Sidekick/Tracker (1 of 2)

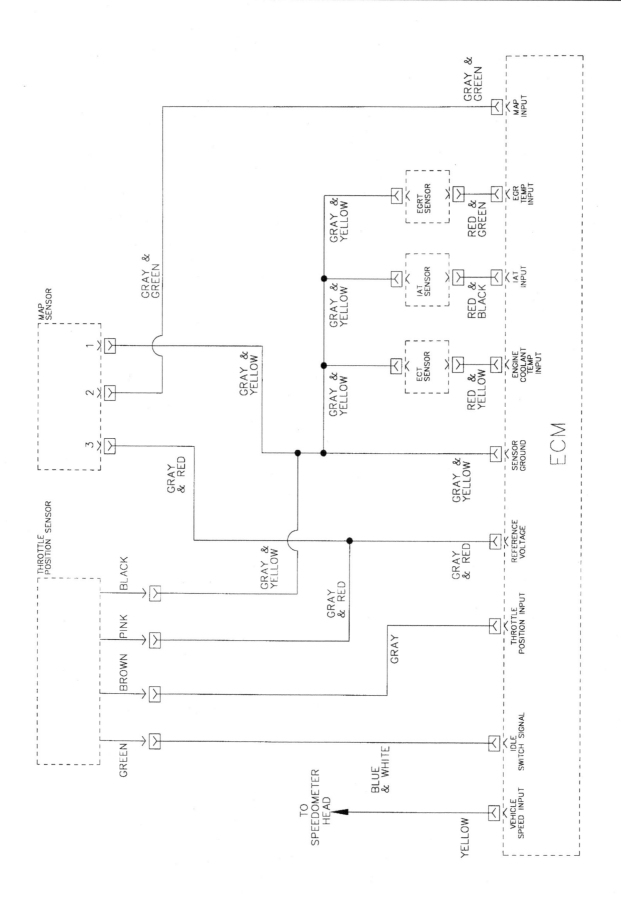

Typical emission and engine control system – 1998 and earlier Sidekick/Tracker (2 of 2)

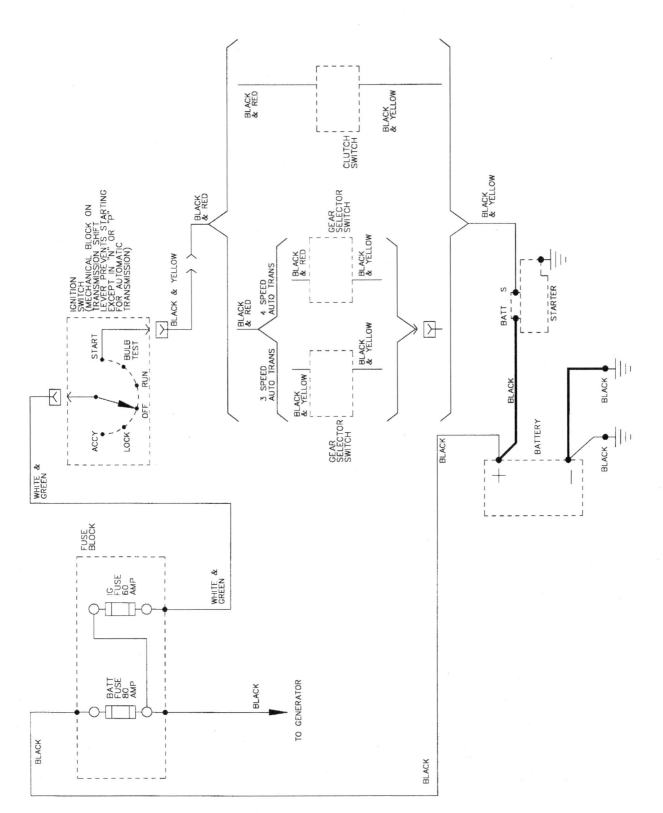

Typical starting system - 1998 and earlier Sidekick/Tracker

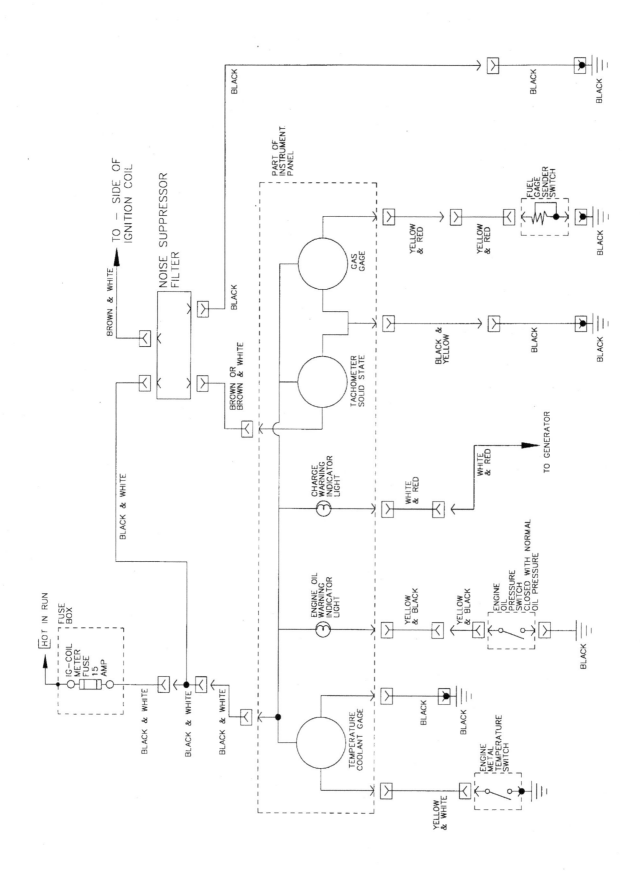

Typical gauges and warning light system - 1998 and earlier Sidekick/Tracker

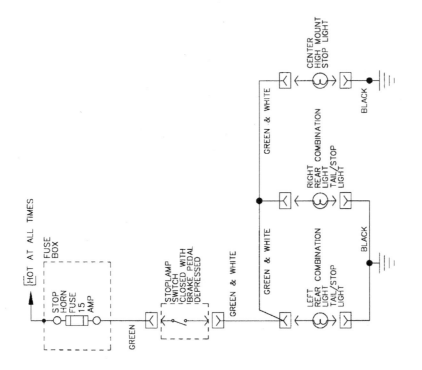

Typical brake light system - 1998 and earlier Sidekick/Tracker

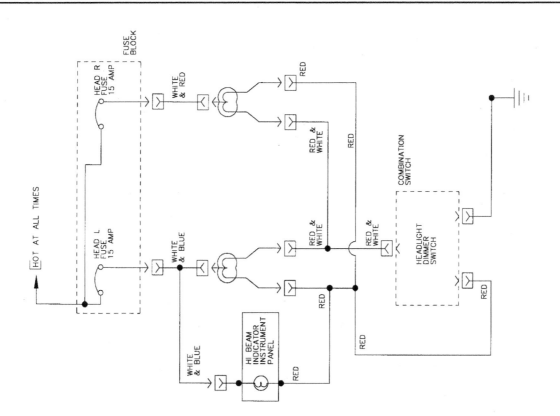

Typical headlight system - 1998 and earlier Sidekick/Tracker

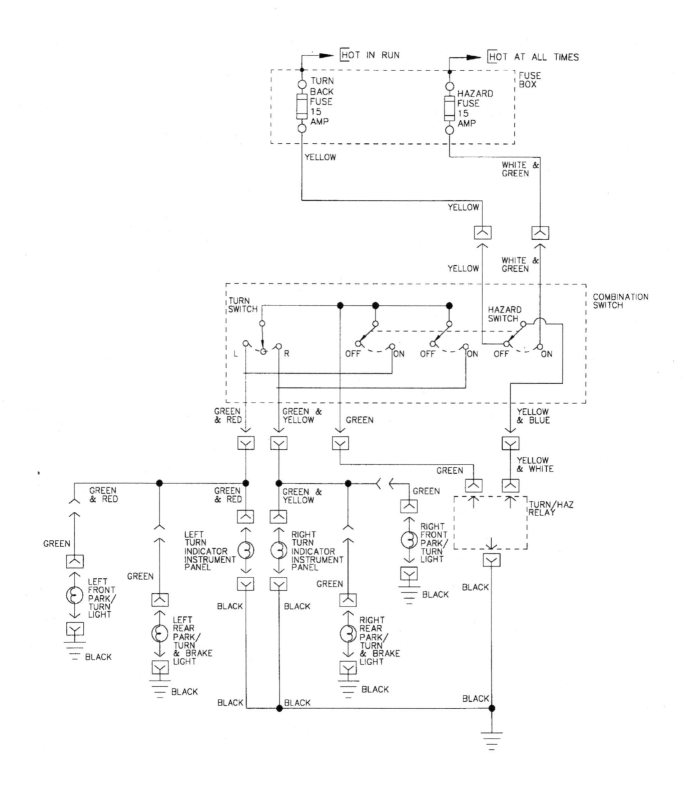

Typical turn signal light system - 1998 and earlier Sidekick/Tracker

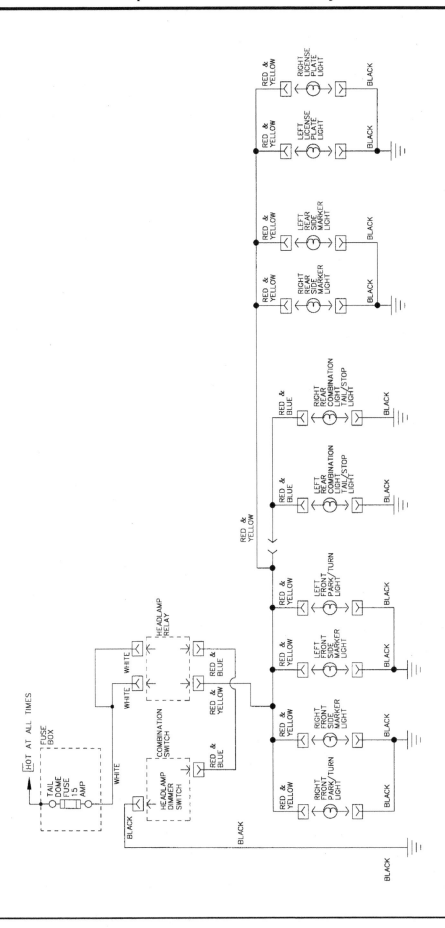

Typical running light system - 1998 and earlier Sidekick/Tracker

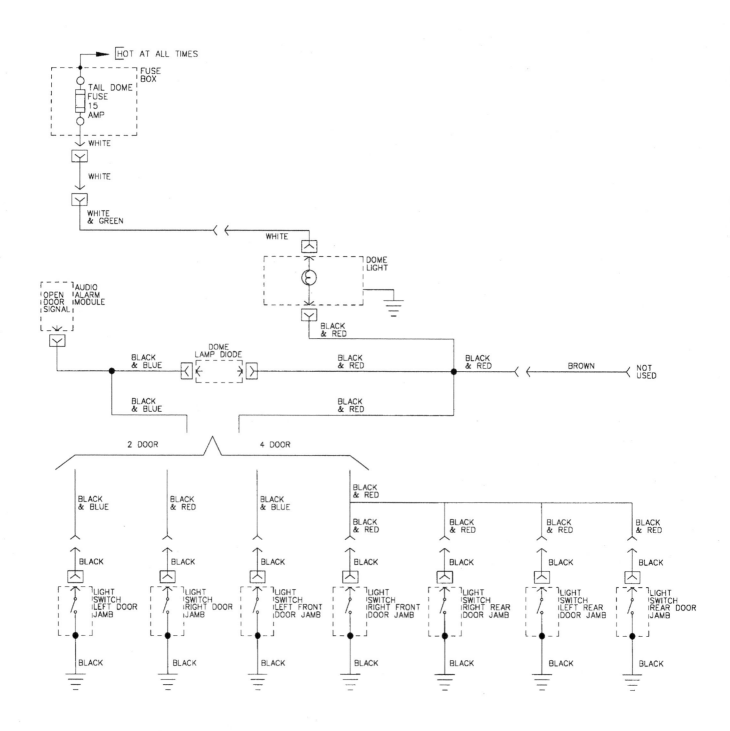

Typical dome light system - 1998 and earlier Sidekick/Tracker

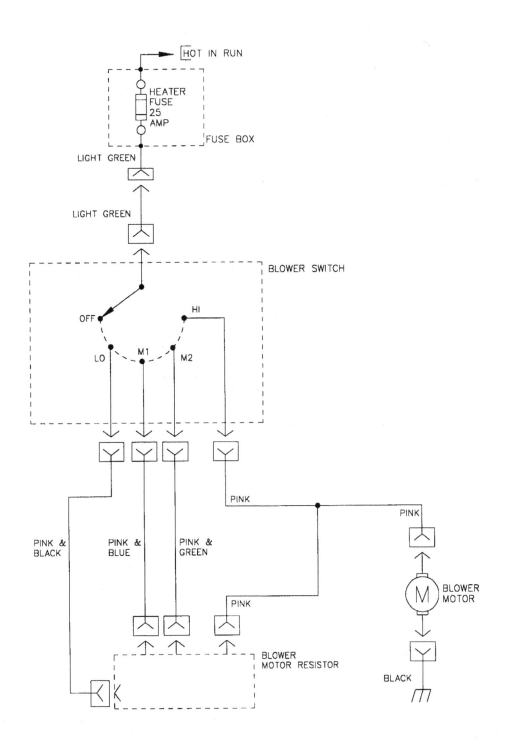

Typical heater blower motor system - 1998 and earlier Sidekick/Tracker

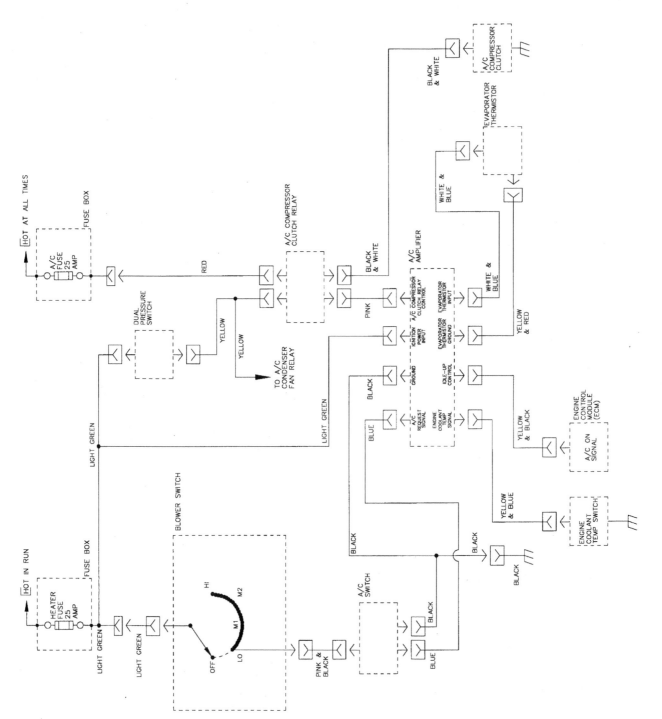

Typical air conditioning system - 1998 and earlier Sidekick/Tracker

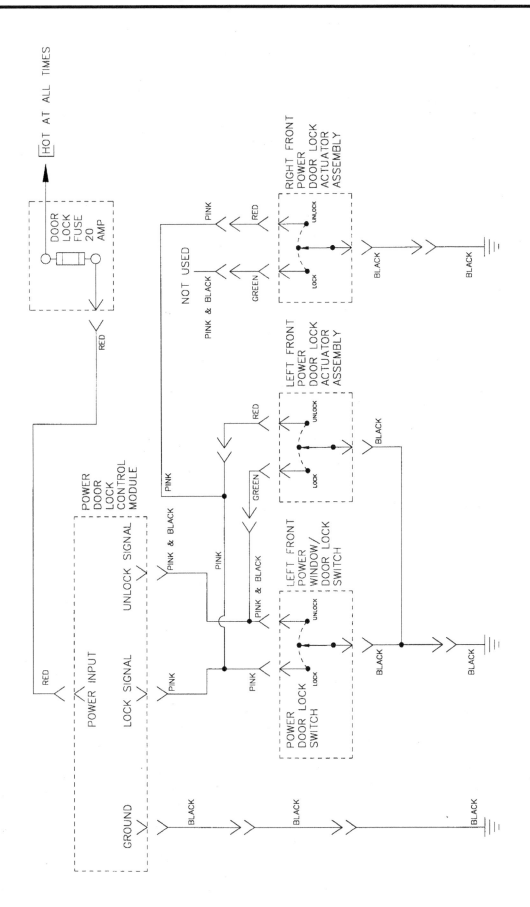

Typical power door lock system (two-door models) - 1998 and earlier Sidekick/Tracker

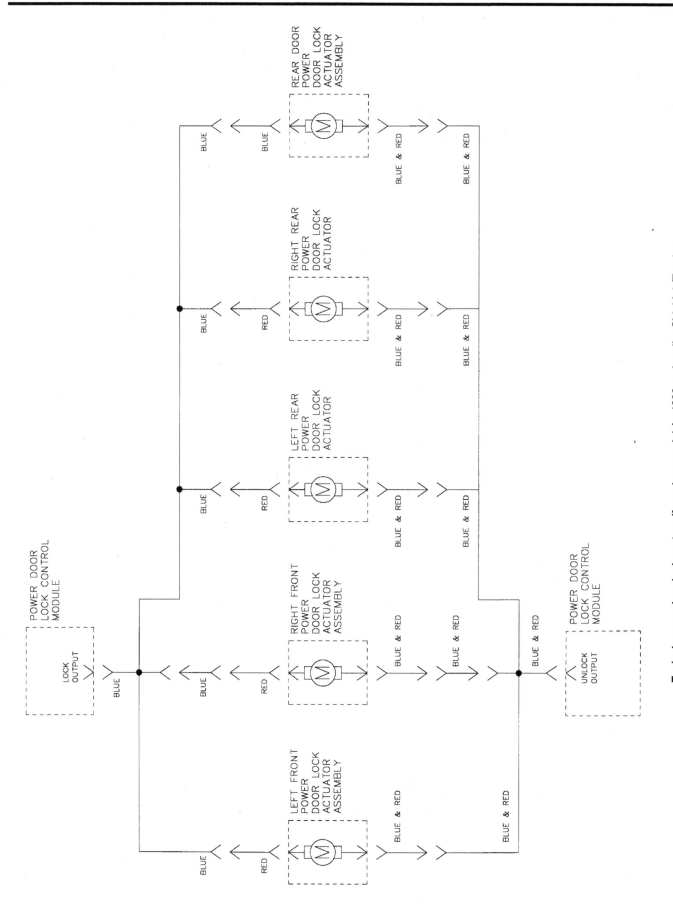

Typical power door lock system (four-door models) - 1998 and earlier Sidekick/Tracker

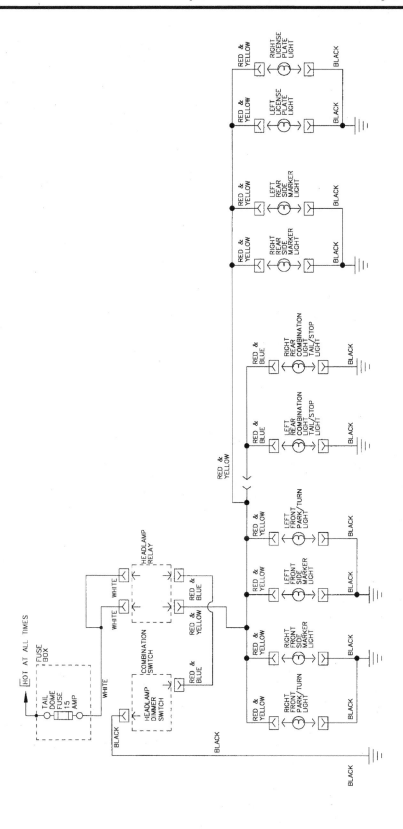

Typical power window system - 1998 and earlier Sidekick/Tracker

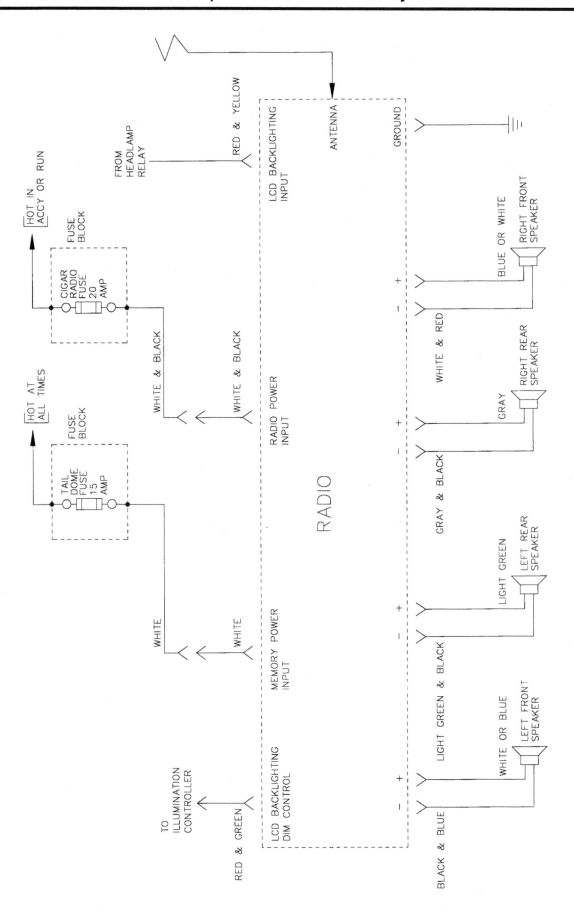

Typical audio system - 1998 and earlier Sidekick/Tracker

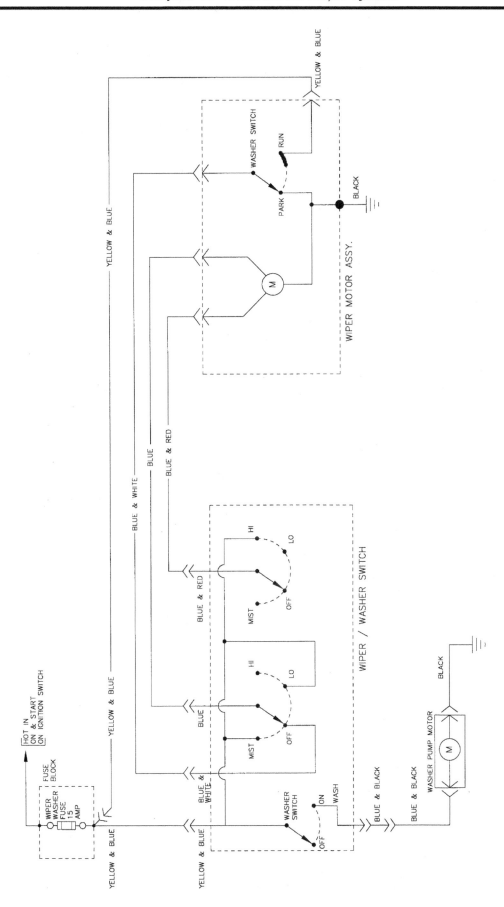

Typical windshield wiper/washer system - 1998 and earlier Sidekick/Tracker

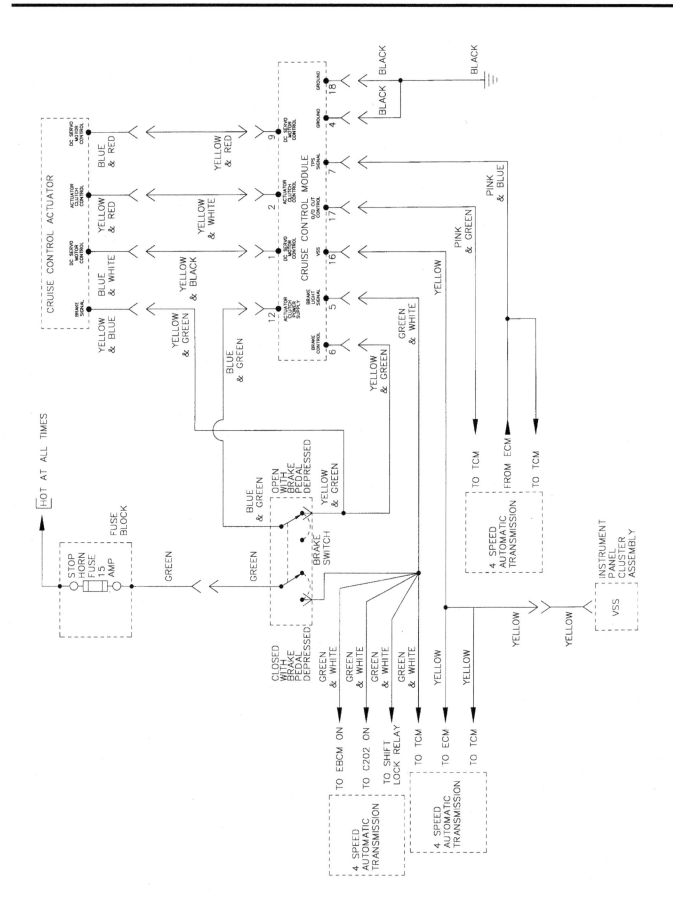

Typical cruise control system (with automatic transmission) - 1998 and earlier Sidekick/Tracker

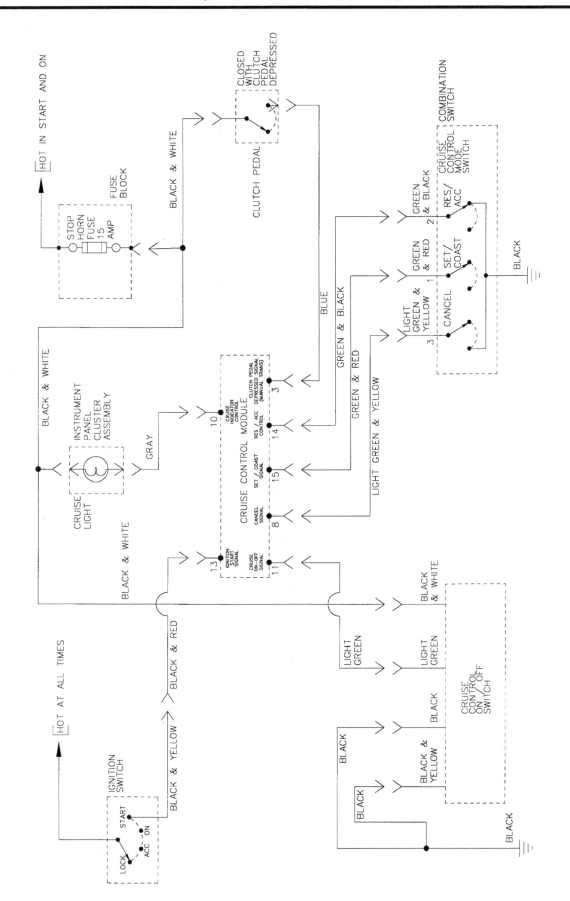

Typical cruise control system (with manual transmission) - 1998 and earlier Sidekick/Tracker

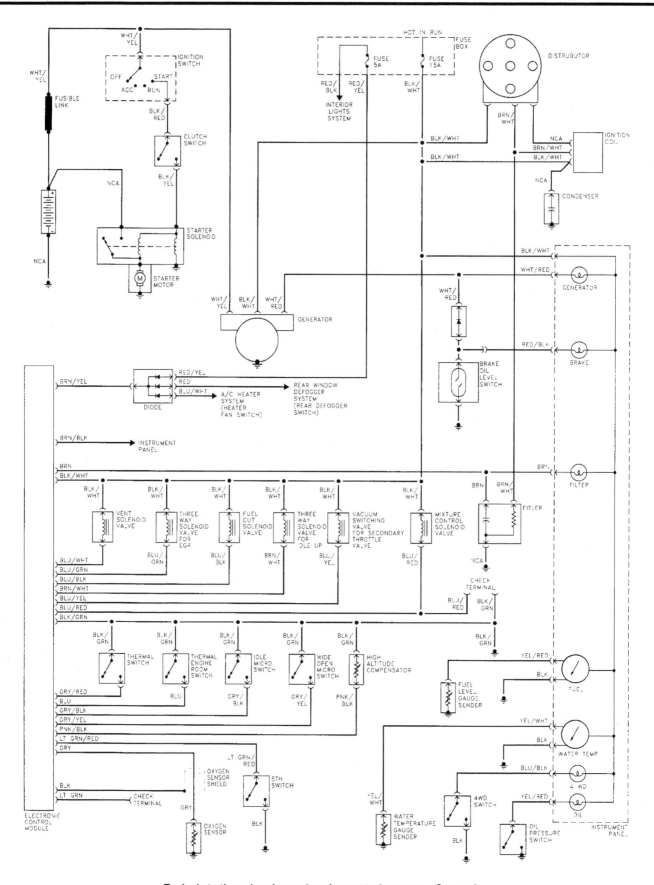

Typical starting, charging and engine control systems - Samurai

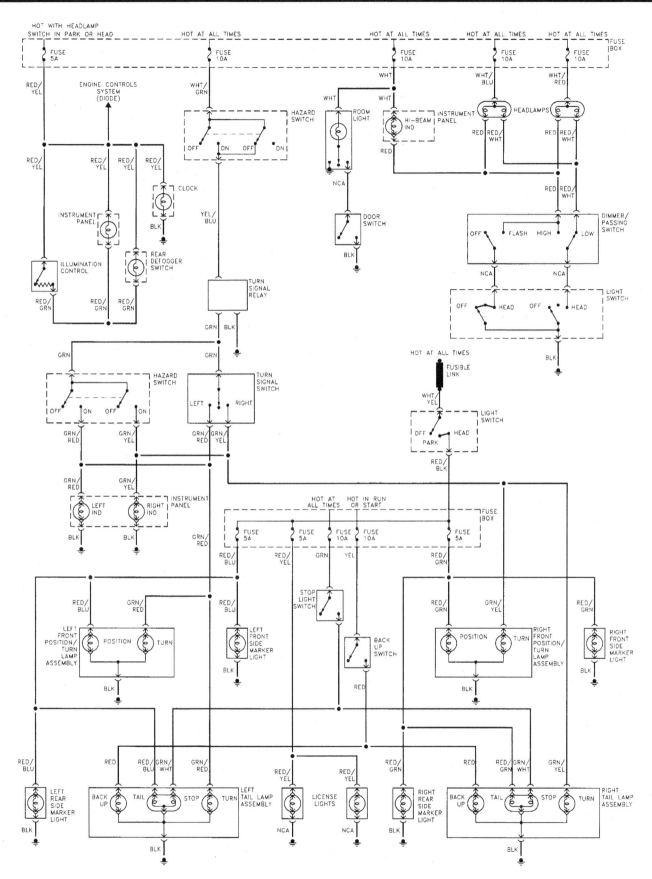

Typical lighting systems - Samurai

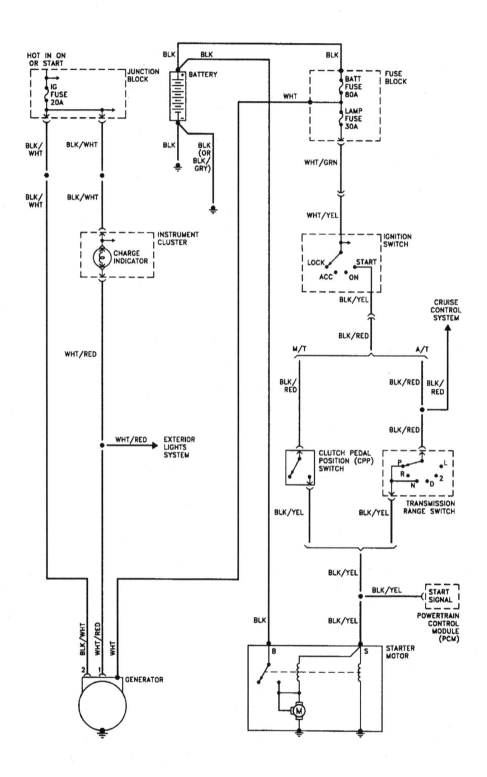

Starting and charging system - 1999 and later models

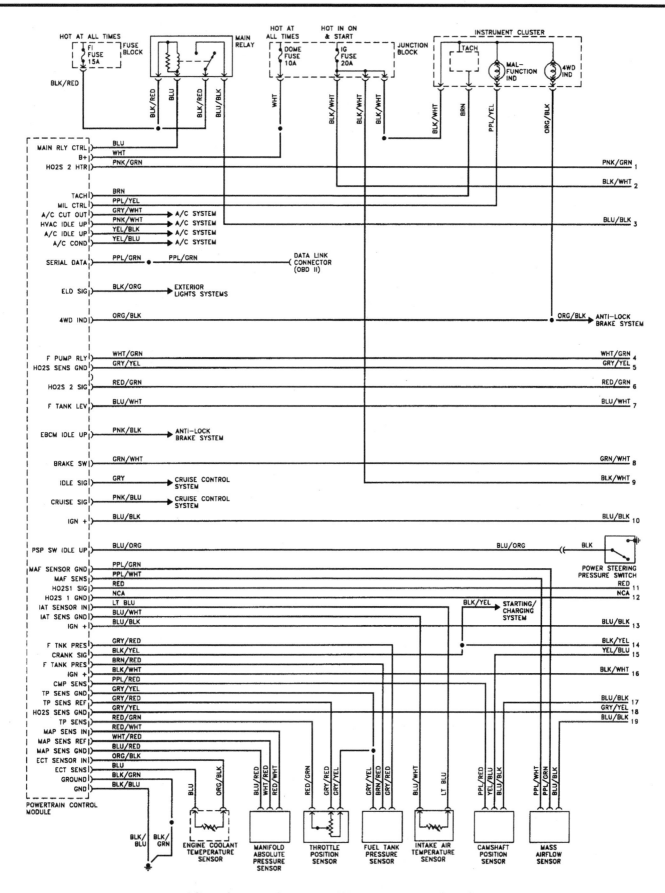

1.6L engine control system - 1999 and later models (1 of 3)

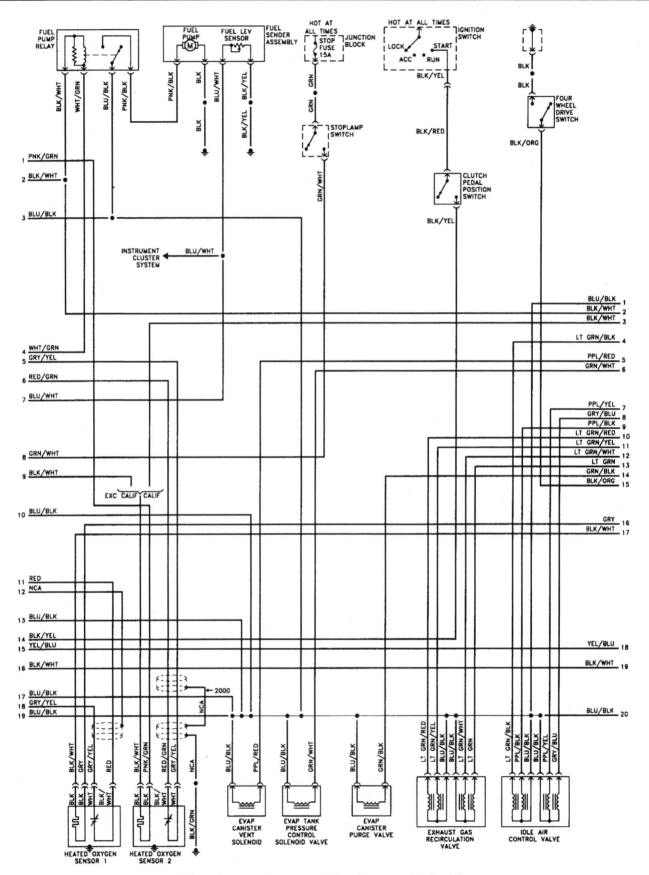

1.6L engine control system - 1999 and later models (2 of 3)

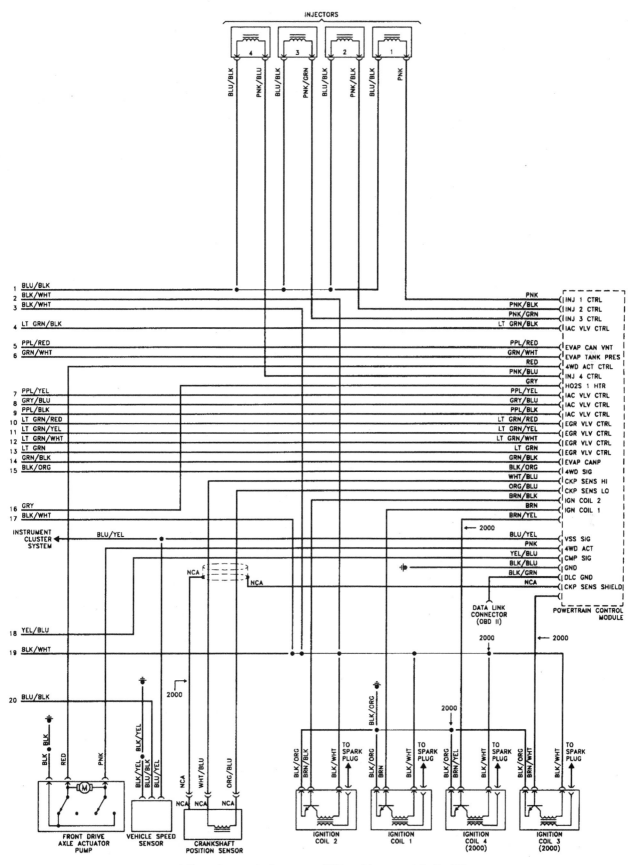

1.6L engine control system - 1999 and later models (3 of 3)

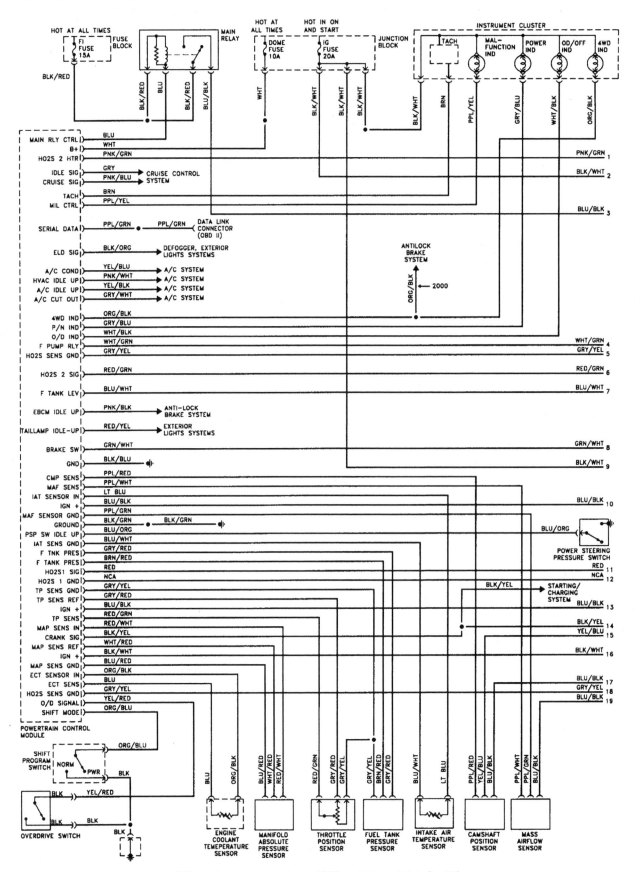

2.0L engine control system - 1999 and later models (1 of 3)

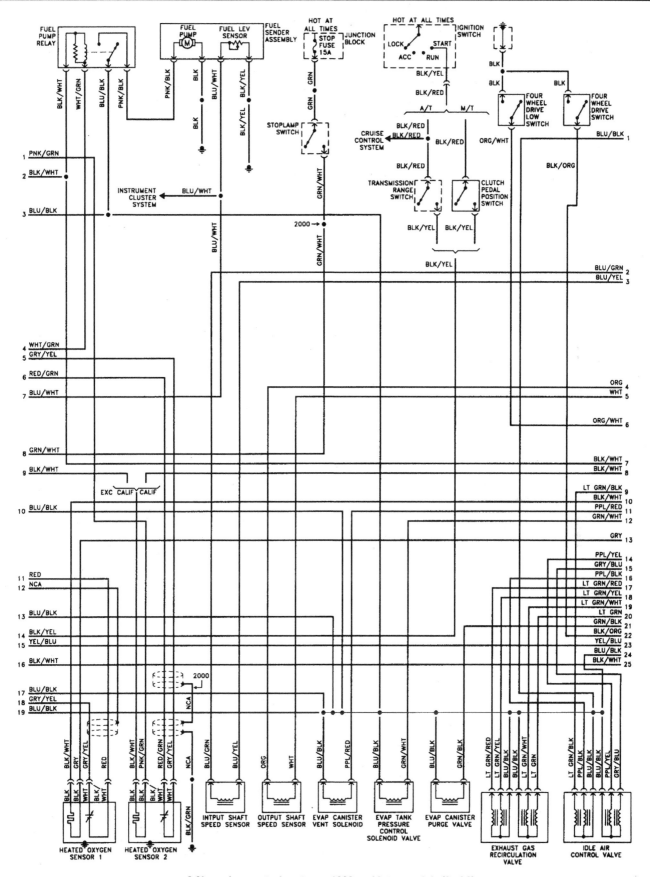

2.0L engine control system - 1999 and later models (2 of 3)

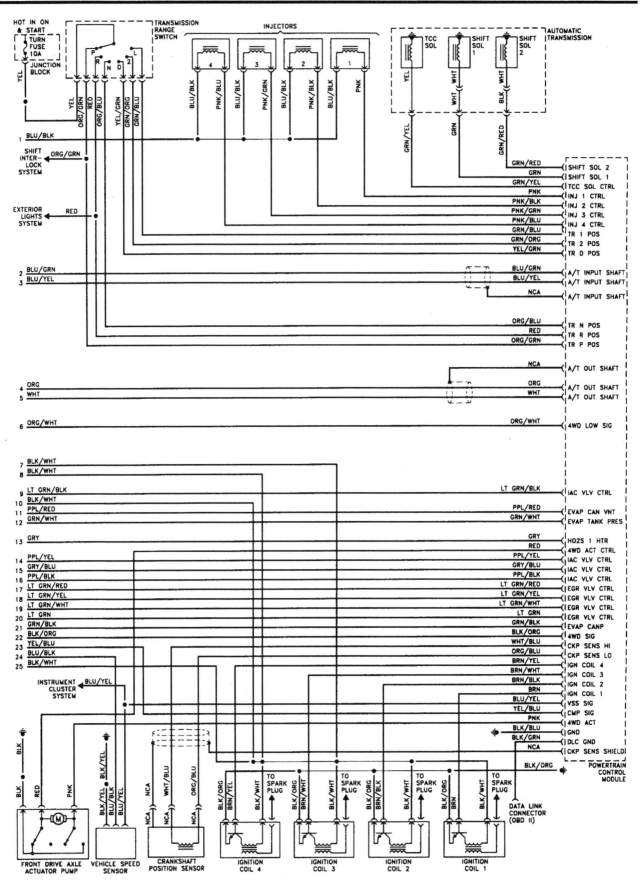

2.0L engine control system - 1999 and later models (3 of 3)

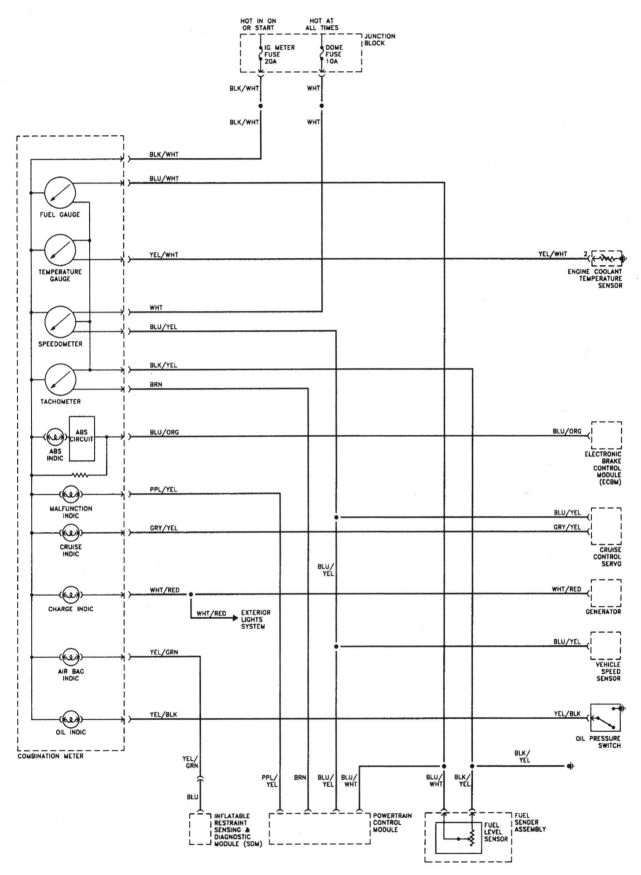

Warning lights and gauges system - 1999 and later models

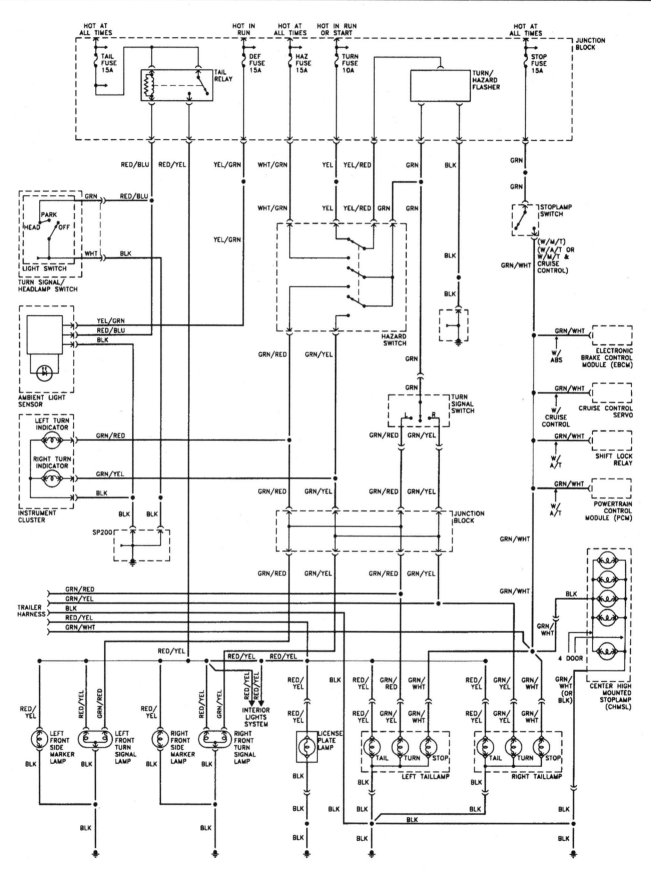

Exterior lighting system - 1999 and later models (1 of 2)

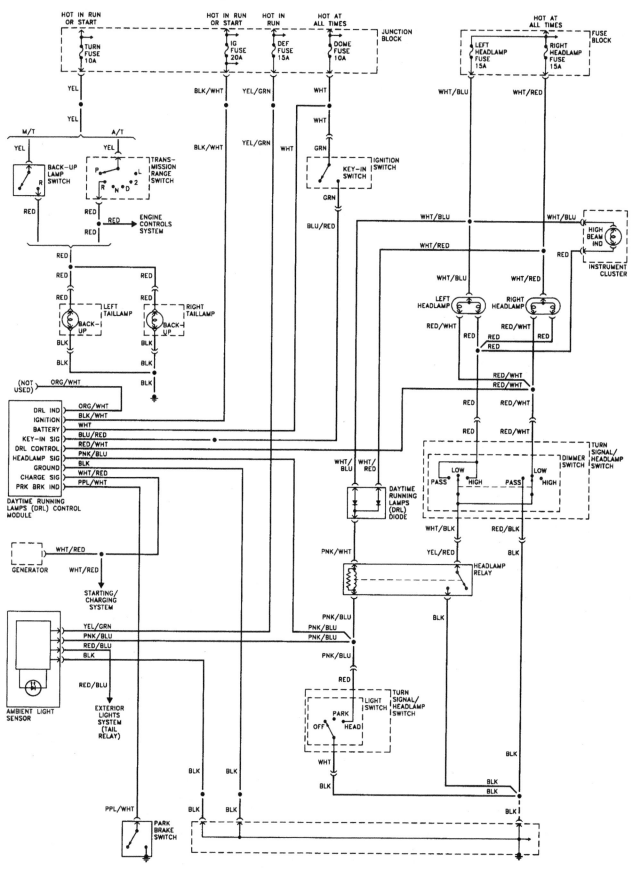

Exterior lighting system - 1999 and later models (2 of 2)

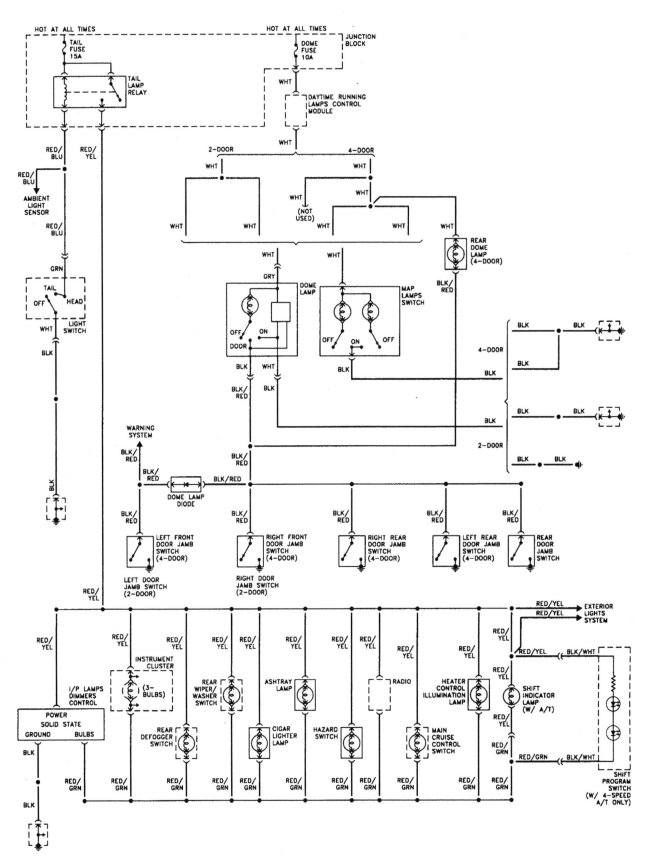

Interior lighting system - 1999 and later models

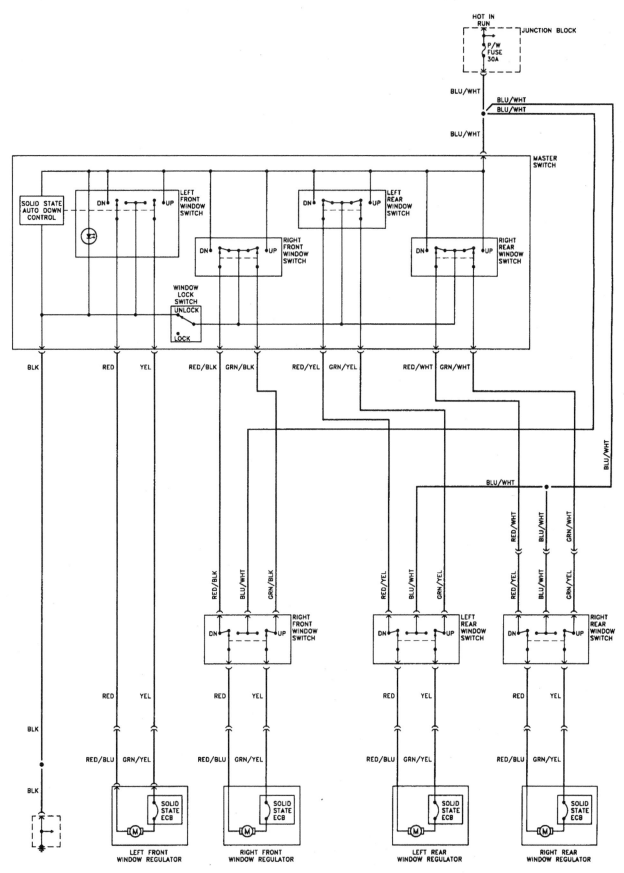

Power window system - 1999 and later models

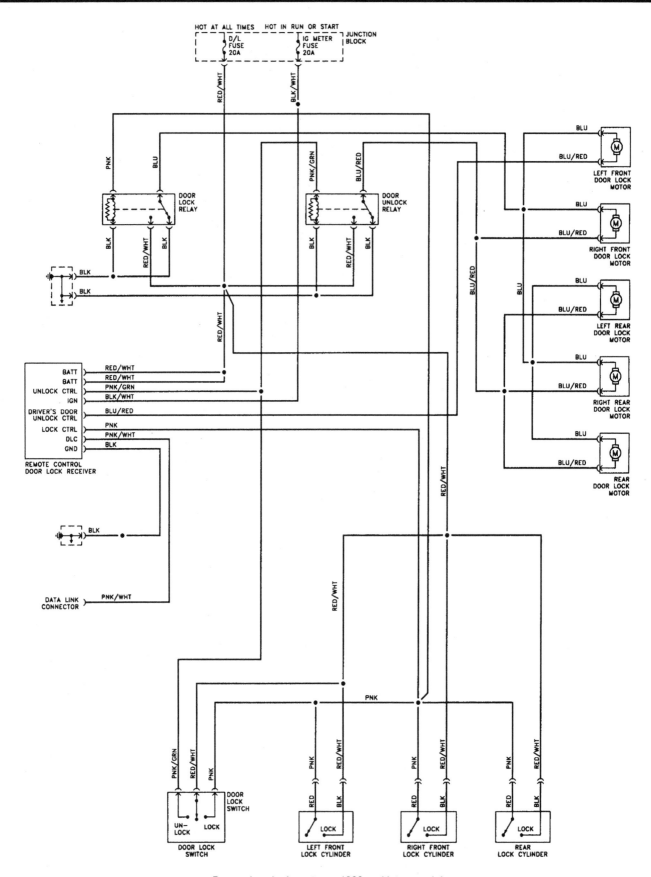

Power door lock system - 1999 and later models

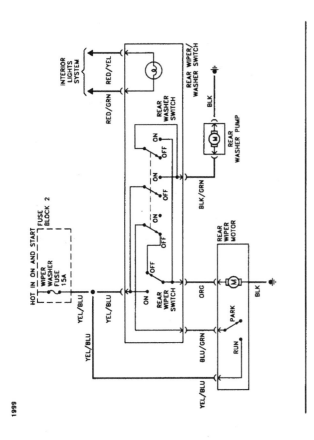

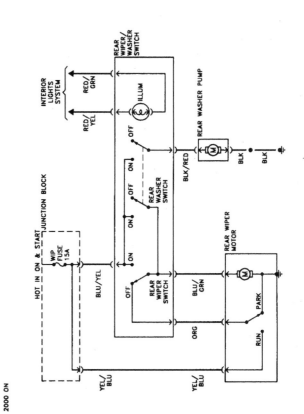

Rear window wiper/washer system - 1999 and later models

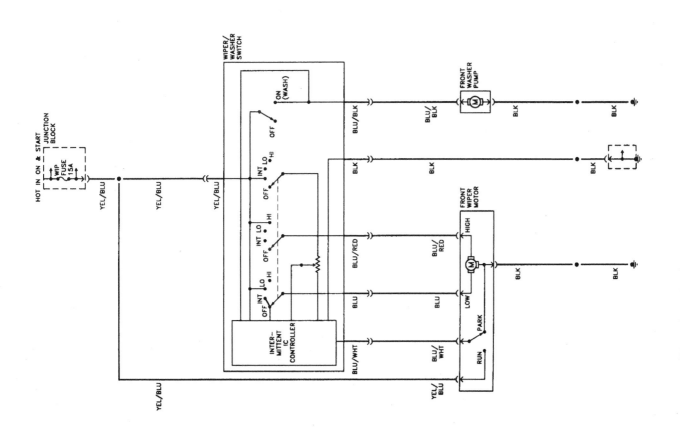

Windshield wiper/washer system - 1999 and later models

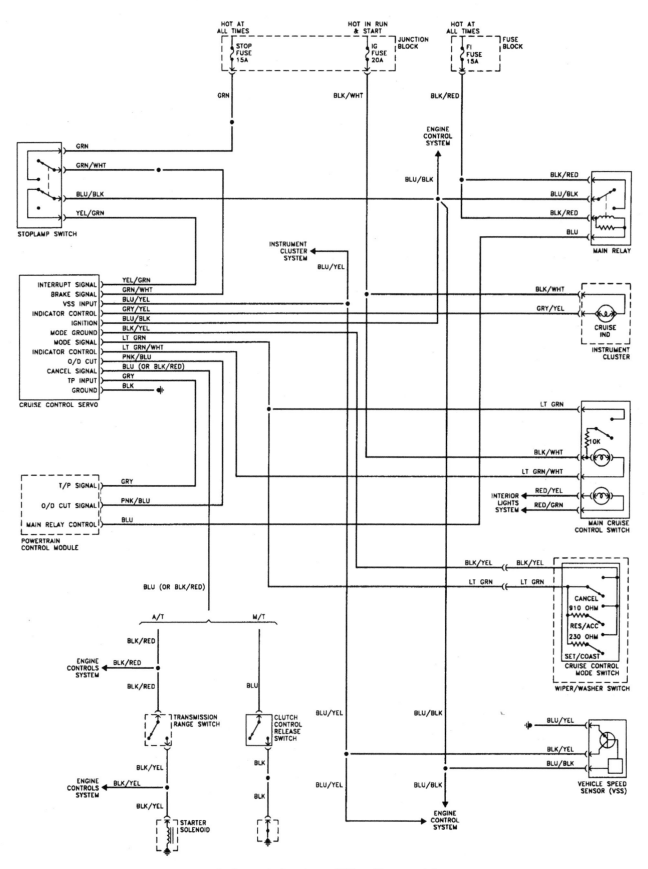

Cruise control system - 1999 and later models

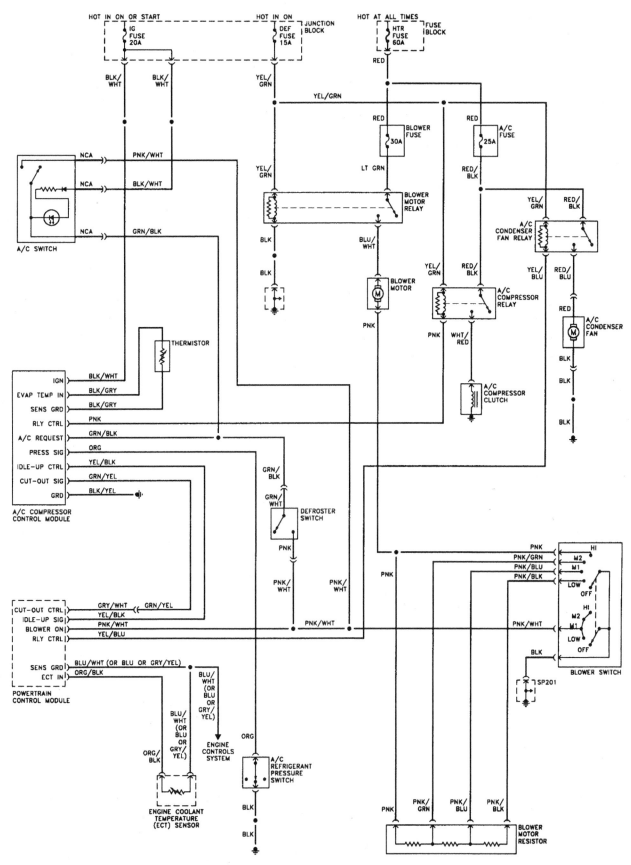

Air conditioning system - 1999 and later models

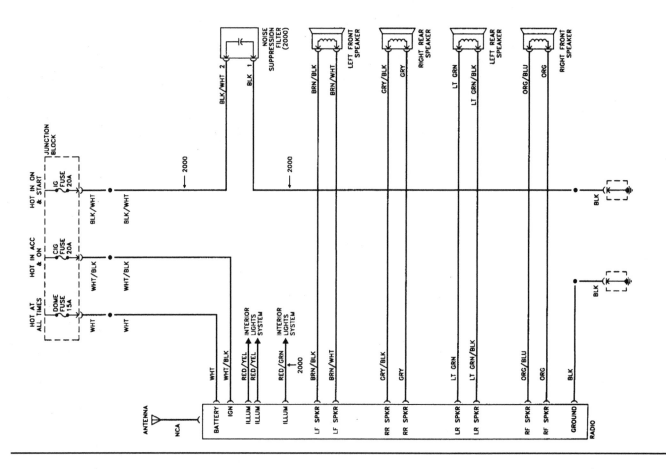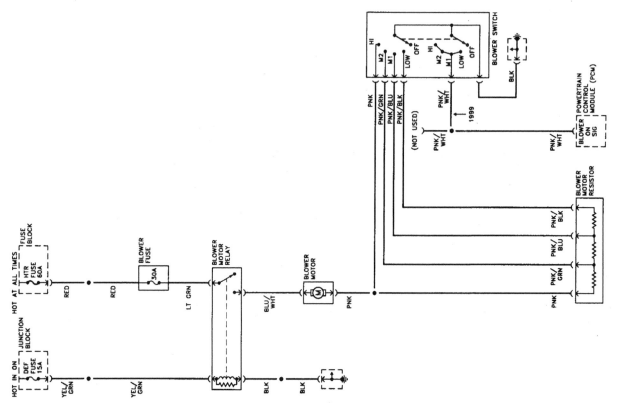

Typical audio system - 1999 and later models

Heater system - 1999 and later models

Index

F

G

H

I

Haynes Automotive Manuals

NOTE: If you do not see a listing for your vehicle, consult your local Haynes dealer for the latest product information.

ACURA
12020 Integra '86 thru '89 & Legend '86 thru '90
12021 Integra '90 thru '93 & Legend '91 thru '95
    Integra '94 thru '00 - see HONDA Civic (42025)
    MDX '01 thru '07 - see HONDA Pilot (42037)
12050 Acura TL all models '99 thru '08

AMC
    Jeep CJ - see JEEP (50020)
14020 Concord/Hornet/Gremlin/Spirit '70 thru '83
14025 (Renault) Alliance & Encore '83 thru '87

AUDI
15020 4000 all models '80 thru '87
15025 5000 all models '77 thru '83
15026 5000 all models '84 thru '88
    Audi A4 '96 thru '01 - see VW Passat (96023)
15030 Audi A4 '02 thru '08

AUSTIN
    Healey Sprite - see MG Midget (66015)

BMW
18020 3/5 Series '82 thru '92
18021 3 Series including Z3 models '92 thru '98
18022 3-Series incl. Z4 models '99 thru '05
18023 3-Series '06 thru '10
18025 320i all 4 cyl models '75 thru '83
18050 1500 thru 2002 except Turbo '59 thru '77

BUICK
19010 Buick Century '97 thru '05
    Century (front-wheel drive) - see GM (38005)
19020 Buick, Oldsmobile & Pontiac Full-size (Front wheel drive) '85 thru '05
19025 Buick, Oldsmobile & Pontiac Full-size (Rear wheel drive) '70 thru '90
19030 Mid-size Regal & Century '74 thru '87
    Regal - see GENERAL MOTORS (38010)
    Skyhawk - see GM (38030)
    Skylark - see GM (38020, 38025)
    Somerset - see GENERAL MOTORS (38025)

CADILLAC
21015 CTS & CTS-V '03 thru '12
21030 Cadillac Rear Wheel Drive '70 thru '93
    Cimarron, Eldorado & Seville - see GM (38015, 38030, 38031)

CHEVROLET
10305 Chevrolet Engine Overhaul Manual
24010 Astro & GMC Safari Mini-vans '85 thru '05
24015 Camaro V8 all '70 thru '81
24016 Camaro all models '82 thru '92
    Cavalier - see GM (38015)
    Celebrity - see GM (38005)
24017 Camaro & Firebird '93 thru '02
24020 Chevelle, Malibu, El Camino '69 thru '87
24024 Chevette & Pontiac T1000 '76 thru '87
    Citation - see GENERAL MOTORS (38020)
24027 Colorado & GMC Canyon '04 thru '10
24032 Corsica/Beretta all models '87 thru '96
24040 Corvette all V8 models '68 thru '82
24041 Corvette all models '84 thru '96
24045 Full-size Sedans Caprice, Impala, Biscayne, Bel Air & Wagons '69 thru '90
24046 Impala SS & Caprice and Buick Roadmaster '91 thru '96
    Impala '00 thru '05 - see LUMINA (24048)
24047 Impala & Monte Carlo all models '06 thru '11
24048 Lumina '90 thru '94 - see GM (38010)
24048 Lumina & Monte Carlo '95 thru '05
    Lumina APV - see GM (38035)
24050 Luv Pick-up all 2WD & 4WD '72 thru '82
    Malibu - see GM (38026)
24055 Monte Carlo all models '70 thru '88
    Monte Carlo '95 thru '01 - see LUMINA
24059 Nova all V8 models '69 thru '79
24060 Nova/Geo Prizm '85 thru '92
24064 Pick-ups '67 thru '87 - Chevrolet & GMC
24065 Pick-ups '88 thru '98 - Chevrolet & GMC
24066 Pick-ups '99 thru '06 - Chevrolet & GMC
24067 Chevy Silverado & GMC Sierra '07 thru '12
24070 S-10 & GMC S-15 Pick-ups '82 thru '93
24071 S-10 & Sonoma & Jimmy '94 thru '04
24072 Chevrolet TrailBlazer, GMC Envoy & Oldsmobile Bravada '02 thru '09
24075 Sprint '85 thru '88, Geo Metro '89 thru '01
24080 Vans - Chevrolet & GMC '68 thru '96
24081 Full-size Vans '96 thru '10

CHRYSLER
10310 Chrysler Engine Overhaul Manual
25015 Chrysler Cirrus, Dodge Stratus, Plymouth Breeze, '95 thru '00
25020 Full-size Front-Wheel Drive '88 thru '93
    K-Cars - see DODGE Aries (30008)
    Laser - see DODGE Daytona (30030)
25025 Chrysler LHS, Concorde & New Yorker, Dodge Intrepid, Eagle Vision, '93 thru '97
25026 Chrysler LHS, Concorde, 300M, Dodge Intrepid '98 thru '04
25027 Chrysler 300, Dodge Charger & Magnum '05 thru '09
25030 Chrysler/Plym. Mid-size '82 thru '95
    Rear-wheel Drive - see DODGE (30050)
25035 PT Cruiser all models '01 thru '10
25040 Chrysler Sebring '95 thru '06, Dodge Stratus '01 thru '06, Dodge Avenger '95 thru '00

DATSUN
28005 200SX all models '80 thru '83
28007 B-210 all models '73 thru '78
28009 210 all models '78 thru '82
28012 240Z, 260Z & 280Z Coupe '70 thru '78
28014 280ZX Coupe & 2+2 '79 thru '83
    300ZX - see NISSAN (72010)
28018 510 & PL521 Pick-up '68 thru '73
28020 510 all models '78 thru '81
28022 620 Series Pick-up all models '73 thru '79
    720 Series Pick-up - see NISSAN (72030)
28025 810/Maxima all gas models '77 thru '84

DODGE
    400 & 600 - see CHRYSLER (25030)
30008 Aries & Plymouth Reliant '81 thru '89
30010 Caravan & Ply. Voyager '84 thru '95
30011 Caravan & Ply. Voyager '96 thru '02
30012 Challenger/Plymouth Saporro '78 thru '83
    Challenger '67-'76 - see DART (30025)
30013 Caravan, Chrysler Voyager, Town & Country '03 thru '07
30016 Colt/Plymouth Champ '78 thru '87
30020 Dakota Pick-ups all models '87 thru '96
30021 Durango '98 & '99, Dakota '97 thru '99
30022 Durango '00 thru '03, Dakota '00 thru '04
30023 Durango '04 thru '09, Dakota '05 thru '11
30025 Dart, Challenger/Plymouth Barracuda & Valiant 6 cyl models '67 thru '76
30030 Daytona & Chrysler Laser '84 thru '89
    Intrepid - see Chrysler (25025, 25026)
30034 Dodge & Plymouth Neon '95 thru '99
30035 Omni & Plymouth Horizon '78 thru '90
30036 Dodge and Plymouth Neon '00 thru '05
30040 Pick-ups all full-size models '74 thru '93
30041 Pick-ups all full-size models '94 thru '01
30042 Pick-ups full-size models '02 thru '08
30045 Ram 50/D50 Pick-ups & Raider and Plymouth Arrow Pick-ups '79 thru '93
30050 Dodge/Ply./Chrysler RWD '71 thru '89
30055 Shadow/Plymouth Sundance '87 thru '94
30060 Spirit & Plymouth Acclaim '89 thru '95
30065 Vans - Dodge & Plymouth '71 thru '03

EAGLE
    Talon - see MITSUBISHI (68030, 68031)
    Vision - see CHRYSLER (25025)

FIAT
34010 124 Sport Coupe & Spider '68 thru '78
34025 X1/9 all models '74 thru '80

FORD
10320 Ford Engine Overhaul Manual
10355 Ford Automatic Transmission Overhaul
11500 Mustang '64-1/2 thru '70 Restoration Guide
36004 Aerostar Mini-vans '86 thru '97
    Aspire - see FORD Festiva (36030)
36006 Contour/Mercury Mystique '95 thru '00
36008 Courier Pick-up all models '72 thru '82
36012 Crown Victoria & Mercury Grand Marquis '88 thru '10
36016 Escort/Mercury Lynx '81 thru '90
36020 Escort/Mercury Tracer '91 thru '02
    Expedition - see FORD Pick-up (36059)
36022 Escape & Mazda Tribute '01 thru '11
36024 Explorer & Mazda Navajo '91 thru '01
36025 Explorer/Mercury Mountaineer '02 thru '10
36028 Fairmont & Mercury Zephyr '78 thru '83
36030 Festiva & Aspire '88 thru '97
36032 Fiesta all models '77 thru '80
36034 Focus all models '00 thru '11
36036 Ford & Mercury Full-size '75 thru '87
36044 Ford & Mercury Mid-size '75 thru '86
36045 Ford Fusion & Mercury Milan '06 thru '10
36048 Mustang V8 all models '64-1/2 thru '73
36049 Mustang II 4 cyl, V6 & V8 '74 thru '78
36050 Mustang & Mercury Capri '79 thru '93
36051 Mustang all models '94 thru '04
36052 Mustang '05 thru '10
36054 Pick-ups and Bronco '73 thru '79
36058 Pick-ups and Bronco '80 thru '96
36059 Pick-ups & Expedition '97 thru '09
36060 Super Duty Pick-up, Excursion '99 thru '10
36061 F-150 full-size '04 thru '10
36062 Pinto & Mercury Bobcat '75 thru '80
36066 Probe all models '89 thru '92
    Probe '93 thru '97 - see MAZDA 626 (61042)
36070 Ranger/Bronco II gas models '83 thru '92
36071 Ford Ranger '93 thru '10 &
    Mazda Pick-ups '94 thru '09
36074 Taurus & Mercury Sable '86 thru '95
36075 Taurus & Mercury Sable '96 thru '01
36078 Tempo & Mercury Topaz '84 thru '94
36082 Thunderbird/Mercury Cougar '83 thru '88
36086 Thunderbird/Mercury Cougar '89 thru '97
36090 Vans all V8 Econoline models '69 thru '91
36094 Vans full size '92 thru '10
36097 Windstar Mini-van '95 thru '07

GENERAL MOTORS
10360 GM Automatic Transmission Overhaul
38005 Buick Century, Chevrolet Celebrity, Olds Cutlass Ciera & Pontiac 6000 '82 thru '96
38010 Buick Regal, Chevrolet Lumina, Oldsmobile Cutlass Supreme & Pontiac Grand Prix front wheel drive '88 thru '07
38015 Buick Skyhawk, Cadillac Cimarron, Chevrolet Cavalier, Oldsmobile Firenza Pontiac J-2000 & Sunbird '82 thru '94
38016 Chevrolet Cavalier/Pontiac Sunfire '95 thru '05
38017 Chevrolet Cobalt & Pontiac G5 '05 thru '11
38020 Buick Skylark, Chevrolet Citation, Olds Omega, Pontiac Phoenix '80 thru '85
38025 Buick Skylark & Somerset, Olds Achieva, Calais & Pontiac Grand Am '85 thru '98
38026 Chevrolet Malibu, Olds Alero & Cutlass, Pontiac Grand Am '97 thru '03
38027 Chevrolet Malibu '04 thru '10
38030 Cadillac Eldorado & Oldsmobile Toronado '71 thru '85, Seville '80 thru '85, Buick Riviera '79 thru '85
38031 Cadillac Eldorado & Seville '86 thru '91, DeVille & Buick Riviera '86 thru '93, Fleetwood & Olds Toronado '86 thru '92
38032 DeVille '94 thru '05, Seville '92 thru '04
    Cadillac DTS '06 thru '10
38035 Chevrolet Lumina APV, Olds Silhouette & Pontiac Trans Sport '90 thru '96
38036 Chevrolet Venture, Olds Silhouette, Pontiac Trans Sport & Montana '97 thru '05
    GM Full-size RWD - see BUICK (19025)
38040 Chevrolet Equinox '05 thru '09
    Pontiac Torrent '06 thru '09
38070 Chevrolet HHR '06 thru '11

GEO
    Metro - see CHEVROLET Sprint (24075)
    Prizm - see CHEVROLET (24060) or TOYOTA (92036)
40030 Storm all models '90 thru '93
    Tracker - see SUZUKI Samurai (90010)

GMC
    Vans & Pick-ups - see CHEVROLET

HONDA
42010 Accord CVCC all models '76 thru '83
42011 Accord all models '84 thru '89
42012 Accord all models '90 thru '93
42013 Accord all models '94 thru '97
42014 Accord all models '98 thru '02
42015 Accord '03 thru '07
42020 Civic 1200 all models '73 thru '79
42021 Civic 1300 & 1500 CVCC '80 thru '83
42022 Civic 1500 CVCC all models '75 thru '79
42023 Civic all models '84 thru '91
42024 Civic & del Sol '92 thru '95
42025 Civic '96 thru '00, CR-V '97 thru '01, Acura Integra '94 thru '00
42026 Civic '01 thru '10, CR-V '02 thru '09
42035 Odyssey all models '99 thru '10
    Passport - see ISUZU Rodeo (47017)

HYUNDAI
43010 Elantra all models '96 thru '10
43015 Excel & Accent all models '86 thru '09
43050 Santa Fe all models '01 thru '06
43055 Sonata all models '99 thru '08

INFINITI
    G35 '03 thru '08 - see NISSAN 350Z (72011)

ISUZU
    Hombre - see CHEVROLET S-10 (24071)
47017 Rodeo, Amigo & Honda Passport '89 thru '02
47020 Trooper '84 thru '91, Pick-up '81 thru '93

JAGUAR
49010 XJ6 all 6 cyl models '68 thru '86
49011 XJ6 all models '88 thru '94
49015 XJ12 & XJS all 12 cyl models '72 thru '85

JEEP
50010 Cherokee, Comanche & Wagoneer Limited all models '84 thru '01
50020 CJ all models '49 thru '86
50025 Grand Cherokee all models '93 thru '04
50026 Grand Cherokee '05 thru '09
50029 Grand Wagoneer & Pick-up '72 thru '91
50030 Wrangler all models '87 thru '11
50035 Liberty '02 thru '07

KIA
54050 Optima '01 thru '10
54070 Sephia '94 thru '01, Spectra '00 thru '09, Sportage '05 thru '10

LEXUS
    ES 300/330 - see TOYOTA Camry (92007) (92008)
    RX 330 - see TOYOTA Highlander (92095)

LINCOLN
    Navigator - see FORD Pick-up (36059)
59010 Rear Wheel Drive all models '70 thru '10

MAZDA
61010 GLC (rear wheel drive) '77 thru '83
61011 GLC (front wheel drive) '81 thru '85
61012 Mazda3 '04 thru '11
61015 323 & Protegé '90 thru '03
61016 MX-5 Miata '90 thru '09
61020 MPV all models '89 thru '98
    Navajo - see FORD Explorer (36024)
61030 Pick-ups '72 thru '93
    Pick-ups '94 on - see Ford (36071)
61035 RX-7 all models '79 thru '85
61036 RX-7 all models '86 thru '91
61040 626 (rear wheel drive) '79 thru '82
61041 626 & MX-6 (front wheel drive) '83 thru '92
61042 626 '93 thru '01, & MX-6/Ford Probe '93 thru '02
61043 Mazda6 '03 thru '11

MERCEDES-BENZ
63012 123 Series Diesel '76 thru '85
63015 190 Series 4-cyl gas models, '84 thru '88
63020 230, 250 & 280 6 cyl sohc '68 thru '72
63025 280 123 Series gas models '77 thru '81
63030 350 & 450 all models '71 thru '80
63040 C-Class: C230/C240/C280/C320/C350 '01 thru '07

MERCURY
64200 Villager & Nissan Quest '93 thru '01
    All other titles, see FORD listing.

MG
66010 MGB Roadster & GT Coupe '62 thru '80
66015 MG Midget & Austin Healey Sprite Roadster '58 thru '80

MINI
67020 Mini '02 thru '11

MITSUBISHI
68020 Cordia, Tredia, Galant, Precis & Mirage '83 thru '93
68030 Eclipse, Eagle Talon & Plymouth Laser '90 thru '94
68031 Eclipse '95 thru '05, Eagle Talon '95 thru '98
68035 Galant '94 thru '10
68040 Pick-up '83 thru '96, Montero '83 thru '93

NISSAN
72010 300ZX all models incl. Turbo '84 thru '89
72011 350Z & Infiniti G35 all models '03 thru '08
72015 Altima all models '93 thru '06
72016 Altima '07 thru '10
72020 Maxima all models '85 thru '92
72021 Maxima all models '93 thru '01
72025 Murano '03 thru '10
72030 Pick-ups '80 thru '97, Pathfinder '87 thru '95
72031 Frontier Pick-up, Xterra, Pathfinder '96 thru '04
72032 Frontier & Xterra '05 thru '11
72040 Pulsar all models '83 thru '86
72050 Sentra all models '82 thru '94
72051 Sentra & 200SX all models '95 thru '06
72060 Stanza all models '82 thru '90
72070 Titan pick-ups '04 thru '10, Armada '05 thru '10

OLDSMOBILE
73015 Cutlass '74 thru '88
    For other OLDSMOBILE titles, see BUICK, CHEVROLET or GM listings.

PLYMOUTH
    For PLYMOUTH titles, see DODGE.

PONTIAC
79008 Fiero all models '84 thru '88
79018 Firebird V8 models except Turbo '70 thru '81
79019 Firebird all models '82 thru '92
79025 G6 all models '05 thru '09
79040 Mid-size Rear-wheel Drive '70 thru '87
    Vibe '03 thru '11 - see TOYOTA Matrix (92060)
    For other PONTIAC titles, see BUICK, CHEVROLET or GM listings.

PORSCHE
80020 911 Coupe & Targa models '65 thru '89
80025 914 all 4 cyl models '69 thru '76
80030 924 all models incl. Turbo '76 thru '82
80035 944 all models incl. Turbo '83 thru '89

RENAULT
    Alliance, Encore - see AMC (14020)

SAAB
84010 900 including Turbo '79 thru '88

SATURN
87010 Saturn all S-series models '91 thru '02
87011 Saturn Ion '03 thru '07

SATURN (cont.)
87020 Saturn all L-series models '00 thru '04
87040 Saturn VUE '02 thru '07

SUBARU
89002 1100, 1300, 1400 & 1600 '71 thru '79
89003 1600 & 1800 2WD & 4WD '80 thru '94
89100 Legacy models '90 thru '99
89101 Legacy & Forester '00 thru '06

SUZUKI
90010 Samurai/Sidekick/Geo Tracker '86 thru '01

TOYOTA
92005 Camry all models '83 thru '91
92006 Camry all models '92 thru '96
92007 Camry/Avalon/Solara/Lexus ES 300 '97 thru '01
92008 Toyota Camry, Avalon and Solara & Lexus ES 300/330 all models '02 thru '06
92009 Camry '07 thru '11
92015 Celica Rear Wheel Drive '71 thru '85
92020 Celica Front Wheel Drive '86 thru '99
92025 Celica Supra all models '79 thru '92
92030 Corolla all models '75 thru '79
92032 Corolla rear wheel drive models '80 thru '87
92035 Corolla front wheel drive models '84 thru '92
92036 Corolla & Geo Prizm '93 thru '02
92037 Corolla models '03 thru '11
92040 Corolla Tercel all models '80 thru '82
92045 Corona all models '74 thru '82
92050 Cressida all models '78 thru '82
92055 Land Cruiser FJ40/43/45/55 '68 thru '82
92056 Land Cruiser FJ60/62/80/FZJ80 '80 thru '96
92060 Matrix & Pontiac Vibe '03 thru '11
92065 MR2 all models '85 thru '87
92070 Pick-up all models '69 thru '78
92075 Pick-up all models '79 thru '95
92076 Tacoma, 4Runner & T100 '93 thru '04
92077 Tacoma all models '05 thru '09
92078 Tundra '00 thru '06, Sequoia '01 thru '07
92079 4Runner all models '03 thru '09
92080 Previa all models '91 thru '95
92081 Prius '01 thru '08
92082 RAV4 all models '96 thru '10
92085 Tercel all models '87 thru '94
92090 Sienna all models '98 thru '09
92095 Highlander & Lexus RX-330 '99 thru '07

TRIUMPH
94007 Spitfire all models '62 thru '81
94010 TR7 all models '75 thru '81

VW
96008 Beetle & Karmann Ghia '54 thru '79
96009 New Beetle '98 thru '10
96016 Rabbit, Jetta, Scirocco, & Pick-up gas models '75 thru '92 & Convertible '80 thru '92
96017 Golf, GTI & Jetta '93 thru '98, Cabrio '95 thru 02
96018 Golf, GTI & Jetta '99 thru '05
96019 Jetta, Rabbit, GTI & Golf '05 thru '11
96020 Rabbit, Jetta, Pick-up diesel '77 thru '84
96023 Passat '98 thru '05, Audi A4 '96 thru '01
96030 Transporter 1600 all models '68 thru '79
96035 Transporter 1700, 1800, 2000 '72 thru '79
96040 Type 3 1500 & 1600 '63 thru '73
96045 Vanagon air-cooled models '80 thru '83

VOLVO
97010 120, 130 Series & 1800 Sports '61 thru '73
97015 140 Series all models '66 thru '74
97020 240 Series all models '76 thru '93
97040 740 & 760 Series all models '82 thru '88

TECHBOOK MANUALS
10205 Automotive Computer Codes
10206 OBD-II & Electronic Engine Management
10210 Automotive Emissions Control Manual
10215 Fuel Injection Manual, '78 thru '85
10220 Fuel Injection Manual, '86 thru '99
10225 Holley Carburetor Manual
10230 Rochester Carburetor Manual
10240 Weber/Zenith/Stromberg/SU Carburetor
10305 Chevrolet Engine Overhaul Manual
10310 Chrysler Engine Overhaul Manual
10320 Ford Engine Overhaul Manual
10330 GM and Ford Diesel Engine Repair
10333 Engine Performance Manual
10340 Small Engine Repair Manual
10345 Suspension, Steering & Driveline
10355 Ford Automatic Transmission Overhaul
10360 GM Automatic Transmission Overhaul
10405 Automotive Body Repair & Painting
10410 Automotive Brake Manual
10415 Automotive Detaling Manual
10420 Automotive Electrical Manual
10425 Automotive Heating & Air Conditioning
10430 Automotive Reference Dictionary
10435 Automotive Tools Manual
10440 Used Car Buying Guide
10445 Welding Manual
10450 ATV Basics
10452 Scooters 50cc to 250cc

SPANISH MANUALS
98903 Reparación de Carrocería & Pintura
98904 Manual de Carburador Modelos Holley & Rochester
98905 Códigos Automotrices de la Computadora
98906 OBD-II & Sistemas de Control Electrónico del Motor
98910 Frenos Automotriz
98913 Electricidad Automotriz
98915 Inyección de Combustible '86 a '99
99040 Chevrolet & GMC Camionetas '67 al '87
99041 Chevrolet & GMC Camionetas '88 al '98
99042 Chevrolet Camionetas Cerradas '68 al '95
99043 Chevrolet/GMC Camionetas '94 a '04
99048 Chevrolet/GMC Camionetas '99 al '06
99055 Dodge Caravan/Ply. Voyager '84 al '95
99075 Ford Camionetas y Bronco '80 al '94
99076 Ford F-150 '97 al '09
99077 Ford Camionetas Cerradas '69 al '91
99088 Ford Modelos de Tamaño Mediano '75 al '86
99089 Ford Camionetas Ranger '93 al '10
99091 Ford Taurus & Mercury Sable '86 al '95
99095 GM Modelos de Tamaño Grande '70 al '90
99100 GM Modelos de Tamaño Mediano '70 al '88
99106 Jeep Cherokee, Wagoneer & Comanche '84 al '00
99110 Nissan Camionetas y Pathfinder '80 al '96
99118 Nissan Sentra '82 al '94
99125 Toyota Camionetas y 4-Runner '79 al '95

Over 100 Haynes motorcycle manuals also available

7-12